S0-CAS-500

Correlation Analysis in Chemistry

Recent Advances

Contributors

R. P. Bell, Department of Chemistry, The University of Stirling, Stirling, FK9 4LA, Scotland

M. Charton, Department of Chemistry, Pratt Institute, Brooklyn, New York, 11205, U.S.A.

C. Duboc, Université de Paris VI, F-75005 Paris, France

D. F. Ewing, Chemistry Department, The University, Hull, HU6 7RX, England

O. Exner, Institute of Organic Chemistry and Biochemistry, Czechoslovak Academy of Sciences, Prague 6, Czechoslovakia

G. P. Ford, Department of Chemistry, University of Texas at Austin, Austin, Texas 78712, U.S.A.

M. Godfrey, Department of Chemistry, The University, Southampton, S09 5NH, England

C. Hansch, Department of Chemistry, Pomona College, Claremont, California 91711, U.S.A.

A. R. Katritzky, School of Chemical Sciences, University of East Anglia, University Plain, Norwich, NR4 7TJ, England

J. Shorter, Department of Chemistry, The University, Hull, HU6 7RX, England

M. Sjöström, Research Group for Chemometrics, Institute of Chemistry, Umeå University, S-901 87 Umeå, Sweden

R. D. Topsom, Department of Chemistry, La Trobe University, Bundoora, Melbourne, Victoria, Australia 3083

S. Wold, Research Group for Chemometrics, Institute of Chemistry, Umeå University, S-901 87 Umeå, Sweden

Correlation Analysis in Chemistry

Recent Advances

Edited by

N. B. Chapman

and

J. Shorter

The University of Hull
Hull, England

Plenum Press · New York and London

Library of Congress Cataloging in Publication Data

Main entry under title:

Correlation analysis in chemistry.

Includes bibliographical references and index.
1. Linear free energy relationship. I. Chapman, Norman Bellamy, 1916- II. Shorter, John, 1926-
QD501.C792 541.39 78-1081
ISBN 0-306-31068-6

A Division of Plenum Publishing Corporation
227 West 17th Street, New York, N.Y. 10011

Printed in the United States of America

Preface

This book, *Correlation Analysis in Chemistry: Recent Advances*, is a sequel to our *Advances in Linear Free Energy Relationships.*† The change in the title is designed to reflect more accurately the nature of the field and the contents of the volume. The term LFER is still widely used, but it is often applied rather loosely to correlation equations that are not LFER in the restricted sense of a relationship involving logarithms of rate or equilibrium constants on each side of the equation. The term "correlation analysis" seems to us more appropriate for the whole subject. The use of this term has compelled us also to introduce "chemistry" into the title; we have preferred not to prefix this with "organic" on the grounds that several areas of interest are not "organic chemistry" as usually understood, although, of course, traditional applications of the basic relationships associated with the names of Hammett and of Taft continue to be of interest.

In the first volume we sought through our authors to provide a series of general articles covering the various aspects of the field as they seemed to us. Since the book was the first international research monograph in its field, each chapter, while giving prominence to recent developments, did not neglect earlier work, so that each article presented a comprehensive account of its own area. The same general idea underlies the present volume, but most of the articles are on much more specialized and restricted topics, and treat in greater depth and detail matters that were dealt with in only a few pages in the earlier volume. Consequently some important topics such as correlation analysis in relation to solvent effects on organic reactivity and to inorganic and organometallic chemistry are treated only incidentally in the present volume. The traditional areas of LFER associated with Hammett and Taft are now covered by articles dealing with LFER as statistical models (Chapter 1), the relationship to quantum chemistry (Chapter 3), multiparameter extensions of the Hammett equation (Chapter 4), applications to polycyclic arenes and heterocyclic compounds (Chapter 5) and to olefinic systems (Chapter 6), and a critical compilation of substituent constants (Chapter 10). Applications to

† N. B. Chapman and J. Shorter, editors, Plenum Press, London and New York (1972).

spectroscopy of various types are of frequent occurrence in some of these chapters, while nmr chemical shifts are the special concern of Chapter 8. The effect of the reagent on organic reactivity is covered in two chapters, one on the Brönsted equation, which is the oldest LFER (Chapter 2), and the other on nucleophilicity (Chapter 7). The subject of biochemical quantitative structure–activity relationships (QSAR) is developing so rapidly that another general treatment (Chapter 9) seemed justified to supplement the chapters on enzymology and drug action in the first volume.

It seems appropriate here to repeat a paragraph that we as Editors wrote in our own defense in our preface to *Advances in Linear Free Energy Relationships*:

> Authors were given only very general guidance as to the length and content of their respective chapters, and they had full freedom to develop their topics as they felt appropriate. We are aware that the ten chapters bridge the extremes of style from the "essay–review," with a relatively low density of literature citations, to the "progress report" from which no signficant contribution is omitted. The chapters also vary widely in the extent to which tables or figures are used. Some authors present correlations that have been carried out specially for this book, and in some cases the approach used reflects the individual views of the author. Accordingly we are prepared for the criticism that the book has the usual failings of multiauthor works. However, in our editorial work we have tried to impress on the book a certain unity, despite the diversity of style and approach. We have, moreover, been mindful that parts of it may be read by scientists from a wide range of disciplines. Authors were asked to remember this and to try to make themselves intelligible to a wide readership.

For the first volume we wrote an Editors' Introduction dealing mainly with *Signs, Symbols, and Terminology for Substituent Effects*, which is reprinted in the present volume. The only significant change in our views is that we would now wish to encourage the use of K (*konjugativ*) for the total delocalization effect.

As editors we are greatly indebted to many people. The authors, like their predecessors in the earlier volume, have made strenuous efforts to meet our requirements and suggestions in many directions and have been very forbearing with the idiosyncrasies of the editors. Several colleagues in Hull have helped us in various ways. The very considerable secretarial work has been excellently done by Christina Pindar, Janet Bailey, and Dorothy Wilson, all of the Chemistry Department. We are also grateful to Jane Fisher of the British Library for work on references.

Finally, as in all our work, we are most grateful to our wives for their support and encouragement.

N. B. Chapman
J. Shorter

Contents

Errata
for
Advances in Linear Free Energy Relationships,
edited by N. B. Chapman and J. Shorter, Plenum Press, London and New York (1972)

p. 18 Footnote ‡: For σ_R, read ρ_R.
p. 76 Table 2.1: For Ph_2C read Ph_2CH, and E_s for $Cl(CH_2)_2$ should be −0.90. In footnote *b*, insert "hydrate" after "acetaldehyde."
p. 110 Line 9 from bottom: For 1 : 3, read 1 : 1.
p. 121 Equation (3.1): Insert c^2 after 3000 in the denominator.
p. 184 Equation (4.14): The symbol in the second term on the right side should be σ_R^0.
p. 253 The last line should read ". . . follow as an . . .".
p. 258 In entry 218, ε should be 11.5 and n_D^{20} should be 1.4659.
p. 372 Section 8.1.2: Insert "energy" after "free" in first line of main text.
p. 396 Line 7 from bottom: The compounds should be "*p*-nitrophenyl benzoates."
p. 408 Table 9.3: The values of σ_m should be as follows: F, 0.34; Cl, 0.37; Br, 0.39; I, 0.35.

Editors' Introduction †

Signs, Symbols, and Terminology for Substituent Effects

Since this is an area of considerable confusion we felt it necessary to give authors some guidance, broadly as follows.

There is no uncertainty as to the sign to be attached to a σ value: an electron-attracting group has a positive σ value, and *vice versa*. There are two possible conventions, however, for the *sign* of a substituent effect. In the Ingold convention the movement of electrons towards a substituent is signified by a negative sign; the movement of electrons away from a substituent by a positive sign. Thus NO_2 has a $-M$ effect, while OMe has a $+M$ effect. *Electronegativity* is thus associated with a *negative* sign, and *electropositivity* with a *positive* sign. In the other convention, originally associated with Robinson, the signs are reversed.

There is little doubt that the Ingold convention is the more widely used the world over. We note, however, that many authors (especially those of textbooks and monographs published in the U.S.A.) avoid using signs and symbols for electronic effects and use phrases such as "a mesomerically electron-attracting group." From an LFER standpoint the Ingold convention has the disadvantage that the signs of the electronic effects and those of the corresponding σ values are opposite. For this reason some writers on LFER have adopted the Robinson sign convention. In deciding not to follow suit in this volume, we were influenced by correspondence with Sir Christopher Ingold a few months before he died, and by the appearance of the second edition of Ingold's *Structure and Mechanism in Organic Chemistry*, and of Hammett's *Physical Organic Chemistry*. Both these books will have the status of major works for many years to come; both of them adhere to the Ingold sign convention.

We therefore asked authors to use the Ingold convention, but in the event signs and symbols to describe electronic effects have not been extensively used.

†This Introduction originally appeared in *Advances in Linear Free Energy Relationships*, N. B. Chapman and J. Shorter, eds., Plenum Press, London and New York (1972).

As to symbols, the use of I for the inductive effect (possibly with a subscript for a special purpose) is universal practice. Delocalization effects are more of a problem, the possibilities being T, R, K, or C for the total effect, with M for the time-permanent and E for the time-variable effect. T was not to be used since Ingold had abandoned it in favor of K, based on the German for "conjugative"; he preferred K to the more obvious C, clearly inconvenient in organic chemistry. However, K seemed to have definite disadvantages in connection with LFER. We suggested that K or C might be used, or R, which has the advantage of fitting in with σ_R, which we felt we could not tamper with, although we dislike the term "resonance." Authors were asked not to use M with reference to conditions under which time-variable effects might be considered important (for separate specification of the latter E remains appropriate). Since, however, some authors probably consider that the analysis into time-permanent and time-variable effects has sometimes been overemphasized, it is likely that this advice has not been strictly adhered to.

As to terminology, we recognized that the difficulties mentioned in the previous paragraph are carried over and others are added. Having alerted the authors to the problems as we saw them, we had hoped to examine very closely the use they made of the various terms with a view to trying to ensure some degree of uniformity of usage. In the event this task has proved beyond us, and we can do no more than lay the problems before the reader.

We consider a reactant molecule RY and an appropriate standard molecule R_0Y. Initially we suppose that Y is not conjugated with either R or R_0. For RY the *polar effect* of the group, R, comprises all the processes whereby a substituent may modify the electrostatic forces operating at the reaction center Y, relative to the standard R_0Y. These forces may be governed by charge separations arising from differences in the electronegativity of atoms (leading to the presence of dipoles), the presence of unipoles, or electron delocalization. Field (or direct), inductive (through-bond polarization involving σ or π electrons), and mesomeric effects may in principle be distinguished. Because of the difficulty of distinguishing between field and through-bond effects in practice, the term *inductive effect* is often used to cover both. It is so used widely in this book, but in some places the more restricted meaning is implied. The term *resonance effect* or *resonance polar effect* is often used in connection with the *mesomeric effect*, e.g., the analysis of σ values into *inductive* and *mesomeric* components is also known as the separation of *inductive* and *resonance effects*, as shown by the use of σ_R as the basic symbol for the delocalization component.

When R and R_0 may be conjugated with Y, the above discussion holds, but additional influences may arise from the more extensive electron-delocalization. The mesomeric part of the polar effect will be modified. There will also be the *resonance effect* of R which is concerned with the extent of conjugation of R and Y, relative to the standard R_0Y, and is not part of the polar effect. This distinction is sometimes not clearly made: "resonance effects" in a wide sense are lumped together and treated as if they are polar in nature; cf. the distinction between σ_R^0 on the one hand and σ_R, σ_R^+, and σ_R^- on the other.

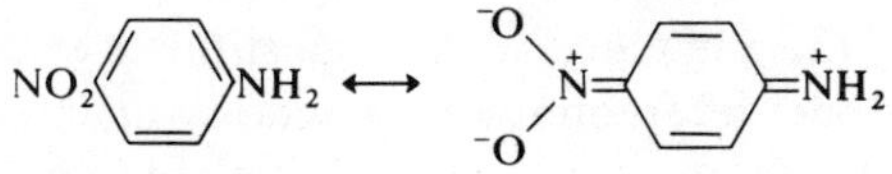

(I)

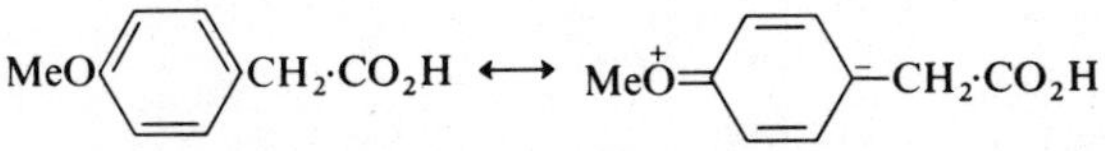

(II)

It thus seems that the terms *resonance effect* and *mesomeric effect* may be used to cover (*a*) conjugation of the substituent with the functional group, mediated by the aromatic or other delocalized system, or (*b*) that of the substituent with the delocalized system, not including the functional group. Structure (I) is an example of (*a*), and (II) of (*b*).

It also seems common practice to use the term *resonance effect* in connection with (*c*), the conjugation of the functional group with the delocalized system, e.g., as in (III)

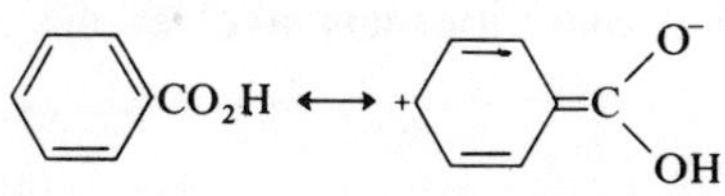

(III)

The precise meaning of *resonance effect* may thus only be understood in context.

Hyperconjugation or the *hyperconjugative effect*, as the name implies, is historically and commonly regarded as a special kind of conjugative effect involving delocalization of electrons in C—H bonds adjacent to an unsaturated system. The actual nature of this substituent effect, however, must be regarded as still in doubt. At most places in this book where the effect is referred to, no particular view of hyperconjugation seems to be implied: The term simply means a special effect of α-hydrogen atoms in the same sense as mesomeric electron release. An analogous effect of C—C bonds under circumstances in which ordinary conjugation cannot occur is also sometimes invoked; this is C—C hyperconjugation, in contrast to CH hyperconjugation.

A word on steric effects is appropriate, although there are no particular problems. *Steric effects* are caused by the intense repulsive forces operating when two nonbonded atoms approach each other so closely that nonbonded compressions are involved. The *primary steric effect* of R is the direct result of compression which arises because R differs in structure from R_0 in the vicinity of the reaction center. A *secondary steric effect* involves the moderation of a polar effect or resonance effect by nonbonded compressions.

In discussing the influence of any substituent effect it is of course necessary to consider differentially interactions in initial and transition states in the case of rate processes, and in initial and final states in the case of equilibria.

References

References are collected at the end of each chapter in the order of first citation. We have provided references to *Chemical Abstracts* for periodicals which, we believe, will be inaccessible to many readers. Up to volume 65 (1966) such references give the column number; from volume 66 (1967) the distinctive number of the abstract is cited.

In the case of Russian journals we have also provided references to English translations, if available. These are usually indicated simply by EE (i.e., English edition), followed by the page number, but amplification is sometimes required. The main journals involved are as follows.

Russian Title	*English Title*
Zhurnal Obshchei Khimii	Journal of General Chemistry of the USSR
Zhurnal Organicheskoi Khimii	Journal of Organic Chemistry of the USSR
Zhurnal Fizicheskoi Khimii	Russian Journal of Physical Chemistry
Uspekhi Khimii	Russian Chemical Reviews
Doklady Akademii Nauk SSR	Doklady Chemistry Doklady Physical Chemistry
Izvestiya Akademii Nauk SSR Seriya Khimicheskaya	Bulletin of the Academy of Sciences of the USSR Division of Chemical Science

Reaktsionnaya Sposobnost Organicheskikh Soedinenii	Organic Reactivity
Kinetika i Kataliz	Kinetics and Catalysis
Zhurnal Strukturnoi Khimii	Journal of Structural Chemistry
Optika i Spektroskopiya	Optics and Spectroscopy

For *Angewandte Chemie*, or its *International Edition*, references to the English edition (EE) or to the German edition (GE) as appropriate are given.

Miscellaneous Symbols

All chapters are concerned with measures of the success of correlations and the following symbols are used throughout:

r correlation coefficient in a simple linear regression,
R correlation coefficient in a multiple regression,
s standard deviation of the estimated value of the dependent variable.

1

Linear Free Energy Relationships as Tools for Investigating Chemical Similarity—Theory and Practice

Svante Wold and Michael Sjöström

Svante Wold and Michael Sjöström • Research Group for Chemometrics, Institute of Chemistry, Umeå University, S-901 87 Umeå, Sweden.

1.1. Introduction

Organic chemistry has always depended much more on empirical rules than on models deduced from theoretical postulates. This is because the complexity of organic reactions in solution makes the relationship between experiment and pure "microscopic" theory based on atomic and molecular concepts extremely difficult to discover. Thus, it is at present impossible to reach a thorough understanding of most phenomena in organic chemistry on the basis of microscopic theory.[1]

The organic chemist is forced to have recourse to approximate treatments, which can in principle start from either of two points of view.[2] In one, microscopic theory is in principle retained, but simplifying approximations are introduced to make calculations and interpretations possible. However, in order to treat organic chemical problems in this way, the approximations that have to be made are at present so great that it is doubtful whether the results are closely connected to the original theory.

The other approach is to start from an empirical, macroscopic analysis of problems and add as much of the microscopic theory as is necessary to allow a discussion of the results in terms of molecular concepts. This latter approach has historically been the only way open to the organic chemist, who has always investigated phenomena experimentally long before theories powerful enough to cope with them have been established.

Consequently, in organic chemistry a number of empirical models for the description of relationships between structure and reactivity have emerged. The most successful and intensively investigated are the linear free energy relationships[3] (LFER), with the Hammett equation[4–7] as the most prominent example.

The term LFER derives from the major use of these relationships as mathematical tools for correlating changes in free energy in different reaction series. Relationships of the same mathematical form as LFER can, however, be used to correlate the change of any property measured on any ensemble of similar processes, systems, or objects. This would make the terms "correlation analysis"[7] or "quantitative analogy models"[8] more appropriate. Nevertheless, we shall henceforth use the term LFER in order to conform with the other articles in this volume. We mean thereby a mathematical relationship of a special form, viz., equation (1.5) or (1.8), used to correlate any type of observed quantities (denoted by y) regardless of whether they are proportional to free energies or not (see p. 6).

Since LFER relate observations on one system (e.g., a chemical reaction or a chemical compound) to observations on other similar systems, one can see LFER as quantifications of the classical chemical principle of analogy.[9,10] To quote Hammett: "From its beginning the science of organic chemistry has depended on the empirical and qualitative rule that like substances react similarly and that similar changes in structure produce similar changes in reactivity."[10] "Linear free energy relationships constitute the quantitative specialisation of this fundamental principle."[11] Hence, one can see LFER as tools of chemical pattern recognition,[12] whereby the regular behavior—the behavioral pattern—of similar chemical systems is given a mathematical representation.

This view of LFER as approximate mathematical relationships describing changes in similar systems will be taken in the present chapter. This will be supported by a theoretical derivation of LFER based on mathematical arguments. Thus, provided that certain continuity assumptions are valid, generalized LFER are shown to describe the variation of any measurement made on similar systems. This property of LFER has certain consequences for their interpretation and applicability in practical cases, which we shall discuss later. The discussion will be illustrated by applications to the Hammett equation.

Regardless of whether one views LFER as exact chemical "laws" or general mathematical similarity models, it is important to use correct mathematical and statistical methods when LFER are investigated. Though pertinent methods are well defined in mathematical statistical literature, we feel that there is a great need to discuss these methods in the present context. Hence, we shall discuss the use of LFER in practice, including the extensions of LFER to methods of pattern recognition.

1.2. Theoretical Foundations of LFER

Palm[13] and Leffler and Grunwald[2] were the first to show that LFER can be seen as mathematical models capable of approximating small changes in a class of similar reactions. Their derivations were, however, valid only for special cases, owing to their use of functions of single variables. Later work based on functions of vector variables led to the more general results presented below.[8]

Though an understanding of the principal results of the theoretical derivation given below is of some importance for the interpretation of chemical data in terms of LFER, the detail of the mathematical formalism in the derivation (i.e., Sections 1.2.1–1.2.4) is, in our view, of less importance. It is sufficient to realize that a simple LFER [equation (1.5)] and a multiple-term LFER [equation (1.8)] have the same formal basis as

Taylor expansions of a function $y = f(x)$. Thus, an arbitrary function $y = f(x)$ can be approximated by a low-degree polynomial, $y = c_0 + c_1x$ or $y = c_0 + c_1x + c_2x^2$, in a small interval around a point $x = x_0$. Analogously, data y_{ik} observed on the systems S_{ik}—say chemical reactions of compounds with reaction centers $i = 1, 2, \ldots, M$ and substituents $k = 1, 2, \ldots, N$—can, under certain circumstances, be approximated by equation (1.5) or (1.8) provided that the variation of the systems with subscript k (i.e., substituent) is sufficiently small.

1.2.1. *Mathematical Derivation, Discrete Case*

Let us consider modifications made in a chemical system S. By a chemical system we shall henceforth mean any dynamic or static ensemble of atoms and molecules of interest to a chemist. Thus, a system might be a particular chemical reaction, a particular compound in solution, or anything that can be thought of as a recognizable, but not necessarily isolated, collection of atoms and molecules that is studied from a chemical point of view.

Let the system S be discretely modified in two distinct ways, for instance by changing the reaction center and a substituent. Letting the two ways of change be denoted by the change of the subscripts i and k, we thus get the systems S_{ik}. We now make one measurement on each modified system, determining, for instance, the logarithm of a rate constant. These measurements, denoted by y_{ik}, can *formally* be described by a function F plus "small" errors of measurement ε, [equation (1.1)]. This function F is

$$y_{ik} = F(\mathbf{Z}_i, \mathbf{X}_k) + \varepsilon_{ik} \tag{1.1}$$

constructed so that everything that changes with the first modification (e.g., the change of reaction center) is incorporated as an element into the vector variable $\mathbf{Z}$, and everything that changes with the second modification (e.g., the change of substituent) is incorporated into the vector variable $\mathbf{X}$. Though the function F may be extremely complicated and involve very many variables—in equation (1.1) represented as elements in either or both of the vector variables—this formalism certainly is in accordance with present physical and chemical theory. Thus, for instance, quantum theory represents any observable property (y) as a solution of a linear operator equation.[14] This, in turn, makes the observable y depend on microscopic variables in a functionlike manner. The residuals (ε_{ik}) henceforth denote the "random" variation not described by the mathematical model. They contain, among other factors, errors of measurement.

In chemical language, the vector variables $\mathbf{Z}$ and $\mathbf{X}$ contain as elements all such *microscopic variables* as charge distributions, dipole moments, solvation, etc. that we use to "explain" the variation of a property between one reaction and another or between one compound and another. What is interesting in the present treatment is that we need not know what all these microscopic variables are. It is sufficient to assume that they exist and have certain properties.

Let us now study some mathematical properties of the function F. First, for the trivial case that for a given i (e.g., a given reaction center), all the systems S_{ik} are virtually identical for all k (e.g., for all substituents), the vector $\mathbf{X}$ will, of course, be identical for all systems and the measurements can be described by equation (1.2). Let then the identity between different

$$y_{ik} = F(\mathbf{Z}_i, \mathbf{C}) + \varepsilon_{ik} = f(\mathbf{Z}_i) + \varepsilon_{ik} = \alpha_i + \varepsilon_{ik} \tag{1.2}$$

k be somewhat relaxed so that for a given i, all systems S_{ik} are very similar. Assume furthermore that the small change with k can be described by some kind of single coordinate analogous to a reaction coordinate (Fig. 1.1). If the change in this coordinate, henceforth called the coordinate of change and denoted by t, is sufficiently small when k is changed, the elements in the vector $\mathbf{X}$ will change linearly in relation to one another (Fig. 1.1), and the vector $\mathbf{X}$ will contain only one independent element, x. In this case, it is easy to show by means of Taylor expansions of the function F [equation (1.3)] around the point ($\mathbf{Z}_0$, x_0), that a second-order

$$\begin{aligned} F(\mathbf{Z}, x) = F_{00} + \sum_r F'_r \,\Delta z_r + F'_x \,\Delta x + \sum_r \sum_s F''_{rs} \,\Delta z_r \,\Delta z_s \\ + \sum_r F''_{rx} \,\Delta z_r \,\Delta x + F''_{xx} \,\Delta x^2 + \text{higher terms} \end{aligned} \tag{1.3}$$

approximation to F is given, after rearrangement, by equation (1.4).[8] Here, the remainder $R(3)$ contains only terms of the third and higher

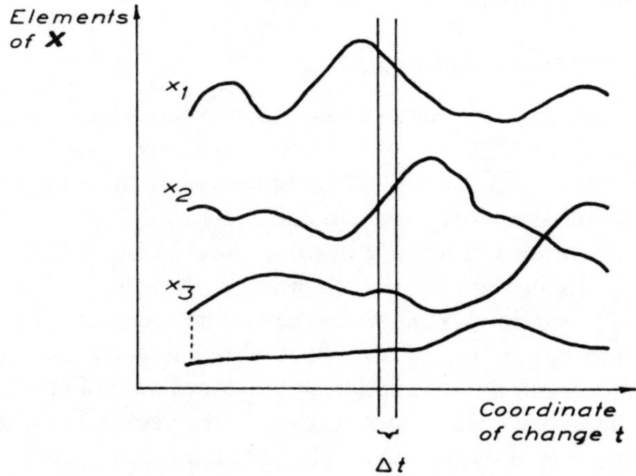

Fig. 1.1. When the change Δt in the coordinate t is small enough, the elements in the vector $\mathbf{X}$ vary linearly with respect to each other.

degree, which all can be made arbitrarily small by keeping the change in the variable x sufficiently small. Terms containing only elements of $\mathbf{Z}$ are

$$F(\mathbf{Z}, x) = f(\mathbf{Z}) + \Delta x\left[F'_x + \sum_r F''_{rx}\ \Delta z_r\right] + \Delta x^2 F''_{xx} + R(3)$$

$$= f(\mathbf{Z}) + \Delta x\left[F'_x + \sum_r F''_{rx}\Delta z_r\right] + \Delta x^2\left[F'_x + \sum_r F''_{rx}\Delta z_r\right]F''_{xx}/F'_x$$

$$- \Delta x^2\left[\sum_r F''_{rx}\Delta z_r\right]F''_{xx}/F'_x + R(3)$$

$$= f(\mathbf{Z}) + g(\mathbf{Z})h(x) + R(3) \tag{1.4}$$

all included in $f(\mathbf{Z})$. Hence, provided that the systems are sufficiently similar for fixed i, $R(3)$ will be negligible and can be incorporated into the residuals ε, and the data y can be well fitted by equation (1.5). Equation

$$y_{ik} = \alpha_i + \beta_i\theta_k + \varepsilon_{ik} \tag{1.5}$$

(1.5), which has the same form as simple LFER such as the Hammett equation ($\alpha_i \approx \log k_{i0}$, $\beta_i = \rho_i$, and $\theta_k = \sigma_k$), the Taft equation, the Brönsted relation, and others, separates the influence of the two modifications (i and k) into separate parameters depending on only one of these modes of change. It should be noted that the residuals ε contain *both* errors of measurement *and* "model errors" owing to the possibly incomplete vanishing of the remainder $R(3)$ in equation (1.4).

One can also obtain a third-degree approximation (1.6) by including a quadratic term in $h(x)$, which leads to the quadratic similarity model (1.7).

$$F(\mathbf{Z}, x) = f(\mathbf{Z}) + g_1(\mathbf{Z})h(x) + g_2(\mathbf{Z})h(x)^2 + R(4) \tag{1.6}$$

$$y_{ik} = \alpha_i + \beta_i\theta_k + \gamma_i\theta_k^2 + \varepsilon_{ik} \tag{1.7}$$

By simple Taylor expansions, one can also derive other similarity models more or less related to LFER, some of which are shown in Table 1.1. The most general of these, equation (1.8), is obtained in the same way as equation (1.4) if the similarity between the systems S_{ik} (for fixed i) is relaxed further, i.e., the variation in t becomes larger (Fig. 1.1). To begin with, quadratic terms in the different elements in $\mathbf{X}$ will start to be important and $\mathbf{X}$ will contain two linearly independent elements, then cubic terms will become larger and $\mathbf{X}$ will have three linearly independent elements, etc. In general, if the variation in the coordinate of change is so large that A elements are linearly independent in the vector $\mathbf{X}$, the data y_{ik} can be well approximated by the model (1.8) containing A product terms.

$$y_{ik} = \alpha_i + \sum_{a=1}^{A} \beta_{ia}\theta_{ak} + \varepsilon_{ik} \tag{1.8}$$

As seen in equation (1.4), the derivations given above are valid only when the derivative F'_x is different from zero; i.e., at least one element in the vector $\mathbf{X}$ must change with the coordinate of change t. If this assumption is not valid, the value of $\mathbf{X}$ is constant and the data are well described by equation (1.2).

The main difference between the derivation sketched above and the derivations of Palm[13] and Leffler and Grunwald[2] is that in our equation (1.4) the vector $\mathbf{Z}$ can contain any finite number of elements and still give the simple LFER (1.5). One consequence is that the reaction center and the substituent are *not* formally symmetric in LFER such as the Hammett equation, contrary to what is often stated (see for instance Ref. 2, p. 141). Since the vector $\mathbf{Z}$ can contain several independent elements, it can incorporate, at least in principle, the variation of solvent, temperature, pressure, etc. The effects of these external variables show only as a change in the parameters α and β. Hence it is, in principle, possible to have a parameter scale θ (e.g., σ in the Hammett equation) which is independent of solvent, pressure, and temperature in addition to its independence of the reaction center. The premise is, as before, that the variation in the variable x is small, i.e., so small that third-degree terms containing both z_r and x are negligible in the Taylor expansion (1.3).

In practice one finds for instance, that for the Hammett equation, the same σ scale [corresponding to θ in equation (1.5)] sometimes cannot be used for both aqueous solvents and pure alcoholic solvents.[15–17] According to the present derivation, this is probably due to changes in the effect of substituents on reactivity when one goes from one solvent system to another.

Finally, we note the connection between equation (1.7) and equation (1.8) with $A = 2$. Both these models represent "breakdowns" of the similarity model (1.5). Thus, when the variation between the systems S_{ik} becomes large so that the data start to show systematic deviations from equation (1.5), the data can at first be described by equation (1.7), but when the deviations become even larger, one has to employ equation (1.8) with $A = 2$, then $A = 3$, and so on. Hence, for borderline cases, it might in practice be difficult to distinguish statistically between model (1.7) and model (1.8). This is a problem since the chemical interpretation of these two models is distinctly different. The first (1.7) is usually interpreted as due to the variation of a "single effect," which causes curvature because of "saturation." The second (1.8) is interpreted as due to the variation of two separate effects. Illustrations are found in Sections 1.3.8 and 1.5.2.

1.2.2. The Case of One Continuous Mode of Change

It is also of some interest to formulate similarity models for the case when one mode of change is continuous in an external variable, for instance temperature. If the mode of change, now denoted by T, affects the vector variable $\mathbf{Z}$, we have from equation (1.4), at the value T, equation (1.9). When the dependence of y on T can be well described by a polynomial in functions $u(T)$—for instance the Arrhenius equation $\log k = A+B/T$—we can write (1.10), which can be simplified to (1.11) since the θ

$$y_{T,k} = f(\mathbf{Z}_T) + g(\mathbf{Z}_T)\theta_k + \varepsilon_{T,k} \tag{1.9}$$

$$y_{T,k} = f_0 + f_1 u(T) + f_2 u(T)^2 + \cdots + (g_0 + g_1 u(T) + \cdots)\theta_k + \varepsilon_{T,k} \tag{1.10}$$

$$y_{T,k} = c_0 + \lambda_k[1 + c_1 u(T) + c_2 u(T)^2 + \cdots] + h_2 u(T)^2 + h_3 u(T)^3 + \cdots + \varepsilon_{T,k} \tag{1.11}$$

scale is arbitrary with respect to (a) a shift of scale, (b) a multiplicative factor. When T is temperature, $u(T) = 1/T$ and quadratic and higher terms are usually negligible. This gives the similarity model (1.12). This

$$y_{T,k} = c_0 + \lambda_k(1 + c_1/T) + \varepsilon_{T,k} \tag{1.12}$$

model, which is the isokinetic model in the formulation of Exner,[18,19] has been derived by assuming only similarity between the systems obtained when the subscript k is varied. Hence, equation (1.12) and analogous models for the case of pressure variation, etc., can be used as probes for chemical similarity in the same way as other LFER. The parameters λ_k are not the same as the parameters θ_k, but they are linearly related to one another.

When analyzing the temperature dependence of a number of reactions one should treat the data directly in relation to equation (1.12) and *not* in relation to other formulations of the isokinetic model obtained by transforming (1.12) under the assumption that this is an *exact* relation—for instance that the variation in $\Delta H^\ddagger$ is linear in $\Delta S^\ddagger$.[2,18] We note that $\Delta H^\ddagger$ and $\Delta S^\ddagger$ contain such terms as $\partial y/\partial(1/T)$, i.e., the derivatives of y with respect to one of the elements in $\mathbf{X}$ in equations (1.3) and (1.4). Hence the degree of approximation is decreased by one. Hence, even when errors of measurement are extremely small, the model errors due to the incomplete vanishing of the remainder $R(3)$ in equation (1.4) makes such transformations highly risky (see Section 1.2.3).

Furthermore, one should be aware of the difference in principle between equations (1.5) and (1.12). It might well happen that data are well described by equation (1.12) but that the same data cannot be well represented by, for instance, the Hammett equation (1.5) at a fixed temperature, say T_1. This means that the variation within the reaction series (variation of the subscript k) at one temperature T_1 is similar to the corresponding variation at another temperature T_2. At the same time the variation within the series at a fixed temperature T_1 is *not* similar to the variation in other series with other reaction centers, which *are* described by the Hammett equation. An example of this is given in Ref. 20, where pK_a values for substituted anilinium ions are well represented by equation (1.12), including the *p*-nitro-substituted compound, which deviates greatly from the simple Hammett equation.

The converse relationship is binding, however. When a reaction series observed at a fixed temperature, T_0, gives data that are well fitted by a simple LFER, such as the Hammett equation, data for the same series when the temperature is varied around T_0 must fit equation (1.12) or its extensions[20] with the same precision. This follows directly from the derivation of equation (1.5), where temperature T can be one of the elements in the vector **Z**. Claims to the contrary in the literature[21–23] are due to erroneous data analysis, as discussed below.

1.2.3. Nontransformability of Approximate Models

We can now discuss a topic that is liable to confusion, viz., the temperature dependence of closely similar reactions.

In Section 1.2.1 it was shown that data y_{ik} for closely similar systems S_{ik} can be described approximately by a simple LFER (1.5). The deviations ε_{ik} in equation (1.5) contain both small model errors and errors of measurement. Writing these separately as δ_{ik} and ε_{ik}, respectively, equation (1.5) becomes (1.13). The temperature dependence of observations

$$y_{ik} = \alpha_i + \beta_i\theta_k + \delta_{ik} + \varepsilon_{ik} \tag{1.13}$$

for similar systems is described by equation (1.12), but one might be interested in deriving the temperature dependence of data for individual reactions from equation (1.13). When the data y_{ik} are logarithms of rate constants, this temperature dependence involves the parameters $\Delta H^\ddagger$ and $\Delta S^\ddagger$, which correspond to the slope and intercept in a linear plot of $\log k$ against $1/T$. $\Delta H^\ddagger$ and $\Delta S^\ddagger$ can be estimated approximately as in (1.14) and (1.15).[24] Assume further that both the substituent scale (θ) and the errors

$$-\Delta H^\ddagger_{ik} = R\frac{\partial y_{ik}}{\partial(1/T)} + RT \quad = R\beta_i + RT \tag{1.14}$$

$$\Delta S^\ddagger_{ik} = Ry_{ik} - \frac{R}{T}\frac{\partial y_{ik}}{\partial(1/T)} = R\alpha_i \tag{1.15}$$

of measurement (ε_{ik}) are independent of temperature (see Section 1.1.2). If, as usual, the temperature dependence of the data y_{ik} is well described by the Arrhenius equation, it follows directly that the parameters α and β in equation (1.13) must be linear in $1/T$. From equations (1.13) to (1.17) we have (1.18), (1.19). Thus, apart from the terms containing δ_{ik}, ε_{ik}, and the temperature derivative of δ_{ik}, we see that $\Delta H^{\ddagger}_{ik}$ and $\Delta S^{\ddagger}_{ik}$ are linearly related for fixed i and given temperature $T = T_0$. The relationship is complicated by

$$\alpha_i = a_{0i} + a_{1i}/T \tag{1.16}$$

$$\beta_i = b_{0i} + b_{1i}/T \tag{1.17}$$

$$\Delta H^{\ddagger}_{ik} = -Ra_{1i} - R\theta_k b_{1i} - R\frac{\partial \delta_{ik}}{\partial(1/T)} - RT \tag{1.18}$$

$$\Delta S^{\ddagger}_{ik} = R\left(a_{0i} + \frac{a_{1i}}{T} + \delta_{ik} + \varepsilon_{ik}\right) - \frac{R}{T}(a_{1i} + \theta_k b_{1i}) - \frac{R}{T}\frac{\partial \delta_{ik}}{\partial(1/T)} + R\theta_k(b_{0i} + b_{1i}/T)$$

$$= \varphi_{1i} + \varphi_{2i}\theta_k + R(\delta_{ik} + \varepsilon_{ik}) - \frac{R}{T}\frac{\partial \delta_{ik}}{\partial(1/T)} \tag{1.19}$$

the fact that in practice $\Delta H^{\ddagger}$ and $\Delta S^{\ddagger}$ are often determined from the same experimental results, by evaluating $\Delta H^{\ddagger}$ as the slope of a plot of $R \ln k$ against $1/T$ and $\Delta S^{\ddagger}$ from the corresponding intercept. This introduces a nonnegligible spurious correlation between $\Delta H^{\ddagger}$ and $\Delta S^{\ddagger}$ as a consequence of additional terms common to both $\Delta H^{\ddagger}$ and $\Delta S^{\ddagger}$. These terms all contain the residuals $(\delta_{ik} + \varepsilon_{ik})$ and can completely dominate the apparent correlation between $\Delta H^{\ddagger}$ and $\Delta S^{\ddagger}$. The problem has been adequately discussed by Petersen,[25] Exner,[18] and others.[26]

It has, however, been claimed that in the absence of errors of measurement, or if these are sufficiently small, one can still use a plot of $\Delta H^{\ddagger}$ against $\Delta S^{\ddagger}$ to investigate the similarity between systems.[21] It is seen from the present treatment that this can be done only when the terms $\partial\delta_{ik}/\partial(1/T)$ are small compared to other terms in equations (1.18) and (1.19). Hammett has, however, shown that these terms sometimes are substantial,[27] which partly explains why plots of $\Delta H^{\ddagger}$ against $\Delta S^{\ddagger}$ cannot be used as similarity probes; data should instead be directly treated by means of the form corresponding to the ordinary LFER, i.e., equation (1.12).

This example shows the difficulties resulting when an approximate model is translated from one coordinate system to another. Only by retaining both residual terms is it possible to investigate whether a particular transformation is feasible or not. It is much safer, and therefore to be recommended, always to treat the data in the same coordinate system, i.e., always relating y_{ik} to other y_{jk} by means of various similarity models and not sometimes comparing y_{ik} with, for instance, $\partial y_{jk}/\partial(1/T)$.

1.2.4. *Summary of the Mathematical Derivation*

In Sections 1.2.1 and 1.2.2 we have discussed some mathematical properties of measurements (y_{ik}) made on the similar systems S_{ik}, obtained

Table 1.1. Mathematical Similarity Models of Differing Approximating Power[a]

Model	Degree of approximation, M	Equation number
$y_{ik} = \gamma + \varepsilon_{ik}$	0	(1.20)
$y_{ik} = \alpha_i + \varepsilon_{ik}$	0	(1.2)
$y_{ik} = \alpha_i + \theta_k + \varepsilon_{ik}$	1	(1.21)
$y_{ik} = \alpha_i + \beta_i\theta_k + \varepsilon_{ik}$	2	(1.5)
$y_{T,k} = c_0 + (1 + c_1/T)\lambda_k + \varepsilon_{Tk}$	2	(1.12)
$y_{ik} = \alpha_i + \beta_i\theta_k + \gamma_i\theta_k^2 + \varepsilon_{ik}$	3	(1.7)
$y_{ik} = \alpha_i + \sum_{a=1}^{A} \beta_{ia}\theta_{ak} + \varepsilon_{ik}$	$2 \leqq M \leqq A$	(1.8)

[a] The subscript i signifies the larger and the subscript k the smaller variation of the systems S_{ik} on which the data y_{ik} are measured. The parameters α, β, and θ are, in principle, estimated from the data in order to make the deviations (residuals) ε_{ik} as small as possible according to some criterion, usually the least-squares criterion. The degree of approximation, M, signifies that the remainder in the Taylor expansion contains terms of degree higher than M only.

by modifying the system S by two separate modifications involving subscripts i and k, respectively. The case when one of the modifications is made in a continuous fashion, giving the system $S_{T,k}$ and the corresponding data $y_{T,k}$, has also been discussed.

The values of these measurements (y) can, if the change of the systems with subscript k is small over all the data, be represented by one of the mathematical relations in Table 1.1. Here, equation (1.5) corresponds to the standard formulation of LFER, e.g., the Hammett equation ($\alpha_i \approx \log k_{i0}$, $\beta_i = \rho_i$, and $\theta_k = \sigma_k$). Equation (1.8) corresponds to multiple-term LFER (A terms) and can result from either a large number of independent "effects" influencing the system when subscript k is changed, or from the large variation of one "effect" between the systems, or from a mixture of these possibilities. Hence, only equations (1.5), (1.7), and (1.12) have a simple and unique interpretation. Equations (1.20) and (1.21) are included in the table to indicate their connection with other similarity models. Equation (1.20) describes the trivial case when all changes in both i and k are so small that all the systems S_{ik} are virtually identical. Equation (1.21), the "additivity relation," is obtained when equation (1.1) is expanded by use of Taylor's method and all terms of degree 2 and higher are negligible, i.e., the variation with both i and k is extremely small.

1.2.5. What is Meant by "Similarity"?

In the mathematical derivation given in Section 1.2.1, the treatment was based on the assumption that the variation of the systems S_{ik} with the subscript k (e.g., variation with substituent) was much smaller than the variation of the same systems with the subscript i (e.g., the reaction center). Hence, when data are well fitted by equation (1.5), with the parameters α_i significantly different from zero, one must conclude that the influences of the two modes of change corresponding to i and k differ in magnitude, the first being larger than the second.

For instance, when a reaction series is well fitted by the Hammett equation, this means that the substituents in this series influence the reactivity (as measured by y) in a manner that is closely similar to the manner in which the same substituents influence the reactivity in other reaction series, which are also well fitted by the Hammett equation, i.e., the change with subscript k for one series is closely similar to the corresponding change with subscript k for other series. In addition, the reaction center behaves very similarly *within* each series, i.e., the reactions within each series including the first series, follow closely similar mechanisms. The mechanism can, however, differ widely from one series to another; the similarity between the first reaction center and the other reaction centers need not be very close. This is well exemplified by the wide range of reactions to which the Hammett equation is applicable, from benzoic acid ionization equilibria to enzymic reactions.[3] At the same time, the range of substituents is not very wide.[7]

One important corollary of the form of the similarity models (1.5) and (1.8) is that similarity is a collective property of the class, not of the individual systems. In order to study whether this class similarity extends to a particular system S_{jm}, one must therefore study the relationship between the corresponding observation y_{jm} and at least two other observations for the same subscript j (say y_{jk} and y_{jl}) and at least another two observations for the same subscript m (say y_{im} and y_{gm}). Hence, by means of LFER, one does not study the similarity between single systems but rather the similarity between series of systems—say S_{ik} and S_{jk} with $k = 1, 2, \ldots, N$. However, if one makes several observations on each system, giving for system k the data y_{jk}, the similarity between a number of systems S_k, $k = 1, 2, \ldots, N$, can be studied by analyzing the data y_{jk} by means of the models (1.5) and/or (1.8). This will be the subject of Section 1.4.

1.2.6. *Normalization of Parameters*

When the models (1.5) and (1.8) are fitted to data y_{ik}, the parameters α, β, and θ are not uniquely defined. Hence, a unique foundation of the models and their parameter scales requires additional constraints on the parameters, so-called normalization constraints. Two constraints are needed in equation (1.5). A multiplication of all β_i's by a factor μ can be compensated by division of all θ_k's by the same factor. Hence a constraint must be imposed on the former parameters, either by fixing one β_i to have a value different from zero, say $\beta_1 = 1$, or by, for instance, keeping $\sum \beta_i^2 = 1$. Moreover, addition of a number to all the parameters θ_k can be compensated by the subtraction of $\Delta\beta_i$ from all the parameters α_i. Hence one constraint must be imposed on the θ_k's. Usually one specifies one θ_k to have a certain value, say $\theta_1 = 0$.[1–3]

In equation (1.8), one therefore needs two normalization conditions for each of the A product terms $\beta_{ia}\theta_{ak}$, and these can be chosen in the same way for each term, as described above for the single term in equation (1.5). The traditional normalization of the parameters in model (1.5) is to set $\beta_i = 1.0$ for one well-known standard reaction (with subscript $i = 1$). For the Hammett equation, the ionization of benzoic acids in water at 298.15 K (25°C) is defined to have a ρ value (corresponding to β_1) of unity. The substituent scale is fixed by arbitrarily giving σ_k (corresponding to θ_1) for hydrogen the value zero. It should be noted that when fitting a single series (with fixed subscript i) to equation (1.5) or (1.8), the parameters θ are already defined in value. Hence, this single series fitting corresponds to the estimation of the parameters α and β only, which does not require any normalization.

1.2.7. Chemical Interpretation and Consequences

So far, we have discussed a mathematical formalism based on the local approximation of a rather general function F of two vector variables. The formalism has some consequences for the interpretation of chemical data related to LFER. Two assumptions were made as a basis for the mathematical treatment, viz.:

(*a*) The measured data y_{ik} are thought of as generated by a function F of two vector variables $\mathbf{Z}$ and $\mathbf{X}$, and the function is assumed to be continuous and several times differentiable.

(*b*) The change with the subscript k (i.e., the substituent in the Hammett equation) can be described by a single coordinate of change, t (see Fig. 1.1).

The first of these assumptions is not especially controversial; it is generally believed that measurements made on a macroscopic chemical system with about 10^{23} molecules in it can be described in this way. The second assumption is, however, both crucial and debatable. In certain cases, as when the change with k corresponds to the change of temperature or solvent composition, this assumption is rather natural. When, however, this change corresponds to a change of substituent (Hammett and Taft equations), solvent or reactant (Grunwald–Winstein, Swain–Scott, and other equations), or a change of other discrete entities,[3] this assumption seems more debatable. It is difficult to imagine a variable of change in which a small and continuous change converts, for instance, a nitro-group into a hydrogen atom. Nevertheless, the fact that LFER such as the Hammett equation correlate well large masses of data shows that, phenomenologically, such a single variable of change is a good approximation to what happens in the processes generating these data.

In our view, this is rationalized most easily if we see a substituent, or any other discrete molecule or molecular part corresponding to the subscript k in LFER (1.5), in terms of a shell model with properties similar to the valence model of an atom (Fig. 1.2). Inside the shell, in the core, any number of variables may operate in any complex pattern, as long as the outside of the shell, analogous to the valence shell of the atom, behaves regularly. Chemists observe only the effects of the substituents on other parts of the molecule such as the reaction center. Hence, operationally they can "see" only the "outside" of the substituent shell and therefore they neither need to nor can they interest themselves in what happens inside the shell. This definition of "substituent" corresponds to that of Palm,[13] which has also been discussed by Taft[28] and Krygowski.[29]

If substituents are "distant" from the reaction center, the replacement of one substituent by another is seen only as the change of one outer shell into another. This can be described phenomenologically as the change of a single variable, t, and the data can be described by equation (1.5). When the substituent interacts strongly with the reaction center, we observe "inner shell" effects such as resonance and steric effects. The strong interaction as such might, however, demand a type of mathematical model other than the LFER model, since strong interactions usually correspond to large changes between systems and therefore possibly also to the breakdown of the assumptions made in the derivation of the similarity models (1.5) and (1.8).

With the shell model, one might ask how far a relation such as the Hammett equation can be applied, i.e., what is the largest domain of substituents and reactions that can be covered by equation (1.5) and a single substituent-parameter scale θ (corresponding to σ in the Hammett equation)? If we see the substituents as in Fig. 1.2, there is nothing in

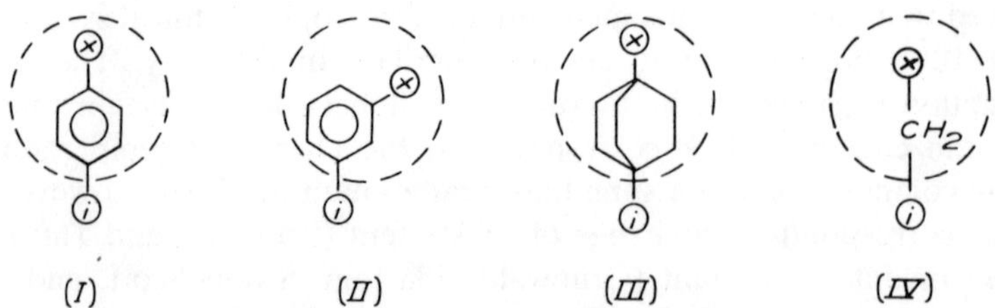

Fig. 1.2. The shell model of a substituent. According to this model, the four substituents in (I)–(IV) are different even if the group x is the same.

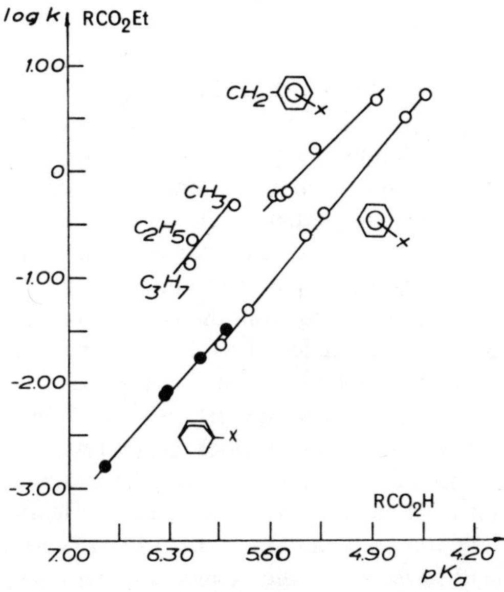

Fig. 1.3. Plot of the logarithms of rate constants for hydrolysis of esters RCO_2Et in 87.8% EtOH–water against the corresponding logarithms of ionization constants of RCO_2H in 50% EtOH–water. Data from Refs. 31–36. The obvious lack of fit of the data to a single line shows that equation (1.5) cannot simultaneously fit data for compounds with substituents as in (I)–(IV) in Fig. 1.2.

principle that prevents the Hammett equation also being applied to alicyclic and to aliphatic systems such as (III) and (IV). A simple way to test whether this is possible is to plot data for two reaction series against one another for a number of substituents of the types in (I)–(IV). This is done in Fig. 1.3, resulting in clearly different lines for substituents of the types in (III) and (IV); in other words, several "effects" vary outside the shell. This shows, as is already well known, that the Hammett equation is limited to *meta*- and *para*-substituted benzene derivatives. The same approach was used by Wells[30] to show that the Hammett equation could not be extended to naphthalene series. The problem of defining the domain of an LFER is further discussed in Section 1.3.2.

This shell model of substituents also resolves the paradox pointed out by Hine[37] (see also a discussion in Ref. 2, p. 193). The paradox can be formulated as follows. The fact that the Hammett equation can be applied to data for both *meta*- and *para*-substituted benzene derivatives shows, according to the similarity interpretation of LFER, that the influence of both types of substituent on the reactivity of these compounds is similar. One would then, assuming the validity of a linear similarity model between similar substituents, expect that a plot of data for *meta*-substituted compounds against the corresponding data for *para*-substituted compounds should be a straight line. This is not so for all substituents that fit the Hammett equation.

However, if we see each substituent as a sphere containing also the aromatic nucleus (cf. Fig. 1.2), we realize that the first test—whether or not a substituent fits the Hammett equation—is a test of whether a single "effect" varies outside the shell when one substituent is replaced by another. The second test—whether a plot of data for *meta*-substituted compounds against corresponding data for *para*-substituted compounds gives a single straight line—is a test of whether a single "effect" operates *inside* the sphere. Thus, it is clear from the nonlinearity of the plot that at least two such "effects" are needed. This does not present any theoretical problem; Katritzky and Topsom[38] have listed at least five such possible "effects." In practice, the result of the first test, that only one "substituent effect" influences reactivity in many cases, is the result of interest to the chemist who uses LFER as a tool to investigate similarities and dissimilarities among his reactions or other systems. He is interested in a simple and efficient tool, and the one-term LFER (1.5) is ideal for this purpose. The role of physical organic chemistry in this context is to provide optimal substituent scales and explicit definitions of reaction types to which LFER (1.5) with these scales are applicable.

The physical organic chemist, however, is also interested in explaining substituent effects in terms of concepts that he or she, at least at present, believes have a deeper foundation than the empirical θ parameter scales of equation (1.5). Concepts currently favored are, for example, effects of electron delocalization and polarization, interactions between unipoles, dipoles, and quadrupoles, and so on. To investigate the influence of these effects on reactivity, LFER-like models, such as equation (1.8), but with an explicit physical interpretation of each term, are often used. The theory and practice of such investigations is conceptually different from the topic dealt with here, however, and will not be further discussed. We note only the difficulties with the interpretation of multiple-term LFER, see for instance recent discussions by Ehrenson *et al.*[39] and Clementi *et al.*[40] However, let us discuss some consequences of seeing LFER as empirical similarity models, able to describe data y_{ik} for systems S_{ik} provided that the variation with subscript k is small. It follows that even if one believes that several effects vary within an ensemble of similar chemical systems, it is difficult to distinguish between them phenomenologically and estimate their separate influence on the systems. If the variation between the systems is kept small so that one can expect linear contributions from each effect, these contributions tend to mix and become indistinguishable—the simple LFER (1.5) will then be a good approximate model for the data. When, on the other hand, the variation between the systems is made larger, one can expect a "decoupling" of the effects and nonlinearities in each of them due to the larger variation. The data will then have to be analyzed in relation to a more complex model, which may be difficult to interpret.

Similarly, when we find that data are adequately approximated by the model (1.8) with A terms, this does *not* necessarily imply that there are A different effects influencing the systems under study. The reason may be that the variation of a single effect is so large that several terms in equation (1.8) are needed to approximate its nonlinear behavior. The reason may also be the limited variation of a large number of effects (larger than A), the small variation of which results in the effective number of independent elements in $\mathbf{X}$ in equation (1.1) being A. Finally, the reason may be anything between these two extremes. Hence, for a large class of systems, such as the class of all aromatic reactions, one might anchor the center of an LFER in different places and get different resulting θ scales (Fig. 1.4). Alternatively, it is self-evident that several one-term LFER can be fused together into a multiple-term LFER (1.8). The interpretation and application of this LFER, however, is more difficult than the use of a number of single-term LFER.

This possibility of slightly different and still overlapping LFER is a source of confusion. The field of aromatic reactivity is a good illustration with its proliferation of different σ scales and multiple-term LFER. Swain and Lupton[41] have made a rough analysis of a large number of substituent scales and concluded that the combination of two independent scales is sufficient to approximate all these scales. In the absence of a more stringent statistical analysis, which should include an investigation of how well the observations are approximated by these two scales, this indicates that, even

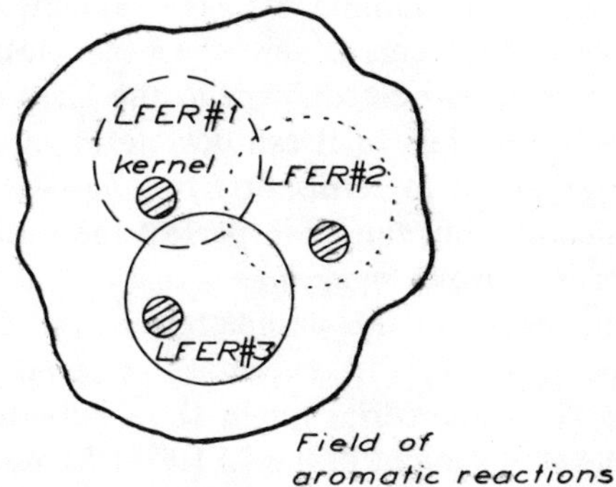

Fig. 1.4. In a large class of systems, such as the class of aromatic reactions, a single-area, one-term LFER (1.5) cannot describe more than parts of the data. Hence, for such a class, there may be several more or less overlapping LFER. Each LFER is defined by a kernel domain and additional series of data that are well fitted by the LFER with the same θ scale as obtained for the kernel data.

that are well fitted by the same LFER are similar, at least in some ways. with alicyclic and aliphatic systems such as (III) and (IV) in Fig. 1.2, the main variation in many series of data is rather well described by a two-term LFER (1.8).

1.3. Using LFER in Practice

LFER can be interpreted as empirical models of similarity. Systems that are well fitted by the same LFER are similar, at least in some ways. Equally important, a system that does *not* fit the LFER is *not* similar to the other processes. Hence, an LFER can be used as a kind of sieve through which most systems under analysis pass. The chemist can then concentrate on the interesting systems that stay in the "sieve," the nontypical cases. The following are the main problems where LFER are useful as tools in data analysis:

(*a*) To determine whether a class of chemical systems, usually reactions or compounds, behave similarly, as discussed in Section 1.2.5. Subproblems: Which systems (if any) behave "abnormally"? Which observations (subscript k) made on these systems behave "abnormally" (if any)?

Thus, by the application of a simple LFER (1.5) to logarithms of rate and equilibrium constants for reactions of *meta*- and *para*-substituted benzene derivatives, it is possible to (i) establish a domain of reaction types and substituents where the simple LFER (1.5), the Hammett equation ($\alpha_i \approx \log k_{0i}$, $\beta_i = \rho_i$, and $\theta_k = \sigma_k$), is applicable with a single substituent scale (σ^0), (ii) determine the values of the parameters in this scale that make the LFER (1.5) fit best the data in this domain.[15–17]

(*b*) To find, once the similar behavior of systems in a class is established, as in the example with the Hammett equation, whether an additional series of systems is similar to this class. This series, usually a reaction series, may either conform to the class behavior as well as the other systems in the class or it can deviate in various ways. These deviations can be analyzed in terms of (i) *outliers*—a single point or points deviate significantly, but the main part of the series conforms well to the class behavior, (ii) *two or more lines*—one can, for instance, get different lines (different values of the parameters α and β) for *meta*- and *para*-substituted compounds, (iii) *curvature*—instead of conforming to the simple LFER (1.5), the series might show curvature, equation (1.7), (iv) *lack of fit*—the series might fit the LFER (1.5) much worse than the other systems in the class or it might show a complete lack of fit, i.e., the LFER does not represent the variation within the series at all.

(*c*) To place the system (e.g., a reaction series) in one of several possible classes. If observations on a large class of systems, say reactions, cannot be fitted by a single LFER (1.5), this means that the variation within part of the class is dissimilar to the variation in other parts of the class. However, it is often possible to divide the class into subclasses so that within each subclass the systems behave similarly, while the subclasses are dissimilar. One can then use the corresponding subclass LFER to classify the new system according to which of these LFER fit the data best.

In the area of the Hammett equation, several examples of this type of classification are found. Thus, for instance, a reaction series better fitting σ^+ constants than σ^0 constants is often interpreted to have direct conjugation between donor-substituents and the reaction center in the transition state.[42]

(*d*) Prediction. By fitting an LFER to a few points for a reaction series, other values in the series can be predicted from the coefficients thus obtained. Though the Hammett equation has been used in this way from time to time, this is apparently not an area where LFER have so far been extensively used; see for instance a discussion by Exner.[7]

(*e*) Explanation. When a particular LFER, such as the Hammett equation, has been applied in various situations for a long time, the experience thus accumulated can often be used to relate parameter values of the LFER to microscopic theory. For the Hammett equation, the size and sign of the parameter ρ [corresponding to β in equation (1.5)] is often given physical significance.[7] Similarly, the values of the parameters θ_k (σ_k in the Hammett equation) are often interpreted in terms of electronic effects.[7] The explanatory use of the LFER is, however, the subject of many of the other articles in the present volume and is better treated in the relevant context.

It is of some importance to keep in mind the following points when the parameters obtained by fitting an LFER to data are used in theoretical discussions: (i) limitations in the interpretability of the parameters in view of the approximate nature of LFER, (ii) the reliability of the parameter values. Information about the latter can be obtained only when statistical methodology is used in the data analysis. This will be the subject of the following sections. We note the connection between, on the one hand, the models (1.5) and (1.8) and, on the other hand, the models for multiple linear regression. The parameters θ in the former models correspond to the independent x variables in the latter. The difference is that the θ's are usually estimated from the data y, while the x values are directly measured. Hence, once the θ's are defined—by estimation from a data matrix y or in terms of measured values in a "standard series"—the two classes of models are equivalent. Their interpretation in terms of cause and effect is therefore similar; the substituents k cause a change of the measured response y, the size of which is estimated by the parameters θ.

1.3.1. The Statistical Approach

Data can be analyzed by means of LFER by various different methods. In the early days of LFER, the only way to analyze data in terms of equation (1.5) was to define θ values by means of a standard process, e.g., a standard reaction. For the Hammett equation, this standard series defining the σ constants was the ionization of benzoic acids in water at 298.15 K.

In the statistical approach, on the other hand, all data are regarded as of equal weight and one tries to find parameters [α, β, and θ in equation 1.5)] that make the particular LFER fit a large body of carefully selected data as well as possible, in some quantitative sense. This θ scale is then used in later analyses of single series. The first attempt to use statistical methods with LFER was made by Jaffé, who made an analysis of the Hammett equation (under some simplifying assumptions) based on 371 data series.[43] The resulting σ scales have formed the basis of most later work in the field. With the advent of large computers it is no longer necessary to have recourse to standard series, or to make simplifying assumptions in the statistical analysis. By means of iterative methods, one can determine a θ scale that fits the data at hand in an optimal way (see below).

It has been argued that basing a θ scale (σ scale in the Hammett equation) on the statistical analysis of a large number of data series has some disadvantages. First, it would make the θ values depend on the body of knowledge at the time of the evaluation, making a revision necessary from time to time. Secondly, the large data basis would tend to obscure possible deviations, since these would tend to be smoothed out. Both arguments are invalid. First, models are *always* approximate and should never be taken to be definitive. Secondly, since there is no evidence that any particular reaction series is theoretically better suited to be a standard than others, the statistical approach is the only reasonable[44] one. By making the analysis carefully, one can avoid the danger of smoothing out the deviations. In fact, *only by the use of objective statistical methods can such deviations be unambiguously detected.* However, it is certainly not adequate to make a purely statistical analysis by trying to obtain parameters that fit best all kinds of data. The data incorporated into the analysis must be carefully chosen so as to correspond to the chemical scope of the LFER. The results are otherwise impossible to interpret and therefore useless.

There is also a practical reason for adopting the statistical approach. If the θ constants are defined by a standard series, these values will contain a certain error, δ, due to errors of measurement and also, if we accept the theoretical arguments in Section 1.2.1, due to model errors resulting from the incomplete vanishing of the remainder $R(3)$ in equation (1.4). Hence we have equation (1.22). When these θ values are used later in correlating other observations by means of equation (1.5), the deviations ε_{ik} will be larger by a factor f, which is given approximately by (1.23). Here S_δ^2 is the

$$\theta_{\text{obs}} = \theta_{\text{true}} + \delta \tag{1.22}$$

$$f = (S_\varepsilon^2 + S_\delta^2)^{1/2}(S_\varepsilon^2)^{-1/2} \tag{1.23}$$

variance of δ in equation (1.22) and S_ε^2 the variance of the observed deviations ε_{ik} in equation (1.5). Hence, the precision of the correlation will be unnecessarily low owing to the choice of θ values as defined by a standard process.

1.3.2. Analysis of Data from a Whole Class

The problem of investigating whether a whole class of systems is similar with respect to the observed data, y_{ik}, can be formulated as an investigation of whether the data y_{ik} are well fitted by equation (1.8), with the number of terms A being "small." Provided that the data matrix **Y** is complete, i.e., the data y_{ik} are defined by observed values for all $i = 1, 2, \ldots, M$ and all $k = 1, 2, \ldots, N$, this is a well-studied problem in multivariate statistics known as *principal components analysis* or *factor analysis*.[45–48] Straightforward procedures have been developed for determining how many terms (A) are needed for equation (1.8) to describe the data adequately.[49]

The situation in chemistry is, however, more difficult insofar as most data in the matrix **Y** usually are undefined. Thus, for the Hammett equation, Jaffé estimated that about 3200 out of 42000 possible data in his statistical analysis were defined by actual observation,[43] i.e., about 92% of the data were missing. The situation has not essentially changed since then, and if a statistical analysis of the applicability of an LFER to a class of chemical systems is to be made, one cannot, therefore, use the standard methods of multivariate analysis. However, methods have been worked out for fitting equation (1.5), i.e., the simple one-term LFER, to incomplete data,[44,50–52] including the estimation of confidence intervals of the θ parameters.[52]

It is important to point out that data analysis by these statistical methods does not rely on the definition of θ values in terms of a standard process. On the contrary, for a given incomplete data matrix **Y**, the parameters α_i, β_i, and θ_k are calculated so as to minimize the total residual sum of squares ($\sum_{i,k} \varepsilon_{ik}^2$). It should also be noted that it is advantageous to treat the similarity model in the form (1.5) and estimate both the α and the β parameters for each modification i (e.g., reaction series) and the parameters θ for each modification k (e.g., substituent). The LFER (1.5) is sometimes rewritten in the form (1.24) by approximating α_i by y_{i0} (usually the value for the unsubstituted compound in the series). The observed

$$y_{ik} - y_{i0} = \beta_i \theta_k + \varepsilon_{ik} \tag{1.24}$$

value y_{i0} contains, as discussed in Section 1.3.1, an error δ, which is spread out over the other members of the series if the data are analyzed in relation to the model (1.24). In addition, one can no longer test whether the unsubstituted compound (or whatever corresponds to y_{i0}) deviates significantly from the LFER. Thus, the treatment of data in the form (1.24) corresponds to the assumption that the unsubstituted compound is always well behaved. Since the estimation techniques do not require this assumption, we recommend that the data be treated according to equation (1.5).

When the model (1.5) does not fit the *whole* data matrix **Y** well, as is the case with the Hammett equation and aromatic reactivity data, the problem is no longer one of simple model fitting, but is one that must be attacked by a strategy rather than by an algorithm. The problem is complicated by LFER being approximate models. Thus, the data can usually be divided in several different ways so that a simple LFER fits well each part of the divided matrix (Section 1.2.7). Hence, a chemically sound strategy should involve the following steps:

(*a*) The definition of a core domain where the actual LFER is thought to have some theoretical foundation. If, for example, we wish to define an area of application of the Hammett equation in which conjugative interaction between substituent and reaction center is absent, reaction series for which the reaction center is structurally isolated from the aromatic nucleus, as in phenylacetic acids, for example, would provide a first approximation to such a core domain.

(*b*) The fitting of equation (1.5) to the core data indicated in (*a*), by using, for instance, the method described in Ref. 15.

(*c*) The screening of resulting residuals for abnormally large individual values (test for outliers, Section 1.3.7) and for abnormally large values for a reaction series and/or a substituent. For the latter, one can use approximate F tests based on (1.25). Here s_i^2 and s_k^2 are the residual

$$F = s_i^2/s_0^2 \quad \text{or} \quad F = s_k^2/s_0^2 \tag{1.25}$$

variance of the ith series and of the kth substituent, based on all defined residuals ε_{ik} for that i or k, respectively, and s_0^2 is the total variance of all defined residuals (for data defined by actual observed values), equation (1.26) below.

(*d*) The deletion of data for those substituents and/or reaction series and/or individual points that show abnormal deviations.

(*e*) Refitting according to step (*b*).

(*f*) Obtaining, after one or two iterations (*b*)–(*e*), in favorable cases, a core-data matrix for which all series and substituents show good behavior.

(*g*) Extending this core-data matrix in various directions, one direction at a time, by the incorporation of a number of series for each direction. For each new direction, the total data matrix is refitted according to equation (1.5). Deviations are tested for as in step (*c*), but only deviating observations that were included in the *last step* are deleted, i.e., the core-data matrix should not be changed.

Thus one obtains the following final results:

(*a*) A "domain" defined by combinations of well-behaved reaction series and substituents, for which the LFER (1.5) fits the data "well"; i.e., no systematic deviations are detected.

(*b*) Deviations, ε_{ik}, for the observed data in the domain. The standard deviation [s_0 in (1.26)] expresses the size of typical deviations. The number of degrees of freedom in (1.26), i.e., the number that the sum of squares is

$$s_0 = \left\{ \sum_i^M \sum_k^N \varepsilon_{ik}^2 \Big/ \left[\sum_{i=1}^M (n_i - P) - A(N-2) \right] \right\}^{1/2} \tag{1.26}$$

divided by, is the total number of observed data, n_i, minus the number of estimated parameters. This number is $P(=1+A)$ for each series (α_i and β_i) plus $A(N-2)$ (θ values but for two constraints).

In some cases the standard deviation of individual series is related to the parameter β. This can be expected if LFER are seen as approximate models and has indeed been found for the Hammett equation[17] (see p. 24). In such a case, outliers in a series should not be tested for by using s_0 in equation (1.26), but rather the typical s_i value for the β value at hand. Figure 1.5 shows the estimated values, s_i, as a function of the corresponding ρ values (corresponding to β_i) obtained in a statistical analysis of the Hammett equation.[17] One sees a clear dependence of s_i on ρ, which means that the fit of the Hammett equation in terms of the residual standard deviation is not constant.

(*c*) Values of the parameters θ_k that fit the data best in the domain, under the assumption that the data used in the analysis are representative of all data in the same domain. Also, one obtains confidence intervals of the θ parameters. This gives, *inter alia*, an indication of which θ values are significantly different from one another and which are not.

1.3.3. Application to the Hammett Equation

The strategy described above has been applied to aromatic reactivity data in order to define a region where the simple Hammett equation (1.5) (with $\alpha_i \approx \log k_{0i}$, $\beta_i = \rho_i$ and $\theta_k = \sigma_k$) is valid with a single σ-parameter scale.[15–17] The data were chosen so as to give a σ scale, σ^0, which should

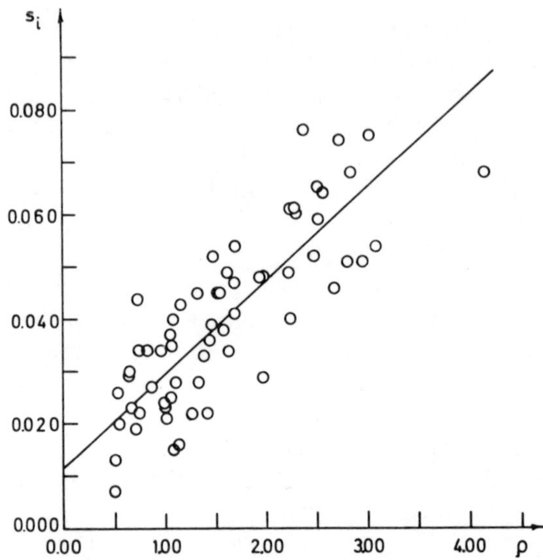

Fig. 1.5. The residual standard deviation s_i [equation (1.37)] as a function of ρ as found in the statistical analysis of the Hammett equation.[17]

correspond to substituent influences excluding cross-conjugation and steric effects. This was accomplished by starting in step (a) with a core domain defined by 21 reaction series ($i = 1, 2, \ldots, 21$) in which the reaction center was isolated from the aromatic nucleus by a CH_2 group or another nonconjugating group.

With these core data, we investigated whether data corresponding to *meta*- and *para*-substituents can be approximated by the same LFER or whether they fit separate scales better. We found that they could indeed be incorporated into the same LFER. For these data, none of the substituents included showed large deviations.

The core-data matrix was then expanded to include reaction series of σ type (typified by ionization of benzoic acids in water), σ^- type (typified by ionization of phenols in water), and σ^+ data (typified by solvolysis of *t*-cumyl chlorides in acetone–water). A number of substituents were found to deviate significantly in parts of the data matrix, and in Table 1.2 the resulting "domain" of the Hammett equation with a σ^0 scale is found together with estimated values of σ^0. The details are in Refs. 15–17.

One should be aware of the requirements the data must meet for a statistical analysis. In order to get reliable estimates of the σ constants, data should be available for the substituent in at least three and preferably five or more series. Similarly, *at least* five or six data points must be defined by observation in each series for the analysis to have a stable solution. This

Table 1.2. σ^0 Constants with 95% Confidence Intervals from Ref. 17[a]

Substituent		*meta*	*para*	3,5-Disubstituted	3,4-Disubstituted
H		0.00 ± 0.03			
Me	(−)	−0.06 ± 0.03	−0.14 ± 0.03	−0.10 ± 0.06	−0.19 ± 0.09
Et	(−)	−0.08 ± 0.12	−0.13 ± 0.09		
Pr^i	(−)	−0.08 ± 0.14	−0.13 ± 0.07		
Bu^t	(−)	−0.09 ± 0.11	−0.15 ± 0.07		
CH_2Ph	(−)	−0.05 ± 0.18	−0.06 ± 0.13		
Ph	(−)	0.04 ± 0.14	0.05 ± 0.10		
F	(− −)	0.34 ± 0.05	0.15 ± 0.06		
Cl	(−)	0.37 ± 0.03	0.24 ± 0.03	0.75 ± 0.13	0.57 ± 0.09
Br	(−)	0.37 ± 0.04	0.27 ± 0.04	0.76 ± 0.10	
I	(−)	0.34 ± 0.04	0.28 ± 0.04		
CF_3		0.46 ± 0.09	0.54 ± 0.11		
COMe	(+ +)	0.36 ± 0.07	0.47 ± 0.10		
COPh	(+ +)	0.36 ± 0.18	0.46 ± 0.10		
$COBu^t$	(+ +)		0.33 ± 0.18		
CHO	(+ +)	0.41 ± 0.13	0.47 ± 0.18		
CO_2R	(+ +)	0.35 ± 0.10	0.44 ± 0.09		
CO_2H	(+ +)	0.36 ± 0.18	0.44 ± 0.18		
SO_2NH_2	(+ +)	0.58 ± 0.13	0.58 ± 0.12		
SO_2Me	(+ +)	0.69 ± 0.10	0.73 ± 0.09		
CN	(+ +)	0.62 ± 0.06	0.71 ± 0.08		
NO_2	(+ +)	0.71 ± 0.04	0.81 ± 0.05	1.42 ± 0.08	
OMe	(− −)	0.10 ± 0.03	−0.12 ± 0.05	0.23 ± 0.11	
OH	(− −)	0.02 ± 0.08	−0.22 ± 0.12		
OPh	(− −)		0.06 ± 0.13		
SMe	(− −)	0.14 ± 0.18	0.06 ± 0.18		
NH_2	(− −)	−0.09 ± 0.05	−0.30 ± 0.11		
NMe_2	(− −)	−0.10 ± 0.09	−0.32 ± 0.12		
NHAc	(− −)	0.14 ± 0.11	0.00 ± 0.13		

[a] These values are applicable in all reactions of *meta*- and *para*-substituted benzene compounds in aqueous solvents, with the following exceptions: (*i*) strong donors in the *para*-position, marked with (− −), show resonance interaction in σ series, typified by the ionization of benzoic acids; (*ii*) strong and weak donors, the latter marked with (−), show resonance interaction in σ^+ series, typified by the solvolysis of *t*-cumyl chlorides; also plots for such reactions often exhibit significant curvature which makes them difficult to include in a multiple series analysis; (*iii*) strong acceptors in *para*-positions, marked with (++), show resonance interaction in σ^- series, typified by the ionization of phenols.

limits rather severely the data that can be included in the analysis. Moreover, in order to estimate a *second* component, for instance a resonance-parameter scale, one would need at least another three points per series for which a significant resonance contribution is expected. To make things worse, there is little agreement on which deviations from the simple Hammett equation are definitely due to resonance effects; other possible causes such as steric effects, solvent effects, changes in mechanism, and so on, can often be proposed (see Section 1.5.2).

These limitations have made it impossible to extend the study of the simple Hammett equation to cope with data that deviate from equation (1.5). One can conclude that, at present, observations do not exist that allow of this type of statistical analysis of data in relation to any model more complex than (1.5). One may well ask if this is an indication also of the limited utility of more complex Hammett-type models.

1.3.4. Single Series

The application of an LFER (1.5) to a class of systems, S_{ik}, to determine which systems in the class behave similarly with respect to the data, y_{ik}, was discussed in the previous section. The principal results of such an analysis were (*a*) a domain defined by combinations of i and k where the LFER (1.5) is applicable with a single parameter scale θ; (*b*) values of these parameters θ; (*c*) a typical size of the deviations between data and model, expressed as the residual standard deviation, s_0 [equation (1.26)]. Alternatively, the residual standard deviation might be related to the parameter β as in Fig. 1.5.

One can now claim with some confidence that if data for a new series of systems are well fitted by the LFER (1.5) with the same θ scale, this is an indication of similarity between this new series and the class as a whole in the way discussed in Section 1.2.5. Equally important, one can conclude that systems that deviate from the pertinent LFER are *not* similar to the class. In order to deal with such matters as what constitutes a good fit and what should be considered to be a deviation, simple statistical techniques are available, as will be discussed below.

A large number of questions can be answered, at least partially, by the analysis of a single series. Some of these questions are discussed below, but others are not discussed for various reasons. Thus, one type of application in which the fitting of an LFER (1.5) to a single series is used is the determination of so-called secondary σ values. Since we strongly believe that σ values never should be based on a single series, the statistical problems involved in this application of LFER will not be dealt with in this chapter.

1.3.5. Plots

The mandatory first step in all data analysis is to make a plot of the data in the most illustrative way possible. In the present case, this is done by plotting the observed y values (e.g., logarithms of rate constants) on the y axis against the pertinent θ values (e.g., σ^0) on the x axis, as exemplified in Fig. 1.6. The line resulting from the least-squares fit of the model (1.5) to the data (see below) can be incorporated into the same plot to show the

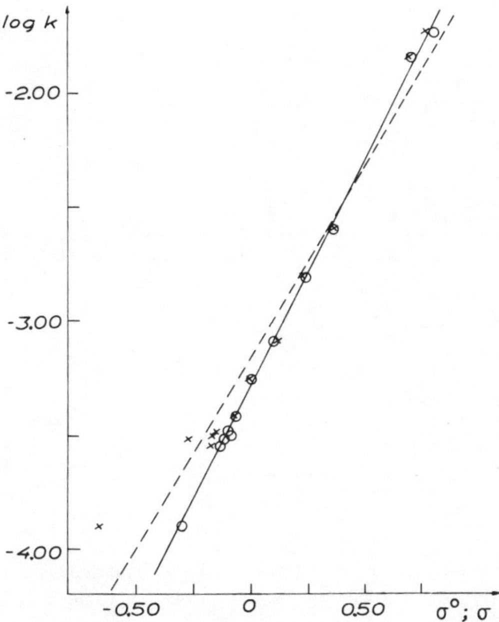

Fig. 1.6. Data for the alkaline hydrolysis of aryl tosylates (313 K, H_2O) from Maremäe and Palm[53] (Table 1.3) plotted against σ^0 from Ref. 17 (circles) and σ from Refs. 7 and 57 (crosses). The two lines are obtained by using the least-squares method (see Section 1.3.6).

extent of the agreement between data and model. In order to check for systematic deviations between data and model, it is recommended that a plot be made of the resulting residuals (ε) against the y values as shown in Fig. 1.7. The latter type of plot readily shows such things as outliers and curvature.

1.3.6. Least Squares

The fitting of the model (1.5) or (1.8) with fixed θ scale(s) to the n observations on a single series, here denoted by y_k, is best made by means of the least-squares method. Thus, the parameters α and β are determined so as to minimize the sum of the squares of the deviations ε_k between model and data. This least-squares condition gives a linear equation system of size $A+1$ [2 for equation (1.5)], which is easily solved by means of a desk calculator or a computer. The technique is well treated in numerous texts[54–56] and here we shall only summarize the results for the simplest case of equation (1.5). The least-squares method is based on a number of assumptions. Some of these are discussed below and shown to be rather natural in the present context.

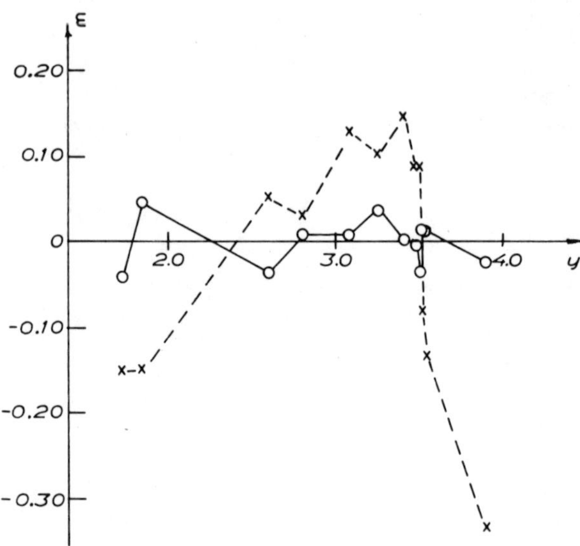

Fig. 1.7. Residuals ε from least-squares analyses (Table 1.3) plotted against the values of y for the two procedures (*i*) and (*ii*) in Table 1.3, which correspond to the lines in Fig. 1.6. Symbols are the same as in Fig. 1.6.

Computationally, the fitting of a straight line, equation (1.5), is particularly simple.[54] It involves the calculation of five sums only, viz., $\sum \theta$, $\sum y$, $\sum \theta^2$, $\sum y^2$, and $\sum \theta y$. By means of these and the number of points in the series, n, the intermediate results in equations (1.30)–(1.34) are calculated. These are then used to calculate the principal parameters of the analysis, viz., the slope, β, and intercept, α, of the line [equations (1.29) and (1.28)] and the residual standard deviation [equation (1.37)]. The individual residuals (ε_k) are then calculated from the data (y_k, θ_k) and the parameters α and β by using the equation of the line (1.27).

In the interpretation of the results, the residuals (ε_k) and the residual standard deviation (s) are of particular importance. The former tell us whether the individual systems behave similarly to the class as a whole. The latter carries information about the overall similarity of the series to the class as a whole.

Some of the technical details of the fitting are given below. The line goes through the point defined by the averages of the y and θ values [for $\bar{y}$, $\bar{\theta}$, see equations (1.33)–(1.34)] and the line can be expressed as in (1.27).

$$y_k = \bar{y} + \beta(\theta_k - \bar{\theta}) + \varepsilon_k \tag{1.27}$$

Hence, the parameter α in equation (1.5) is calculated from (1.28). The slope of the line, β, is given by (1.29). The sums S_{xy}, S_{xx}, and S_{yy} have their usual meanings, (1.30)–(1.32). Confidence intervals of $\bar{y}$ and β can be calculated from equations (1.35) and (1.36) under the assumption that the deviations ε_k are approximately normally distributed—a realistic assumption in most LFER applications; the residual standard deviation s is

$$\alpha = \bar{y} - \beta\bar{\theta} \tag{1.28}$$

$$\beta = S_{xy}/S_{xx} \tag{1.29}$$

$$S_{xy} = \sum_k (\theta_k - \bar{\theta})(y_k - \bar{y}) = \sum_k \theta_k y_k - \frac{1}{n}\sum_k \theta_k y_k \tag{1.30}$$

$$S_{xx} = \sum_k (\theta_k - \bar{\theta})^2 = \sum_k \theta_k^2 - \frac{1}{n}\Big(\sum_k \theta_k\Big)^2 \tag{1.31}$$

$$S_{yy} = \sum_k (y_k - \bar{y})^2 = \sum_k y_k^2 - \frac{1}{n}\Big(\sum_k y_k\Big)^2 \tag{1.32}$$

$$\bar{y} = \sum_k y_k/n \tag{1.33}$$

$$\bar{\theta} = \sum_k \theta_k/n \tag{1.34}$$

$$\bar{y} \pm ts_{\bar{y}} = \bar{y} \pm ts/n^{1/2} \tag{1.35}$$

$$\beta \pm ts_\beta = \beta \pm ts/(S_{xx})^{1/2} \tag{1.36}$$

defined in equation (1.37) and the t value is usually chosen to correspond to 95% confidence. Equation (1.36) and the second formulation of equation (1.37) are valid only for the similarity model (1.5). The number of parameters P is two in that case and $A+1$ with model (1.8).

$$s = \Big[\sum_k \varepsilon_k^2 \Big/ (n-P)\Big]^{1/2} = \left[\frac{(S_{yy} - S_{xy}^2/S_{xx})}{(n-P)}\right]^{1/2} \tag{1.37}$$

The equation for the slope given above (1.29) is based on the assumption that the values of θ_k are known exactly. This is, of course, never fulfilled in reality, but the ordinary least-squares formula can be used when the inaccuracy of the y values is more than three times the inaccuracy in the θ values (in terms of their estimated standard error). This ratio is well fulfilled when the θ values are based on the statistical analysis of several data series, but less well when the θ values are defined in terms of a standard process. When the inaccuracy in the y and θ values is of the same magnitude, the slope β is calculated instead by equation (1.38), where the parameter λ is the ratio of the variances of y and θ (1.39).[54] The least-squares formulas given above, (1.27)–(1.39), are based on the assumption that the values of y have the same accuracy (constant variance) within a single series. This condition is usually well fulfilled when the y values are logarithms of rate or equilibrium constants. When some points are

$$\beta = \{S_{yy} - \lambda S_{xx} + [(S_{yy} - \lambda S_{xx})^2 + 4\lambda S_{xy}^2]^{1/2}\}/2S_{xy} \tag{1.38}$$

$$\lambda = \sigma_y^2/\sigma_\theta^2 \tag{1.39}$$

considerably less accurate than others—significantly according to an F test—one can use analogous formulas giving each point a weight w_k inversely proportional to the variance of the point,[54] or one can simply delete the less accurate points before the analysis.

Applying these formulas to the example shown in Fig. 1.6 (data are given in Table 1.3 and are taken from Ref. 47), we get the results in Table 1.4. Even if we assume that the inaccuracy is the same in y as in θ, the slope does not change so much that it falls outside the confidence interval calculated under the assumption that the θ values are exact. Since the calculated value of β is usually not used in an absolute sense, but rather in comparison with other β values, the error made in using equation (1.29) instead of (1.38) is not large if all β values are calculated from the same equation, i.e., (1.29).

To conclude this section, the fitting of equations (1.5) and (1.8) to the data is most easily done by using least-squares methods. The ensuing bias is likely to be small.

1.3.7. Detection of Outliers

One of the questions one wishes to have answered by the application of an LFER to a data series is whether one (or a few) points deviate significantly from the others. Also, it is important for the questions dealt with in the next section (goodness of fit) that the fit is not severely distorted by a few outliers. The problem of finding outliers within the framework of least squares has been treated by several authors (see Refs. 54–56 and references therein). All approaches are based on the assumption that the nondeviating residuals [ε in equation (1.5)] are approximately normally distributed. This is a most reasonable assumption in connection with data emerging from physical processes, e.g., the processes studied in chemistry. Davies and Goldsmith[54] and others have published simple procedures based on normalized residuals R_n:

$$R_n = |\varepsilon|/s \tag{1.40}$$

Here ε is the suspected deviation and s the computed standard deviation of the residuals including the suspected point [equation (1.37)]. Critical values of R_n according to Davies and Goldsmith for equation (1.5) are given in Table 1.5. Analogous tables for use in multiple regression with

Table 1.3. Logarithms of Rate Constants (y) for the Alkaline Hydrolysis of Aryl Tosylates[53] *Fitted by the Hammett Equation (1.5) with (i) $\theta = \sigma^0$, (ii) $\theta = \sigma$, (iii) $\theta = \sigma$, Twelfth Point Deleted, and (iv) $\theta = \sigma$, Tenth and Twelfth Points Deleted. Values of ε are the Resulting Deviations*

	Substituent											
	1	2	3	4	5	6	7	8	9	10	11	12
	p-NO_2	*m*-NO_2	*m*-Cl	*p*-Cl	*m*-OMe	H	*m*-Me	*m*-NMe_2	*m*-NH_2	*p*-OMe	*p*-Me	*p*-NH_2
$y = \log k$ (Ref. 53)	−1.733	−1.845	−2.604	−2.812	−3.089	−3.260	−3.419	−3.490	−3.506	−3.519	−3.547	−3.903
(*i*) σ^0 (Ref. 17)	0.814	0.713	0.365	0.242	0.101	0.000	−0.062	−0.096	−0.088	−0.120	−0.135	−0.296
(*ii*) σ (Refs. 57 and 7)	0.78	0.71	0.37	0.23	0.12	0.00	−0.07	−0.15	−0.16	−0.27	−0.17	−0.66
(*i*) ε (y vs σ^0; $n = 12$)	−0.041	0.046	−0.026	0.008	0.008	0.036	0.001	−0.005	−0.036	0.014	0.017	−0.024
(*ii*) ε (y vs σ; $n = 12$)	−0.152	−0.154	0.049	0.028	0.125	0.100	0.145	0.085	0.084	−0.082	−0.136	−0.336
(*iii*) ε (y vs σ; $n = 11$)	−0.049	−0.066	0.068	0.020	0.102	0.044	0.074	−0.002	−0.004	−0.194	0.018	
(*iv*) ε (y vs σ; $n = 10$)	−0.023	−0.045	0.064	0.004	0.070	0.012	0.037	−0.045	0.048	(−0.246)	−0.026	(−0.608)

Table 1.4. *Resulting Parameters (See Text) for the Four Different Fittings of the Data in Table 1.3 to the Hammett Equation*

Fitting	$\bar{y}$	$\bar{\theta}$	S_{xx}	S_{xy}	S_{yy}	β	s_ε	$\beta_{\lambda=10}$	$\beta_{\lambda=1}$	ψ	r	C_ρ
(*i*)	−3.060583	0.11983	1.341680	2.643978	5.218627	1.971 ± 0.025	0.029	1.972	1.973	0.043	0·9992	0.030
(*ii*)	−3.060583	0.060833	1.862692	3.044506	5.218627	1.634 ± 0.114	0.156	1.651	1.693	0.236	0.976	0.156
(*iii*) $n = 11$	−2.984000	0.126364	1.295855	2.382060	4.444446	1.838 ± 0.075	0.085	1.845	1.859	0.134	0.993	0.092
(*iv*) $n = 10$	−2.9305	0.1660	1.123040	2.1488	4.129599	1.913 ± 0.045	0.048	1.916	1.920	0.074	0.998	0.054

Table 1.5. Critical Values of R_n [Equation (1.40)] at 95% Confidence Level[a]

Sample size n	R_n	Sample size n	R_n
3	1.15	10	2.29
4	1.48	11	2.36
5	1.71	12	2.41
6	1.89	15	2.55
7	2.02	20	2.71
8	2.13	30	2.91
9	2.21		

[a] After Ref. 54, p. 460.

equation (1.8) are given by Prescott.[58] Looking at the resulting residuals in Table 1.3, we see that the value for p-NH_2 in row (*ii*) is considerably larger than the others in the same row. The value of R for this residual is 2.154 (0.336/0.156), which is smaller than the critical value of 2.41 ($p = 0.05$) in Table 1.5. When the deviations ε are clearly nonnormal owing to distortion by an outlier, however, the standard deviation s is best calculated by refitting the LFER to the series with the suspected point deleted. In the example in Table 1.3 (see also Fig. 1.7) the deviations are shown to be nonrandom and hence nonnormal by a simple run test (see Section 1.3.8). When the line is refitted without the p-NH_2 point, the residual standard deviation s becomes 0.085. The value of R for p-NH_2 then becomes 3.953 (0.336/0.085), which is clearly larger than the critical value, and the point should therefore be deleted. When equation (1.5) is refitted to the remaining eleven points [row (*iii*) in Tables 1.3 and 1.4], the residual for p-OMe is seen to be large. The R_n value based on s, including this point, is 2.28, which is smaller than the critical value in Table 1.5, and when the residuals are tested for nonrandomness the outcome is negative (see Section 1.3.8). Hence the point corresponding to p-OMe should not be considered to be an outlier. The above test for outliers is used when there is no theoretical reason *in advance* for suspecting the validity of a particular point. When applying LFER the situation is sometimes different. For example, in aromatic reaction series, donor and/or acceptor substituents can be suspected to show significant deviations in certain types of reactions, because of resonance interactions. In this case, the suspected points are best tested by fitting the LFER to the well-behaved points, and then calculating predicted values for the suspected points from the resulting values of α and β and the given value of θ. In the example shown in Tables 1.3 and 1.4, the row (*iv*) shows the results when the Hammett equation with the σ scale in row (*ii*) is fitted to all points but p-OMe and p-NH_2. The calculated deviations for the latter two substituents (−0.246 and −0.608) are more than 4 times the residual standard deviations calculated by equation (1.41) (0.054 and 0.063, respectively), where s is 0.048, based on

$$s_{\text{calc}} = s[1 + 1/n + (\theta - \bar{\theta})^2/S_{xx}]^{1/2} \tag{1.41}$$

the ten "well-behaved" points. The corresponding critical t-value [$p =$ 0.05, 8 degrees of freedom (d.f.)] is 2.31, showing that the suspected points do deviate significantly at the 95% level. It must be noted that in the treatment of outliers, the rules must be followed strictly in order not to delete points prematurely. The computation of s, to be used in equation (1.40), can be based on the reduced series (minus the suspected points) *only* when it is shown by a stringent test that the residuals are not random. This can be done only when there are at least ten points in the series (including the suspected ones); when there are fewer points, the outlier must therefore be very large in order to be detected. The last-mentioned procedure, which was based on the initial deletion of suspected point(s), can be used only when these points are from the beginning suspected from a theoretical point of view; in such a case they should probably not be included in the reaction series anyway, i.e., they do not belong to the proper domain of the LFER.

1.3.8. Systematic Deviations from Linearity

By using the test for outliers discussed in the previous section, one considers whether a few points deviate from the linear model. Another issue, for which tests for outliers are rather inefficient, is whether the linear model is inadequate to describe the data series as a whole. Thus, a number of small but systematic deviations from linearity, each of which might go undetected by tests for outliers, could add up to a significant deviation from linearity for the whole series. The straightforward way to test for curvature is to fit both equation (1.5) and (1.7) to the data and test whether the difference in residual variance [s^2 with s computed according to equation (1.37)] is significant according to an F test.[54] The results of such a test for $\log k$ vs. σ (data in Table 1.3) are shown in Table 1.6. Manifestly, according to this test, the curvature is highly significant. One might also wish to test whether the data are better described by a two-term LFER (1.8). This is done in the same way by fitting the two equations (1.5) and (1.8) separately to the data and then testing the difference in residual variance by an F test.

Another simple and useful test, which is based on testing the residuals ε for nonrandomness, is the run test.[59] This test can also be used to determine whether a single point distorts the fit, so that the residuals become nonrandom (see Section 1.3.7). In this test one calculates the

Table 1.6. Analysis of Variance of Significance of Quadratic Term When Models (1.5) and (1.7) Are Fitted to the 12 Points y, σ in Table 1.3[a]

Model	Sum of squares	Number of degrees of freedom	Mean square	Standard deviation
Straight line (1.5)	0.245321	10 (=12−2)	0.02453	0.156
Parabola (1.7)	0.024288	9 (=12−3)	$a=0.002699$	0.052
Difference	0.221033	1	$b=0.221033$	

[a] Ref. 54, p. 380. F value: $b/a=0.221033/0.002699=81.9 \gg F_{crit}$. Critical F value ($p=0.01$, 1, and 9 d.f.) = 10.56.

number of runs of consecutively positive and consecutively negative residuals. This is the same as $(1+q)$, where q is the number of times the residuals change sign when they are plotted against y as in Fig. 1.7. If the numbers of positive and negative residuals are denoted by n_+ and n_-, respectively, the expected value of the number of runs, U_{theor}, is given by equation (1.42) and is approximately normally distributed, with variance given by equation (1.43), where $n=(n_+ + n_-)$. Hence, an actual U value

$$U_{theor}=1+\frac{2n_+n_-}{(n_+ + n_-)}\approx 1+\frac{n}{2} \tag{1.42}$$

$$V(U)=\frac{2n_+n_-(2n_+n_- - n_+ - n_-)}{(n_+ + n_-)^2(n_+ + n_- - 1)}\approx\frac{n}{4} \tag{1.43}$$

that is smaller than $U_{theor}-2[V(U)]^{1/2}$ signifies a nonrandomness of the residuals ε. In the example shown in Table 1.3, the numbers of runs of ε in rows (*i*)–(*iv*) are 7, 3, 4, and 5, respectively. The theoretical values according to equation (1.42) are 6.83, 6.83, 6.45, and 5.80 with the corresponding variances according to equation (1.43) being 2.56, 2.56, 2.43, and 2.03. This gives minimum values for U $[U_{theor}-2V(U)^{1/2}]$ of 3.63, 3.63, 3.34, and 2.95, which shows that series (*ii*) is significantly nonrandom. This, in turn, is a criterion for systematic deviations from linearity due to curvature or for other reasons. In the present case, the two substituents *p*-OMe and *p*-NH_2 probably cause this deviation because of resonance interactions in the series defining the σ values.

For the example in Table 1.3, statistical tests cannot distinguish between whether the *p*-NH_2 point is an outlier or whether the similarity model is curved. This is a problem particularly in the Hammett equation, where deviating substituents usually have large positive or large negative σ values, deviations that can also be described as due to a curved Hammett plot. Only the analysis of several series, and/or independent chemical information, can serve to distinguish between the two "explanations."

One should also note that negative answers to the statistical tests above do not imply an absence of curvature. They imply that possible inadequacies in fit due to curvature are small compared to the "random" scatter around the line.

1.3.9. Several Lines

Sometimes it seems as if two or more lines would fit the data better than a single line, for instance, one line for *meta*- and one line for *para*-substituted compounds.[60] The question as to whether the fit is significantly better for two lines must be tested on the basis of the residual sums of squares. It is *not* sufficient to say, for example, that the correlation coefficient for a single line is 0.96 but for two lines 0.99 and 0.99. Such reasoning based on correlation coefficients is misleading and has no statistical or chemical rationale whatsoever. Table 1.7 shows a correct way to make the test based on the residual sums of squares, exemplified with the data in Table 1.3. As seen from the insignificant F value, two lines do not give a better fit in this case. Also here, possible outliers should be deleted before attempting the fittings in order to make the statistical test efficient.

1.3.10. Goodness of Fit

After coping with systematic deviations between data and the linear model (1.5) due to outliers, curvature, several lines, or other possibilities, it is of interest to see how well the remaining data fit the model, i.e., how large the "random" scatter is around the line. Quantitative criteria for this are usually called criteria of goodness of fit (CGOF). Jaffé,[43] being the first

Table 1.7. Analysis of Variance[61] of the Fit of Two Separate Lines for Meta and Para Substituents (Hydrogen is Treated as a Para Substituent) versus the Fit of a Single Line[a]

Model	Sum of squares	Numbers of degrees of freedom	Mean square	r
Single line	0.008279	10		0.9992
Two lines	0.005509(= 0.002561 +0.002948)	8 = 4 + 4	$a = 0.000689$	0.9994(m) 0.9995(p)
Difference	0.002770	2	$b = 0.001385$	

[a] Data from Table 1.3 (y and σ^0). F value $= b/a = 2.01 \ll F_{\text{crit}}$. Critical F value ($p = 0.05$, 2 and 8 d.f.) = 4.46.

to use statistical techniques extensively in connection with LFER, unfortunately advocated the use of the correlation coefficient r [cf. (1.44)] as

$$r = S_{xy}(S_{xx}S_{yy})^{-1/2} \tag{1.44}$$

a CGOF. The correlation coefficient r has some very undesirable properties as a CGOF; this has been pointed out by numerous authors.[15,62,63] The most serious drawback is that it automatically gets larger with a smaller sample size (n). Therefore one cannot use r for comparing the fit of two series with different numbers of points in them. This increase of r with smaller sample size also means that a large value, e.g., 0.95, which is highly significant if $n = 10$, is even nonsignificant for $n = 4$.

Several other CGOF have been proposed, but before dealing with them, let us discuss the purposes one might have in applying a CGOF. These might be as follows:

(*a*) Does the LFER explain anything at all of the variation in the series—is the fit statistically significant? To answer this question, any one of three criteria r [equation (1.44)], ψ [equation (1.45), Ref. 63], or C

$$\psi = [n(1 - r^2)/(n - 2)]^{1/2} \tag{1.45}$$

$$C = t_{n-2}s_\beta/\beta = t_{n-2}s_\varepsilon S_{xx}^{1/2}/S_{xy} \tag{1.46}$$

[equation (1.46), Ref. 15] can be used. If r or ψ is used, it must be tested for statistical significance. This can be done in various ways,[54,63] for instance by using the values in Table 1.8. The coefficient C contains a statistical test already, viz., by means of the t value corresponding to the desired level of confidence, usually 95% ($p = 0.05$), and is significant if smaller than 1.0, giving a better fit the closer C is to zero. A positive answer to this first question means that the LFER explains some of the variation within the series. In such a case we proceed to question (*b*). A negative answer can mean one of several things:

(i) The fit is obscured by outliers, curvature, or other systematic deviations; this should, however, have been corrected before looking at the CGOF (see previous sections).

(ii) Too few data points have been collected; thus the fit is obscured because of large errors of measurement.

(iii) The wrong θ scale has been used.

(iv) The LFER is of little relevance for explaining the variation in the series, i.e., the β value is small relative to the standard deviation in y.

On all occasions, the plot of y against θ (σ in the Hammett equation) will provide useful information about the possible cause of the negative answer to question (*a*). (See also a discussion by Exner[60] for a thorough coverage of the subject.)

Table 1.8. Lower Values of r [Equation (1.44)] and ψ [Equation (1.45)] at 95% and 99% Levels of Confidence[54]

Confidence level	Criteria	Sample size n 3	4	5	6	7	8	9	10	15	20
$(1-p)=0.95$	r	0.997	0.950	0.878	0.811	0.754	0.707	0.666	0.632	0.514	0.444
	ψ	0.134	0.442	0.618	0.717	0.777	0.817	0.846	0.866	0.921	0.944
$(1-p)=0.99$	r	1.00	0.990	0.959	0.917	0.875	0.834	0.798	0.765	0.641	0.561
	ψ	0.00	0.199	0.366	0.489	0.573	0.637	0.683	0.720	0.824	0.873

(*b*) Is the series at hand so similar to other series that are well described by the LFER that no additional "effects" (causal factors) are needed to explain the variation within the series? In order to answer this question, one should compare the residual standard deviation of the present series, s, equation (1.37), with the average standard deviation for well-fitting series, s_0, equation (1.26). If the standard deviation for well-fitting series depends on the slope β, one should instead compare s with the appropriate s_0 value for the series; for the Hammett equation this is obtained from Fig. 1.5. The two values are compared by means of an F

$$F = s^2/s_0^2 \tag{1.47}$$

value (equation (1.47). An F value larger than a critical value (recommended level, $p=0.01$) with $n-2$ and infinite degrees of freedom will indicate that the answer to the question is negative.

(*c*) Do the data fit one θ scale better than another? This is also tested by an F value, this time computed by taking the quotient of the squared residual standard deviations obtained when fitting the data to the two scales, with the smaller value in the denominator. The critical F value has $n-2$ degrees of freedom in each respect of F.

Example. The data in Table 1.3 give significant values of r, ψ, and C (only one need be used) in all four cases when fitted to equation (1.5); see Tables 1.4 and 1.8. Hence, the Hammett equation explains at least some of the variation in the series. For a β value of 1.97 (ρ in the Hammett equation), Fig. 1.5 gives a typical standard deviation of about 0.050. The F values according to (1.47) are then 0.34, 9.73, 2.89, and 0.92 for (*i*)–(*iv*), respectively. The critical F values ($p=0.01$) for $n-2$ and infinite degrees of freedom are 2.32, 2.32, 2.41, and 2.51, respectively, which shows that there is a lack of fit for (*ii*) and (*iii*). When finally the fit in (*i*) is compared with the fit in (*iii*) and (*iv*), the F values 8.59 ($0.085^2/0.029^2$) and 2.74 ($0.048^2/0.029^2$) are to be compared with the critical values 3.14 (10, 9 d.f.)

and 3.35 (10, 8 d.f.). Hence, the fit in case (*i*) is significantly better than in case (*iii*) but not better than in case (*iv*).

1.3.11. Demands on Data

The incentive for fitting data to an LFER (1.5) or (1.8) is basically to study whether the present series behaves similarly to a class of systems studied earlier. Depending on how the data are chosen, i.e., how the experiments are performed, the efficiency of this study can vary greatly. Most chemists realize that in order to investigate whether a reaction series is well fitted by the Hammett equation, it is not sufficient to measure the relevant property (e.g., the rate constant) of the compounds with substituents *m*-Cl, *p*-Cl, *m*-Br, and *p*-Br only, inasmuch as the range of the corresponding σ values is too narrow to permit of any conclusions. Thus, in order to study a reaction series efficiently, substituents with large, intermediate, and small θ values should be used. Similarly one must choose data for a multiple-term LFER so that the different θ scales really are different for as many substituents as possible. Strict rules for the number of points necessary in various cases cannot be given, since this number depends on the precision of the data, the differences in parameter scales, and other factors,[64] but it is probably safe to say that the minimum number of different substituents (with *significantly* different θ values) is 5 when equation (1.5) is used and 8 when equation (1.8) with $A = 2$ is used. In the latter case at least 3 points must differ substantially in value for θ_{1k} and θ_{2k}.

1.3.12. Summary of the Analysis of a Single Series

To conclude the present section, we may summarize the data analysis of a single series in the following scheme:

(*a*) Plot the data by using a pertinent θ scale (σ^0 or σ for the Hammett equation).

(*b*) Fit the model (1.5) by least squares to the data, using the same θ scale(s). Calculate the residual standard deviation, *s*, and the value and the confidence interval of the slope β.

(*c*) If the plot shows points which deviate *much* more than the others, test for outliers. Delete significant outliers and refit.

(*d*) Plot the resulting residuals against the *y* values.

(*e*) If the plot in step (*d*) indicates systematic patterns in the residuals, test for curvature. Look also for systematic deviations for certain types of substituents. If they are seen, test for several lines and/or a multiple-term LFER.

(*f*) Test for significance using *r*, ψ, or *C*. If a two-term LFER is employed, use instead the multiple correlation coefficient and calculated confidence intervals of the coefficients β_1 and β_2.

(*g*) Test for similarity with the class by using *s*.

(*h*) Make a *chemical* interpretation of the results.

The last step, (*h*), must not be forgotten; chemical data are after all collected to get answers to chemical questions, not to be the basis for a mathematical exercise.

1.4. Extension to Pattern Recognition

Chemists have always used classification schemes as tools to acquire knowledge about chemical systems. The periodic system, the classification of organic compounds according to functional groups or as aromatic or aliphatic, and the classification of reaction mechanisms into such types as S_N1 and S_N2 are important examples for the physical organic chemist. The key to a successful classification is to find classes such that the systems display similarity with respect to the pertinent properties within the classes but dissimilarity between classes. In the past, chemical systems have been classified mainly on a qualitative basis. With the increased number of physical instruments in the chemical laboratory, ir, nmr, uv and mass spectrometers, gas chromatographs, etc., the chemist gets more quantitative information about chemical systems. This, in turn, makes for a demand for classification schemes based on quantitative data.

The similarity models (1.5) and (1.8) are not restricted to applications where the data y are proportional to free energy changes. The derivation in Section 1.2.1 was based only on the assumption that the data y were measurements made on similar systems, and on certain continuity properties of these measurements. Hence, if we have several disjoint classes of systems, where each class contains similar systems, the models (1.5) and (1.8) can describe the data structure of each class separately. These data structures can then be used for the classification of new systems on the basis of measurements. This problem area is called *pattern recognition* in chemistry,[65–67] classification and discrimination in statistics,[68] and other names in other branches of science.

In earlier discussion of LFER, we have assumed that one observation only was made on each system S_{ik}. In pattern recognition, one studies similarities and dissimilarities between single systems, say S_k. To accomplish this, a number of observations (subscript i) are made on each system, giving the data y_{ik}. Just as before, we can argue that the systems in a single class q are similar if their data y_{ik} are well described by the similarity models (1.5) or (1.8) with few terms. This is the basis of the method of pattern recognition described below.

1.4.1. Example of Pattern Recognition

Chemical pattern recognition is best understood in terms of an example. The following treatment concerns a problem of structure determination and is based on the results of Mecke and Noack, who made one of the first applications of heuristic pattern recognition in chemistry.[69] They were interested in the conformation of α,β-unsaturated carbonyl compounds in solution. Two planar conformations—(V) = "*trans*" and (VI) = "*cis*" (Fig. 1.8)—and a number of twisted conformations can be proposed. Current chemical knowledge held the stable conformation to be "*trans*" (V) when the groups R_1 and R_3 were small, say hydrogen. Mecke and Noack were particularly interested in what happened when these substituents were large. They attacked this problem in the way organic chemists have always favored, basing the approach on the assumption that all "*trans*" compounds of type (V) behave more similarly to one another than to "*cis*" compounds of type (VI) and vice versa. They synthesized a number of compounds with known or presumed conformations, (Va), (V) with $R_1 = R_3 = H$ ("*trans*"), and (VIa) ("*cis*") and then recorded ir and uv spectra of these compounds and of other compounds with unknown conformations. They extracted seven variables from these spectra—ir frequencies and intensities for the carbonyl and C=C absorptions, half-bandwidth of the former, and wavelength and intensity of the maximal absorption in the uv (see Tables 1.9 and 1.10)—and analyzed the similarities and dissimilarities between the data for each compound by means of a number of two-dimensional plots.

The analysis can also be done quantitatively by means of the similarity models (1.5) or (1.8)[49] or other similarity models.[65] The steps in the analysis are essentially as follows.

(*a*) Define the classes of interest in terms of systems that are "known" to belong to these classes. At least five systems are needed for each class. These systems are called the reference sets of the classes. In the carbonyl example the compounds 1–6 and 7–13 (Table 1.9) constitute the reference

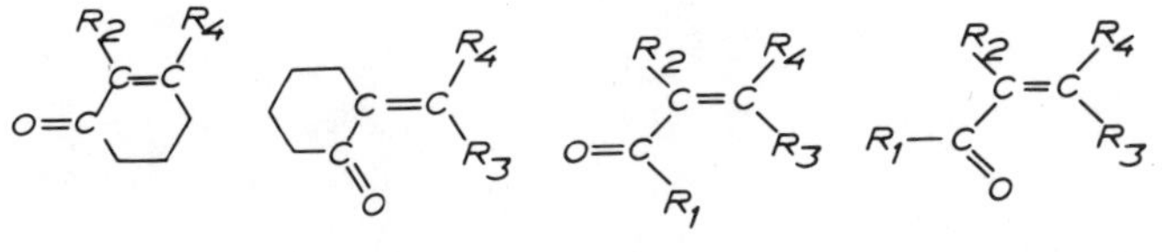

(Va) "trans" (VIa) "cis" (V) "trans" (VI) "cis"

Fig. 1.8. α,β-Unsaturated carbonyl compounds with known conformations and the two possible planar conformations for open-chain compounds. The designations "*cis*" and "*trans*" refer to the conformational relationships of the C=O bond and the C=C bond.

Table 1.9. Spectroscopic Data for the Carbonyl Compounds in Reference Set 1 ("trans," Nos. 1–6) and Reference Set 2 (Sterically Hindered, Nos. 7–13) after Mecke and Noack[69]

Number	Compound[a]	$\nu_{C=O}$ (cm^{-1})	$\int \varepsilon\, d\lambda$ ($cm^3\, mol^{-1}$)	$\Delta\nu_{1/2}$ (cm^{-1})	$\nu_{C=C}$ (cm^{-1})	$\int \varepsilon\, d\lambda$ ($cm^3\, mol^{-1}$)	λ_{max} (Å)	ε_{max} ($cm^{-1}\, mol^{-1}$)	Class	$s_k^{(1)}$	$s_k^{(2)}$	F value
1	CH_2=CHCHO	1703	7.25	7.0	1620	0.19	2080	9000	(V)	0.043	0.196	20.9
2	MeCH$\overset{t}{=}$CHCHO	1700	8.40	5.5	1644	1.37	2160	14400	(V)	0.101	0.347	11.7
3	(CH_2COCH=$CHCH_2CH_2$)	1691	8.90	9.0	1621	0.03	2180	11200	(V)	0.081	0.276	11.7
4	(CH_2COCH=$CMeCH_2CH_2$)	1680	9.45	25.0	1635	1.53	2250	15000	(V)	0.103	0.391	14.4
5	(CH_2COCH=$CMeCH_2CMe_2$)	1674	9.00	16.0	1637	1.52	2260	13900	(V)	0.045	0.342	58.6
6	(CH_2COCMe=$CMeCH_2CH_2$)	1673	8.50	11.5	1635	1.87	2340	13200	(V)	0.127	0.302	5.6
7	MeCOCH$\overset{c}{=}$CHMe	1699	3.61	7.0	1618	2.30	2230	9300	S	0.257	0.016	243.9
8	MeCOCH$\overset{c}{=}$CHPr	1697	3.99	7.5	1616	2.87	2240	9900	S	0.265	0.034	60.4
9	MeCOCH=CMe_2	1693	3.88	5.5	1622	6.28	2320	12300	S	0.391	0.064	36.8
10	EtCOCH=CMeEt	1694	4.13	13.0	1625	5.26	2320	12900	S	0.339	0.076	19.7
11	MeCOCMe=CHMe[b]	1696	4.45	11.5	1626	1.70	2280	5700	S	0.260	0.051	25.5
12	MeCOCMe=CMe_2	1689	4.98	14.5	1622	2.66	2400	6600	S	0.297	0.041	53.1
13	MeCOCMe=CMeEt[c]	1690	4.87	15.0	1625	2.65	2390	5300	S	0.312	0.021	211.5

[a] For each compound are given (*a*) frequency, intensity, and half-bandwidth for the C=O absorption in the ir, (*b*) frequency and intensity for the C=C absorption in the ir, and (*c*) wavelength and intensity for the maximal uv absorption. The residual standard deviations obtained when the data are fitted to each class model, "*trans*" and sterically hindered, are denoted by $s_k^{(1)}$ and $s_k^{(2)}$, respectively, and the *F* value is their squared ratio.

[b] H and Me are *cis*.

[c] Me and Me are *cis*.

sets for the two classes "*trans*" and *sterically hindered*, designated (V) and S, respectively, in Table 1.9.

(*b*) Normalize the data so that each variable has zero mean and unit variance over all systems (autoscaling). This is done to give each measured property equal weight in the classification. If there is prior knowledge about the importance of the variables, they should instead be weighted according to this importance. In the carbonyl example there is no such prior knowledge and the data are autoscaled by subtracting $\bar{y}_i$ and dividing by s_i for each variable (see Table 1.11).

(*c*) Find the typical "data structure" of each class by fitting model (1.8) to the data for the reference set of each class separately, as described in Section 1.3.2. In the carbonyl example, the data matrix is complete, and this fitting is a simple problem, giving the parameters in Table 1.11.

(*d*) Use the data structures from step (*c*) to classify the nonassigned systems. This is done by fitting the data of each nonassigned system to each class model, with the α and β values fixed as those determined in step (*c*). Each such fitting corresponds to a linear multiple regression and gives as one result, among others, a residual standard deviation, $s_k^{(q)}$ (for system k and class q) [equation (1.37)]. A nonassigned system is classified as belonging to the class for which this standard deviation, $s_k^{(q)}$, is smallest, provided that it is of the same magnitude as the "typical" residual standard deviation for that class, s_0, equation (1.26). Tables 1.9 and 1.10 show these $s_k^{(q)}$ values for all the 35 carbonyl compounds. The systems in the reference sets (Table 1.9) are much closer to their own class than to the other—the two classes are well separated. The three compounds with known "*cis*" conformation (VIa) (Nos. 33–35 in Table 1.10) are seen to be much closer to class S (sterically hindered), and the residual standard deviations for the first two are of the same magnitude as s_0 for that class (not significantly larger according to an F test). Hence one can conclude: (i) Sterically hindered compounds indeed do have a conformation other than "*trans*" and (ii) this conformation is probably "*cis.*"

One can get an indicator of which variables display analogy (similarity) within a class by comparing, for each class and variable separately, the residual variance of the compounds in the reference set with the variance of the corresponding data y_{ik}. A small ratio between these two values means that much of the variation in the data is taken care of by the similarity model (1.8); the variable behaves similarly to the general data structure of the class. In the carbonyl example, these ratios are shown in Table 1.11 and it is seen that variable 2 has the least and variable 5 the most similarity.

Pattern recognition can also be used to find which variables really distinguish one class from another. The residuals in the analysis above can also be used for this problem. By comparing, for a single variable i, the

Table 1.10. Spectroscopic Data for the Unassigned Compounds in the Carbonyl Example (the Test Set)[a]

Number	Compound	(a)			(b)		(c)		Assigned class[b]	$s_k^{(1)}$	$s_k^{(2)}$	F value
14	HCOCMe=CH_2	1702	5.70	5.0	1638	0.60	2150	13700	(V)	0.129	0.254	3.8
15	HCOCMe=CHEt	1693	7.90	7.0	1645	1.76	2250	16200	(V)+	0.071	0.342	23.1
16	HCOCH=CMe_2	1686	7.60	6.0	1638	1.43	2280	14200	(V)+	0.067	0.293	19.1
17	HCOCH=CMe_2	1686	7.60	6.0	1621	1.43	2280	14200	(V)	0.151	0.252	2.8
18	MeCOCH=CH_2	1706	4.93	8.0	1618	0.69	2080	8200	S, (V)	0.148	0.113	1.7
19	MeCOCH=CH_2	1686	4.93	8.0	1618	0.69	2080	8200	[S, (V)]	0.190	0.160	1.4
20	EtCOCH=CH_2	1707	4.80	11.0	1619	1.24	2120	7000	[S, (V)]	0.190	0.123	2.4
21	EtCOCH=CH_2	1690	4.80	11.0	1619	1.24	2120	7000	[S, (V)]	0.196	0.125	2.5
22	MeCOCMe=CH_2	1684	6.56	9.5	1630	0.68	2140	7900	(V)	0.148	0.204	1.9
23	MeCOCH=CHMe	1701	6.30	8.0	1634	1.91	2160	11600	(V)	0.121	0.200	2.8
24	MeCOCH$\overset{t}{=}$CHMe	1682	6.30	8.0	1634	1.91	2160	11600	(V)	0.135	0.231	2.9
25	MeCOCH$\overset{t}{=}$CHEt	1701	6.51	9.0	1629	2.45	2170	9800	(V), S	0.146	0.165	1.3
26	MeCOCH$\overset{t}{=}$CHEt	1682	6.51	9.0	1629	2.45	2170	9800	(V), S	0.159	0.195	1.5
27	MeCOCH$\overset{t}{=}$CHPr^n	1702	7.30	16.0	1631	2.07	2180	11000	(V)	0.155	0.236	2.3
28	MeCOCH$\overset{t}{=}$CHPr^n	1681	7.30	16.0	1631	2.07	2180	11000	(V)+	0.110	0.244	4.9

29	MeCOCH$\overset{t}{=}$CHPr^i	1702	6.68	9.0	1626	2.57	2180	12700	(V)	0.148	0.182	1.5
30	MeCOCH$\overset{t}{=}$CHPr^i	1683	6.68	9.0	1626	2.57	2180	12700	(V)	0.145	0.207	2.0
31	MeCOCMe$\overset{t}{=}$CHMe	1675	6.64	7.0	1646	1.44	2250	13000	(V)+	0.134	0.327	6.0
32	MeCOCEt$\overset{t}{=}$CHMe	1675	6.06	8.0	1642	1.34	2245	13400	(V)	0.149	0.302	4.1
33	$\overline{CH_2CH_2CH_2CH_2COC}$=CHMe	1693	4.40	11.0	1622	4.08	2380	7400	S+	0.336	0.053	39.6
34	$\overline{CH_2CH_2CH_2CH_2COC}$=CHPr^n	1692	4.51	11.5	1620	4.04	2400	8800	S+	0.331	0.051	42.6
35	$\overline{CH_2CH_2CHMeCH_2COC}$=CMe$_2$	1685	4.98	16.5	1611	3.51	2450	7900	[S]	0.345	0.139	6.2

[a] Compounds with two carbonyl bands in the ir are tabulated as two separate compounds; notation in headings as in Table 1.9.

[b] A compound is assigned to a class q if the residual standard deviation $s_k^{(q)}$ is not significantly larger than $s_0^{(q)}$ according to the F test described in Section 1.3.10, equation (1.47). The values of $s_0^{(1)}$ and $s_0^{(2)}$ are 0.0995 and 0.0537, respectively, giving the limits 0.21 and 0.11, respectively ($F_{5,17} = 4.34$ and $F_{5,23} = 3.94$ for $p = 0.01$). A plus sign indicates that there is a significant difference in fit for the two classes (the F value in the column on the extreme right above exceeds the value 5, which is the 5% value of $F_{5,5}$). Assignments within square brackets are doubtful owing to the large values of $s_k^{(q)}$.

Table 1.11. Resulting Parameter Values When Model (1.8) with A = 2 is Fitted to the Two Reference Sets (Table 1.9) Separately[a]

		1	2	3	4	5	6	7
α_i	class 1	−0.068	0.253	0.086	0.079	−0.127	−0.032	0.120
β_{i1}	class 1	0.526	−0.163	−0.652	−0.230	−0.193	−0.336	−0.261
β_{i2}	class 1	0.052	0.009	0.585	−0.668	−0.270	−0.208	−0.304
α_i	class 2	0.060	−0.202	0.015	−0.110	0.170	0.141	−0.111
β_{i1}	class 2	0.056	−0.133	−0.355	−0.060	0.682	−0.118	0.609
β_{i2}	class 2	0.295	−0.182	−0.547	−0.253	−0.478	−0.537	0.021
$\bar{y}_i$		1691	6.18	10.2	1628	2.07	2230	10740
s_i		9.62	1.62	4.21	9.07	1.33	99.0	2920
	(*i*)	0.66	0.56	0.88	0.69	0.96	0.72	0.83
	(*ii*)	4.0	206	2.8	14.3	163	20.6	36.7

[a] The data have first been autoscaled by subtracting $\bar{y}_i$ and dividing by s_i (rows 7 and 8). The last two rows show the relevance of the variables (*i*) for describing the variation within the classes (low values, low power; values close to 1.0, high power), (*ii*) for the discrimination between the classes (large values, high relevance; low values, low relevance).

residuals of all systems in the reference sets when fitted to their own class, with the residuals for the same systems when their data are fitted to all class models but their own, one gets a quantitative measure of the discriminating power of the *i*th variable. These measures for variables 1–7 in the carbonyl example are shown in Table 1.11; variables 1 and 3 are seen to be much less discriminating than the others.

1.4.2. Graphical Illustration

The data for a single system can be thought of as a point in the *M*-dimensional space obtained by letting each of the *M* variables define one coordinate axis. Pattern recognition thus can be seen as a set of mathematical methods for finding structures in this *M* space. In order to visualize data and results, however, it is necessary to project the *M*-space onto a *2*-space. Different methods for doing this are described in the literature.[66,67] As an example, we show an eigenvector projection of the carbonyl data and the simplest class models [equation (1.5)] in Fig. 1.9. The graph preserves much of the structure of the problem; one can see the two reference sets well distinguished, the good fit of the class models to these reference sets and the small distance between the points for the "*cis*" compounds and those for the sterically hindered compounds.

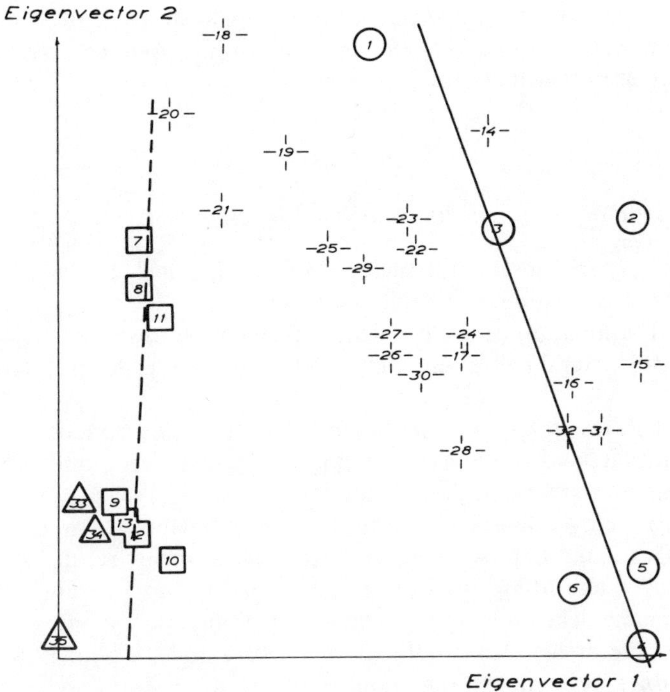

Fig. 1.9. Eigenvalue projection of the data in Tables 1.9 and 1.10. Circles correspond to compounds with known "*trans*" conformations [(V) and (V*a*)], triangles to compounds with "*cis*" conformations (VI*a*), and squares correspond to compounds with large substituents R_1 and R_3 (Fig. 1.8).

1.4.3. Conclusion

LFER are quantitative models of similarity that can be used in chemical pattern recognition for (*a*) finding the typical data structures for groups of similar systems—reactions, compounds, spectra, mixtures, etc., (*b*) utilizing these data structures either for the classification of new, unassigned, systems or for constructing new systems with desired properties. In the last case one of the groups in step (*a*) contains systems having the desirable properties, say being active against some kind of disease, and the analysis gives as its result the typical data structure for such compounds.

Pattern recognition is based solely on empirical data and relationships between them. The result of the analysis therefore depends on the quality and relevance of the data as well as an intelligent selection of the systems in the reference sets. Though pattern recognition in quantitative terms is rather new in chemistry, one can see many applications of these methods, especially since they correspond closely to the qualitative analogical reasoning used by generations of chemists.

1.5. Discussion

1.5.1. The Place of LFER in a Chemical Investigation

The experimental investigation of a chemical problem can be described as consisting of the following steps.

(*a*) Formulation of the problem in *chemical* terms, e.g., "what is the structure of a certain chemical compound?" or "what is the mechanism of a certain reaction?"

(*b*) (i) Design of the experiment—choice of instruments, reagents, concentration and temperature ranges, solvent, etc; and points of measurement in terms of these factors.

(ii) Reformulation of the problem in mathematical and statistical terms. This latter step is necessary since measurements result in numbers which, to make things worse, are not exact, but contain errors of measurement. The only way to deal with imprecise numbers is to use statistical methods. These methods might be simple, e.g., making use of mean values, or complex, e.g., nonlinear regression, but they must explicitly or implicitly treat data as consisting of a systematic part and an "error" part.

Steps (*b*)(i) and (*b*)(ii) are intimately connected. When, for example, one investigates the mechanism of a chemical reaction, one tries to design experiments that (i) provide information about mechanistic details, e.g., kinetic experiments, (ii) give data that can be analyzed without too much difficulty; one chooses conditions so that the observed kinetics are first order or another simple form.

(*c*) Performing the experiment, collecting results.

(*d*) Data analysis. This is naturally described as consisting of the following two steps.

(i) *Data reduction*, in which the raw data obtained in the experiment are reduced and translated into data of chemical relevance. In a kinetic experiment, this corresponds to fitting a kinetic model, e.g., an exponential function in first-order kinetics, to the (y_i, t_i) pairs. Here y_i is an instrument reading (e.g., conductance) which is related to the concentration of one of the reactants or products, and t_i is the time at which this reading is made. As a result of the data reduction, one gets a rate constant and possibly some other parameters as well.

(ii) *Data interpretation*, in which the chemical data obtained in step (*d*)(i) are related to the original problem formulated in steps (*a*) and (*b*)(ii).

This step is usually taken qualitatively, e.g., by saying that a rate constant obtained in the present experiment is much larger than a rate constant for a model system, which indicates, say, a neighboring group effect.

The present article concerns the use of LFER as quantitative tools in step (*d*)(ii), i.e., as tools for relating chemical data to chemical questions. This use of mathematical and statistical methods is still in its infancy in organic chemistry, but the advantages of these methods will probably make for an increased use in the future.

With the quantitative similarity models (1.5) and (1.8) discussed in the present article, the obvious advantage is to give the chemist *quantitative* support for the analogical reasoning which is the basis of so much chemical theory. In particular, we think that it is valuable to get information about which chemical interpretation is feasible for a certain body of data—i.e., consistent with the structure of the data—and which interpretation is not. If, for instance, one finds that a one-component model (1.5) fits well the logarithms of rate constants for a number of reaction series, a chemical interpretation that postulates five different effects varying between the reactions would be a clearly unacceptable overinterpretation.

Another valuable type of information that LFER can provide is evidence as to whether there really is a difference in behavior between two classes of systems, and, if so, where this difference lies. One example is in connection with the Hammett equation, where one gets solid evidence for the *absence* of differences between *meta*- and *para*-substituents in aromatic reaction series;[15–17] a joint body of data with both *meta*- and *para*-substituted compounds gives an equally good fit to equation (1.5) as does the disjoint fitting of data for *meta*- and *para*-substituted compounds when cross-conjugation is absent. Another example is provided by the carbonyl problem of Mecke and Noack (Section 1.4.1), where the quantitative models established that there really is a difference between the "*trans*"-compounds and the compounds with large R_1 and R_3 substituents.

In this context it is perhaps relevant to discuss the widely misused concept of statistical significance. To take a concrete example, say that one is interested in whether the slope β [in equation (1.5)] is different for two reaction series. By means of the standard formulas of linear regression, one has the standard error of the computed slope as $s/S_{xx}^{1/2}$ (see Section 1.3.6), and the confidence interval for the computed slope of a line is given by equation (1.36). The two slopes are different at a statistical significance level slightly higher than that corresponding to the chosen t value if the two confidence intervals do not overlap.

This does not mean, however, that the slopes are necessarily different from a chemical point of view. Two slopes can always be made statistically different by including sufficient data—S_{xx} grows with the number of points—but a chemist would, in the specific example, perhaps demand a difference of, say, 25% in the slopes before he considered the two slopes as different in chemical significance. Hence, in order to establish a chemically significant difference, the difference must also be statistically significant, but the latter does not always imply the former.

With the advent of computers one can use statistical methods for estimating the parameters α, β, and θ in equation (1.5), without having recourse to simplifying assumptions or to the definition of the θ values in terms of a standard series. Hence one can get an *objective* picture of the complexity inherent in the data, in particular whether equation (1.5) is adequate to represent the data or whether a more complex model is needed. In our view, each area where multiple-term LFER are now used would benefit much from such an analysis; consequently the empirical relations now used might be considerably simplified.

1.5.2. Concluding Remarks

LFER are found in a large number of areas in chemistry.[1–3] One interesting question is whether each LFER is a special kind of "natural law," which can be derived for the particular application from underlying principles such as quantum mechanics, or whether the abundance of LFER indicates, rather, that they exhibit some general principle in the behavior of chemical systems.

The first approach, in which individual LFER are interpreted in detail in terms of microscopic theory, has been made by several authors, e.g., Marks and Drago[70] for their EC relation, and Godfrey,[71] Ponec,[72] and others for the Hammett equation. We note that according to the derivation given in Section 1.2, an LFER can be derived from *any* continuous function *F*. Hence, the possibility of deriving LFER from specific quantum mechanical formulations can be seen as a special case of the general formulation (1.1) and, in our view, provides no evidence for or against the similarity interpretation of LFER proposed in this chapter.

We have discussed the possibility of using LFER for the purpose of microscopic interpretation of chemical phenomena. One-term LFER (1.5) can be well interpreted both in terms of microscopic theory—a single "effect" varies between the systems in a single series and the same "effect" is found in all series—and in terms of empirical similarity. For multiple-term LFER (1.8), however, we are more pessimistic. The main cause of this pessimism is that a statistical approach is necessary with LFER, not only because the data themselves are inexact owing to errors of measurement, but also because the models (1.5) and (1.8) are approximate. However, it is presently impossible to analyze multiple-term LFER by proper statistical methods because of the lack of suitable data (see Section 1.3.). The published attempts, most recently by Ehrenson *et al.*[39] and by Swain and

Lupton,[41] are *not* based on nonrestricted statistical analysis. Thus, the analysis of Ehrenson *et al.* is based on an arbitrarily fixed "inductive" scale; i.e., the first θ scale is fixed as σ_I, obtained from aliphatic standard series. Moreover, on vague arguments they exclude the parameter α from the analysis by subtracting log k for the unsubstituted compound from other log k values in each series. This is not to be recommended; see Section 1.3.2 and a similar discussion by Clementi *et al.*[40] Swain and Lupton make the same simplification and other dubious assumptions, and this is aggravated by their neglect to analyze directly the experimental results; instead they analyze the correlations between various σ scales in the hope of separating field and resonance effects.

It is difficult to select data suitable for an analysis by means of a multiple-term LFER (1.8) without forcing an "explanation" onto the behavior of the data. If, for instance, we wish to extend the single-term Hammett equation to a two-term LFER with the second term describing resonance effects, we must select data that (*i*) deviate from the one-term LFER (1.5), (*ii*) deviate *for certain* because of resonance interaction between substituent and reaction center.

The reasons for deviations from the simple equation (1.5) other than resonance effects can, however, be many; probably the most common are as follows.

(*a*) A large variation in reactivity within a single series, which may lead to curvature in the Hammett plot. The deviations from linearity due to this curvature are naturally most pronounced for substituents at the ends of the σ scale, i.e., strong donors and strong acceptors. This makes for confusion with resonance effects.

(*b*) Steric effects, which occur with large substituents (typically iodo and phenyl), in particular in the *meta*-position.

(*c*) Several mechanisms in the series due to change in, for instance, solvation. The substituents most likely to cause a change in mechanism are again the strong donors and/or acceptors, owing to their large affinity for various reactants and solvents.

(*d*) Several steps in the mechanism, the rates of which change greatly with substituent. This can be seen as a special case of (*a*).

(*e*) Others; the deviations are too large to be acceptable, but no particular pattern can be seen; see a discussion by Hammett[73] and by Exner.[60]

To summarize: One can be virtually certain of the interpretation of the points that are well fitted by the simple LFER (1.5), but the deviations can be due to a large number of causes. Thus, we agree with Exner[74] when he states: "A procedure which does not seem to be desirable is the extending of existing relationships by additional terms or the defining of further types of constants, without regard to the approximate character of all the relationships, and on an insufficient experimental basis. On the contrary, in fundamental studies the use of sigma constants should be restricted, and replaced by direct correlations between experimental quantities."

For operational use in studying the similarities between chemical systems, however, LFER with one or several terms are ideally suited. They have simple mathematical form—they are linear once the θ values are determined—which makes them simple to apply, especially since this can be done within the simple statistical least-squares framework. The important point for the chemist is probably the close connection of LFER to classical chemical reasoning in terms of classification and analogies within classes, which makes LFER uniquely suitable for the analysis of chemical data. LFER can handle information that is contained in several variables, while in a qualitative data analysis one has great difficulty in reasoning in terms of more than one variable at a time. Hence LFER can extract much more information out of a given body of data than is possible by a qualitative analysis. With rising experimental costs, the virtues are obvious.

In all data analysis one must remember that there is never one "best" way to perform the analysis. Statistical methods can be applied to data only after a specific question is formulated, sometimes dressed up as a statistical hypothesis, and different questions posed lead to different methods of analysis. Thus, LFER are tools that can provide answers to questions formulated in terms of similarity and analogy. When other answers are sought from the data, other models might be better, with the corresponding use of different statistical methods. One should also be aware of the possibility that the data under analysis do not contain sufficient information to give unambiguous answers concerning the original problem (Section 1.5.1). In such cases, only more experiments, perhaps better designed, can avoid the predicament; no method of data analysis can extract information that is not already in the data. To conclude, LFER are by no means substitutes for chemical knowledge and intelligence, but rather a quantitative support for chemical reasoning, helping chemists efficiently to extract certain types of information from chemical data.

Acknowledgments

We thank the Swedish Natural Science Research Council and the Swedish Institute for Applied Mathematics for supporting grants.

References and Notes

1. L. P. Hammett, *Physical Organic Chemistry*, 2nd edition, p. 347 (McGraw-Hill, New York, 1970).
2. J. E. Leffler and E. Grunwald, *Rates and Equilibria of Organic Reactions*, p. 128 (Wiley, New York, 1963).

3. N. B. Chapman and J. Shorter, eds., *Advances in Linear Free Energy Relationships* (Plenum, London, 1972).
4. L. P. Hammett, *Physical Organic Chemistry*, Chap. 7 (McGraw-Hill, New York, 1940).
5. Reference 1, p. 355.
6. Reference 2, p. 172.
7. O. Exner, in Ref. 3, p. 2.
8. S. Wold, *Chem. Scripta*, **5**, 97 (1974).
9. The analogy principle was clearly formulated in organic chemistry in the middle of the 19th century; see for instance A. Kekulé, *Lehrbuch der Organischen Chemie*, Band I, pp. 124–132 (Verlag von Ferdinand Enke, Erlangen, 1861).
10. Reference 1, p. 348.
11. L. P. Hammett, Foreword to Ref. 3.
12. G. S. Hammond, *J. Chem. Educ.*, **51**, 559 (1974).
13. V. A. Palm, *Osnovy Kolichestvennoi Teorii organicheskikh Reaktsii* (Izdatelstvo Khimiya, Leningrad, 1967); German translation, *Grundlagen der Quantitativen Theorie Organischer Reaktionen* (Akademie Verlag, Berlin, D.D.R., 1971).
14. H. Eyring, J. Walter, and G. E. Kimball, *Quantum Chemistry*, p. 28 (Wiley, New York, 1944).
15. S. Wold and M. Sjöström, *Chem. Scripta*, **2**, 49 (1972).
16. M. Sjöström and S. Wold, *Chem. Scripta*, **6**, 114 (1974).
17. M. Sjöström and S. Wold, *Chem. Scripta*, **9**, 200 (1976).
18. O. Exner, *Progr. Phys. Org. Chem.*, **10**, 411 (1973).
19. O. Exner, *Nature*, **227**, 366 (1970).
20. S. Wold and O. Exner, *Chem. Scripta*, **3**, 5 (1973).
21. C. D. Ritchie and W. F. Sager, *Progr. Phys. Org. Chem.*, **2**, 363 (1964).
22. Reference 2, p. 376.
23. L. G. Hepler, *Canad. J. Chem.*, **49**, 2803 (1971).
24. Reference 1, p. 107.
25. R. C. Petersen, *J. Org. Chem.*, **29**, 3133 (1964).
26. K. B. Wiberg, *Physical Organic Chemistry*, p. 379 (Wiley, New York, 1964).
27. Reference 1, p. 401.
28. R. W. Taft, *J. Phys. Chem.*, **64**, 1805 (1960).
29. T. M. Krygowski, *Bull. Acad. polon. Sci., Sér. Sci. chim.*, **19**, 61 (1971).
30. P. R. Wells and W. Adcock, *Austral. J. Chem.*, **19**, 221 (1966).
31. J. D. Roberts, E. A. McElhill, and R. Armstrong, *J. Amer. Chem. Soc.*, **71**, 2923 (1949).
32. J. D. Roberts and W. T. Moreland Jr., *J. Amer. Chem. Soc.*, **75**, 2167 (1953).
33. K. Kindler, *Justus Liebigs Ann. Chem.*, **450**, 1 (1926).
34. A. J. Hoefnagel and B. M. Wepster, *J. Amer. Chem. Soc.*, **95**, 5357 (1973).
35. K. Kindler, *Justus Liebigs Ann. Chem.*, **452**, 90 (1927).
36. E. Grünwald and B. J. Berkowitz, *J. Amer. Chem. Soc.*, **73**, 4939 (1951).
37. J. Hine, *J. Amer. Chem. Soc.*, **81**, 1126 (1959).
38. A. R. Katritzky and R. D. Topsom, *J. Chem. Educ.*, **48**, 427 (1971).
39. S. Ehrenson, R. T. C. Brownlee, and R. W. Taft, *Progr. Phys. Org. Chem.*, **10**, 1 (1973).
40. S. Clementi, F. Fringuelli, P. Linda, and G. Savelli, *Gazz. Chim. Ital.*, **105**, 281 (1975).
41. C. G. Swain and E. C. Lupton, *J. Amer. Chem. Soc.*, **90**, 4328 (1968).
42. L. M. Stock and H. C. Brown, *Adv. Phys. Org. Chem.*, **1**, 35 (1963).
43. H. H. Jaffé, *Chem. Rev.*, **53**, 191 (1953).
44. S. Ehrenson, *Progr. Phys. Org. Chem.*, **2**, 241 (1964).
45. T. W. Anderson, *An Introduction to Multivariate Statistical Analysis* (Wiley, New York, 1958).
46. H. H. Harman, *Modern Factor Analysis* (University of Chicago Press, Chicago, 2nd edn., 1967).

47. P. H. Weiner, E. R. Malinowski, and A. R. Levinstone, *J. Phys. Chem.*, **74**, 4537 (1970).
48. P. H. Weiner and E. R. Malinowski, *Chemical Applications of Factor Analysis* (to be published, 1977).
49. S. Wold, *Pattern Recognition*, **8**, 127 (1976).
50. J. W. Al and B. M. Wepster, Report, January 1958, Technische Hogeschool, Delft.
51. H. Wold, in *Multivariate Analysis*, P. R. Krishnaiah, ed. (Academic Press, New York, 1967).
52. A. Christoffersson, *The One-Component Model with Incomplete Data* (Thesis, Uppsala University, Uppsala, Sweden, 1970).
53. V. M. Maremäe and V. A. Palm, *Reakts. spos. org. Soedinenii*, **1**(2), 85 (1964).
54. O. L. Davies and P. L. Goldsmith, eds., *Statistical Methods in Research and Production* (Oliver and Boyd, Edinburgh, 4th edn., 1972).
55. P. D. Lark, B. R. Craven, and R. C. L. Bosworth, *The Handling of Chemical Data*, p. 182 (Pergamon Press, Oxford, 1968).
56. N. R. Draper and H. Smith, *Applied Regression Analysis* (Wiley, New York, 1966).
57. D. H. McDaniel and H. C. Brown, *J. Org. Chem.*, **23**, 420 (1958).
58. P. Prescott, *Technometrics*, **17**, 129 (1975).
59. I. M. Chakravarti, R. G. Laha, and J. Roy, *Handbook of Methods of Applied Statistics*, Vol. 1, pp. 395–403 (Wiley, New York, 1967).
60. O. Exner, Ref. 3, p. 12.
61. Reference 54, p. 261.
62. C. K. Hancock, *J. Chem. Educ.*, **42**, 608 (1965).
63. O. Exner, Ref. 3, p. 18.
64. G. E. P. Box and W. G. Hunter, *Technometrics*, **7**, 23 (1965).
65. B. R. Kowalski and C. F. Bender, *J. Amer. Chem. Soc.*, **94**, 5632 (1972).
66. B. R. Kowalski and C. F. Bender, *J. Amer. Chem. Soc.*, **95**, 686 (1973).
67. B. R. Kowalski, in *Computers in Chemical and Biological Research*, Vol. 2, C. E. Klopfenstein and C. L. Wilkins, eds. (Academic Press, New York, 1974).
68. T. Cacoullos, ed., *Discriminant Analysis and Applications* (Academic Press, New York, 1973).
69. R. Mecke and K. Noack, *Chem. Ber.*, **93**, 210 (1960).
70. A. P. Marks and R. S. Drago, *J. Amer. Chem. Soc.*, **97**, 3324 (1975).
71. M. Godfrey, *J. Chem. Soc. Perkin II*, 1016 (1975).
72. R. Ponec and V. Chvalovský, *Coll. Czech. Chem. Comm.*, **39**, 3091 (1974).
73. Reference 1, p. 365.
74. Reference 3, p. 52.

2

The Brönsted Equation—Its First Half-Century

R. P. Bell

2.1. Introduction

It is now just over 50 years since the Brönsted relation† between catalytic power and acid–base strength was proposed. Since then it has been very widely used, and constitutes the first example of a linear free energy relationship, a term that was not introduced until considerably later: In fact, the Brönsted equation preceded by some ten years the sigma–rho correlations introduced by Hammett and often regarded as the first example of this type of relationship.[1] Further, there has recently been a

† In writing about this topic a choice has to be made between the two spellings Brønsted and Brönsted. The former is correct in Danish, but the letter ø is peculiar to the Danish and Norwegian languages, and Brönsted himself used the latter spelling in all his publications in English, French, and German. Moreover, it provides most readers with a clue to the correct pronunciation of his name, and has therefore been adopted here.

R. P. Bell • Department of Chemistry, The University of Stirling, Stirling FK9 4LA, Scotland.

revival of interest in the Brönsted equation in connection with reaction mechanisms and molecular pictures of kinetic processes. A review of the way in which our knowledge and interpretation of this relation have developed may therefore be of interest.

2.2. *The Discovery of the Brönsted Relation in the Decomposition of Nitramide*

In 1924 Brönsted and Pedersen[2] published a study of the kinetics of the decomposition of nitramide in aqueous solution according to the equation $NH_2NO_2 \rightarrow H_2O + N_2O$. This is a very remarkable paper, and a large proportion of its 50 pages repays reading even today. Since it will be quoted repeatedly in the present chapter it will be referred to simply as BP. It was already known[3] that nitramide decomposes very rapidly in alkaline solution, and the original object of the investigation was to find a convenient method of measuring hydroxide ion concentrations in buffer solutions, with a view to testing current theories about salt effects and activity coefficients. The nitramide decomposition is experimentally suitable for this purpose, since it produces only neutral and unreactive compounds and can be conveniently followed by measuring the volume or pressure of the nitrous oxide evolved, at that time probably the most convenient nonanalytical method of following the course of a reaction.

In the event, the reaction turned out to be unsuitable for measuring hydroxide ion concentrations, but its actual behavior led to findings of much greater interest. The observed first-order rate constant, k, in buffer solutions was found to be independent of pH over a wide range, but to depend upon the concentration of one of the buffer components: Thus in acetate buffers the results could be represented by equation (2.1), where k_0

$$k = k_0 + k_{AcO}[AcO^-] \tag{2.1}$$

depends only on the temperature, and k_{AcO} (in general k_B) only on the temperature and the nature of the buffer component. Similar behavior was observed in buffer solutions prepared from a number of weak acids and from aniline.

These findings were related by BP to Brönsted's 1923 definition of acids and bases[4] as species having a tendency to lose or gain a proton, respectively, as expressed in the (hypothetical) scheme $A \rightleftharpoons B + H^+$. It had previously been supposed that acid–base catalysis was the exclusive property of hydrogen or hydroxide ions (or their analogs in other solvents), but this became an unreasonable view once it was realized that the hydrogen ion in aqueous solution exists essentially as H_3O^+, and that the hydroxide ion is only one of a large class of bases, comprising the anions of weak acids. A more natural assumption was that under suitable circumstances any species satisfying the Brönsted acid–base definition might act as a catalyst: This was termed *general acid–base catalysis,* in contrast to the specific catalysis by hydrogen or hydroxide ions which had previously been assumed. Equation (2.1) is clearly consistent with the occurrence of general catalysis by bases, and although much earlier work by Dawson and his collaborators[5,6] on the acetone–iodine reaction had produced good evidence for general catalysis, in this instance by acids, the general importance of this finding had not been realized.

Even when catalysis is actually being effected by a general base B, this does not necessarily lead to observed relations of the type (2.1), as may be shown by analyzing the reaction scheme (2.2), where SH is the substrate

$$\begin{aligned} \mathrm{SH + B} &\underset{k_{-1}}{\overset{k_1}{\rightleftharpoons}} \mathrm{S^- + A} \\ \mathrm{S^-} &\underset{k_2}{\rightarrow} \text{products} \end{aligned} \tag{2.2}$$

and A the acid related to B by $\mathrm{A \rightleftharpoons B + H^+}$. It is easily shown that the observed reaction rate will be proportional to [B] only if $k_2 \gg k_{-1}[\mathrm{A}]$, when k_1 refers to the rate-limiting step. If on the other hand, $k_2 \ll k_{-1}[\mathrm{A}]$ the observed rate will always be proportional to $[\mathrm{OH^-}]$: The first stage of the reaction is then at equilibrium, and k_2 becomes rate-limiting. This was clearly realized by BP, who stated that the decomposition of nitramide must involve a slow proton transfer to the base from nitramide or an isomer of nitramide. The same theme was subsequently treated more generally for both acid and base catalysis by Brönsted,[7] and extended by Pedersen[8] to reactions such as keto–enol transformations which involve two successive proton transfers. The resulting equations have been rediscovered many times by later authors in connection with specific reactions.

In equation (2.1) k_0 refers to catalysis by the solvent, and k_{AcO} (more generally k_{B}) to catalysis by the basic component of the buffer. (In this particular reaction, catalysis by $\mathrm{OH^-}$ is negligible over a considerable range of pH, though this will not always be true). BP found that for a series of related catalysts, k_{B} was related to the acid–base strength of the catalyst by equation (2.3), where K_{B} is the basic strength of the catalyst B, K_{A} the acid

$$k_{\mathrm{B}} = G'_{\mathrm{B}} K_{\mathrm{B}}^{\beta} = G_{\mathrm{B}} K_{\mathrm{A}}^{-\beta} \tag{2.3}$$

strength of the corresponding acid A, and G_{B} (or G'_{B}) is a constant for this particular reaction, temperature, solvent, and series of catalysts. Under the same conditions β is also a constant, which had the value 0.83 for the decomposition of nitramide catalyzed by a series of anionic bases. Equation (2.3) implies that a plot of log k_{B} against log K_{A} should be a straight

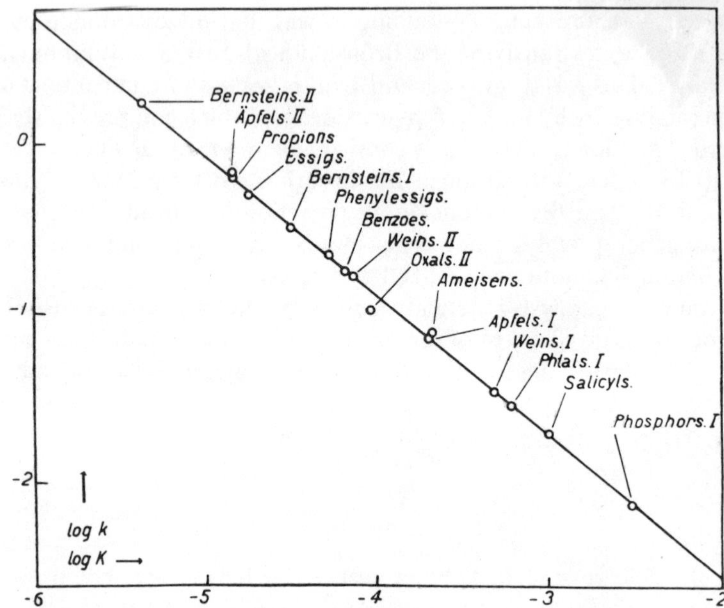

Fig. 2.1. The first Brönsted relation. Anion catalysis in the decomposition of nitramide. (Reproduced from Ref. 2.)

line of slope $-\beta$, as illustrated in Fig. 2.1, reproduced from the original paper of BP.†

Equation (2.3) needs some modification for statistical factors. If the basic catalyst contains q equivalent sites at which a proton can be accepted, and the corresponding acid p equivalent acidic protons, the correct relation for base catalysis is (2.4), and the corresponding expression for acid catalysis§ is (2.5).

$$k_B/q = G_B(p/qK_A)^\beta \tag{2.4}$$

$$k_A/p = G_A(qK_A/p)^\beta \tag{2.5}$$

Statistical factors were introduced in an incomplete form by BP and later derived correctly by Brönsted.[7] More sophisticated derivations have been given recently by a number of authors.[9–12]

† This plot should strictly be slightly modified, since BP used an incomplete form of statistical correction [see (2.4)]; however, the resulting plot differs very little from Fig. 2.1.

§ Many authors have represented the Brönsted exponents for acid and base catalysis by α and β, respectively. Since the distinction between substrate and catalyst is an artificial one from a molecular point of view, and disappears completely when proton-transfer rates are measured directly, we have preferred to use β (for Brönsted) throughout.

In the following decade Brönsted and his co-workers showed that equation (2.4) was also applicable to catalysis of the nitramide decomposition by substituted anilines[13] and cationic bases,[14] though the values of G and β were somewhat different for each class of base. Similar behavior was observed when the decomposition was studied in isoamyl alcohol (3-methylbutan-1-ol)[15] or *m*-cresol,[16] better agreement being obtained if the values of k_B and K_A referred to the same solvent. It is noteworthy that in all these cases a precise correlation was obtained, as illustrated by Fig. 2.1; in fact the precision is higher than for most subsequent examples of the Brönsted relation, or for other classes of LFER. This can be attributed to two factors. First, in contrast to most LFER, the two quantities correlated by the Brönsted relation refer to the *same reaction*.† Secondly, because of the small size of the proton, deviations due to steric effects will be at a minimum for proton-transfer reactions, especially for a small molecule such as nitramide.

2.3. Applicability of the Brönsted Relation to Other Reactions

A relation equivalent to the Brönsted relation with $\beta = \frac{1}{2}$ was proposed at an early date for a number of reactions,[17] but most of the evidence on which it was based would not now be regarded as valid. The observations of Dawson and his collaborators[5,6] on general acid catalysis in the acetone–iodine reaction do in fact conform to such a relation,[18] though this was not shown by the authors. Brönsted's work on the nitramide reaction was soon followed by a study of the mutarotation of glucose,[19] in which general catalysis by both acids and bases was found to obey equations (2.4) and (2.5), though the results were less extensive and accurate than for the nitramide decomposition.

In the following 40–50 years the validity of the Brönsted relation has been established for very many examples of general acid–base catalysis: In fact, any considerable deviations would now be regarded as casting doubt on the reaction mechanism, or reflecting some abnormal properties of the species concerned. No attempt will be made to list or document these very numerous investigations, but some of the most studied classes of reaction will be mentioned; a more extensive account has been given elsewhere.[20] The ionization or enolization of ketones and similar substances exhibits general catalysis by both acids and bases and has been followed by using halogens as scavengers to remove the enol or enolate ion, by the racemization of optically active substrates, or by exchange of hydrogen isotopes.

† In a catalyzed reaction such as (2.2) the value of the equilibrium constant K for the reaction $SH + B \rightleftharpoons S^- + A$ is frequently unknown, but if SH is kept constant, variations in $\log K$ will be equal to those in $\log(1/K_A)$.

Many hydrolytic reactions (for example, those of ortho-esters or vinyl ethers) are subject to general catalysis and the same is true of the reversible addition of nucleophiles to carbonyl groups, of which the ring–chain interconversion involved in the mutarotation of glucose was the earliest example to be studied. Although the decomposition of ethyl diazoacetate has long been known as an example of specific catalysis by hydrogen ions, for several other diazo-compounds the decomposition exhibits general acid catalysis, and there are several examples in the interesting borderline range, i.e., in the scheme (2.6) [analogous to (2.2) for base

$$\mathrm{SH} + \mathrm{A} \underset{k_{-1}}{\overset{k_1}{\rightleftharpoons}} \mathrm{SH}_2^+ + \mathrm{B}$$

$$\mathrm{SH}_2^+ \underset{k_2}{\rightarrow} \text{products} \tag{2.6}$$

catalysis] $k_{-1}[\mathrm{B}]$ and k_2 are of comparable magnitude. Finally, isotopic hydrogen exchange in aromatic compounds is usually catalyzed by general acids, and because of the sensitivity of methods for detecting tritium the Brönsted relation can be studied over a very wide range of acid strengths.

In catalyzed reactions the proton transfer is nearly always thermodynamically unfavorable or "up-hill" (i.e., ΔG^0 has a large positive value) and the rate can be studied only by making use of subsequent changes in the reaction products. A thermodynamically favorable or "down-hill" proton transfer can in principle be observed directly, though its rate is usually too high to be measured by conventional means. However, the development of relaxation methods for following fast reactions, mainly by Eigen and his school, has largely removed this limitation: for example, values were obtained for the rates of reaction of the keto, enol, and anionic forms of acetylacetone with a wide variety of acids and bases.[21] In general the relationship between rate and equilibrium constants resembles that found for catalyzed reactions, though, as will be shown later, the wide range of accessible rates and chemical structures reveals new phenomena of interest.

As first pointed out by Horiuti and Polanyi,[22] a close analogy with the Brönsted relation is provided by the well-known Tafel relation between current and overvoltage in the cathodic evolution of hydrogen. This is expressed by (2.7), where i is the current density, F the Faraday, V the

$$\ln i = \alpha \mathsf{F} V/RT + \text{const} \tag{2.7}$$

overvoltage, and α usually has a value of about 0.5. Whereas in the usual Brönsted relation the free energy change in the process is varied by altering the chemical nature of the reactants, in the cathodic discharge of hydrogen the variation is caused by changing the overvoltage. The latter can often be varied over a range of 2 V, which would correspond to a change of 10^{34} in an equilibrium constant; hence the study of electrode processes provides a valuable extension of the conditions usually available for studying proton-transfer reactions.

2.4. Molecular Interpretation of the Brönsted Relation

Early workers in this field regarded the Brönsted relation as essentially empirical, but some years later plausible molecular interpretations were suggested. For this purpose it is convenient to write the relation in the form (2.8), where ΔG^0 is the standard free energy change for the overall

$$\delta(\Delta G^{\ddagger}) = \beta\delta(\Delta G^0) \tag{2.8}$$

reaction, $\Delta G^{\ddagger}$ the free energy of activation (both corrected for statistical factors), and the operator δ represents the effect of chemical modifications in the catalyst. Since in general $\Delta G^0 = -RT \ln K$, the integration of (2.8) with the assumption that β remains constant leads to the familiar form of the Brönsted relation. As already mentioned, if the substrate remains the same and there is no association between the solute species, $\delta(\Delta G^0)$ for the proton-transfer reaction being considered will be identical with $\delta(\Delta G^0)$ for any equilibrium with a standard acid–base pair (in water usually H_3O^+–H_2O), which is used as a measure of the acid–base strength of the catalyst. Equation (2.8) is readily recognizable as an LFER, in which β is the ratio of the ρ values for the ρ–σ relations representing the effect of substituents on the rates and equilibria, respectively.

A molecular interpretation of equation (2.8) was first given by Horiuti and Polanyi in 1935.[22] They supposed that the energy of the transition state corresponded to the intersection of the potential energy curves representing the binding of the proton in its initial and final states, as illustrated in Fig. 2.2, and that the effect of modifying the catalyst could be represented by a vertical displacement of one of these curves, without changing its shape or the horizontal distance between the two curves. If this displacement is small, and changes in potential energy are identified with the changes in free energy, simple geometry leads directly to equation (2.8), with $\beta = s_1/(s_1+s_2)$, where s_1 and s_2 are the slopes of the two curves at their point of intersection.

Figure 2.2 implies that identical behavior would result from the vertical displacement of either curve, since $\Delta G^{\ddagger}$ depends only upon their relative positions. This means that in a catalyzed reaction Brönsted relations with identical values of β should be generated by variations in the acid–base strength of either the catalyst or the substrate. This is not a thermodynamic necessity, as may be shown by the following consideration of a general acid–base reaction $A_1 + B_2 \rightleftharpoons A_2 + B_1$ for which $\Delta G^{\ddagger}_{+}$, $\Delta G^{\ddagger}_{-}$, and ΔG^0 refer

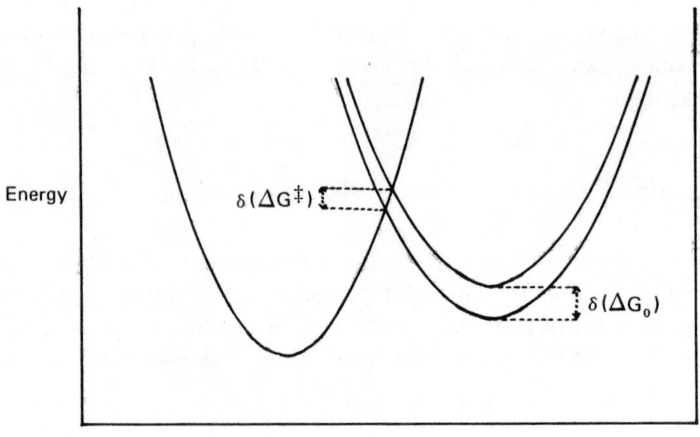

Fig. 2.2. A simple molecular basis for the Brönsted relation.

to the forward, reverse, and overall reaction, respectively. We can then, in general, write (2.9), where δ_1, β_1 and δ_2, β_2 refer to changes in

$$\begin{aligned}\delta(\Delta G_+^\ddagger) &= \delta_1(\Delta G_+^\ddagger)+\delta_2(\Delta G_+^\ddagger) = (\beta_1+\beta_2)\delta\Delta G^0\\ \delta(\Delta G_-^\ddagger) &= \delta_1(\Delta G_-^\ddagger)+\delta_2(\Delta G_-^\ddagger) = -[(1-\beta_1)+(1-\beta_2)]\delta\Delta G^0\end{aligned} \tag{2.9}$$

the pairs A_1–B_1 and A_2–B_2, respectively. This is consistent with the necessary condition $\delta(\Delta G^0) = \delta(\Delta G_+^\ddagger) - \delta(\Delta G_-^\ddagger)$ even when β_1 and β_2 have different values, and we shall see later that such differences are found experimentally.

Apart from this limitation, Fig. 2.2 as it stands obviously represents a highly specialized model. It is therefore of interest that a quite different model, in which the proton moves in the field of two negatively charged centers, leads to exactly the same conclusion.[23] In fact, very general considerations[24] show that the validity of equation (2.8) depends only on the assumption that for the series of substances considered, both ΔG^0 and $\Delta G^\ddagger$ are functions of a single parameter ξ: β is then simply $(\partial\Delta G^\ddagger/\partial\xi)/(\partial\Delta G^0/\partial\xi)$, which may be regarded as a constant over limited intervals of $\Delta G^\ddagger$ and ΔG^0. The last statement stresses the fact that all realistic theoretical derivations lead to the differential form of the Brönsted relation, i.e., equation (2.8). The range over which it is legitimate to treat β as a constant, and hence to integrate (2.8) to give (2.3), will depend upon the details of the model: Thus in Fig. 2.2, it is the range over which the intersecting potential energy curves may be regarded as linear. In practice this question must be answered by reference to experiment, which shows that β is usually effectively constant over at least several powers of ten in k

or K. The question of variations in β over very large ranges, and their interpretation in terms of molecular models, will be considered in Section 2.5.3.

The identification of model potential energies with observed free energies of activation or reaction raises a point of general interest that has not been satisfactorily solved. It might be thought that Fig. 2.2 would predict relations between changes in activation energy and enthalpies of reaction, rather than between $\Delta G^{\ddagger}$ and ΔG^0. However, at finite temperatures all these quantities represent averages over a large number of energy states, which also involve the solvent, the averaging process being different for ΔH and ΔG. The quantities that occur in models usually ignore both the spread of energy states and interaction with the solvent and therefore do not correspond directly with any measured quantities in solution at finite temperatures. However, there is some theoretical justification (related to the frequent cancellation of energy and entropy effects) for supposing that the effect of substituents is reflected more closely by changes in ΔG^0 and $\Delta G^{\ddagger}$ than by the corresponding enthalpy quantities,[25] and in the few instances where systematic measurements have been made, the correlation between activation energies for catalysis and the enthalpy of dissociation of the catalyst is considerably worse than that between the rate constants and dissociation constants.[26]

In their molecular interpretation of the Brönsted relation, Horiuti and Polanyi[22] pointed out that in any quantitative treatment of proton transfers it would almost certainly be necessary to take into account the tunnel effect, i.e., the nonclassical behavior of particles of low mass. Although there is now a good deal of evidence, especially from hydrogen isotope effects, that tunneling corrections are important in many proton-transfer reactions,[27] they do not appear to distort regular behavior in terms of the Brönsted relation. This means that the magnitude of tunneling corrections must either remain effectively constant in reactions of a series of similar acids or bases, or that it must vary in a smooth manner with ΔG^0. The latter type of behavior does in fact emerge from theoretical treatments of two different models.[23,28] Tunneling behavior will therefore not be considered further in this review, though it should be remembered that its inclusion might well modify some of the quantitative conclusions arrived at. At the end of Section 2.5.3 reference is made to recent theories of proton transfer in which tunneling is allotted a dominant role.

2.5. *The Range of Validity of Brönsted Relations*

As often happens when a quantitative relationship has become firmly established, interest has gradually shifted to the range of validity of the

Brönsted relation and the explanation of deviations from it. Two main questions may be asked here: (*a*) how drastic are the structural variations that are permissible, i.e., how wide a meaning has the operator δ in equation (2.8); (*b*) what range of values of ΔG^0 (or reaction rates) corresponds to an effectively constant value of β, so that equation (2.8) can be integrated to give equation (2.3); i.e., will Brönsted plots be curved rather than linear if extended over a wide range? It is convenient to discuss these topics under three headings.

2.5.1. *Deviations within a Series of Similar Acids and Bases*

It was anticipated by BP that the relations which they established for the decomposition of nitramide would be valid only for a series of catalysts that were chemically very similar, and most of the early work on catalyzed reactions employed series such as carboxylic acids (or their anions), ring-substituted anilines, and the like. Much less information was available about the effect of varying the substrate, partly because its acid–base strength was often too low to be measured. When such variation was attempted, as for example in prototropic reactions of ketones, it often involved substitution close to the seat of reaction, so that the condition of chemical similarity was less strictly observed. Recently, however, several series of remotely substituted substrates have been studied, notably the ionization of ring-substituted phenylnitromethanes and phenylnitropropanes,[29–33] and of the series $XC_6H_4CH_2CH(COMe)CO_2Et$ and $XC_6H_4CH_2CH(COMe)_2$.[34] For a given catalyst each of these series obeys a Brönsted relation, but the values of β differ considerably from those applying to variation of catalyst, i.e., β_1 and β_2 in equation (2.9) are different. For the nitroalkanes the values of β (substrate) show marked anomalies, as discussed in Section 2.6.1.

When we revert to series of similar catalysts, individual deviations from the Brönsted relation can often be attributed to specific effects. *Steric hindrance* (which will affect rates much more than equilibria) is not prominent, because of the small size of the proton, but the low rates sometimes observed when 2- and 6-substituted pyridines are used as catalysts[35–37] can presumably be attributed to steric repulsions. An effect in the opposite direction has been observed in the anion-catalyzed halogenation of various ketones and esters.[38] When both the catalyst and the substrate contain a large group near the seat of reaction, the observed rate constant is greater (by up to 300%) than that predicted by the Brönsted relation obeyed by other species. This is probably because when the two large groups are close together, they cause the liberation of fewer water molecules than when they are separated: This type of interaction, under the name of *hydrophobic bonding*, is now commonly invoked to account for the configuration and association of macromolecules.

It was realized at an early date[16,39] that in the decomposition of nitramide, catalysis by *primary*, *secondary*, *and tertiary amines* could not be accommodated on a single Brönsted plot, and subsequent more detailed study[40,41] showed that the three classes of amine give rise to three separate but parallel plots. This behavior is almost certainly due to the decreasing solvation of the cations in the order $RNH_3^+ > R_2NH_2^+ > R_3NH^+$. This sequence will distort the measured values of pK, especially in aqueous solution, but will have less effect on the transition states, where the charge is more dispersed.[42] Similar considerations serve to explain the anomalous order of acid–base strengths of the alkylamines in water, an anomaly that disappears in solvents with little or no power of hydrogen bonding, and in the gas phase.[43]

The hydration of the amine cations depends upon electrostatic forces or hydrogen bonding, but in some instances hydration to give definite covalent species occurs, notably in the system $CO_2 + H_2O \rightleftharpoons H_2CO_3 \rightleftharpoons H^+ + HCO_3^-$. The ratio $[CO_2]/[H_2CO_3]$ is about 270 at 298 K,[44] and this has an important bearing on the catalytic activity of the bicarbonate ion. The conventional pK of carbonic acid (6.35) does not differentiate between the species CO_2 and H_2CO_3, while the "true" basic strength of HCO_3^- is determined by the equilibrium $H_2CO_3 \rightleftharpoons H^+ + HCO_3^-$, for which p$K = 3.89$. Hence if the conventional pK is used in a Brönsted plot, the point for catalysis by HCO_3^- should show a large negative deviation from the line defined by other bases, while the use of p$K = 3.89$ should place it approximately on the line. BP predicted this behavior but were not able to verify it for the nitramide decomposition; however, deviations of the expected magnitude have been observed in other base-catalyzed reactions.[45,46]

These examples involving hydration represent a special case of deviations from the Brönsted relation, which occur when the acid–base equilibrium of the catalyst is accompanied by subsidiary equilibria, which may or may not involve the solvent. BP gave a full treatment of this problem, the results of which are as follows. Consider the general equilibrium (2.10),

$$\begin{array}{ccc} HX & \rightleftharpoons & X^- + H^+ \\ \downarrow\uparrow & & \downarrow\uparrow \\ HY & \rightleftharpoons & Y^- + H^+ \end{array} \tag{2.10}$$

and let x_1 and x_2 represent the equilibrium fractions of the acid and its anion which are in the forms HY and Y^-, respectively. If X^- and Y^- can act separately as basic catalysts for a given reaction, with catalytic constants k_X and k_Y obeying the same Brönsted relation, then it is easily shown that the observed catalytic constant for the equilibrium mixture is given by (2.11), where K_A is the acid dissociation constant of the equilibrium mixture measured by conventional means. A similar expression can be derived for acid catalysis by HX and HY. The expression in square brackets in (2.11)

$$k_B = (1-x_2)k_X + x_2 k_Y = GK_A^{-\beta}[(1-x_1)^{\beta}(1-x_2)^{1-\beta} + x_1^{\beta}x_2^{1-\beta}] \quad (2.11)$$

represents the deviation from the Brönsted relation caused by the presence of the subsidiary equilibria in (2.10). In the general case a measurement of this deviation for a single reaction is insufficient to determine both x_1 and x_2, though in principle they could be derived by measuring the deviations for two reactions with different values of β. However, if, as is often the case, either x_1 or x_2 is close to zero, equation (2.11) is simplified, and the study of a single reaction is sufficient.

BP observed that the anions of nitrourethane and nitramide itself had an unexpectedly low catalytic effect in the nitramide decomposition, and used the above equations to estimate the equilibrium proportions of the isomers $HN{=}NO_2H$ and $EtO_2CN{=}NO_2H$. However, their values are of doubtful significance, since these species are structurally very different from the carboxylate ions on which the standard Brönsted relation was based. Since 1924 no use appears to have been made of this principle for investigating isomerization equilibria in solution until some recent studies of ring–chain equilibria in keto-acids, which can exist in the three forms shown in (2.12).[47–50] The overall dissociation constants were determined by conventional means, and the "true" dissociation constants of the

$$\underset{\text{lactol}}{\underbrace{RC(OH)O \cdot CO}} \rightleftharpoons \underset{\text{keto-acid}}{\underbrace{RCO \quad CO_2H}} \rightleftharpoons \underset{\text{carboxylate ion}}{\underbrace{RCO \quad CO_2^-}} + H^+ \quad (2.12)$$

keto-acid forms by measuring the catalytic effect of the carboxylate ions in the decomposition of nitramide, and for some compounds also in the mutarotation of glucose. The ratio [lactol]/[keto-acid] was determined for 27 aliphatic or aromatic keto-acids, and in the few cases where comparison is possible, the values obtained agree well with those from other sources.

Two further points are worth mentioning here. First, the above treatment implies some restriction on the time scale for the interconversion of the species involved. This must be sufficiently fast for conventional pK measurements to yield a true equilibrium value. On the other hand it must be sufficiently slow for interconversion to be unlikely during the lifetime of the encounter complex between catalyst and substrate: Otherwise it is not meaningful to attribute separate catalytic constants to X^- and Y^- in scheme (2.10), or to relate the catalytic power of the carboxylate ion in (2.12) to the dissociation constant of the keto-acid form rather than to that of the equilibrium mixture. The rates of interconversion of CO_2 and H_2CO_3 are well known and certainly fulfil these conditions, and relaxation measurements on solutions of keto-acids[47,51] show that the same is true of these systems; however, there may well be systems for which this assumption cannot be made. Secondly, the use of deviations from the Brönsted relation for obtaining information about subsidiary equilibria depends in no way on a theoretical basis or molecular picture for this relation, but may be regarded as the application of a law established by experiment. BP suggested several other applications of this kind, of which two examples will be given. The first relates to the dissociation of an unsymmetrical dibasic acid according to the scheme $H^+ + {}^-XYH \rightleftharpoons HXYH \rightleftharpoons HXY^- + H^+$, where the measurement of base catalysis by an equilibrium mixture of the ions ^-XYH and HXY^- offers a method of determining the ratio of their concentrations. The second refers to the structures in solution of oxyacids and their anions: For example, phosphorous acids might exist as $P(OH)_3$, $OPH(OH)_2$, OPH_2OH, or OPOH. BP show that the different statistical factors that these formulations imply lead to different kinetic consequences, and hence to the possibility of distinguishing between them. Although modern methods of structure determination obviously diminish the usefulness of such applications, it seems worthwhile to investigate their potentialities further.

2.5.2. The Effect of Extensive Structural Changes

Brönsted and his collaborators[2,13,14] showed that in the nitramide decomposition, catalysis by anionic bases, substituted anilines, and cationic bases generated three separate but approximately parallel Brönsted plots. At least part of these differences can be attributed to differences in charge, a view supported by theoretical considerations,[8,52] but it is difficult to separate such effects from those due to differences in chemical nature. It was also found at an early stage that acid–base catalysis by the species H_3O^+, H_2O, and OH^- could not usually be represented by the Brönsted relations valid for other species, and a recent review by Kresge[53] tabulates deviations for 27 examples of catalysis by H_3O^+ and OH^-: Such deviations are usually, but not always, negative. It is again difficult to separate the effects of charge and of chemical nature, and there is the added difficulty that in the conventional formulation of the acidity constants of these species, viz.,

$$K(H_3O^+) = [H_3O^+][H_2O]/[H_3O^+] = [H_2O]$$
$$K(H_2O) = [H_3O^+][OH^-]/[H_2O] = 10^{-14}/[H_2O]$$

it is usual to write $[H_2O] = 55.5$, and to calculate the catalytic constant of H_2O by dividing the observed rate constant by 55.5, which is not a satisfactory basis for comparison with acid–base systems present in dilute solution. An alternative procedure is to regard $[H_2O]$ as the concentration of monomeric water molecules in liquid water,[54] but there is no unequivocal method for estimating this quantity, and the procedure is open to question. Moreover, the values of K_A and k_A or k_B for the species H_3O^+, H_2O, and OH^- nearly always lie well outside the range covered by the other catalysts studied, and deviations may therefore be related to the curvature of Brönsted plots discussed in Section 2.5.3.

There are, however, many examples of individual deviations from Brönsted plots by solute catalysts that bear the same charge and that are not involved in subsidiary equilibria of the type discussed in Section 2.5.1. For example, in the reaction $CH_3CH(OH)_2 \rightleftharpoons CH_3CHO + H_2O$, catalysis by 52 phenols and carboxylic acids conforms well to a Brönsted relation; on the other hand the catalytic coefficients for oximes are anomalously high and those for nitroalkanes and urethane anomalously low[55] by factors of up to 100.

Even larger anomalies appear when substrates in catalytic reactions undergo drastic structural variations close to the site of reaction, or in a direct study of the rate of reaction of such species with acids or bases. For example, phenol, acetylacetone, and nitromethane are acids of similar strength, yet the rate constants for their reactions with hydroxide ions are approximately 10^{10}, 10^5, and $10\ \text{dm}^3\ \text{mol}^{-1}\ \text{s}^{-1}$, respectively; obviously such behavior implies enormous deviations from any hypothetical Brönsted relation.

The slow reaction of nitroalkanes with bases was originally accounted for by Hantzsch[56] by supposing that the "pseudoacid" $R^1R^2CHNO_2$ isomerized slowly into the "true acid" $R^1R^2C{=}NO_2H$, which then reacted rapidly with base, and on the basis of optical evidence he regarded many other types of acid as pseudoacids. This explanation would not now be regarded as tenable, since the interconversion of the two isomers in the presence of base is believed to involve the intermediate formation of the anion, but the reasons for the large kinetic differences between different classes of acid–base systems are still not fully understood. These differences are probably related to the extent of structural change and redistribution of charge accompanying the interconversion of the acid–base pair, in the sense that a high degree of reorganization leads to a slower reaction. This suggestion can be rationalized in terms of potential energy curves[55,57]: It is qualitatively in accord with the slowness of the proton-transfer reactions of ketones and similar compounds, and especially of nitroalkanes, and is consistent with the decreasing order of catalytic power in the following series: oximes > carboxylic acids, phenols > nitrourethane > nitroalkanes, found in the reversible hydration of acetaldehyde,[55] and with a number of other observations. In an extreme form of this hypothesis,[58] the structural change is assumed to involve an intermediate (or "virtual intermediate"), but this suggestion has not yet been generally accepted.

It was at one time believed[59] that, *ceteris paribus*, the rate of proton transfer depended primarily on the nature of the atom bearing the proton. Thus for acids of equal strength reacting with a given base the order would be OH ≈ NH > SH ≫ CH, this being related to the decreasing tendency to form hydrogen bonds in this series. However, it now seems likely that the extent of structural reorganization is the more important factor, since the degree of charge delocalization needed to produce an acid of given strength increases in traversing the same series. This view is strengthened by the recent discovery that some classes of carbon acids (cyanocarbons,[60] chloroform,[61] phenylacetylene,[62] and disulfones[63]) behave kinetically just like oxygen acids. It must, however, be admitted that in spite of much discussion there is no general agreement about the reasons for the very large deviations from the Brönsted relation that arise when drastic structural alterations are made close to the seat of reaction, and it seems likely that the observed behavior is a consequence of several different effects.

A special type of positive deviation is sometimes encountered in reactions involving *bifunctional catalysts*, i.e., species that can both add and remove a proton.[64] These effects, which are relevant to enzyme action, are particularly marked for catalysts in which the successive addition and removal of a proton at different sites produces a chemically identical species, as in imidazole, pyrazole, the enol form of acetylacetone, and ions such as HPO_4^{2-} and $H_2PO_4^-$. However, such effects are found only for some classes of reaction and do not always appear when they might be expected[65]; moreover, a very general treatment of concerted proton transfers[66] indicates that simple bifunctional acid–base catalysis is likely to be of importance only under very restricted conditions.

2.5.3. Nonlinear Brönsted Plots

It was recognized by BP that the Brönsted relation is fundamentally a differential one, as expressed by equation (2.8), and that integrated forms such as (2.3) represent approximations that will be valid only over a limited range. In other words, the exponent β is only a "local" constant, and conventional Brönsted plots may be expected to exhibit curvature if they can be extended over a sufficiently wide range of rates and acid–base strengths. In particular, this is evident from any interpretation in terms of intersecting energy curves, such as Fig. 2.2, since β will remain constant only if these curves remain linear over their region of intersection. It is true that many experimental Brönsted plots show no signs of curvature over several powers of ten in k or K, but it must be remembered that a factor of 10 in a rate or equilibrium constant corresponds at 298 K to a free energy difference of only about 6 kJ mol^{-1}, compared with average bond energies of about 400 kJ mol^{-1}; thus variations of several powers of ten represent relatively small displacements of energy curves such as those in Fig. 2.2.

Attempts to study Brönsted relations over even larger ranges often run into experimental difficulties, partly because the reactions become too fast or too slow for convenient study, and partly because of interference by equilibria or rate processes involving solvent species. Moreover, a large extension of the range of catalysts (or substrates) is often incompatible with the requirement that they should be chemically similar, so that it becomes uncertain whether observed deviations represent a true curvature of the Brönsted plot or are due to structural factors. However, during the last decade such curvature has been fairly definitely established for some ten reactions, being always in the sense that the slope β decreases as the reaction becomes faster. A list of such reactions up to 1973 has been compiled by Kresge,[53] and further examples were given subsequently.[67–69]†

For simple proton-transfer reactions, two essentially different (but not mutually exclusive) explanations have been given for the observed curvature of Brönsted plots, the first of which dates back to the paper of BP. They pointed out that in a general acid–base reaction $A_1+B_2 \rightleftharpoons A_2+B_1$, if the acid strength of A_1 or the basic strength of B_2 is progressively increased by substitution, a point will eventually be reached at which the forward reaction takes place at every encounter, so that its rate cannot be further increased by further modification of the reactant: This corresponds to $\beta=0$. As previously pointed out [equation (2.9)] this necessarily implies that $\beta=1$ for the reverse reaction. Similarly, there will be another region in which the reverse reaction takes place at every encounter ($\beta=0$), and hence $\beta=1$ for the forward reaction. The complete picture for the dependence of rate upon equilibrium constant will therefore resemble Fig. 2.3, reproduced from the paper of BP. It is clear that a Brönsted relation with a constant slope between zero and unity represents an approximation to part of the central section of these curves, and that curvature (in the sense of decreasing β with increasing rate constant) is to be expected if the range is sufficiently extended.

Essentially the same argument was put forward much later by Eigen,[59,71] who illustrated it by a diagram almost identical to Fig. 2.3. He also treated the problem more quantitatively, paying particular attention to the case in which there is no chemical activation barrier to the transfer of the proton. Under these circumstances the curved section of Fig. 2.3 extends effectively over only about two powers of ten in K or k, and the

† It should be noted that in some instances the curvature of the Brönsted plot barely exceeds the scatter of points about the line. Moreover, in the frequently quoted example of the reactions of a variety of ketonic compounds with bases it has been recently suggested[70] that the experimental results are better represented by two straight lines, one for monocarbonyl and the other for β-dicarbonyl compounds. However, there can be no doubt that the curvature is real in several of the instances quoted.

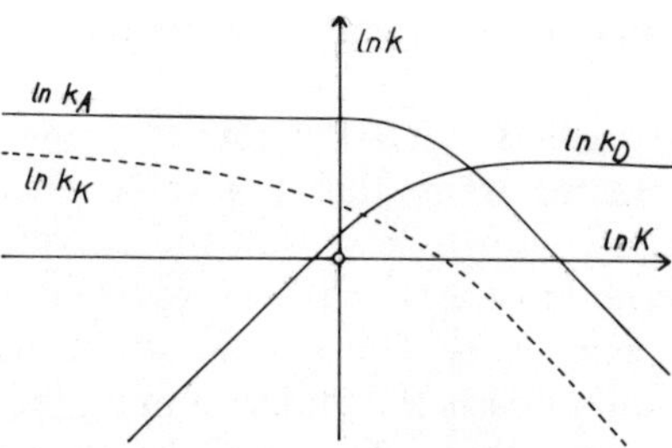

Fig. 2.3. Relations between rate constants and equilibrium constants of proton-transfer reactions over a large range. (Reproduced from Ref. 2.)

maximum deviation from the straight lines of slope zero or unity amounts to ln 2; this is approximately in accord with kinetic measurements on reactions between certain classes of acids and bases.[72,60–63] This type of effect (often referred to as *Eigen curvature*) must always be present, but it cannot account for the gradual curvature often found experimentally, even when neither the forward nor the reverse reaction rate is approaching the diffusion-controlled limit, nor for the fact that β often remains effectively constant but less than unity over a wide range of rates.

The second type of explanation offered for nonlinear Brönsted plots, often referred to as *Marcus curvature*, focuses attention on the chemical activation barrier. Such barriers must frequently be present, since proton-transfer reactions are often slow even when they are thermodynamically favorable, especially for many carbon acids. There is no general reason why $\Delta G^\ddagger$ should be a linear function of ΔG^0 over an extended range, and we have already seen that in the simplest interpretation of the Brönsted relation in terms of intersecting potential energy curves (Fig. 2.2), this assumption is valid only when these curves can be approximated by straight lines. Predictions of Brönsted curvature therefore demand some assumption about how the energy profile changes as ΔG^0 changes, and this is what is done in Marcus's theory, originally developed for electron transfers,[73] and subsequently applied to proton transfers.[74] In the simplest form of this theory it is assumed that the curves in Fig. 2.2 are parabolas of identical curvature, and that the only effect of varying ΔG^0 is to displace these curves vertically with respect to one another. This leads to the relations (2.13), where Z is a collision number and $\Delta G_0^\ddagger$ (assumed constant for a given type of reaction) is the so-called "intrinsic barrier" corresponding to $\Delta G^0 = 0$; for the general reaction, $A_1 + B_2 \rightleftharpoons B_1 + A_2$, $\Delta G_0^\ddagger$ is often taken

$$\begin{aligned} \Delta G^\ddagger &= -RT \ln (k/Z) = (1 + \Delta G^0/4\Delta G_0^\ddagger)^2 \Delta G_0^\ddagger \\ \beta &= d\Delta G_0^\ddagger/d\Delta G^0 = \tfrac{1}{2}(1 + \Delta G^0/4\Delta G_0^\ddagger) \\ d\beta/d\Delta G^0 &= \tfrac{1}{8}\Delta G_0^\ddagger \end{aligned} \tag{2.13}$$

as the mean of the values for the identity reactions, $A_1 + B_1 \rightleftharpoons B_1 + A_1$ and $A_2 + B_2 \rightleftharpoons B_2 + A_2$.

Equation (2.13) predicts that β will have the value one-half when ΔG^0 is zero, but will be greater or less than this value when ΔG^0 is positive or negative, respectively; it also follows that the curvature of the Brönsted plot in the neighborhood of $\Delta G^0 = 0$ will be greater for fast reactions, which have low values of $\Delta G_0^{\ddagger}$. There is no doubt that these expressions give a qualitatively sound basis for understanding the curvature of Brönsted plots, but they naturally suffer from the limitations of the model employed, and care must be taken in drawing quantitative conclusions from them. Thus it was shown by Koeppl and Kresge[75] that if the curvatures and separation of the intersecting parabolas are allowed to vary within a series, a plot of β against ΔG^0 is no longer linear, as demanded by (2.13), but becomes a sigmoid curve. Similar deviations from (2.13) occur when calculations are made in terms of intersecting Morse curves,[76] or if the picture of intersecting curves is replaced by an energy profile computed by the semiempirical bond-energy–bond-order method.[74] In all these cases the relationship between β and ΔG^0 may be effectively linear over a limited region, but the slope will no longer be $\frac{1}{8}\Delta G_0^{\ddagger}$, and hence values of $\Delta G_0^{\ddagger}$ obtained by applying equation (2.13) to experimental results will lose any clear significance. Further, the assumption that $\Delta G_0^{\ddagger}$ remains constant within a series can hardly be justified. In the usual type of catalyzed reaction it is especially likely to break down for variations of substrate, since structural variations near the seat of reaction will certainly affect the rate of the identity reaction $A + B \rightleftharpoons B + A$ for the substrate.

A more serious defect of equation (2.13) (or its equivalent for any other form of energy profile) lies in the assumption that the measured values of $\Delta G^{\ddagger}$ and ΔG^0 refer to the process in which the proton is actually being transferred. It is now generally believed that after the two reactants have diffused together, some reorganization of the surrounding solvent, and in some cases also structural reorganization of the reacting species, must take place before transfer of the proton occurs, followed by analo-

Diffusion:	$XH + Y \rightleftharpoons XH\|Y$	(encounter complex)	
Reorganization:	$XH\|Y \overset{w^r}{\rightleftharpoons} XH \cdot Y$	(reaction complex)	
Proton transfer:	$XH \cdot Y \overset{\Delta G\ddagger}{\rightleftharpoons} X \cdot HY$	(reaction complex)	(2.14)
Reorganization:	$X \cdot HY \overset{w^p}{\rightleftharpoons} X\|HY$	(encounter complex)	
Diffusion:	$X\|HY \rightleftharpoons X + HY$	(products)	

gous reorganizations before the products separate.† All these reorganizations will involve free energy changes, which are not likely to depend upon ΔG^0 for the actual proton-transfer step. The overall scheme for a proton-transfer reaction then becomes as in (2.14), charges being omitted, where w^r and w^p are the reorganization energies for the reactants and products, respectively. If this scheme is combined with the model of intersecting identical parabolas for the proton-transfer step, we obtain (2.15)

$$(\Delta G^{\ddagger})_{\text{obs}} = w^r + \{1 + [(\Delta G^0)_{\text{obs}} - w^r + w^p]/4\Delta G_0^{\ddagger}\}^2 \Delta G_0^{\ddagger} \tag{2.15}$$

instead of (2.13), where $\Delta G_0^{\ddagger}$ refers only to the actual process of proton transfer.

This type of equation was derived by Marcus but not employed by him in his treatment of experimental data.[74] Equation (2.15) has, however, been widely used recently in the quantitative interpretation of nonlinear Brönsted plots[53,67–69,77–79] with the assumption that the intrinsic barrier, $\Delta G_0^{\ddagger}$, and the two reorganization energies, w^r and w^p, remain constant within a series. If the absolute values of $(\Delta G^0)_{\text{obs}}$ are available, all three constants can in principle be obtained by analyzing the experimental data, but if (as is frequently true for catalyzed reactions) these absolute values are unknown, then only w^r and $\Delta G_0^{\ddagger}$ can be obtained. The values obtained for these parameters should be treated with caution, in view of the numerous assumptions underlying (2.15) and the uncertainty with which the departures from linearity are defined experimentally. Nevertheless, it is significant that the values obtained for $\Delta G_0^{\ddagger}$ are quite small, being usually less than 20 kJ mol^{-1}, while the reorganization energy, w^r, varies between 35 and 70 kJ mol^{-1} and thus contributes a large proportion of the observed reaction barrier. The qualitative effect of this is to produce Brönsted plots that are more curved than would be expected if the observed values of $\Delta G^{\ddagger}$ could be attributed mainly to the proton-transfer step. The importance of w^r is shown in a particularly convincing manner by recent work on the decomposition of diazodiphenylmethane[68] catalyzed by 19 acids with strengths covering a range of 10^{11}. For the stronger acids the catalytic constant becomes independent of acid strength but has a value of only about 1 l mol^{-1} s^{-1}, compared with 10^9–10^{10} expected for a diffusion-controlled reaction; this finding is of course independent of any details of Marcus theory. Even if no considerable reorganization terms are present it is necessary in a full treatment to combine the considerations on which Eigen curvature is based with those depending on the form of the energy profile, and a recent analysis shows[80] that the effect of this is to increase the curvature of Brönsted plots even when neither the forward nor the reverse rate is close to the diffusion-controlled value.

† The timing of the reorganization and proton-transfer steps has been the subject of much discussion, largely inconclusive. The sequence described here represents the majority view.

Furthermore two other lines of argument have led to equations essentially equivalent to (2.13) or (2.15). The first of these[80] depends upon a quantitative application of the Leffler principle[81] and the Hammond postulate.[82] The second forms part of a general treatment of processes in polar media put forward by Dogonadze and his collaborators[83] and differs fundamentally from the generally accepted picture of proton transfer. According to these authors, transfer takes place mainly by a tunneling process from the lowest vibrational level of the X—H bond (as previously proposed by Weiss[84]), and the thermal activation involved is that involved in reorienting the solvent molecules or reorganizing the heavy-atom framework of the reactants. While this picture is probably correct for electron transfer, it has not been generally accepted for the transfer of the much heavier proton, and the question must remain open.

The reactions so far considered are believed to involve simple proton transfers as the rate-determining step. For more complex reactions, in which proton transfer is accompanied by other processes involving the making and breaking of covalent bonds, it is not uncommon to encounter quite abrupt changes of the Brönsted slope β over a small range of pK values. This is notably the case for the reactions of nucleophiles with carbonyl compounds, investigated particularly by Jencks and his collaborators.[85] The reason for this may be illustrated by a recent example, the addition of methoxyamine to *p*-chlorobenzaldehyde, which shows general catalysis by buffer acids.[86] If we write the aldehyde as $\rangle C{=}O$, the

$$\begin{aligned} &\mathrm{MeONH_2} + \rangle\mathrm{C{=}O} \underset{k_{-1}}{\overset{k_1}{\rightleftharpoons}} \mathrm{Me\overset{+}{O}NH_2{-}\overset{|}{\underset{|}{C}}{-}O^-} \\ &\mathrm{Me\overset{+}{O}NH_2{-}\overset{|}{\underset{|}{C}}{-}O^-} + \mathrm{A} \xrightarrow[k_A]{} \mathrm{Me\overset{+}{O}NH_2{-}\overset{|}{\underset{|}{C}}{-}OH} + \mathrm{B} \end{aligned} \tag{2.16}$$

sequence of reactions is as in (2.16), and the usual steady-state treatment gives for the observed reaction rate the expression (2.17). If the catalyzing

$$v = k_1[\mathrm{MeONH_2}]\left[\rangle\mathrm{C{=}O}\right]\frac{k_A[\mathrm{A}]}{k_{-1}+k_A[\mathrm{A}]} \tag{2.17}$$

acid is sufficiently weak, $k_A[\mathrm{HA}] \ll k_{-1}$, we have (2.18), i.e., the proton transfer is rate-limiting, and a Brönsted plot will be obtained with a slope β

$$v = k_1/k_{-1}[\mathrm{MeONH_2}]\left[\rangle\mathrm{C{=}O}\right]k_A[\mathrm{A}] \tag{2.18}$$

appropriate to the constant k_A. At the other extreme, for a sufficiently strong acid, $k_A[\mathrm{HA}] \gg k_{-1}$, and (2.17), becomes (2.19), i.e., the formation

$$v = k_1[\mathrm{MeONH_2}]\left[\rangle\mathrm{C{=}O}\right] \tag{2.19}$$

of the intermediate is rate-determining, and the observed rate becomes independent of the nature and concentration of the catalyst, giving a Brönsted plot of zero slope. For acids of intermediate strength there will be a region in which the slope changes from β to zero, and the rate is a nonlinear function of catalyst concentration, which is what is observed. In other systems there may be more than one intermediate and consequently more complex behavior, but in general the nonlinear Brönsted plots observed for this class of reaction are now fairly well understood.

2.6. The Significance of Brönsted Exponents

The exponent β was originally regarded as an empirical constant lying between zero and unity, but in recent years much attention has been given to its molecular interpretation and its relation to other characteristics of the reaction.[87] We have already seen that β must approach zero for reactions that are thermodynamically very favorable, and unity for those that are very unfavorable. These two classes of reaction have transition states closely resembling the initial state and the final state, respectively, and it is therefore natural to regard intermediate values of β as an index of the *extent of proton transfer* or the *degree of symmetry* in the transition state. This can be illustrated by examining the slopes of the energy curves in Fig. 2.2 and receives a more quantitative expression from the related equation (2.13), which predicts that $\beta = \frac{1}{2}$ when $\Delta G^0 = 0$, and is, respectively, greater or less than one-half when ΔG^0 is positive or negative.

There have therefore been many attempts to compare the observed values of β with estimates of transition state symmetry arrived at by other means, especially by a study of *kinetic isotope effects*. There is now much experimental evidence that in a series of similar proton-transfer reactions, especially for catalysis of a given reaction by a series of similar catalysts, k^H/k^D has a maximum value (usually 7–10) when ΔG^0 is close to zero and decreases when ΔG^0 assumes considerable positive or negative value.[88–95] Even when the maximum is not experimentally accessible, the direction in which k^H/k^D changes with increasing reactivity has frequently been used as a guide to the extent of proton transfer in the transition state. These correlations were originally attributed to the part played by unsymmetrical vibrations of the transition state,[96] but recent model calculations[28,97] suggest that this explanation is insufficient to account for the facts and that the tunnel effect may be involved. A correlation between the magnitudes of the isotope effect and of β is also predicted by the Marcus theory, since on the reasonable assumptions that both $\Delta G_0^{\ddagger}$ and the reorganization terms w^r and w^p are unaffected by isotopic substitution, either (2.13) or (2.15) leads to the expression (2.20), which represents a parabolic decrease of k^H/k^D on either side of the maximum value at $\beta = \frac{1}{2}$.

$$\ln (k^H/k^D) = [1-(2\beta-1)^2] \ln (k^H/k^D)_{max} \tag{2.20}$$

Whatever the theoretical interpretation, there is some experimental evidence for a correlation between the magnitude of the isotope effects in a series of reactions and the corresponding values of β, for example in the ionization of carbon acids,[98,99] though individual deviations are large. Another use of isotope effects involves the secondary effects produced by the isotopic substitution of hydrogen atoms not directly involved in the proton transfer, which has the advantage that it will have a negligible effect on the potential energy surface. Thus it is possible to measure the dissociation constants K^H and K^D of the species CH_3CO_2H and CD_3CO_2H in the same solvent, and also their catalytic effects k^H and k^D for a given reaction. An "isotopic Brönsted exponent," β_i, is then defined by equation (2.21), and can be compared with the conventional β determined by

$$\ln (k^H/k^D) = \beta_i \ln (K^H/K^D) \tag{2.21}$$

varying the strength of the acid. Since these effects are small, considerable experimental precision is required, but some interesting results have been obtained. Thus in the decomposition of 3-diazobutan-2-one[100] catalyzed by the pairs HCO_2H–DCO_2H and CH_3CO_2H–CD_3CO_2H, the values of β_i were, respectively, 0.64 ± 0.07 and 0.64 ± 0.06, compared with the conventional value 0.61 ± 0.03, based upon the same two acids. However, this good agreement may be exceptional, since in an investigation[101] of rates and equilibria in the reactions of $(CH_3)_2CHNO_2$ and $(CD_3)_2CHNO_2$ with hydroxide and acetate ions, no such agreement was found.

For reactions in which the rate-determining step is the transfer of a proton from hydronium ions to the substrate, measurements in H_2O–D_2O mixtures make it possible to compare the rates of two processes such as those in (2.22). Since the equilibria between the various isotopic solvent

$$\begin{aligned} &H_2O^+{-}H + S \rightarrow [H_2O \cdots H \cdots S]^+ \rightarrow H_2O + HS^+ \\ &D_2O^+{-}H + S \rightarrow [D_2O \cdots H \cdots S]^+ \rightarrow D_2O + HS^+ \end{aligned} \tag{2.22}$$

species in these mixtures are known quite accurately, it is possible to obtain values of β_i for the secondary effect of deuterium substitution in the hydronium ion and to compare these with the exponents based on general acid catalysis by other species. This has been done for about a dozen reactions, but although there is a rough correspondence between the general magnitudes of these exponents, statistical analysis indicates[87] that there is no significant correlation between their variations from one reaction to another.

Thus although the interpretation of the Brönsted exponent as a measure of the extent of proton transfer in the transition state seems qualitatively reasonable, it receives little quantitative support from comparisons with similar interpretations of hydrogen isotope effects. This is perhaps not surprising, since the concept of "degree of proton transfer" (or transition-state symmetry) is not well defined: For example, it could be related to the geometrical position of the proton, the orders or force constants of the two bonds holding the proton, the distribution of charge, or the free-energy change in the overall reaction. These quantities need not vary in parallel in a series of reactions, and each of them may be important in different contexts. These problems become particularly acute when we deal with anomalous Brönsted exponents, as described in the next section.

2.7. *Anomalous Brönsted Exponents*

This term is usually applied to exponents that do not lie between the limits zero and unity. It was tacitly assumed by early workers that β must be positive and less than unity, and the same limitation follows from the picture of intersecting potential energy curves (Fig. 2.2) or from the Marcus equations (2.13) and (2.15), provided it is assumed that the quantities $\Delta G_0^\ddagger$, w^r, and w^p remain constant throughout the series, and that ΔG^0 conforms with $-4\Delta G_0^\ddagger < \Delta G_0 < +4\Delta G_0^\ddagger$, as is demanded by the physical basis of the equation. In more general terms, if the operator δ represents the effect of substitution, and r, p, and $\ddagger$ refer to the reactants, products, and transition state, respectively, we can write equation (2.23), and the

$$\beta = \frac{\delta \Delta G^\ddagger}{\delta \Delta G^0} = \frac{\delta G^\ddagger - \delta G^r}{\delta G^p - \delta G^r} \tag{2.23}$$

condition $0 < \beta < 1$ will be satisfied provided that $\delta G^\ddagger$ lies between δG^r and δG^p, as is implied by the Leffler principle.[81] It is also obvious that values of β outside these limits are meaningless if β is to be regarded as a measure of the extent of proton transfer.

The great majority of systems investigated yield positive values of β less than unity, but recently a number of exceptions to this have been found, almost all involving the ionization of nitrocompounds. Thus Bordwell and his collaborators[29–33] investigated the reactions of the compounds $XC_6H_4CH_2NO_2$, $XC_6H_4CHMeNO_2$, $XC_6H_4CH_2CHMeNO_2$, and $XCH_2CH_2CH_2NO_2$ with a number of bases in each case. They found that when the base is held constant, the effect of varying X conformed to Brönsted relations with slopes varying from 1.14 ± 0.04 to 1.61 ± 0.11. Since the Brönsted exponents for the forward and reverse reactions must always add up to unity, this implies values of β between -0.14 and -0.61 for the protonation of the anions, i.e., the effect of a substituent upon the rate constants of the protonation is in the direction opposite to its effect upon the equilibrium constants. The same authors pointed out that published values for the acidity constants and rate constants for reactions with hydroxide ion of CH_3NO_2, $MeCH_2NO_2$, and Me_2CHNO_2 showed similar behavior, in that methyl substitution gave a stronger acid but a slower reaction[102–105]; the results correspond to a β value of about -0.5. In this series β is therefore about 1.5 for the transfer of a proton from water to the corresponding anions. It should be noted that for all these examples the deprotonation of a given nitroalkane by a series of bases gives a "normal" value of β of about 0.5.

A number of different explanations of this anomalous behavior have been put forward, and since these are often involved and sometimes not very convincing, no attempt will be made to describe them in detail. The Marcus equation can be formally adapted to include these anomalous values, if we remove the condition that the intrinsic barrier $\Delta G_0^\ddagger$ remains constant within a series.[74] Equation (2.13) then gives (2.24), which can in

$$\beta = \frac{d\Delta G^\ddagger}{d\Delta G^0} = \frac{1}{2}\left(1 + \frac{\Delta G^0}{4\Delta G_0^\ddagger}\right) + \left[1 - \left(\frac{\Delta G^0}{4\Delta G_0^\ddagger}\right)^2\right]\frac{d\Delta G_0^\ddagger}{d\Delta G^0} \tag{2.24}$$

principle give values of β outside the usual limits even when the condition $-4\Delta G_0^\ddagger < \Delta G^0 < +4\Delta G_0^\ddagger$ (demanded by the physical basis of the equation) is satisfied. The same conclusion follows from the more complete equation (2.15), though if the terms w^r and w^p in (2.15) contain a considerable contribution from the reorganization of the reacting species, which will also vary within a series, it is no longer legitimate to regard these terms as constants. This treatment does not of course yield any direct molecular basis for the anomalies, but it predicts that in a reaction between a simple acid–base system (such as those normally used as catalysts) and a system involving a considerable structural reorganization (as is the case for most substrates), anomalies are more likely to arise when the latter system is varied by substitution. This follows from the assumption that $\Delta G_0^\ddagger$ represents some kind of mean between the values for the identity reactions for the two systems, and this mean will be dominated by the large barrier that characterizes pseudosystems.

From another point of view it is clear from equation (2.23) that anomalous exponents can arise when the effect of a substituent upon the transition state does not lie between its effects upon the initial and the final

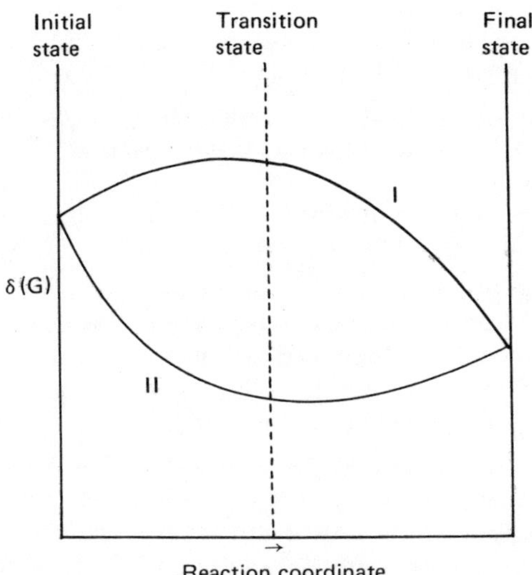

Fig. 2.4. Conditions leading to the occurrence of abnormal Brönsted exponents.

state. This is illustrated in Fig. 2.4, in which the effect of the substituent is plotted schematically against the reaction coordinate, or the extent of proton transfer: Curve I would give $\beta<0$, and curve II $\beta>1$. This behavior can easily arise when the free energy of the system contains a large component due to solvation, as stressed particularly by Cox and Gibson.[106] In a proton-transfer reaction between a charged and an uncharged species, the charge is usually more highly dispersed in the transition state than in either the reactants or the products, so that solvation (and the effect of substituents upon it) will be at a minimum in the transition state. Conversely, in a proton transfer between two uncharged species the transition state will be more polar than either the initial or the final state, corresponding to a maximum degree of solvation. Cox and Gibson[106] illustrated the importance of solvation in this respect by showing that in mixtures of water and dimethyl sulfoxide the values of β for the reactions of several nitroalkanes became more normal as the proportion of dimethyl sulfoxide was increased, thereby decreasing the solvating power of the medium.† An extreme case of specific solvation of the initial state occurs in the work of Jones and Patel[110] on the effect of successive fluorine

† There are several investigations in which a "solvent β" has been derived by measuring the effect of solvent composition on the equilibrium constants and rate constants of proton-transfer reactions, and in some instances their values are close to those obtained by varying the acid–base strength of the reactants by substitution.[78,107–109] However, this behavior cannot be general, since Cox and Gibson[106] found several instances in which the rate was very sensitive to solvent composition, while the equilibrium constant was little affected.

substitution in acetylacetone on acidities and rates of detritiation. The anomalous values of β observed were attributed to covalent hydration of the carbonyl group according to the equation $R^1R^2CO+H_2O \rightleftharpoons R^1R^2C(OH)_2$, which will affect the measured acidities, but cannot occur in the transition state.

Apart from solvation, a maximum substituent interaction in the transition state can be accounted for in terms of the displacement of charge that takes place during the ionization of a nitroalkane.[101,111] Thus in the transition state for reaction with a negatively charged base, the negative charge will be divided between the base, the carbon atom, and the oxygen atoms of the nitro-group. That part of it that is on carbon will be particularly close to substituents and hence able to interact strongly with them. In the final state the nitronate ion will of course bear a full negative charge, but this will be located almost completely on the oxygen atoms, and therefore may interact less strongly with the substituent than the partial negative charge on carbon in the transition state. In an extreme form of this hypothesis, Bordwell and Boyle[58] have proposed that the reaction takes place through two intermediates, the first being a true carbanion, and the second a nitronate ion, bearing the negative charge on carbon and oxygen respectively; but this explanation seems to demand that electrons move more slowly than protons, and the picture of a continuous shift of charge as the proton is transferred seems preferable.

Curves like those in Fig. 2.4 can also result[112] if substituents interact with the reaction zone by more than one mechanism, even if each interaction separately is a monotonic function of the degree of proton transfer. The conditions for producing an abnormal value of β are (*a*) that the two interactions should vary in opposite directions with the extent of proton transfer, (*b*) that they should develop at different rates as proton transfer proceeds. This idea has been applied particularly to the effect of methyl substitution on the ionization of the nitroalkanes. The usual acid-weakening effect of methyl groups† will increase smoothly as proton transfer proceeds, but is opposed by hyperconjugation between the methyl group and the C=N bond of the nitronate ion. This latter effect (which is responsible for the fact that methyl substitution increases the acid strength of the nitroalkanes) will not come into play until the carbon–nitrogen bond has acquired considerable double-bond character, and this difference between the rates at which the two effects develop could easily produce values of β greater than unity.

There is thus no lack of explanations for the existence of anomalous Brönsted exponents, and the problem is rather to choose between them

† This effect is normally described as inductive, but in view of recent work on acid–base reactions in the gas phase[113] it appears that polarizability and inhibition of solvation are more important factors.

and to explain why anomalous values occur so rarely, being almost entirely confined to nitro-compounds.† It seems likely that the factors outlined in this section may distort the quantitative behavior of other types of compound, such as ketones, without being sufficiently marked to produce anomalous exponents.

2.8. Conclusions

The development of the Brönsted relation over the last fifty years has followed a familiar course. Starting as an empirical relation between catalytic power and acid–base strength, it was later found to have much wider implications, and increasing attention was paid to its relationship to general theories of reaction kinetics, and to molecular pictures. For some years after its discovery it remained the province of specialists, but the growth of general interest in linear free energy relationships soon had repercussions, and the last ten or fifteen years have seen a great increase in both experimental and theoretical investigations of Brönsted relations, and especially of deviations from the simple law. In conjunction with isotope effects, a study of Brönsted relations constitutes the most powerful method for elucidating the detailed mechanism of proton-transfer reactions, one of the most important classes of reaction in chemistry and biology. Finally, it is worth emphasizing again the remarkable scope and insight of the original paper of Brönsted and Pedersen,[2] which has been quoted no less than twelve times in the course of this survey.

Acknowledgments

This review owes a great deal to two articles by Professor A. J. Kresge,[53,57] and much relevant material will be found in a Faraday Symposium[115] and in a recent book.[116] The author also wishes to express his thanks to the Fysisk-kemisk Institut, Technical University of Denmark, whose hospitality he enjoyed while writing this chapter.

References and Notes

1. See L. P. Hammett, Foreword to *Advances in Linear Free Energy Relationships*, N. B. Chapman and J. Shorter, eds. (Plenum Press, London, 1972).
2. J. N. Brönsted and and K. Pedersen, *Z. phys. Chem.*, **108**, 185, (1924).

† A value of $\beta = 1.2 \pm 0.1$ has been reported for the reaction of a series of internally hydrogen-bonded phenols with hydroxide ions,[114] but the compounds do not form a very homogeneous set and the value given for the exponent depends essentially on the result for a single substance. These compounds are not pseudoacids, and the observed rates are very high, so that an anomalous value of β would not be anticipated.

3. J. Thiele and A. Lachmann, *Justus Liebigs Ann. Chem.*, **288**, 267 (1895).
4. J. N. Brönsted, *Rec. Trav. chim.*, **42**, 718 (1923).
5. H. M. Dawson and F. Powis, *J. Chem. Soc.*, 2135 (1913).
6. H. M. Dawson and C. K. Reiman, *J. Chem. Soc.*, 1426 (1915).
7. J. N. Brönsted, Om Syre- og Basekatalyse (University of Copenhagen, 1926); English translation, *Chem. Rev.*, **5**, 231 (1928).
8. K. J. Pedersen, Den almindelige Syre- og Basekatalyse, Doctoral thesis (Copenhagen, 1932); *J. Phys. Chem.*, **38**, 581 (1934); *Trans. Faraday Soc.*, **34**, 237 (1938).
9. S. W. Benson, *J. Amer. Chem. Soc.*, **80**, 5151 (1958).
10. E. W. Schlag, *J. Chem. Phys.*, **38**, 2480 (1963); E. W. Schlag and G. L. Haller, *J. Chem. Phys.*, **42**, 584 (1965).
11. D. M. Bishop and K. J. Laidler, *J. Chem. Phys.*, **42**, 1688 (1965); J. N. Murrell and K. J. Laidler, *Trans. Faraday Soc.*, **64**, 371 (1968).
12. V. Gold, *Trans. Faraday Soc.*, **60**, 739 (1964).
13. J. N. Brönsted and H. C. Duus, *Z. phys. Chem.*, **117**, 299 (1925).
14. J. N. Brönsted and K. Volqvartz, *Z. phys. Chem.*, *A*, **155**, 211 (1931).
15. J. N. Brönsted and J. E. Vance, *Z. phys. Chem.*, *A*, **163**, 240 (1933).
16. J. N. Brönsted, A. L. Nicholson, and A. Delbanco, *Z. phys. Chem.*, *A*, **169**, 379 (1934).
17. H. S. Taylor, *Z. Elektrochem.*, **20**, 201 (1914).
18. R. P. Bell, *Acid–Base Catalysis*, p. 91 (Oxford University Press, 1941).
19. J. N. Brönsted and E. A. Guggenheim, *J. Amer. Chem. Soc.*, **49**, 2554 (1927).
20. R. P. Bell, *The Proton in Chemistry*, Chap. 9 (Chapman & Hall, London, 2nd edn., 1973).
21. M.-L. Ahrens, M. Eigen, W. Kruse, and G. Maass, *Ber. Bunsengesellschaft Phys. Chem.*, **74**, 380 (1970).
22. J. Horiuti and M. Polanyi, *Acta Physicochim. U.R.S.S.*, **2**, 505 (1935).
23. R. P. Bell, *Proc. Roy. Soc.*, *A*, **154**, 414 (1936).
24. M. G. Evans and M. Polanyi, *Trans. Faraday Soc.*, **32**, 1333 (1936).
25. Reference 20, pp. 79–82.
26. E. C. Baughan and R. P. Bell, *Proc. Roy. Soc.*, *A*, **158**, 464 (1937); see also Ref. 18, pp. 189–192.
27. R. P. Bell, *Chem. Soc. Rev.*, **3**, 513 (1974).
28. R. P. Bell, W. H. Sachs, and R. L. Tranter, *Trans. Faraday Soc.*, **67**, 1995 (1971).
29. F. G. Bordwell, W. J. Boyle, J. A. Hautala, and K. C. Yee, *J. Amer. Chem. Soc.*, **91**, 4002 (1969).
30. F. G. Bordwell, W. J. Boyle, and K. C. Yee, *J. Amer. Chem. Soc.*, **92**, 5926 ((1970).
31. M. Fukuyama, P. W. K. Flanagan, F. T. Williams, L. Frainier, S. A. Miller, and H. Schechter, *J. Amer. Chem. Soc.*, **92**, 4689 (1970).
32. F. G. Bordwell and W. J. Boyle, *J. Amer. Chem. Soc.*, **93**, 511 (1971); **94**, 3907 (1972).
33. F. G. Bordwell, *Faraday Symposia*, **10**, 100 (1975).
34. R. P. Bell and S. Grainger, *J.C.S. Perkin II*, 1367 (1976).
35. R. P. Bell, M. H. Rand, and K. M. A. Wynne-Jones, *Trans. Faraday Soc.*, **52**, 1093 (1956).
36. F. Covitz and F. H. Westheimer, *J. Amer. Chem. Soc.*, **85**, 1773 (1963).
37. J. A. Feather and V. Gold, *J. Chem. Soc.*, 1752 (1965).
38. R. P. Bell, E. Gelles, and E. Möller, *Proc. Roy. Soc.*, *A*, **198**, 308 (1949).
39. H. L. Pfluger, *J. Amer. Chem. Soc.*, **60**, 1513 (1938).
40. R. P. Bell and A. F. Trotman-Dickenson, *J. Chem. Soc.*, 1288 (1949).
41. R. P. Bell and G. L. Wilson, *Trans. Faraday Soc.*, **46**, 407 (1950).
42. A. F. Trotman-Dickenson, *J. Chem. Soc.*, 1293 (1949).
43. For references see Ref. 20, p. 219.

44. D. Berg and A. Patterson, *J. Amer. Chem. Soc.*, **75**, 5197 (1953); K. F. Wissbrunn, D. M. French, and A. Patterson, *J. Phys. Chem.*, **58**, 693 (1954).
45. F. J. W. Roughton and V. H. Booth, *Biochem. J.*, **32**, 2048 (1938).
46. A. R. Olson and P. V. Youle, *J. Amer. Chem. Soc.*, **62**, 1027 (1940).
47. R. P. Bell, B. G. Cox, and B. A. Timimi, *J. Chem. Soc.*, (B), 2247 (1971).
48. R. P. Bell, B. G. Cox, and J. B. Henshall, *J.C.S. Perkin II*, 1232 (1972).
49. R. P. Bell, D. W. Earls, and J. B. Henshall, *J.C.S. Perkin II*, 39 (1976).
50. R. P. Bell and A. D. Covington, *J.C.S. Perkin II*, 1343 (1975).
51. R. P. Bell and B. G. Cox, *J.C.S. Perkin II*, 1349 (1975).
52. A. J. Kresge and Y. Chiang, *J. Amer. Chem. Soc.*, **95**, 803 (1973).
53. A. J. Kresge, *Chem. Soc., Rev.*, **2**, 475 (1973).
54. R. P. Bell, *Trans. Faraday Soc.*, **39**, 253 (1943).
55. R. P. Bell and W. C. E. Higginson, *Proc. Roy. Soc., A*, **197**, 141 (1949).
56. A. Hantzsch, *Ber. Dtsch. Chem. Gesell*, **32**, 575 (1899).
57. R. P. Bell, *J. Phys. Chem.*, **55**, 885 (1951).
58. F. G. Bordwell and W. J. Boyle, *J. Amer. Chem. Soc.*, **97**, 3447 (1975).
59. M. Eigen, *Angew. Chem. Internat. Edn.*, **3**, 1 (1964); GE, **75**, 489 (1963).
60. E. A. Walters and F. A. Long, *J. Amer. Chem. Soc.*, **91**, 3733 (1969); F. Hibbert, F. A. Long, and E. A. Walters, *J. Amer. Chem. Soc.*, **93**, 2829 (1971); F. Hibbert and F. A. Long, *J. Amer. Chem. Soc.*, **94**, 2647 (1972).
61. Z. Margolin and F. A. Long, *J. Amer. Chem. Soc.*, **94**, 5108 (1972); **95**, 2757 (1973).
62. A. J. Kresge and A. C. Lin, *J.C.S. Chem. Comm.*, 761 (1973).
63. R. P. Bell and B. G. Cox, *J. Chem. Soc. (B)*, 652 (1971); F. Hibbert, *J.C.S. Perkin II*, 1289 (1973).
64. For references see P. R. Rony, *J. Amer. Chem. Soc.*, **90**, 2824 (1968); **91**, 6090 (1969); also Ref. 20, pp. 155–157.
65. G. Lienhard and F. H. Anderson, *J. Org. Chem.*, **32**, 2229 (1967).
66. J. E. Critchlow, *J.C.S. Faraday I*, **68**, 1774 (1972).
67. E. H. Baughman and M. M. Kreevoy, *J. Phys. Chem.*, **78**, 421 (1974).
68. A. I. Hassid, M. M. Kreevoy, and T.-M. Laing, *Faraday Symposia*, **10**, 69 (1975).
69. C. E. Bannister, D. W. Margerum, J. M. T. Raycheba, and L. F. Wong, *Faraday Symposia*, **10**, 78 (1975).
70. D. S. Kemp and M. L. Casey, *J. Amer. Chem. Soc.*, **95**, 6670 (1973).
71. M. Eigen, *Z. phys. Chem. (Frankfurt)*, **1**, 176 (1954).
72. M.-L. Ahrens and G. Maass, *Angew. Chem. Internat. Edn.*, **7**, 818 (1968); GE, **80**, 848 (1968).
73. R. A. Marcus, *J. Chem. Phys.*, **24**, 966 (1956); *Discuss. Faraday Soc.*, **29**, 21 (1960); *J. Phys. Chem.*, **67**, 853, 2889 (1963); *Ann. Rev. Phys. Chem.*, **15**, 155 (1964); *J. Chem. Phys.*, **43**, 679 (1965).
74. R. A. Marcus, *J. Phys. Chem.*, **72**, 891 (1968); *J. Amer. Chem. Soc.*, **91**, 7224 (1969); A. O. Cohen and R. A. Marcus, *J. Phys. Chem.*, **72**, 4249 (1968).
75. G. W. Koeppl and A. J. Kresge, *J.C.S. Chem. Comm.*, 371 (1973).
76. R. P. Bell, *Faraday Symposia*, **10**, 1 (1975).
77. M. M. Kreevoy and D. E. Konasewich, *Adv. Chem. Phys.*, **21**, 243 (1971); M. M. Kreevoy and S.-W. Oh, *J. Amer. Chem. Soc.*, **95**, 4805 (1973).
78. A. J. Kresge, S. G. Mylonakis, Y. Sato, and V. P. Vitullo, *J. Amer. Chem. Soc.*, **93**, 6181 (1971).
79. W. J. Albery, A. N. Campbell-Crawford, and J. S. Curran, *J.C.S. Perkin II*, 2206 (1972).
80. J. R. Murdoch, *J. Amer. Chem. Soc.*, **94**, 4410 (1972).
81. J. E. Leffler, *Science*, **117**, 340 (1953).
82. G. S. Hammond, *J. Amer. Chem. Soc.*, **77**, 334 (1955).

83. E. D. German, R. R. Dogonadze, A. M. Kuznetsov, V. G. Levich, and Yu. I. Kharkats, *J Res. Inst. Catal. Hokkaido Univ.*, **19**, 115 (1971) (in English); M. V. Vol'kenshtein, R. R. Dogonadze, A. K. Madumarov, and Yu. I. Kharkats, *Dokl. Akad. Nauk SSSR, Ser. Fiz. Khim.*, **199**, 124 (1971); EE (Phys. Chem.), 569; R. R. Dogonadze, J. Ulstrup, and Yu. I. Kharkats, *Dokl. Akad. Nauk SSSR, Ser. Fiz. Khim.*, **207**, 640 (1972); EE (Phys. Chem.), 965; *J.C.S. Faraday II*, **70**, 64 (1974).
84. J. J. Weiss, *J. Chem. Phys.*, **41**, 1120 (1964).
85. For a summary and references, see W. P. Jencks, *Chem. Rev.*, **72**, 705 (1972); W. P. Jencks and J. M. Sayer, *Faraday Symposia*, **10**, 41 (1975).
86. S. Rosenberg, S. M. Silver, J. M. Sayer, and W. P. Jencks, *J. Amer. Chem. Soc.*, **96**, 7986 (1974).
87. For a recent review see A. J. Kresge in *Proton-Transfer Reactions*, E. F. Caldin and V. Gold, eds. (Chapman & Hall, London, 1975).
88. A. J. Kresge, *Discuss. Faraday Soc.*, **39**, 48 (1965).
89. R. P. Bell and D. M. Goodall, *Proc. Roy. Soc., A*, **294**, 273 (1966).
90. J. E. Dixon and T. C. Bruice, *J. Amer. Chem. Soc.*, **92**, 905 (1970).
91. F. G. Bordwell and W. J. Boyle, *J. Amer. Chem. Soc.*, **93**, 512 (1971).
92. D. J. Barnes and R. P. Bell, *Proc. Roy. Soc., A*, **318**, 421 (1970).
93. S. B. Hanna, C. Jermini, and H. Zollinger, *Tetrahedron Lett.*, 4415 (1969).
94. D. Cook, R. E. J. Hutchinson, J. K. MacLeod, and A. J. Parker, *J. Org. Chem.*, **39**, 534 (1974).
95. R. P. Bell and S. Grainger, *J.C.S. Perkin II*, 1606 (1976).
96. F. H. Westheimer, *Chem. Rev.*, **61**, 265 (1961).
97. A. V. Willi, *Helv. Chim. Acta*, **54**, 1220 (1971).
98. R. P. Bell, *Discuss. Faraday Soc.*, **39**, 16 (1965).
99. R. A. More O'Ferrall, in *Proton-Transfer Reactions*, E. G. Caldin and V. Gold, eds. (Chapman & Hall, London, 1975).
100. W. J. Albery, J. R. Bridgeland, and J. S. Curran, *J.C.S. Perkin II*, 2203 (1972).
101. M. H. Davies, *J.C.S. Perkin II*, 1018 (1974).
102. R. G. Pearson and R. L. Dillon, *J. Amer. Chem. Soc.*, **75**, 2439 (1953).
103. D. Turnbull and S. H. Maron, *J. Amer. Chem. Soc.*, **65**, 212 (1943).
104. G. W. Wheland and J. Farr, *J. Amer. Chem. Soc.*, **65**, 1433 (1943).
105. S. H. Maron and V. K. La Mer, *J. Amer. Chem. Soc.*, **60**, 2588 (1938).
106. B. G. Cox and A. Gibson, *J.C.S. Chem. Comm.*, 638 (1974); *Faraday Symposia*, **10**, 107 (1975).
107. A. J. Kresge, R. A. More O'Ferrall, L. E. Hakka, and V. P. Vitullo, *Chem. Comm.*, 46 (1965).
108. R. P. Bell and B. G. Cox, *J. Chem. Soc.* (*B*), 194 (1970).
109. R. P. Bell and B. G. Cox, *J. Chem. Soc.* (*B*), 783 (1971).
110. J. R. Jones and S. P. Patel, *J. Amer. Chem. Soc.*, **96**, 574 (1974).
111. A. J. Kresge, *Canad. J. Chem.*, **52**, 1897 (1974).
112. A. J. Kresge, *J. Amer. Chem. Soc.*, **92**, 3210 (1970).
113. For a summary see R. W. Taft in *Proton-Transfer Reactions*, E. F. Caldin and V. Gold, eds. (Chapman & Hall, London, 1975).
114. M. C. Rose and J. Stuehr, *J. Amer. Chem. Soc.*, **93**, 4350 (1971); **94**, 5532 (1972).
115. *Faraday Symposia*, **10** (Proton Transfers) (1975).
116 E. F. Caldin and V. Gold, eds. *Proton-Transfer Reactions* (Chapman & Hall, London, 1975).

3

Theoretical Models for Interpreting Linear Correlations in Organic Chemistry

Martin Godfrey

3.1. Introduction

No general theory of the relationship between molecular structure and molecular properties in organic chemistry can reasonably be regarded as satisfactory unless it provides a sound basis for predicting and interpreting linear relationships among molecular quantities. In this chapter the basic factors involved in the design of a satisfactory theoretical model for interpreting linear correlations will be discussed, and evidence, including the results of molecular orbital calculations, which indicates that popular

Martin Godfrey • Department of Chemistry, The University, Southampton, SO9 5NH, England.

treatments are unsound, will be reviewed. In order to illustrate that the basic postulates in popular theoretical models for linear correlations are arbitrary and could reasonably be replaced, an alternative theoretical model is described, which is superior in its ability to account for the results of molecular orbital studies of electronic structures and the results of statistical analyses of substituent effects. If the basic postulates of this alternative model are essentially correct, the overall pattern of physical and chemical behavior, and not merely the pattern of substituent effects, must be significantly different from that which would be expected by applying the popular theory, but not significantly more complicated.

3.2. *On the Design of Theoretical Models*

From the designer's point of view a satisfactory theoretical model for linear correlations in organic chemistry should allow reliable predictions to be made as easily as possible concerning both the circumstances in which correlations should occur (e.g., between which properties and for which compounds) and the magnitudes of the regression coefficients. The concepts used in the model (e.g., analysis into polar, resonance, and steric effects) should be defined in such a way that the knowledge gained through the interpretation of the linear correlations can be readily used in other areas of organic chemistry (e.g., in elucidating reaction mechanisms).

A theoretical model must relate *empirical* linear correlations to linear correlations among certain fundamental properties of chemical compounds. A common feature of all the theoretical models that have so far been proposed is the assumption that empirical linear correlations are reflections of linear relationships between changes in the localized electronic structure in one part of a molecular system and consequent changes in the localized electronic structure in another part of the same system. The changes in each of the quantities involved in the empirical correlations should therefore be related in some characteristic way to the relevant changes in the electronic structure. In this chapter a molecular system is considered to consist of a substrate molecule and, where appropriate, reagent, solvent, and catalyst molecules too. Each molecular system may be formally subdivided into a "source" region of electronic disturbances, a "detector" region for the electronic disturbances, and a "core" region.

There are three fundamental problems to be tackled in designing a theoretical model. The first involves the establishment of the physical nature (e.g., Coulombic) of the significant types of electronic disturbance, and of the mechanisms (e.g., induction) by which these disturbances are transmitted from the source region to the detector region in a molecular system. The second involves the relationship between the electronic disturbances and the causative chemical or physical processes, e.g., addition, elimination, or substitution of atoms or groups of atoms, and the gain, loss, or excitation of electrons. The third problem involves the establishment of direct relationships between changes in the electronic structures of molecular systems and variations in measurable quantities (e.g., equilibrium constants). The fundamental problems have to be solved in such a way that the model is as satisfactory as possible when judged against the criteria mentioned in the first paragraph of this section.

An important factor to be considered by the model designer is the degree of parameterization required. An underparameterized model will fail to predict the existence of certain significant experimentally detectable features in the pattern of chemical behavior. An overparameterized model will suggest that a more complex pattern of chemical behavior exists than is true. It is not better to err on the side of overparameterization since the more parameters that are involved in a model, the harder it is to apply that model in making predictions. Statistical analyses of experimental data are a valuable guide when searching for the optimum degree of parameterization. Swain and Lupton,[1] and Ehrenson *et al.*,[2] have analyzed extensively data concerning the effects of substituents on molecular properties. In the former study the various empirical substituent parameters used in the Hammett equation[3,4] and its extensions (e.g., σ, σ', σ^*, σ_R^0) are analyzed in terms of a substituent field-parameter, $\mathfrak{F}$, and a substituent resonance parameter, $\mathfrak{R}$ [equation (3.1)], cf. Chapter 4. Ehrenson *et al.* express the raw experimental data for substituent effects on the molecular properties, δP, in terms of a substituent inductive/field-parameter, σ_I, and a substituent resonance-parameter, σ_R [equation (3.2)]. Swain and Lupton's expression for δP is equation (3.3).

$$\sigma = f\mathfrak{F} + r\mathfrak{R} \tag{3.1}$$

$$\delta P = \rho_I\sigma_I + \rho_R\sigma_R \tag{3.2}$$

$$\delta P = (\rho f)\mathfrak{F} + (\rho r)\mathfrak{R} \tag{3.3}$$

From the results of their statistical analysis, Swain and Lupton concluded that a dual-term equation (3.3) with universally applicable substituent parameters is sufficiently precise to make any provision for additional flexibility in the analysis superfluous. However, Ehrenson *et al.* concluded that the precision of correlation in their dual-term equation (3.2) is significantly increased if the substituent resonance-parameter is chosen from a set of four parameters, each of which is applicable within a broad but limited range of situations. The existence of dual-term correlations implies that there must be at least two distinct (i.e., not simply related) significant physical sources of electronic disturbance.

There are three general results of dual-term statistical analyses that have to be accounted for in a satisfactory theoretical model. First, whether or not a given set of changes in a molecular system leads to empirical linear correlations sometimes depends on the *position* of the detector region relative to the source region (e.g., linear correlations sometimes exist when the detector region is remote from the source region but not when it is close to it). Secondly, in certain cases, one of the terms in the correlation equation is directly proportional to the other term, i.e., a single-term correlation exists. In principle, a single-term correlation can exist only accidentally unless, for each of the molecular systems involved, one kind of electronic disturbance cannot be induced in at least part of the core. The third result is that dual-term correlations exist when one or more of the measured properties is *itself* associated with the creation of electronic disturbances. This situation occurs, for example, in correlations involving substituent effects on equilibrium constants: the electronic structure of the molecular system changes not only with the nature of the substituent group (source-region disturbance) but also in going from one side of the equilibrium to the other (detector-region disturbance). The measured substituent effect reflects not only the ability of the substituent to cause electronic disturbances at the reaction site, but also its ability to respond to (i.e., to support or to oppose) electronic changes produced at the substituent site by the changes in molecular structure at the reaction site.

It was mentioned above that the existence of dual-term correlations implies that there must be at least two distinct physical sources of electronic disturbance. The problem arises as to whether or not one kind of physical source is significantly affected by the other. In other words, are there *independent* mechanisms for the transmission of each kind of physical disturbance? It has usually been assumed that there are.[5] However, this assumption is not necessary in order to obtain dual-term correlations, provided that there is direct proportionality between the ability of source-region disturbances to cause *and to respond to* disturbances elsewhere in the molecular system.

Another assumption that is usually made is that there are no discontinuous changes in the structure of the core region during a perturbation of the molecular system. However, such changes might sometimes occur when the size of the perturbation reaches some critical value.[6] Such changes would lead to the nonuniversality of any dual-substituent-parameter correlation.

3.3. Popular Interpretations of Linear Correlation Equations

A basic postulate in popular interpretations of linear correlation equations for substituent effects is that there are two independent electronic effects, or at least, two independent sets of related electronic effects.[5] According to Taft,[7] the two sets are inductive effects and resonance effects, with direct field effects considered to be part of the inductive set. Taft and his co-workers have kept open minds on the detailed description of the modes of transmission of the two sets of effects. On the other hand Dewar and his co-workers[6,8] have attempted to describe the transmission mechanisms in detail. According to them, electrostatic effects are transmitted in such a way that propagation by successive polarization of σ bonds (the classical inductive effect) is unimportant at sites separated from the substituent by more than one or two bonds. Resonance effects are the result of electrostatic interactions between the detector and the changes in the core that result from mesomeric interactions between the core and the substituents. The latter changes are taken to be proportional to those calculated for a CH_2^- substituent by using Hückel molecular orbital (HMO) theory. The polarization of π bonds due to electrostatic interactions between the core and the substituent is considered to be operationally indistinguishable from that due to the corresponding mesomeric interactions. The interaction between the detector site and the rest of the molecular system depends on the nature of the property under investigation. Thus, for example, substituent effects on ^{19}F chemical shifts should not be simply related to substituent effects on chemical reactivity.[6] Dewar's model does not allow for the effects of conjugation between the substituent group and the detector group (through-conjugation), but Dewar *et al.*[6] have boldly claimed that the errors in quantitative predictions made by using the most recent version of their model provide the best estimate currently available of the magnitude of through-conjugation effects.

The results obtained in investigations of the dissociation constants of carboxylic acids of known geometry and free from resonance effects and steric effects indicate that the inductive effect (Taft sense) does not depend on either the number of, or the nature of, the atoms between the substituent and the detector site, nor does it depend on the number of paths available for transmission.[9] On the other hand the results are in very good accord with predictions from an electrostatic model for the transmission of the effects of dipolar substituents.

The field and resonance type of model does not work well universally. In some cases observed physical and chemical behavior is more complicated than predicted. This fact by itself does not imply that the basic principles of this type of model are incorrect: Possibly there are a number of additional complicating factors, each of which is infrequently significant. The validity of the basic principles is, however, seriously challenged by instances in which observed physical and chemical behavior is *more simple* than predicted. The following results are pertinent.

Eaborn *et al.*[10] found that substituent effects on the rates of cleavage of compounds $YC_6H_4(C\equiv C)_nMEt_3$ ($n = 1, 2,$ or 3; $M = Si$ or Ge) are correlated very well by Hammett σ constants, which are measures of substituent effects on the acidity of benzoic acid in aqueous solution. The value of ρ_n/ρ_{n-1} (i.e., the "attenuation factor" for the transmission of electronic disturbances) is independent of the value of n. These results imply that field and resonance effects show identical distance and angular dependence, which, of course, they should not.

Excellent correlations with the Hammett σ constants have also been found[11–13] for substituent effects on ^{13}C and ^{19}F chemical shifts, monitored in the benzene ring further from the substituent in substituted biphenyls, and in certain substituted bridged biphenyls, and for substituent effects on the ^1H and ^{13}C chemical shifts monitored at the H_C and C_β atoms [structure I*a*] of the vinyl group in substituted styrenes,[14] cf. Chapter 8. These

$R(H_A)C_\alpha{=}C_\beta(H_B)(H_C)$

(*a*) R = phenyl— (*c*) R = CH_3—

(*b*) R = $CH_2{=}CH$— (*d*) R = bicyclo[2.2.2]octyl—

(I)

results are in obvious disagreement with the prediction of Dewar *et al.*[6] mentioned above, concerning the relationship between substituent effects on chemical shifts and on chemical reactivity.

Yanovskaya *et al.*[15] measured the effects of dipolar substituents on various physical and chemical properties, including calculated charge distributions, in substituted polyvinylenes [structure (II)] containing varying

$$Y{-}(CH{=}CH)_n{-}X$$

(II)

numbers, n, of vinylene units. They analyzed their results in terms of the Hammett equation by using various sigma scales, each of which should involve considerable non-polar contributions. They found ρ_n/ρ_{n-1} to be a constant for all values of n. This result shows that the apparent direct relationship between substituent field-effects and substituent resonance-effects is not confined to benzene derivatives.

The theoretical basis of the popular interpretations of linear correlation equations appears to be in some way unsound. In the next section the results of molecular orbital calculations of electronic structures will be used to throw light on the sources of the weakness.

3.4. Molecular Orbital Calculations of Electronic Structures

Certain molecular orbital methods have been developed to the level at which they permit the calculation of some observable properties of organic compounds with fair accuracy. They include *ab initio* calculations with small basis sets,[16] and the CNDO/2[17] and MINDO[18–20] approximations. The information that these methods give concerning the detailed electronic structures of organic molecules can be regarded as acceptable for use in the design and testing of theoretical models for linear correlations. The methods have been successful in predicting the most stable conformation for a given set, in many organic molecules. However, they do not allow for changes in the geometry of a particular molecular system that may accompany an electronic disturbance (e.g., the change from a regular hexagonal to an irregular hexagonal geometry for a benzene ring). When the energies of two or more geometric forms of the same molecular system are similar, it is possible that certain electronic disturbances (e.g., those associated with substitution) may change the order of stability of the geometric forms, by comparison with the order for the unperturbed system.

Fliszár *et al.*[21,22] have recently drawn attention to a pitfall to be avoided in interpreting the results of molecular orbital calculations. Electron distributions within molecules are usually determined by using the Mulliken[23] population analysis, in which the overlap population associated with a pair of bonded atoms is arbitrarily divided equally between the two bonded atoms in calculating the gross electronic charges for atoms. The real distribution of electronic charge between two dissimilar bonded atoms may be very different from the calculated one.

The results of certain molecular orbital studies appear to support the validity of the basic theory used in the popular interpretations of linear correlation equations mentioned in the previous section. First, the Dewar–Grisdale field model[8] for Coulombic interactions between source groups and detector groups is supported by the results of calculations by Baird and Dewar[24] of the effects of a variable Coulombic perturbation, applied at one hydrogen atom in a benzene ring, on the charge densities at the other hydrogen atoms. The pattern of changes in the hydrogen-atom charge densities does not match the pattern of changes in the corresponding carbon-atom charge densities, as it would if the classical inductive model were satisfactory. Secondly, Brownlee and Taft[25] have calculated that substituent effects on the σ-electronic charge of the fluorine atom in *para*-substituted fluorobenzenes, and of the fluorine atom in 4-bicyclo-octyl-1-fluorides [here and elsewhere in this chapter the bicyclo-octane is the (2.2.2) isomer] are linearly related, with a slope close to unity. This result appears to indicate that through-bond substituent effects are negligible. Thirdly, Hehre *et al.*[26] have concluded that conjugative substituent effects can reasonably be held responsible for the preferred conformations of substituted benzenes. In the most stable conformations the phenyl group is invariably stabilizing with respect to the methyl group. This result gives support to the validity of the concept of resonance effects.

The through-space versus through-bond controversy is illuminated in a CNDO/2 study by Hermann[27] of substituent effects on the energy of the electrostatic interaction between the proton and the remainder of the molecule in 4-bicyclo-octane carboxylic acids in the gas phase. The calculated effects can be interpreted in terms of a direct interaction between the substituent and the proton. However, the effective dielectric constant for the interaction is well below that for a pure through-space effect in the gas phase ($\varepsilon = 1$), indicating that polarization of the core is an important effect. Further evidence for this conclusion comes from the results of CNDO/2 calculations on styrene derivatives by Reynolds and his co-workers.[28]

There are also many instances of the results of molecular orbital studies *not* being in accord with the predictions of popular theories. The following examples are particularly interesting.

(*a*) Saturated hydrocarbons appear to be electronically disturbed by substituents in a qualitatively similar manner to that for unsaturated hydrocarbons. The polarization of saturated- as well as of unsaturated-chain hydrocarbons induced by substituents produces charge redistributions that do not fall off monotonically with increasing distance from the substituent.[16,29] Also, the polarization of bicyclo-octane by any substituent is qualitatively similar to the polarization of benzene by the same substituent[25] (see, for example, Fig. 3.1). Topsom has pointed out (to the authors of Ref. 25) that the substituent effect on the electron density at the β-, γ- and δ-carbon atoms in the bicyclo-octane system is approximately equal to a linear combination of the σ- and π-electron densities at the *ortho*-,

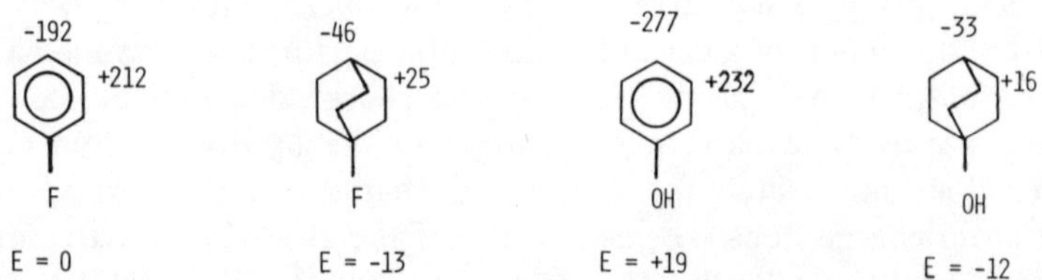

Fig. 3.1. A comparison of the calculated[25,34] substituent effects on the carbon-atom electron densities (q_i, in units of 10^{-4} electron), and on the potential energies of interaction with a unit Coulombic field originating near the δ carbon atom [$E = q_\delta + (4/9)\sum_i q_\gamma$], in benzene and in bicyclo-octane.

meta-, and *para*-carbon atoms in a corresponding monosubstituted benzene system. These results show the inadequacy of the classical inductive model.

(*b*) Relative to the hydrogen atom, alkyl groups usually appear to be attractors of electrons in the principal bond to the core: They have previously been considered to be electron-repelling groups. Fliszár *et al.*[21] have, however, pointed out that the calculated polarity of the C—H bonds in alkanes depends on the method of calculation used. Moreover, depending on the nature of the core, alkyl groups are either weak net donors or weak net acceptors of electrons that are not formally associated with the principal bond to the core (e.g., C—H bonding electrons)[30]: They have usually been considered to be net hyperconjugative donors only. When the core is a neutral unsaturated hydrocarbon, they are weak net donors of these electrons: This behavior is the resultant of moderately strong donation and smaller acceptance (e.g., in C—H antibonding orbitals). Alkyl groups polarize unsaturated hydrocarbon cores markedly in spite of their weak net electron-donating character.[30]

Methyl groups[31] and other alkyl groups[32,33] stabilize both positive ions and negative ions relative to neutral molecules in acid–conjugate-base and base–conjugate-acid pairs. Their ability to stabilize the negative ions was not generally expected.

(*c*) The ability of substituents to participate in charge-transfer with the core is far from small in bicyclo-octanes and in alkanes, although it is less pronounced than in benzene derivatives, alkenes, and alkynes.[25,34] The magnitude of this effect in saturated systems was not previously thought to be so great.

(*d*) The calculated substituent effects on the atomic populations of π electrons in the rings of various monosubstituted benzenes are not directly proportional to one another.[26] With many substituents there is a marked effect on the population at the *meta*-carbon atom of the benzene ring and a marked difference between the populations at the *ortho*- and *para*-carbon atoms; hence the calculated sets of populations are not usually proportional to the set calculated for the benzyl anion by the HMO method. The common treatments of substituent resonance-effects are therefore inadequate.

The results of the molecular orbital studies show that the usual concepts of the inductive effect and of the resonance effect are often inadequate, but there is nothing against the usual concept of the field effect. In the next section a theoretical model for linear correlations is considered, which is based on concepts of interactions between groups, and their modes of transmission, that are in accord with all the results mentioned in this section and in Section 3.3.

3.5. An Alternative Interpretation of Linear Correlation Equations

3.5.1. Introduction

Godfrey[35,36] has proposed an alternative treatment of changes due to electronic disturbances, in the interactions between the source regions and the detector regions in molecular systems. This treatment provides a basis for an interpretation of linear correlation equations that differs radically from the popular treatments discussed in Section 3.3. Although it again involves more than one type of electronic disturbance (in fact it involves three types), it involves only one, albeit complex, important transmission mechanism.

3.5.2. Description of the Theoretical Model

The main features of Godfrey's theoretical model for linear correlations will be brought out by considering its application in typical cases. The types of intergroup interaction are considered for substituted ethylenes, and the mechanism of the transmission of intergroup interactions is considered for substituted buta-1,3-dienes.

Types of Intergroup Interaction

Following Murrell *et al.*[37,38] any substituent is considered to exert both Coulombic and non-Coulombic influences on the core. The Coulombic field falls off with distance from the substituent and for reasons of operational simplicity it is considered to be zero at the third and subsequent atoms of the core from the source site. It also tends to polarize the core. Hence, for example, the π-molecular orbitals of the core should have one of the forms shown qualitatively in Fig. 3.2.† The non-Coulombic influences are of two types. The first is charge transfer, which is the result of electron delocalization between the source site and the core due to the overlapping of occupied orbitals of one moiety with vacant orbitals of the other moiety. It is illustrated for π-molecular orbitals in Fig. 3.3. The consequence of the charge transfer for the ground electronic state of the molecular system is to introduce contributions from the electron configurations shown in the right-hand diagrams of Fig. 3.3.

†In Figs. 3.2, 3.3, 3.4, 3.7, and 3.8, the size of a lobe reflects qualitatively the magnitude of the LCAO coefficient in the Hückel molecular orbital. When the molecular orbital is occupied, the distribution of the electron(s) is given qualitatively by the relative sizes of the lobes, and hence the diagrams indicate the polarization (actual or potential) of the molecular orbitals. The orthogonality condition requires the antibonding π orbital of a vinylene unit (e.g., in ethylene) to be polarized in the opposite sense to the bonding π orbital.

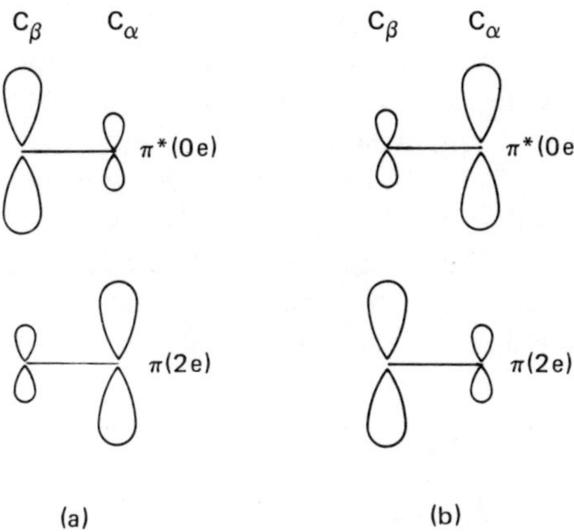

Fig. 3.2. The polarizations of the π-molecular orbitals of ethylene caused by a Coulombic field due to a substituent on C_α that is (*a*) electron attracting, (*b*) electron repelling.

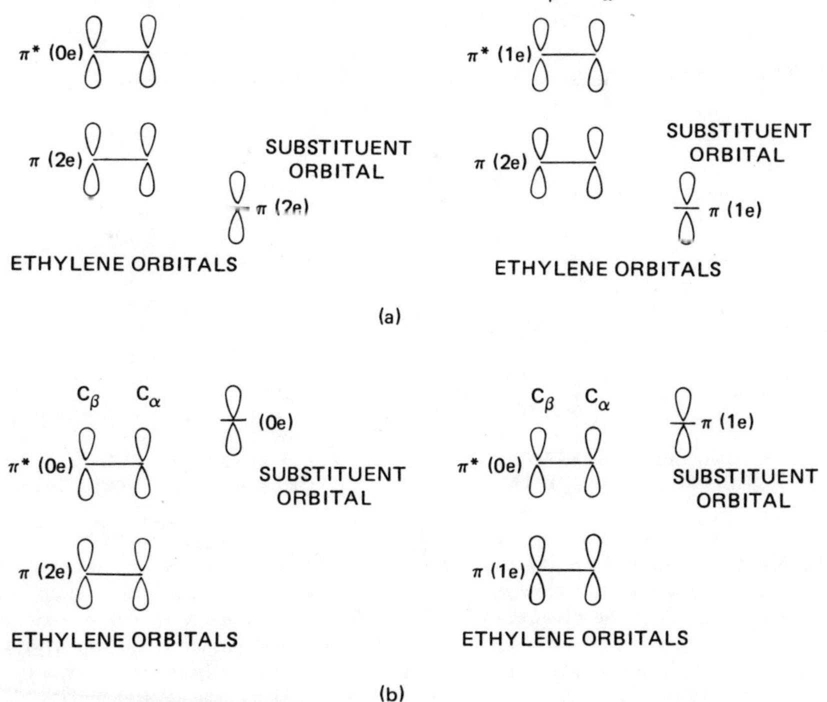

Fig. 3.3. The processes of (*a*) donor, (*b*) acceptor charge transfer. Charge transfer corresponds to the process of going from the electron configuration in the left-hand diagram to the electron configuration in the right-hand diagram (see text).

Charge transfer is a stabilizing process. On the other hand the overlapping of occupied orbitals of one moiety with occupied orbitals of the other moiety is destabilizing.[39] This latter interaction will be called "*overlap repulsion*" in this chapter; its role in organic chemistry is currently attracting much attention.[40–43] In Godfrey's model it constitutes the second type of non-Coulombic influence. The two types of non-Coulombic influence are considered to be the source of an "*overlap field*," which is, by definition, an imaginary field that falls to zero beyond the α atom of the core. The overlap field represents the consequences of drives to maximize

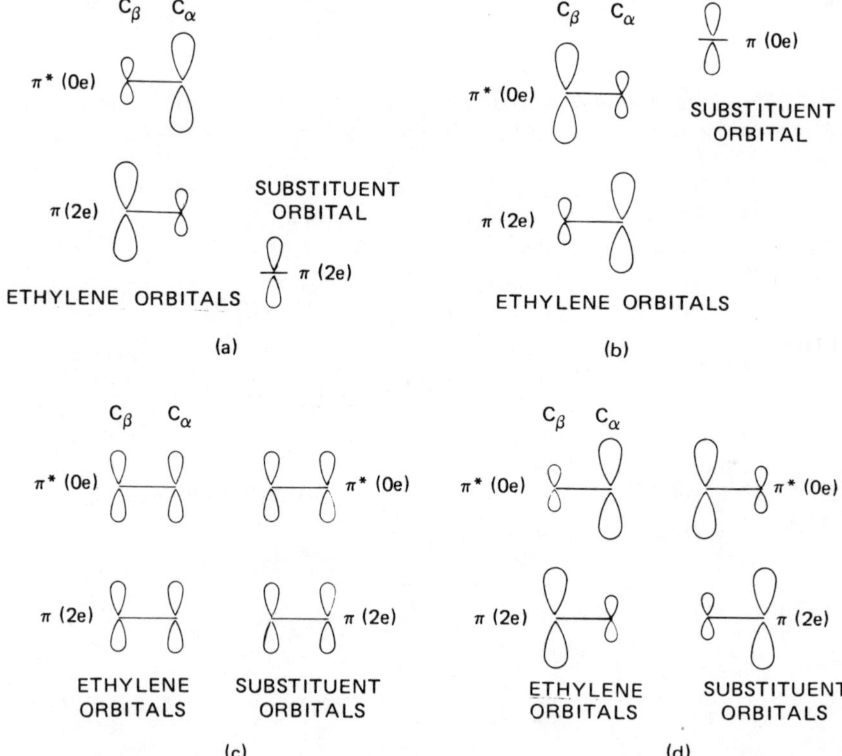

Fig. 3.4. The polarizations of the π-molecular orbitals of ethylene due to terminal overlap fields. (*a*) The drives to maximize donor charge transfer and to minimize overlap repulsion are in concert. The overlap field is electron repelling (positive). (*b*) The sole drive is to maximize acceptor charge transfer. The overlap field is electron attracting (negative). (*c*) The drives to maximize donor charge transfer and to maximize acceptor charge transfer are in opposition, and the drive to minimize overlap repulsion is weak. The overlap field is electron undisturbing (zero). (*d*) In comparison with (*c*), the drives to maximize donor charge transfer and to maximize acceptor charge transfer are again in opposition, but the drive to minimize overlap repulsion is strong. The overlap field is electron repelling (positive).

charge transfer and to minimize overlap-repulsion interactions by polarizing orbitals. Its effects on the electron distribution in the π system of the core are illustrated in Fig. 3.4. When the substituent is primarily an electron donor, the two drives tend to polarize the core in the same direction [Fig. 3.4(*a*)]. Hence donor charge transfer is facilitated even when the principal cause of the overlap field is the destablizing overlap repulsion. When the substituent is primarily an electron acceptor, there is no significant overlap-repulsion and hence the overlap field should be attractive [Fig. 3.4(*b*)]. Whereas Figs. 3.4(*a*) and 3.4(*b*) illustrate the effects of hypothetical substituents that are nonpolarizable and monatomic, Figs. 3.4(*c*) and 3.4(*d*) illustrate the effects of the overlap field on the electron distribution *within* a hypothetical substituent that is polarizable and diatomic. When the substituent is both an electron donor *and* an electron acceptor the drives to maximize donor charge transfer and acceptor charge transfer are in opposition. The magnitude of the overlap field in this case should be determined mainly by the overlap-repulsion contribution, and consequently the amount of net charge transfer†— viz., the electronic charge transferred from a group to its neighbor—is expected to be smaller relative to the magnitude of the overlap field than in the other two cases. The overlap repulsion facilitates donor charge transfer and inhibits acceptor charge transfer. The methyl group and the ammonio (NH_3^+) group are important members of this third class of substituent, because for these purposes they may be regarded as effectively diatomic, and each of them has both a bonding and an antibonding orbital of π symmetry that extends over more than one atom.

For all types of substituent, the magnitudes and directions of the overlap field and of the net charge transfer need not be simply related to the electronegativity of the substituent.

The Transmission Mechanism

How do the interactions between substituents and the neighboring vinylene group affect the interactions between the two vinylene groups in substituted butadienes (see Fig. 3.5)? First, consider the interactions between the two vinylene π systems in *trans*-buta-1,3-diene in the absence of substituents. There should be neither Coulombic field nor net charge transfer, and there should be no overlap field either, unless the drive to minimize overlap repulsion is sufficiently strong to overcome the inherent resistance to perturbation of the π bonding within each group [see Figs. 3.4(*c*) and 3.4(*d*)]. It will be assumed that, in fact, no overlap field is

†Net charge transfer does not include redistribution of electronic charge within a group in order to maximize total charge transfer.

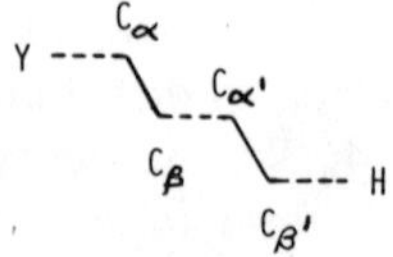

Fig. 3.5. The π system of Y-substituted *trans*-butadiene as a Y,H-substituted bis-vinylene.

established. Next, consider substituent effects. The polarization of the vinylene group adjacent to the substituent should be the source of a new Coulombic field, which in turn causes a polarization of the second vinylene group. Also, the Coulombic and the non-Coulombic source-site disturbances should affect the *energies* of the orbitals of the first vinylene group, and hence the ionization potential and the electron affinity of the group, thereby affecting the ease of charge transfer between the two vinylene groups. Finally, the magnitudes of the drives to establish an overlap field between the two vinylene groups are automatically affected by the fields due to the substituent. This might in certain cases, but need not in general, result in the establishment of an overlap field between the two vinylene groups.

Note that non-Coulombic as well as Coulombic interactions between the source group and the nearer vinylene group lead to Coulombic interactions between the two vinylene groups, and that Coulombic as well as non-Coulombic interactions between the substituent and the nearer vinylene group could, in principle, lead to non-Coulombic interactions between the two vinylene groups. In this sense Coulombic and non-Coulombic disturbances are not transmitted by independent mechanisms. In suitable circumstances it should be possible for a *stabilizing* interaction between one pair of groups, to cause a *destablizing* interaction between a second pair of groups, and vice versa.

The interactions between the two vinylene groups in *cis*-buta-1,3-diene, and the effects of substituents on those interactions, should not be exactly the same as for *trans*-butadiene. The Coulombic interactions will differ because of the angular dependence of the Coulombic fields of dipoles, and the non-Coulombic interactions will differ because of the introduction of a degree of orbital overlap involving atomic centers that are not nearest neighbors.

Generalization of the Theoretical Model

The types of inter-group interaction and the transmission mechanism have been illustrated for delocalized π systems. They should in principle also apply in σ systems and in nonplanar systems for which σ/π analysis is limited. Polarization and charge transfer should occur in all directions but should not in general be isotropic. Since direct field effects of all kinds, including Coulombic, are in this theory considered to be relatively unimportant except between neighboring groups, the structure of the core in three dimensions should be of major importance in determining the magnitudes of the interactions between source groups and detector groups.

The model will, of course, be unsatisfactory when applied in circumstances in which through-space (Coulombic) effects are important relative to through-bond effects. Such circumstances may by no means be as common as suggested by Dewar and Grisdale.[8]

3.5.3. Qualitative Results of Applying the Theoretical Model to General Cases

DOF-Free Regions

In those substituted polyvinylenes in which overlap fields between neighboring vinylene groups do not exist, the third and subsequent vinylene groups will not be directly subject to overlap-field effects either with respect to polarization or with respect to net charge transfer. Hence there is no reason why the relevant changes in the electron populations should not be directly proportional to one another, and why the distributions of the electron populations should not be identical in the third and subsequent vinylene groups. The presence in a molecular system of a region that is not directly subject to overlap field effects will be said to establish a "DOF-free" region (i.e., a region free from Direct Overlap Fields) within the molecular system.

The changes in electron density at every atom within a DOF-free region should be proportional to the same fixed combination of substituent parameters, i.e., they should be proportional to a single *independent* parameter (hereafter called the "DOF-free substituent parameter"). For any given pair of substituents Y_1 and Y_2, the relative change in charge densities, Δq, in the nth vinylene unit ($n \geqslant 3$) from the substituent, should depend only on whether the atom concerned is an α atom or a β atom (see Fig. 3.5). Furthermore, the changes in the *potential energy* of the interaction between the substituent and the detector will be proportional to the same single independent substituent parameter, provided that (*a*) the detector lies within a given DOF-free region, (*b*) there is direct proportionality between the abilities of different substituents to induce *and to respond to* disturbances elsewhere in the molecular system, (*c*) the substituent and the detector are attached to sites that would be electronically indistinguishable in the undisturbed core. It follows that, under the specified conditions, *single-term linear correlations should exist even between different types of measurable property when the detector sites are within a common DOF-free region.* The absolute value of any particular molecular property should decrease exponentially, i.e., according to a negative power law, as the number of vinylene units between the substituent and the detector increases.

In order to illustrate how the efficiency of transmission of substituent effects, and how the scale for the DOF-free substituent parameters, varies with the nature of the DOF-free core, the consequences of replacing the polyvinylene core by a polyphenylene core will now be discussed. For convenience, a phenylene unit will be regarded as a union of two "semiphenylene" units as illustrated in Fig. 3.6 (cf. Fig. 3.5, in which the bis-vinylene system in *trans*-butadiene is shown as a union of two vinylene units).

The Hückel π-molecular orbitals for two isolated semiphenylene units are compared with those for two isolated vinylene units in Fig. 3.7. In each isolated unit there is one bonding orbital and one antibonding orbital. The basis orbitals are closely related in that, for example,

$$\phi_\alpha(\text{vinylene}) = \phi_\alpha(\text{semiphenylene})$$

$$\phi_\beta(\text{vinylene}) = (1/\sqrt{2})(\phi_{\beta 1} + \phi_{\beta 2})(\text{semiphenylene})$$

but the energy separation of the molecular orbitals is greater for semiphenylene than for vinylene. Semiphenylene has also a nonbonding orbital that has no counterpart for vinylene.

The charge transfer drive towards an overlap field should be stronger between two semiphenylene units than between two vinylene units because of the presence in the former of overlapping nonbonding molecular orbitals, each of which contains one electron: The overlap-repulsion drive should be the same in the two systems. It can be assumed that resistance to disturbances within the π bonds of the individual units prevents the creation of an overlap field within phenylene, since benzene has a regular hexagonal structure in its ground electronic state. Hence, the more remote phenylene group in a substituted biphenylene ought to be in a DOF-free region, except when the substituent permits the creation of an overlap field in the nearer phenylene group.

In DOF-free systems, substituent-induced polarization of core units should be more difficult with polysemiphenylenes than with polyvinylenes

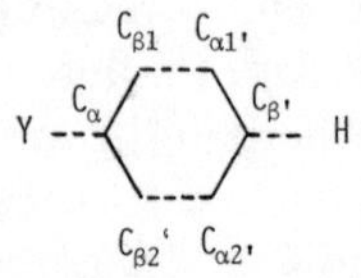

Fig. 3.6. The π system of a Y-substituted benzene as a Y,H-substituted bis-semiphenylene.

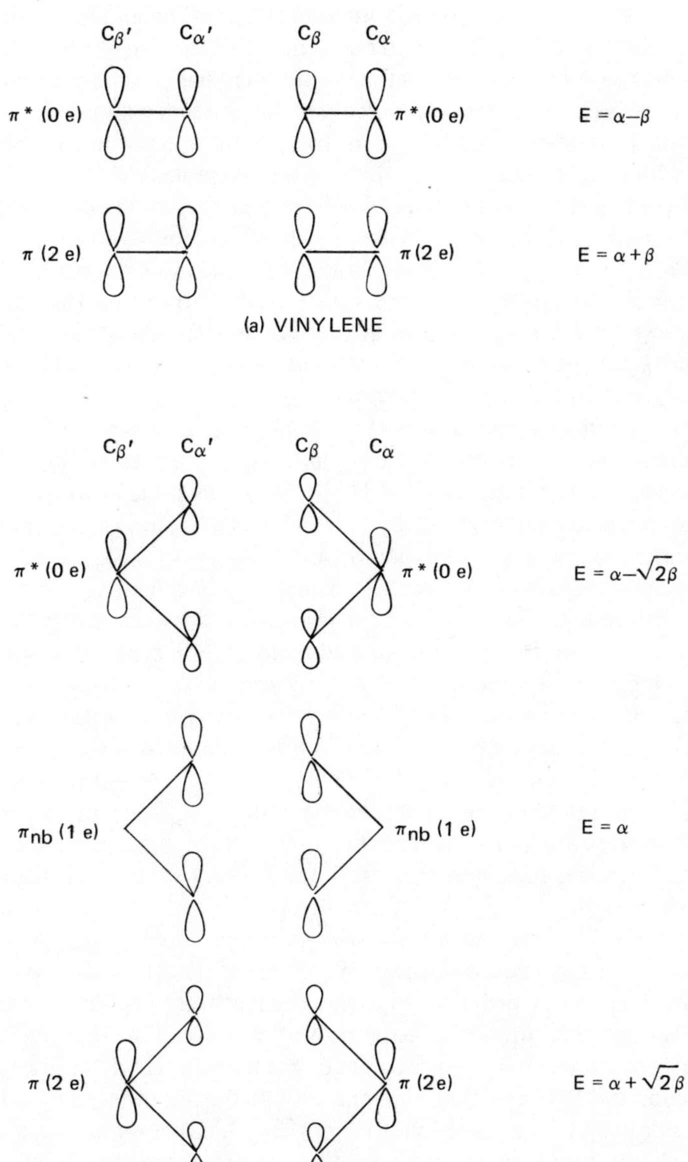

Fig. 3.7. A comparison of the Hückel π-molecular orbitals for (*a*) two isolated vinylene units, (*b*) two isolated semiphenylene units. In the equations, α = Coulomb integral, β = resonance integral.

because of the larger energy separation of the bonding and the antibonding π-molecular orbitals in the former systems. This feature tends to make the rate of decrease of substituent effects with increasing number of core units larger in polysemiphenylenes. However, this tendency should be opposed by the substituent-induced charge transfer between core units being much easier with polysemiphenylenes than with polyvinylenes.

The extra substituent-induced charge transfer between the core units in polysemiphenylenes cannot depend at all on the overlap field of the substituent since there is a node in the relevant molecular orbital of the core unit at the site to which the substituent is attached. Therefore, it is predicted that *the relative importance of the contribution of the overlap-field substituent parameter to the DOF-free substituent parameter will be smaller with polysemiphenylenes than with polyvinylenes.*

The qualitative conclusions reached in the comparison of bis-vinylene and phenylene π-electron systems should, theoretically, apply in the comparison of the ethano- (—CH_2—CH_2—) and the 1,4-cyclohexano-valence-electron systems and, by an obvious extension of the method, in the comparison of the cyclohexano- and bicyclo-octano-valence-electron systems. The results of molecular orbital calculations suggest that electronic disturbances are transmitted less easily through saturated hydrocarbon cores than through unsaturated hydrocarbon cores of similar size.[25] If this conclusion is valid in general, *the variation in the DOF-free substituent parameter scale between different systems ought to be less pronounced within a set of saturated systems than within an analogous set of unsaturated systems.* The data given in Fig. 3.1 clearly indicate that the relative ease of transmission of electronic disturbances through different cores is not directly proportional to the relative influence of different cores on the potential energies for changes in the interactions between substituents and detector groups.

So far in this section transmission through cores made up of sets of identical units has been discussed. We consider next transmission through the styrylene core in order to bring out any features peculiar to cores made up of nonidentical units. Consider, with the aid of Figs. 3.7 and 3.8, the interactions between a semiphenylene group and a vinylene group on the one hand, and between two vinylene groups on the other. The additional molecular orbital in the semiphenylene group has a node through the atom to which the vinylene is attached. Hence, there can be no substituent-induced charge transfer between the units, since this involves this orbital. The same degree of polarization of the semiphenylene unit in the first system, and of the vinylene unit nearer to the substituent in the second system, should induce the same degree of polarization in the more remote core unit. The amount of substituent-induced charge transfer between the two units in the first system should be a little smaller than that between the

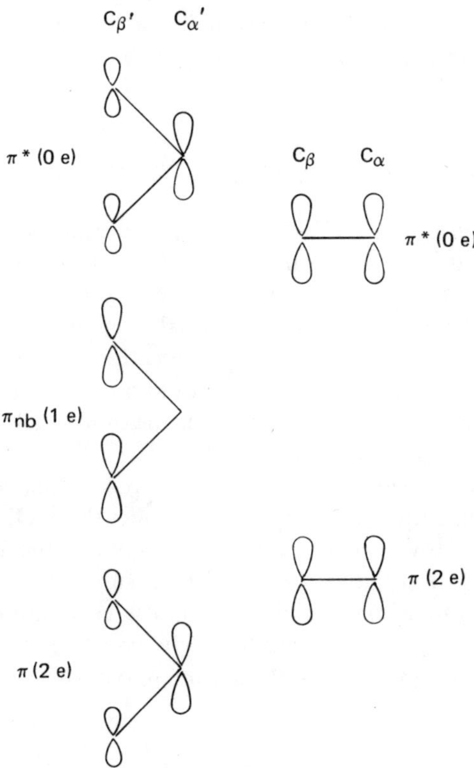

Fig. 3.8. The Hückel π-molecular orbitals for a vinylene/semiphenylene pair.

two units in the second system, because of the higher ionization potential and the lower electron affinity of the semiphenylene unit compared with those for a vinylene unit. Hence, the attenuation factor due to the vinylene group in a styrylene core will be almost as large as that due to the vinylene groups in a tris-vinylene core. The DOF-free substituent parameter scale for styrylene should therefore be almost identical with that for poly-semiphenylenes.

The behavior of the carboxylate group (which is π-isoelectronic with the anion of the semiphenylene group) is of special interest since two important empirical substituent scales are based on the dissociation constants of carboxylic acids. The substitution of a carboxylate group for the vinylene group in styrylene is expected to have very little effect qualitatively. The nonbonding π-molecular orbital of the carboxylate group should not interact with the phenylene molecular orbitals because of the node at the carbon atom. The energies of the bonding and antibonding molecular orbitals should be reduced relative to those of semiphenylene and vinylene by the high electronegativity of the oxygen atoms, but this effect should be opposed by the presence of the additional electron in the nonbonding orbital.

The Conversion of DOF-Free Regions into DOF Regions

The insertion into a DOF-free region of a group that allows overlap fields to be produced between itself and its neighbors will affect the magnitude of substituent effects on molecular properties. The individual unit groups within the resulting new DOF region cannot be associated with a characteristic attenuation factor, and the value of the attenuation factor for the entire new DOF region should not be similar to the product of the values of the attenuation factors for the individual unit groups when they were within a DOF-free region. It may be either larger or smaller. If the insertion still leaves a small DOF-free region between the inserted group and the substituent, this should not affect the DOF-free substituent parameter scale. However, if the insertion converts the DOF-free region totally into a DOF region, the DOF-free substituent parameter scale should be inoperative, and single-term linear correlations are not to be expected either between the same property in different molecular systems or between different properties of the same molecular system.

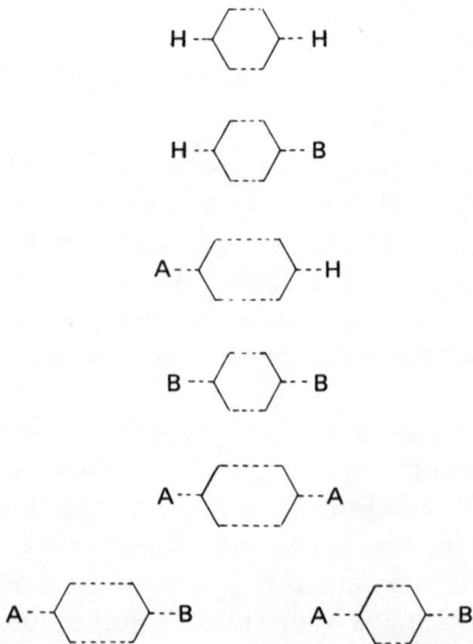

Fig. 3.9. The possible effects of electron-stabilizing substituents A and of electron-destabilizing substituents B on the geometry of the benzene ring in mono- and *para*-di-substituted benzenes.

The insertion of saturated hydrocarbon groups into unsaturated hydrocarbon cores, or of unsaturated hydrocarbon groups into saturated hydrocarbon cores, are important examples of cases in which DOF-free regions are converted into DOF regions.

In favorable circumstances DOF-free regions can be converted into DOF regions simply by changing the substituent. For example, in phenylene the resistance to the creation of an overlap field between the two semiphenylene units might well be just sufficiently large to prevent it from happening. The introduction of a substituent that stabilizes an electron-rich environment by more than a critical amount (hereafter called an "electron-stabilizing" substituent) might well allow the overlap field to be generated. Thus, for example, the introduction of an electron-stabilizing substituent into the *para*-position of a monosubstituted benzene, which has a regular hexagonal carbon ring-structure, may cause the carbon ring-structure to become an irregular hexagon (see Fig. 3.9). On the other hand, the introduction of a substituent that destablizes an electron-rich environment ought to increase the resistance to overlap field generation. The identities of the substituents that favor the generation of overlap fields between identical core units will depend on the nature of the core and on the natures of other groups substituted in the core.

Reagent Effects and Specific Solvent Effects

Whether or not chemical processes within the detector group induce an overlap field between the functional group and the core (or else change the magnitude of an existing overlap field between these groups) should determine the complexity of linear correlations involving reagent effects or specific solvent effects. Single-term linear correlations, such as the Hammett equation and the Brönsted equation, should not be found whenever both Coulombic-field and overlap-field effects are significant. However, dual-term correlations are expected to be much more widespread.

In general, the relative effects of Coulombic and non-Coulombic interactions between the functional group and the reagent on the one hand, and between the functional group and the core on the other, need not be related to one another in a simple way. Consequently there is no reason for a general direct proportionality between reactivity, as measured by the logarithms of the relevant rate constants, and any particular linear combination of detector parameters.

The Cumulative Effects of Substituents

Since in Godfrey's model the magnitudes of interactions between any one pair of groups in a molecular system depend on the magnitudes of the interactions between all pairs of groups in that system, the net effect of any one substituent should not be strictly independent of the presence of other substituents. A molecular orbital study[44] has shown that the total electronic effect of a substituent in a hydrocarbon is not, in general, independent of the nature of other substituents in the hydrocarbon, although the magnitude of the interdependence is usually small compared with the absolute magnitudes of the effects of the substituents. There are two situations in which the "nonadditivity" of substituent effects is expected to be particularly significant. One has just been mentioned: It is when a second substituent modifies the effect of the first substituent on the generation of overlap fields between pairs of core units. The second situation is when marked charge transfer between a substituent and a core unit occurs within the range of powerful Coulombic and overlap fields exerted by another substituent.

3.5.4. Quantification of the Theoretical Model

Definition of Unit Interaction Parameters

Unit overlap field, unit Coulombic field, and unit net charge transfer could, in principle, be defined in a variety of ways. We now describe just one of them. Unit overlap field is defined as the total field operative at the α carbon atom of toluene. Unit Coulombic field is chosen to be equal to unit overlap field (so defined) at the atom of the core that is adjacent to the substituent (the α atom). Unit net charge transfer is defined so as to make the ratios of the values of the net charge-transfer parameters and the overlap-field parameters for a particular set of substituent groups, viz., those that are either predominantly electron donors or predominantly electron acceptors, as close to unity as possible.

The definition of unit overlap field can be seen to be reasonable by considering the results of molecular orbital calculations[26] on the π-electron populations in toluene (Fig. 3.10). The small net charge-transfer

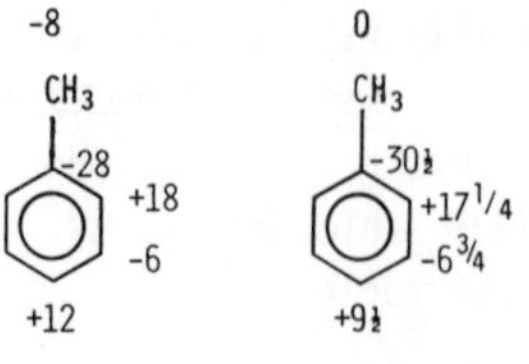

Fig. 3.10. The changes in π-electron populations (in units of 10^{-3} electron) due to the interaction of a methyl and a phenyl group: (*a*) total,[26] (*b*) Coulombic and overlap field contributions only.

contributions to the π-electron populations (obtained from the results of the calculations), given their distribution (determined theoretically[45]), can be subtracted from the total, leaving the combined Coulombic-field and overlap-field contributions. The resulting pattern of the combined contributions shows only a very small net charge transfer between the two semiphenylene units. In terms of the theory described in Section 3.5.2 this result indicates that the Coulombic field contribution must be very small. More specifically, it indicates a slightly electron-attracting Coulombic field.

Evaluation of Substituent Parameters

The Coulombic field, the overlap field, and the net charge-transfer substituent parameters will be denoted by F, S, and T, respectively. The calculated π-electron population distribution for orthogonal nitrobenzene[26] is assumed to be very close to that produced by a pure Coulombic field: The calculations show no net charge transfer between the nitro group and the phenyl group. The calculated distribution can be reproduced by assuming that the Coulombic field at each β atom of the core is approximately 4/9 of that at the corresponding α atom, and that the magnitude of the field at the α atom is 4.5 units.[35] The patterns for π-electron population redistributions due to unit overlap field and to unit Coulombic field are shown in Figs. 3.11(a) and 3.11(b). With the aid of these patterns, values for the substituent overlap-field and Coulombic-field parameters can be obtained from the results of calculations[26] on the π-electron populations in monosubstituted benzenes. The pattern for the π-electron population changes due to unit net charge transfer is shown in Fig. 3.11(c). The pattern for the π-electron population changes calculated by the HMO method, at the *ortho*, the *meta*, and the *para* carbon atom in the benzyl anion (i.e., Dewar and Grisdale's mesomeric pattern[8]) is closely reproduced by a combination of a unit overlap field, a unit net charge transfer, and (15/7) of a unit Coulombic field.

The values of the substituent parameters F and S for several common groups are listed in Table 3.1. They are reasonable in view of the likely electronegativities and molecular orbital characteristics of the groups.

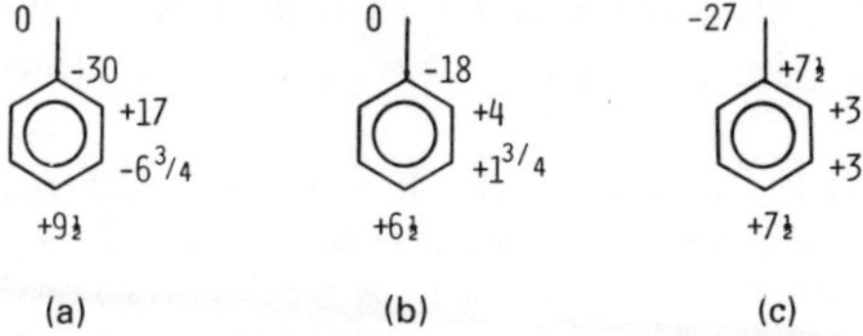

Fig. 3.11. The π-electron redistributions for (a) unit overlap field, (b) unit Coulombic field, (c) unit net charge-transfer perturbations to the benzene ring, in units of 10^{-3} electron.

Table 3.1. Values of the F *and* S *Parameters*

Group	F	S	Group	F	S
NH_2(planar)	−3.8	+5.5	NO(orthogonal)	−3.1	+1.5
NH_2(pyramidal)	−3.0	+4.3	NO_2(planar)	−2.9	−1.6
OH	−4.5	+4.6	NO_2(orthogonal)	−4.5	0.0
OCH_3	−4.2	+4.2	C≡CH	−1.1	−0.2
F	−4.3	+3.2	CN	−2.3	−0.7
$CH{=}CH_2$	0.0	+0.1	NC	−3.4	+0.6
CHO	−0.2	−0.9	CF_3	−1.5	−0.5
CO_2H	−0.2	−1.7	CH_3[a]	−0.7	+1.1
NO(planar)	−0.9	−1.8	Cl[a]	−3.2	+1.8

[a]Calculated from σ^- values.

For easily polarizable substituents it is unreasonable to assume that the values of F and S depend only on the nature of the substituent. Thus, for example, it is no surprise that the values of F and S are by no means identical in different conformations of nitrobenzene and of nitrosobenzene.† However, for other types of substituent (not easily polarizable) it does not seem unreasonable to assume that the values of F and S are approximately independent of the nature of the core whenever the orbital pattern of the core unit neighboring the substituent is qualitatively similar to that in the semiphenylene group.

Electronically Excited States

The patterns of the electron redistributions due to unit substituent effects should depend very much on the nature of the electronic state of the unsubstituted core. Consequently, no single-term correlation is expected between substituent effects on any electronic spectral property and substituent effects on any purely ground-state property, even when both properties are monitored within a DOF-free region. The traditional concept of the resonance substituent effect has been shown to fail even qualitatively in predicting substituent effects on electronic spectra.[45]

The magnitudes of substituent Coulombic fields and of substituent overlap fields are not expected to vary greatly with the nature of the electronic state of the core when the core is an alternant hydrocarbon,

†The value of F depends, *inter alia*, on the distribution of electrons within the substituent group, which in turn depends on overlap-dependent interactions between the substituent group and the core. Since the latter interactions are conformationally dependent in certain molecules, the value of F for the substituent group must also be conformationally dependent in such molecules. This conformational dependence of the value of F is too great to ignore when the substituent group is easily polarizable (e.g., the nitro group in nitrobenzene).

except when there is considerable charge transfer between the substituent and the core in the relevant electronic states. It is not surprising, therefore, to find that substituent effects on the intensity ($\varepsilon^{1/2}$) of the first $\pi \to \pi^*$ transition in benzene are well correlated with a linear combination of the ground-state F and S substituent parameters.

3.5.5. Interpretation of Empirical Dual Substituent-Parameter Equations

If the ratio T : S can be taken to be unity, the theoretical equation for substituent effects on molecular properties, δP, can be written in the form (3.4). Since Coulombic and non-Coulombic disturbances are not transmitted through the molecular system independently of one another (see p. 98), the parameters f and s each reflect both Coulombic and non-Coulombic disturbances in the molecular system.[35] The plot of $\delta P/f$ against s/f generates the theoretically possible patterns of substituent effects (Fig. 3.12).

$$\delta P = f\mathsf{F} + s\mathsf{S}, \qquad \delta P/f = \mathsf{F} + (s/f)\mathsf{S} \tag{3.4}$$

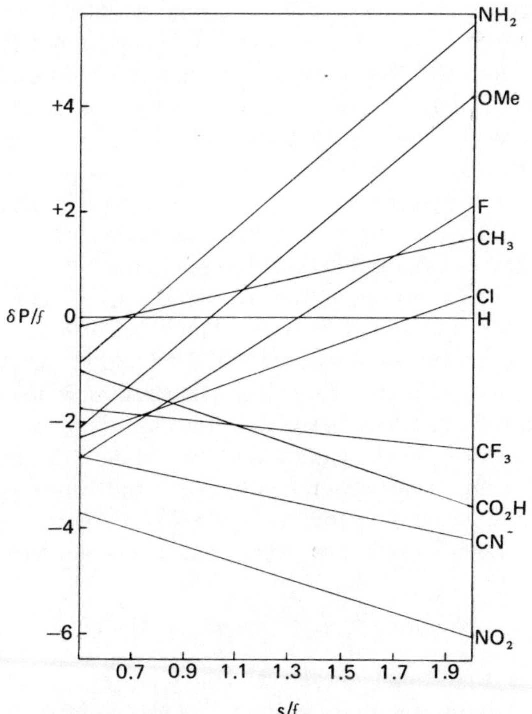

Fig. 3.12. The theoretical patterns of substituent effects.

In order to interpret any quantitative empirical relationship theoretically, it is necessary to assume an equivalence between certain parameters in the empirical equation and certain parameters in the theoretical equation. In using equation (3.4) to interpret the empirical dual substituent-parameter equations of Ehrenson *et al.* and of Swain and Lupton, it will be assumed that each of the empirical substituent parameters is a measure of some linear combination of the theoretical substituent parameters, F and S. Hence, equations (3.5)–(3.8) can be written. The empirical resonance parameters are evaluated by using data for benzene derivatives. In evaluating the regression coefficients (A, a etc.), data for compounds in which ring distortion is a strong possibility have been excluded.

$$\sigma_I = A(\mathsf{F}+a\mathsf{S}), \qquad \mathfrak{F} = A'(\mathsf{F}+a'\mathsf{S}) \tag{3.5}$$

$$\sigma_R = B(\mathsf{F}+b\mathsf{S}) \qquad \mathfrak{R} = \mathrm{B}'(\mathsf{F}+b'\mathsf{S}) \tag{3.6}$$

$$\rho_I = (f-\rho_R B)/A, \qquad \rho f = (f-\rho_R B')/A' \tag{3.7}$$

$$\rho_R = (s-\rho_I Aa)/Bb, \qquad \rho r = (s-\rho_I A'a')/B'b' \tag{3.8}$$

In the equation of Ehrenson *et al.*, $A \approx -0.20$ and $a \approx 0.70$. The values of B and b depend on which substituent resonance scale is used; for σ_R and σ_R^-, $B \approx -0.04$ and $b \approx 4.2$; for σ_R^+, $B \approx -0.20$ and $b \approx 2.1$; and for σ_R^0, $B \approx -0.06$ and $b \approx 2.7$. No common substituent deviates from the σ_R^- correlation, but powerful electron-stabilizing substituents deviate from the other correlations. The physical significance of the ρ parameters follows automatically from the assumptions made about the σ parameters. In each case it is complex. When the σ_R and the σ_R^- scales are used, ρ_I is approximately proportional to f, but it must be remembered that f reflects non-Coulombic as well as Coulombic disturbances in the molecular system.

For the Swain–Lupton equation, a' is equal to a, A' is directly proportional to A, $b' \approx 4.0$, and $B' \approx -0.04$. The value of b' is determined by Swain and Lupton's decision to regard the trimethylammonio group as exerting no resonance effect in the process that is used in the evaluation of $\mathfrak{R}$. If this group were assumed to exert no resonance effect in *any* process, the value of b', and hence the values of $\mathfrak{R}$, would depend on the nature of the process. The important difference between the Ehrenson, Brownlee, and Taft and the Swain–Lupton treatments is that the effects of ring distortion are implicitly taken into account in the former treatment.

3.5.6. Comparison with the Popular Theoretical Models

In both the traditional models and in Godfrey's model, molecular systems in general are regarded as being composed of two types of region with respect to their abilities to transmit electronic disturbances. In the popular models the distinguishing feature is whether or not resonance transmission is permitted within the region, while in Godfrey's model the distinguishing feature is whether or not the region is directly affected by overlap-field disturbances (see Section 3.5.3). If Godfrey's model is valid, the popular models ought to predict too simple a pattern of behavior in resonance-free regions that are also DOF regions, and too complex a pattern of behavior in regions involving resonance that are also DOF-free regions. The popular models should be most satisfactory when applied to behavior in resonance-free regions that are also DOF-free regions.

The results of applying the theoretical dual substituent-parameter equation (3.4) to a variety of experimental data are given in Table 3.2. They will be used in a discussion of correlations among chemical properties, and between chemical properties on the one hand and spectroscopic or calculated electronic properties on the other.

Correlations among Chemical Properties

The concept of the DOF-free region provides a successful basis for interpreting the observed existence (see Section 3.3) of single-term linear correlations in systems that are not resonance-free, as well as in systems that are resonance-free.

The effects of *para*-substituents on the dissociation of benzoic acid in aqueous solution (i.e., the Hammett substituent constants, σ_p) are well correlated with F and S, except for substituents that are likely to be powerful electron stabilizers. The empirical value of s/f (1.3) is very close to the estimated theoretical value[35] for a process in which there is no change in the overlap field between the detector and the core. Hence substituent effects on the side-chain reactivity in any *para*-substituted benzene in which there is no overlap field between the side-chain and the phenylene group, should be well correlated with σ_p, except for substituents that are likely to be powerful electron stabilizers. Deviations from correlation with σ_p are expected for all types of substituents in processes in which an overlap field is generated between the side chain and the phenylene group. A correlation with F and S is still expected but the value of s/f will differ from 1.3. The empirical substituent scales σ_p^0, σ_p^+, and σ_p^- are all well correlated with F and S, when groups that are likely to cause core distortion are omitted. For σ_p^- (measured by the dissociation of anilinium ions in aqueous solution) it is not necessary to omit any common groups.

The effects of *meta*-substituents in benzene derivatives also ought to be correlated with F and S. The value of s/f for processes in which there is no overlap field between the side-chain and the phenylene group should be somewhat less than 1.3. Empirically, σ_m corresponds to an s/f value of

Table 3.2. Empirical Values of f and s/f

δP	f	s/f[a]	Reference
$\sigma_p^{+\,b}$	-0.40[a]	1.40	c
$\sigma_p^{\ b}$	-0.24	1.30	4
σ_p^{-}	-0.24	1.30	d
$\sigma_p^{0\,b}$	-0.26	1.15	4
σ_m, $\sigma_m^{-\,e}$, σ_m^0	-0.22	0.85	4, d
σ_I	-0.20	0.70	4
$\sigma^*_{YCH_2} - \sigma^*_{HCH_2}$	-0.50	0.70	7
log K_d values for anilinium acetates in acetic acid:			
para-substituents[b]	0.12	1.10	f
meta-substituents	0.20	1.15	f
pK_a values for carboxylic acids in 50% w/w aqueous ethanol; bridgehead substituents:			
bicyclo-oct-2-ene[g]	-0.40	0.60	h
bicyclo-octane[g]	-0.33	0.70	h
dibenzobicyclo-octa-2,5-diene[g]	-0.27	0.80	h
pK_a values in aqueous solution:			
pyridinium ions, 4-substituents	-1.7	1.20	i
quinuclidinium ions, 4-substituents	-1.2	0.65	i
3-substituents	-1.1	0.70	j
2,3-dehydroquinuclidinium ions, 3-substituents	-1.25[a]	1.10	j
^{13}C s.c.s in 4-substituted biphenyls in acetone:			
4′-C[b]	0.48	1.35	11
1-C[b]	2.0	1.90	11
Substituent solvent effects (CCl_4 to DMF) on ^{19}F s.c.s.:			
in bridged biphenyls 4′-F	-0.2	0.95	51
in benzenes *p*-F	0.4	0.95	52

[a] To the nearest 0.05 unit.
[b] Assuming ring distortion effects (see text).
[c] L. M. Stock and H. C. Brown, *Adv. Phys. Org. Chem.*, **1**, 35 (1963).
[d] C. L. Liotta, E. M. Perdue, and H. P. Hopkins, *J. Amer. Chem. Soc.*, **95**, 2439 (1973).
[e] Based on the dissociation of anilinium ions in aqueous solution, pK_a/3.
[f] G. W. Ceska and E. Grunwald, *J. Amer. Chem. Soc.*, **89**, 1371 (1967).
[g] The (2.2.2) bicyclic hydrocarbon.
[h] F. W. Baker, R. C. Parish, and L. M. Stock, *J. Amer. Chem. Soc.*, **89**, 5677 (1967).
[i] R. W. Taft and C. A. Grob, *J. Amer. Chem. Soc.*, **96**, 1236 (1974).
[j] C. A. Grob, A. Kaiser, and E. Renk, *Chem and Ind.*, 598 (1957).

0.85. It has been predicted[35] that the introduction of a given overlap field between the side-chain and the phenylene group will have *opposite* effects on the value of s/f in *meta* and *para* compounds. The empirical value of s/f is higher for *meta*-substituents than for *para*-substituents in the case of K_d for anilinium acetates in acetic acid.

The value of s/f for the DOF-free substituent scale for substituents attached to a saturated carbon atom is 0.7, i.e., approximately one-half of the value for the DOF-free substituent scale for substituents attached to an unsaturated carbon atom. The concept of the DOF region provides a reasonable basis for interpreting observed deviations from correlation with σ^* or σ_I. Note that the values of f (see Table 3.2) are not independent of the nature of the hydrocarbon core. Unfortunately there are very few data available on DOF systems in which the substituent is attached to a saturated carbon atom.

The methyl group has a value of F/S such that $\delta P/f \approx 0$ at $s/f = 0.70$. Hence the value of s/f for a process is important in determining the direction of the effect of a methyl group attached to a saturated carbon atom. When the core is a saturated hydrocarbon chain, the unsubstituted system does not obey the correlation equations for substituted systems.[46] This could reasonably be attributed to the existence of an internal overlap field within the core, which is "switched off" by the introduction of most substituent groups.

It has been suggested[35] that solvent molecules play an important role in determining the magnitude of variations in the overlap field between the detector and the core in acid–base equilibria. For aniline, for example, the s/f values for substituent effects suggest that no significant change in the overlap field between the amino group and the phenyl group is produced by protonation in aqueous solution, but that a significant change in this overlap field *is* produced by protonation in a solution in which the solvent has a lower value of the solvent polarity parameter, E_T.[47] The largest change in this overlap field should occur in a gas-phase protonation. In the absence of any change in the overlap field, aniline is predicted to be *less* basic than ammonia, but the production of a change in the overlap field is expected to tend to make aniline *more* basic than ammonia. As expected, the changes in solvent also result in changes in the value of f. For an amine in which any solvent-dependent change in the overlap field between the amino group and the core leaves a DOF-free region between the substituent and the detector, the value of s/f should not be solvent dependent and the value of f should not be very sensitive to the nature of the solvent.

In the light of this discussion on the basicity of amines, it is not surprising that the apparent enhancement of ρ values for substituent effects on the dissociation of carboxylic acids when an alcohol replaces water as the solvent depends very much on the nature of the core.[47]

If Godfrey's model is valid, there should not, in general, be a single-term relationship between the relative rates of two chemical reactions of unsubstituted compounds and the relative f or ρ values for the corresponding reactions of the series of substituted compounds. In other words *there should be no general single-term relationship between reactivity and selectivity* (i.e., the Hammond postulate[48] should not be valid except within a limited range of situations). C. D. Johnson[49] has recently argued that experimental observations are not in accord with a general single-term reactivity–selectivity relationship.

Substituent effects on reactivity are in many cases not additive. For example, Coombes *et al.*[50] have shown departures from additivity of one or two orders of magnitude in the nitration of a series of polychlorobenzenes. With Godfrey's model, marked departures from additivity are expected whenever charge transfer between the core and the functional group is both easy and within the range of the Coulombic and overlap fields exerted by each substituent.

Correlations between nmr Spectroscopy and Chemical Behavior

The concept of the DOF-free region provides a satisfactory basis for explaining the observed single-term correlations between substituent chemical shifts and empirical substituent-reactivity-parameters reported in Section 3.3. The correlations involving chemical shifts observed in the further ring relative to the substituents in the substituted biphenyls give way to dual-term correlations when certain bridging groups are inserted between the phenylene groups. The values of both f and s/f (see Table 3.3) vary somewhat with the nature of the bridging group, so that for certain substituents the substituent effect has been observed to vary more than twentyfold in traversing a series of small bridging groups.[51] When the chemical shifts are monitored in the nearer ring in substituted biphenyls, the values of f and of s/f are markedly different from those observed for the further ring. (The values of f and s/f for substituted benzenes differ little from those for the nearer ring in substituted biphenyls.[11]) These deviations from single-term linear correlations are consistent with the introduction of DOF effects in the molecular cores. Phillips and his coworkers[12,13] have also used the results of observed substituent effects on ^{19}F chemical shifts in bridged biphenyls to question the validity of Taft's

Table 3.3. Values of s/f and Relative Values of f for ^{19}F s.c.s. in 4,4′-Bridged Biphenyls[a]

Bridging group	s/f[b]	f/f(biphenyl)[b]
CH—Ph	1.10	0.45
CH=N	1.25	0.55
N=CH	1.25	0.55
CH=CH	1.25	0.60
N=N	1.30	0.80
O	1.30	1.00
S	1.35	1.20

[a] The solvent is CCl_4. The data are taken from Ref. 51.
[b] To the nearest 0.5 unit.

separation of the transmission of substituent effects across π-electronic systems into inductive/field and resonance components.

The effects of solvent polarity on the values of f and of s in the empirical analyses of substituent effects on chemical shifts are approximately identical (see last two entries in Table 3.2; $s/f \approx 1$). For benzene derivatives, the nitro group and other powerful electron stabilizers do not deviate from the general correlation for solvent polarity effects, even though they do deviate from the general correlation for absolute chemical shifts. Since the sum of the F and the S value for the nitro group is the same for planar and for orthogonal nitrobenzene (Table 3.1), the twisting through 90° of the nitro group ought to have no marked effect on the magnitude of the solvent polarity effect. This is indeed the case.[52]

Correlations between Calculated Electronic Structures and Chemical Behavior

Substituent effects on the polarization of the calculated π-electronic charge remaining after the subtraction of charge-transfer effects, in the terminal vinylene groups of styrene[14] and of butadiene,[53] are proportional to one another and closely follow the empirical σ_p scale. Furthermore, in each compound this π-electronic polarization is directly related to the polarization of the σ-electronic charge in the C_β—H_B bond[14] [see structure (I)]. This behavior is expected on the basis of Godfrey's model since the observations relate to DOF-free regions.

The substituent-induced polarization of the σ-electronic charge in the in-plane C—H bond of the methyl group in *trans*-substituted propenes[54] closely resembles the corresponding polarization of the electronic charge in the C_β—H_B bond of butadiene. On the other hand, substituent effects on the polarization of the C_β—H_B bonds in vinylbicyclo-octane[14] and orthogonal styrene are not obviously related to one another[28] [cf. (I), p. 90]. The inference to be drawn from these results is that the presence of a DOF-free region in a molecular system leads to direct correlation between properties, whereas the presence of a resonance-free region does not.

The pattern of the polarization of the π-electronic charge induced in the further benzene ring by the ammonio group in structure (III)[55] is the same as that for an overlap-independent substituent, but the corresponding

$$H_3\overset{\oplus}{N}—C_6H_4—CH_2—C_6H_5$$

(III)

pattern in the nearer ring indicates an overlap-dependent effect. On replacing the CH_2 group by a CH=CH group the pattern of the polarization is changed in both rings. These results again make sense in terms of the DOF concept but not in terms of the resonance concept.

Conclusion

Although there are insufficient experimental data available to test Godfrey's model thoroughly, there are grounds for hoping that it might well be useful in situations in which the popular models are inadequate.

3.6. Comments on the Theory of Structure–Property Relationships in General

The most important conclusion regarding the general theory of structure–property relationships to be drawn from the studies reported in this chapter is that the concept of the resonance effect does *not* provide the basis for a satisfactory treatment of overlap-dependent interactions. The fact that such a variety of excellent dual-term linear correlations exists encourages the hope that a relatively simple alternative treatment of overlap-dependent interactions that *is* satisfactory might be designed. It has been shown by an example that treatments that allow a degree of interdependence between overlap-dependent and overlap-independent interactions are not necessarily too complicated to be of practical value. The full value of the particular example used still has to be assessed, but it does seem that the interdependence of overlap-dependent and overlap-independent interactions might well be a very important matter.

References

1. C. G. Swain and E. C. Lupton, *J. Amer. Chem. Soc.*, **90**, 4328 (1968).
2. S. Ehrenson, R. T. C. Brownlee, and R. W. Taft, *Progr. Phys. Org. Chem.*, **10**, 1 (1973).
3. L. P. Hammett, *J. Amer. Chem. Soc.*, **59**, 96 (1937).
4. O. Exner in *Advances in Linear Free Energy Relationships*, Chap. 1, N. B. Chapman and J. Shorter, eds. (Plenum, London, 1972).
5. J. E. Leffler and E. Grunwald, *Rates and Equilibria of Organic Reactions* (Wiley, New York, 1963).
6. M. J. S. Dewar, R. Golden, and J. M. Harris, *J. Amer. Chem. Soc.*, **93**, 4187 (1971).
7. R. W. Taft, in *Steric Effects in Organic Chemistry*, Chap. 13, M. S. Newman, ed. (Wiley, New York, 1956).
8. M. J. S. Dewar and P. J. Grisdale, *J. Amer. Chem. Soc.*, **84**, 3548 (1962).
9. L. M. Stock, *J. Chem. Educ.* **49**, 400 (1972).
10. C. Eaborn, R. Eastmond, and D. R. M. Walton, *J. Chem. Soc.* (*B*), 127 (1971).
11. E. M. Schulman, K. A. Christensen, D. M. Grant, and C. Walling, *J. Org. Chem.*, **39**, 2686 (1974).
12. I. R. Ager, L. Phillips, and S. J. Roberts, *J.C.S. Perkin II*, 1988 (1972).
13. P. J. Mitchell and L. Phillips, *J.C.S. Perkin II*, 109 (1974).
14. G. K. Hamer, I. R. Peat, and W. F. Reynolds, *Canad. J. Chem.*, **51**, 897 (1973).
15. L. A. Yanovskaya, G. V. Kryshtal, I. P. Yakovlev, V. F. Kucherov, B. Ya. Simkin, V. A. Bren, V. I. Minkin, O. A. Osipov, and I. A. Tumakova, *Tetrahedron*, **29**, 2053 (1973).
16. W. J. Hehre and J. A. Pople, *J. Amer. Chem. Soc.*, **92**, 2191 (1970).

17. J. A. Pople and G. A. Segal, *J. Chem. Phys.*, **44**, 3289 (1966).
18. N. C. Baird and M. J. S. Dewar, *J. Chem. Phys.*, **50**, 1262 (1969).
19. M. J. S. Dewar and E. Haselbach, *J. Amer. Chem. Soc.*, **92**, 590 (1970).
20. R. C. Bingham, M. J. S. Dewar, and D. H. Lo, *J. Amer. Chem. Soc.*, **97**, 1285 (1975).
21. S. Fliszár, G. Kean, and R. Macaulay, *J. Amer. Chem. Soc.*, **96**, 4353 (1974).
22. S. Fliszár, A. Goursot, and H. Dugas, *J. Amer. Chem. Soc.*, **96**, 4358 (1974).
23. R. S. Mulliken, *J. Chem. Phys.*, **23**, 1833, 1841, 2338, 2343 (1955).
24. N. C. Baird and M. J. S. Dewar, *J. Amer. Chem. Soc.*, **91**, 352 (1969).
25. R. T. C. Brownlee and R. W. Taft, *J. Amer. Chem. Soc.*, **92**, 7007 (1970).
26. W. J. Hehre, L. Radom, and J. A. Pople, *J. Amer. Chem. Soc.*, **94**, 1496 (1972).
27. R. B. Hermann, *J. Amer. Chem. Soc.*, **91**, 3152 (1969).
28. G. K. Hamer, I. R. Peat, and W. F. Reynolds, *Canad. J. Chem.*, **51**, 915 (1973).
29. J. A. Pople and M. Gordon, *J. Amer. Chem. Soc.*, **89**, 4253 (1967).
30. L. Libit and R. Hoffmann, *J. Amer. Chem. Soc.*, **96**, 1370 (1974).
31. W. J. Hehre and J. A. Pople, *Tetrahedron Lett.*, 2959 (1970).
32. L. Radom, J. A. Pople, and P. v. R. Schleyer, *J. Amer. Chem. Soc.*, **94**, 5935 (1972).
33. L. Radom, *Austral. J. Chem.*, **27**, 231 (1974); **28**, 1 (1975).
34. R. T. C. Brownlee and R. W. Taft, *J. Amer. Chem. Soc.*, **90**, 6537 (1968).
35. M. Godfrey, *J.C.S. Perkin II*, 1016 (1975).
36. M. Godfrey, *J. Chem. Soc. (B)*, 1534, 1540 (1971).
37. J. N. Murrell, M. Randić, and D. R. Williams, *Proc. Roy. Soc., A*, **284**, 566 (1965).
38. J. N. Murrell and D. R. Williams, *Proc. Roy. Soc., A*, **291**, 224 (1966).
39. L. Salem, *J. Amer. Chem. Soc.*, **90**, 543 (1968).
40. R. C. Bingham and M. J. S. Dewar, *J. Amer. Chem. Soc.*, **95**, 7180, 7182 (1973).
41. W. J. Hehre and J. A. Pople, *J. Amer. Chem. Soc.*, **97**, 6941 (1975).
42. N. D. Epiotis and R. L. Yates, *J. Amer. Chem. Soc.*, **98**, 461 (1976).
43. R. C. Bingham, *J. Amer. Chem. Soc.*, **98**, 535 (1976).
44. G. P. Ford, T. B. Grindley, A. R. Katritzky, and R. D. Topsom, *J.C.S. Perkin II*, 1569 (1974).
45. J. N. Murrell, *The Theory of the Electronic Spectra of Organic Molecules* (Methuen, London, 1963).
46. C. D. Ritchie and W. F. Sager, *Progr. Phys. Org. Chem.*, **2**, 323 (1964).
47. I. A. Koppel and V. A. Palm, in *Advances in Linear Free Energy Relationships*, Chap. 5, N. B. Chapman and J. Shorter, eds. (Plenum, London, 1972).
48. G. S. Hammond, *J. Amer. Chem. Soc.*, **77**, 334 (1955).
49. C. D. Johnson, *Chem. Rev.*, **75**, 755 (1975).
50. R. G. Coombes, D. H. G. Crout, J. G. Hoggett, R. B. Moodie, and K. Schofield, *J. Chem. Soc. (B)*, 347 (1970).
51. J. Fukunaga and R. W. Taft, *J. Amer. Chem. Soc.*, **97**, 1612 (1975); S. K. Dayal and R. W. Taft, *J. Amer. Chem. Soc.*, **95**, 5595 (1973).
52. R. T. C. Brownlee, S. K. Dayal, J. L. Lyle, and R. W. Taft, *J. Amer. Chem. Soc.*, **94**, 7208 (1972).
53. O. Kajimoto and T. Fueno, *Tetrahedron Lett.*, 3329 (1972).
54. A. Atkinson, A. C. Hopkinson, and E. Lee-Ruff, *Tetrahedron*, **30**, 2023 (1974).
55. D. A. Dawson and W. F. Reynolds, *Canad. J. Chem.*, **53**, 373 (1975).

4

Multiparameter Extensions of the Hammett Equation

John Shorter

John Shorter • The University, Hull, HU6 7RX, England.

4.1. Introduction: The Scope of the Chapter[1]

4.1.1. Deviations from the Hammett Equation

The Hammett equation (1937)[2,3,4] takes forms (4.1), (4.2), where k or K is the rate or equilibrium constant for a side-chain reaction of a *meta*- or

$$\log k = \log k^0 + \rho\sigma \tag{4.1}$$

$$\log K = \log K^0 + \rho\sigma \tag{4.2}$$

para-substituted benzene derivative. The symbol k^0 or K^0 denotes the statistical quantity approximating to k or K for the "unsubstituted" or "parent" compound. The *substituent constant*, σ, measures the polar effect relative to hydrogen of a substituent (in the *meta*- or *para*-position) and is, in principle, independent of the nature of the reaction. The *reaction constant*, ρ, depends on the nature of the reaction (including conditions such as solvent or temperature) and measures the susceptibility of the reaction to polar effects. The ionization of benzoic acids in water at 25°C is the standard process for which ρ is defined as 1.000. The value of σ for a given substituent is thus $\log (K_a/K_a^0)$, where K_a is the ionization constant of the substituted benzoic acid and K_a^0 that of benzoic acid itself.

It is difficult to say what proportion of the relevant reactions conform reasonably well to the Hammett equation in the form described above. Jaffé (1953)[5] examined its application to about 400 reaction series, and on the basis of correlation coefficients he concluded that about 70% of the correlations were "satisfactory" or "excellent."† For reaction series in these categories the Hammett equation has a mean precision of about 15% for k or K, although in some reaction series certain substituents may show large deviations. About 30% of the correlations, however, were classified by Jaffé as "fair" or "poor," and this suggests that many rate or equilibrium data are outside the scope of the Hammett equation in its original form and mode of application.

Deviations are commonly shown by *para*-substituents of considerable $+K$ or $-K$ effect. Hammett himself found that p-NO_2 showed deviations in the correlation of reactions of anilines and phenols. The deviations were systematic in that a σ value of *ca.* 1.27 seemed to apply, compared with 0.78 based on the ionization of p-nitrobenzoic acid. Other examples were soon discovered and it became conventional to treat them similarly in terms of a "duality of substituent constants."

†$r = 0.99$–1.00, excellent; $r = 0.95$–0.99, satisfactory; $r = 0.90$–0.95, fair; $r < 0.90$, poor.

4.1.2. The Duality of Substituent Constants

When σ values based on the ionization of benzoic acid are used, deviations may occur with $-K$ *para*-substituents for reactions involving $+K$ electron-rich reaction centers, and with $+K$ *para*-substituents for reactions involving $-K$ electron-poor reaction centers. The explanation of these deviations is in terms of "cross-conjugation," i.e., conjugation involving substituent and reaction center. In each case "exalted" σ values are required for conformity to the Hammett equation. The special substituent constants for $-K$ substituents (e.g., NO_2, CN, CO_2H, CO_2Me, and SO_2Me) are denoted† by σ^-, and those for $+K$ substituents (e.g. OMe, Me, OH, NH_2, SMe, and Hal) are denoted by σ^+.[7] Values of the former may be based either on the ionization of anilinium ions or of phenols in water. (For some substituents the values based on the two systems are slightly different.) In each case the Hammett line is established with *meta*-substituents and $+K$ *para*-substituents. Values of σ^+ were based by Brown and Okamoto (1958)[7] on the rates of solvolysis of t-cumyl chlorides in 90% acetone-water at 25°C. The Hammett line is established with *meta*-substituents and $-K$ *para*-substituents. (In establishing the Hammett line for σ^-, strong $-K$ substituents are not used in the *meta*-position, and for σ^+, strong $+K$ substituents are similarly not used, in case there is a relayed resonance effect. For such substituents σ_m^- or σ_m^+ values may subsequently be obtained. Brown obtained σ_m^+ or σ_p^+ for *all* substituents from the Hammett line, but the validity of the small corrections applied to substituents already used to establish the line is doubtful.) Tables of sigma values are in Chapter 10.

Values of σ^- have been applied extensively to the correlation of data for the reactions of aniline and phenol derivatives, of carbanions, and of radicals, and have also been applied to nucleophilic aromatic substitution, i.e., where there is no side chain, and reaction occurs at a ring carbon.[4] Values of σ^+ have been applied extensively to solvolyses, aldehyde reactions, basicities of carbonyl compounds and amides, and radical reactions, and most notably to electrophilic aromatic substitution.[4]

4.1.3. The Variability of Cross-Conjugation

The use of σ^+ and σ^- greatly extends the range of applicability of the Hammett equation. However, the "duality of substituent constants" and the attempt to deal with cross-conjugation by selecting σ^+, σ, or σ^- in any given case is somewhat artificial. The contribution of the resonance effect

†The symbol σ^* was at one time used[5] but was abandoned to avoid confusion with Taft's polar substituent constant for aliphatic systems.[6]

of a substituent relative to its inductive effect must in principle vary continuously as the electron-demanding quality of the reaction center is varied, i.e., whether it is electron-rich or electron-poor. A sliding scale of substituent constants would be expected for each substituent having a resonance effect and not just a pair of discrete values: σ^+ and σ for $+K$, or σ^- and σ for $-K$ substituents.

In one way or another this idea was recognized by various authors in the late 1950s. The important question was, how was the idea to be put into practice by modifying the Hammett equation or the way it is used? Obviously great care was needed in finding answers to this question because the value of the Hammett equation would be reduced if there was a proliferation of different constants characteristic of a given substituent. In the limit, σ would become characteristic of the reaction as well as the substituent and the whole system of empirical correlation would then be greatly reduced in value.†

There are, however, several types of treatment, mainly involving multiparameter extensions of the Hammett equation. Before these are introduced, however, we must mention the work of van Bekkum *et al.* (1959),[9] who strongly criticized the duality of sigma constants as an artificial device to describe "mesomeric *para*-interaction," as they termed cross-conjugation, and they proposed a new method of applying the Hammett equation.

The Sliding Scale of Sigma Values. According to van Bekkum *et al.*[9] the ionization of benzoic acids should be used only to evaluate "primary" σ values which can be considered unambiguously "normal", i.e., free from the effects of cross-conjugation. Only such σ values are to be used in calculating ρ for a given reaction series. For other substituents the Hammett line is then used to obtain σ values relevant to the particular reaction series. The authors applied this method to about eighty reaction series and confirmed the hypothesis of a "sliding scale" of σ values arising from cross-conjugation. From the multiplicity of σ values recorded for each substituent the authors selected a normal (unexalted) value, designated σ^n. It was assumed that $+K$ reaction centers give normal σ values for $+K$ substituents, and $-K$ centers for $-K$ substituents, a small spread of calculated σ^n values being averaged for each substituent.

For calculating ρ, eight primary σ values of *meta*-substituents (including H, $\sigma = 0.00$) are of general applicability, and *p*-COMe and *p*-NO_2 are considered acceptable for systems in which cross-conjugation can be ruled out. All other *para*-substituents are unacceptable, along with *meta*-substituents such as OH and OMe, whose strong $+K$ effects may cause

†Certain features of what has happened in the field in the past 20 years are held by some authors to indicate that these worst fears have been realized.[8] However, a greater understanding of structure–reactivity relationships has undoubtedly emerged.

anomalies even from the *meta*-position. The authors claimed that these restrictions led to striking improvements in the success of Hammett correlations, but this conclusion was not entirely secure because insufficient attention was paid to the increase in correlation coefficient due to reducing the number of degrees of freedom of the regression. There is, however, no doubt that the authors' main chemical ideas are sound.

This work of van Bekkum *et al.*[9] is important, but it tends towards the situation in which the "substituent constant" becomes reaction dependent, and correlation analysis is devalued. Therefore we now turn to multiparameter extensions of the Hammett equation.

4.1.4. Multiparameter Extensions of the Hammett Equation

The Yukawa–Tsuno Equation

Yukawa and Tsuno (1959)[10] proposed a method for dealing with the influence of $+K$ substituents on reactions that are more electron-demanding than the ionization of benzoic acid. They suggested that values of $(\sigma^+ - \sigma)$ would provide a scale of enhanced resonance effects, and they modified the Hammett equation to incorporate this,† as shown in (4.3). (A

$$\log k = \log k^0 + \rho[\sigma + r(\sigma^+ - \sigma)] \qquad (4.3)$$

corresponding equation for equilibria may also be written.) The equation implies the multiple linear correlation of $\log k$ with σ and $(\sigma^+ - \sigma)$. (For $-K$ *para*-substituents and for all *meta*-substituents except those of powerful $+K$ effect, simple linear correlation with σ is effectively involved.) The quantity r gives the contribution of the enhanced resonance effect of $+K$ substituents. If $r = 0$ the equation reduces to the simple Hammett equation, and if $r = 1$ it corresponds to straightforward correlation with σ^+. Values of $r > 1$ are also conceivable if there are reactions that are more electron-demanding than the solvolysis of t-cumyl chlorides (used to define σ^+).

A corresponding equation, equation (4.4), with σ^- constants to deal with the influence of $-K$ substituents on reactions that are less electron-demanding than the ionization of benzoic acid was formulated by Yoshioka *et al.* (1962).[11]

$$\log k = \log k^0 + \rho[\sigma + r(\sigma^- - \sigma)] \qquad (4.4)$$

Both these equations have since been modified (Yukawa *et al.*,[12] 1966) to use σ^0 values instead of σ values (see below), but the essential principles are unaltered. All these equations are discussed in Section 4.2.

†The authors did not write the equation exactly as in (4.3), but (4.3) is convenient in the present section; see Section 4.2.1.

The Separation of Inductive and Resonance Effects; Taft's Dual Substituent-Parameter Equation

Hammett σ values measure the resultant of inductive and resonance effects. Taft and Lewis[13,14] (1958–1959, but see also Taft,[6] 1956) suggested that the resultant effects should be quantitatively separable into inductive and resonance contributions through equations (4.5) and (4.6).

$$\sigma_m = \sigma_I + \alpha\sigma_R \tag{4.5}$$

$$\sigma_p = \sigma_I + \sigma_R \tag{4.6}$$

The inductive effect, given by σ_I, is assumed to operate equally from the *meta*- and the *para*-position. The resonance effect, given by σ_R, contributes to σ_m indirectly, α being the "relay coefficient."

Taft and Lewis set up a σ_I scale based on alicyclic and aliphatic reactivities.[6] For ordinary Hammett σ values, based on the ionization of benzoic acid, a value for α of 0.33 was suggested. The authors[15] agreed with van Bekkum *et al.*,[9] however, in regarding the ionization of benzoic acids as a not entirely satisfactory standard process. They proposed the use of "insulated" reaction series, i.e., reactions in which the reaction center is incapable of conjugation with the ring at any stage. The ionization of phenylacetic acids is such a series. Sigma values based on such processes were designated σ^0, and they should correspond to the σ^n of van Bekkum *et al.*, although derived by a different procedure. The analysis into inductive and resonance effects may also be performed for σ^0 constants, giving $\overset{0}{\sigma}_R$ values ($\alpha = 0.5$)[16] or with σ^+ and σ^- constants (giving $\overset{+}{\sigma}_R$ and $\overset{-}{\sigma}_R$, respectively).

The importance of the separation of sigma parameters into σ_I and σ_R-type contributions is that it suggests the possibility of a "dual substituent-parameter" treatment for reaction series through an equation of the form (4.7).[17] Provided that the various σ_R-type scales distinguished

$$\log(k/k^0) = \rho_I\sigma_I + \rho_R\sigma_R \tag{4.7}$$

above are linearly related to each other, it should be satisfactory to characterize each substituent by σ_I and, say, $\overset{0}{\sigma}_R$, and apply equation (4.7) to *meta* and *para* reaction series separately; i.e., with the Hammett equation we have, for each substituent, position-dependent sigma values σ_m and σ_p (arbitrarily becoming σ^+ or σ^- on occasion) but a single ρ value for each reaction series, while in Taft's treatment each substituent is characterized by position-independent σ_I and $\overset{0}{\sigma}_R$ values, but the susceptibility to inductive and to resonance effects (variability of cross-conjugation) is to be expressed separately through position-dependent ρ_I and ρ_R values.

The treatment will be discussed in detail in Section 4.4.

The Swain–Lupton Treatment

The treatment proposed by Swain and Lupton[8] (1968) is closely related to Taft's. It was a reaction against the proliferation of scales of polar substituent constants. According to these authors, the polar effect of any given substituent may be expressed in terms of just two basic characteristics: a field constant, $\mathfrak{F}$, and a resonance constant, $\mathfrak{R}$. All the various sigma scales are held to be linear combinations of $\mathfrak{F}$ and $\mathfrak{R}$ of the form (4.8), where f and r are the field and resonance weighting factors. Correlation analysis of other reactivity data may be carried out in terms of an analogous equation, *meta* and *para* series being dealt with separately.

$$\sigma = f\mathfrak{F} + r\mathfrak{R} \tag{4.8}$$

The Swain–Lupton treatment is discussed in Section 4.3.

4.1.5. The Scope of the Chapter

The three treatments outlined in Section 4.1.4 constitute the main body of the chapter. It has proved convenient to discuss the Yukawa–Tsuno equation first (Section 4.2), then the Swain–Lupton treatment (Section 4.3), and finally Taft's dual substituent-parameter equation (Section 4.4), although the last-mentioned has the longest history.

There will be incidental mention of other treatments bearing an obvious relationship to the above, e.g., those of Knowles *et al.*[18] and of Exner.[19] Certain recent developments, which may be regarded as multiparameter extensions of the Hammett equation, will also be briefly mentioned, e.g., Krygowski's work.[20] These, with a conclusion, are in Section 4.5.

We have deliberately excluded from the present chapter extensions of the Hammett equation designed to deal with the effects of multiple substitution[4] or the *ortho*-effect (with slight exceptions);[21] nor have we paid much attention to extensions to polycyclic or heterocyclic systems (see Chapter 5). We are thus concerned almost entirely with the area for which the Hammett equation was originally devised, i.e., the reactions of *meta*- and *para*-substituted benzene derivatives.

The Dewar–Grisdale[22] treatment is sometimes regarded as generating multiparameter extensions of the Hammett equation. Indeed the F and M parameters of this treatment are sometimes used in multiple regression as if they were analogous to Taft's σ_I and σ_R-type constants. In our view the Dewar–Grisdale treatment is primarily concerned with the analysis of σ_m and σ_p (as originally defined) into field and mesomeric contributions on the basis of a definite model of polar substituent effects, and the use of this model to predict σ values for the various substituent–reaction-center dispositions in polycyclic systems such as naphthalene. The treatment is not much concerned with the variability of cross-conjugation, although there is some reference to this in a more recent paper.[23] We have therefore not included the Dewar–Grisdale treatment in the present chapter; it is dealt with in Chapter 5.

4.2. *The Yukawa–Tsuno Equation*

4.2.1. *The Work of Y. Yukawa and Y. Tsuno, 1959*

Tsuno *et al.*[24] reported that rate constants for the acid-catalyzed decomposition of ω-diazoacetophenones in glacial acetic acid at 40°C did not conform well with the Hammett equation. The use of Brown and Okamoto's[7] σ^+ values instead of σ led to considerable improvement but appeared to overcorrect for the effect of "transition-state resonance": "This is not surprising, because the contribution of resonance would not always be constant but would vary from reaction to reaction. Such failures of the linear relationship using the set of σ^+ are frequently observed;" cf. van Bekkum *et al.*[9] (Section 4.1.3).

A second paper[25] dealt with the ω-diazoacetophenone decomposition in 75% aqueous acetic acid, and with various possible defining systems for electrophilic substituent constants. It was argued that whatever system was used to define σ^+ values, then $(\sigma^+ - \sigma)$ was the proper measure of the resonance effect of a substituent as exerted in an "electrophilic reaction." Moreover, various scales of $(\sigma^+ - \sigma)$ were linearly related to each other.

In the third and best-known paper[10] "The modified Hammett relationship for electrophilic reactions," they suggested that results for such reactions should be correlated by equation (4.9).† The quantity r^+

> . . . is a reaction constant describing the degree of the transition state resonance§ or measuring the magnitude of positive charge to be necessarily stabilized at the transition state,§ and $\Delta\sigma_R^+$, which corresponds to a proper set of $(\sigma^+ - \sigma)$ values, is a substituent constant suggesting the resonating capacity of a substituent.

†Equation (4.9) was originally written with r instead of r^+. Possible confusion of r with correlation coefficient has troubled many authors. Various other symbols have been used, including R[26] (which may be confused with multiple correlation coefficient), γ,[27] and A.[28] Recently Yukawa and Tsuno[29] have themselves introduced the symbol r^+, which seems very suitable, particularly since it suggests the use of r^- in the complementary equation for "nucleophilic reactions" (Section 4.2.4). We shall use r^+ and r^- without regard to symbols used by individual authors who have applied the equation. Also we shall pay no regard to minor variations in the form of the equation, such as writing $(\sigma^+ - \sigma)$ explicitly. The separation of $\log(k/k^0)$ into $\log k = \log k^0 + \cdots$ is in principle a more fundamental matter. Most authors write the equation with a combined log term without necessarily intending to mean that k^0 is the observed rate constant for the parent system (rather than an intercept

The σ^+ scale chosen was H. C. Brown's[7] (Section 4.1.2), and σ values were taken from Jaffé's review,[5] i.e., r^+ is unity for solvolysis of t-cumyl

$$\log(k/k^0) = \rho(\sigma + r^+\Delta\sigma_R^+) \tag{4.9}$$

chlorides in 90% acetone–water at 25°C, and zero for the ionization of benzoic acids in water at 25°C.

Yukawa and Tsuno[10] suggested two methods of correlation analysis with their equation. The most general involves rewriting the equation in the form (4.10). Multiple regression of $\log k$ on σ and $\Delta\sigma_R^+$ gives $a(=\rho)$ and $b(=r^+\rho)$, and $\log k^0$ is the intercept corresponding to $\log k$ for the parent system. Alternatively, when adequate data for suitable *meta*-substituents are available, ρ may be determined by simple regression of $\log k$ on σ, $\Delta\sigma_R^+$ being zero. (This is essentially the procedure of van Bekkum *et al.*[9]) Simple regression of $(\log k - \rho\sigma)$ on $\Delta\sigma_R^+$ for the remaining substituents then gives b and $\log k^0$.¶

$$\log k = \log k^0 + a\sigma + b\Delta\sigma_R^+ \tag{4.10}$$

Yukawa and Tsuno examined the application of equation (4.9) to 35 reactions, mainly rate processes, but also a few equilibria. Nuclear and side-chain processes were variously involved. Values of ρ, r^+, and s were tabulated, and in selected cases the better conformity of $\log(k/k^0)$ to $(\sigma + r^+\Delta\sigma_R^+)$ rather than to σ^+ or σ was demonstrated graphically. Selected data are in Table 4.1. Values of r^+ ranged from 0.19 to 2.29, but were mainly 1.0 ± 0.5; ρ varied from -12 to -0.6, and no clear relationship between ρ and r^+ emerged (see Section 4.2.2). It was admitted that for many reactions having $r^+ = 1.0 \pm 0.3$, simple correlation with σ^+ was reasonably successful.

The authors suggested that r^+ might be a better indicator of the nature of the transition state than ρ. For instance, the ρ values for the two

¶When *relative reactivities* only are available, $\log(k/k^0)$ is the dependent variable, but the regressions should not be forced through the origin. An intercept should be permitted in the least-squares treatment, but in the ordinary way the intercept should emerge as indistinguishable from zero. If it does not, there is some effect connected with the replacement of H by any other substituent. This is possible in some situations (cf. Toyne's paper on the correlation analysis of pmr data[31]), and erroneous conclusions can be drawn from correlation analysis that obscures it.

term) and that the correlation is of *relative* reactivity. (See footnote ¶ above.) Equation (4.9) is commonly referred to as the Yukawa–Tsuno equation, although its originators have not encouraged this. Indeed they have recently devised their own name for the equation[30]: the LASR relationship≡Linear Aromatic Substituent Reactivity relationship. Most physical organic chemists, however, will regard the association of the equation with the originators' names as entirely fitting.

§This must be taken to mean "... the transition state *relative* to the initial state ..." and to include 'final state' and 'initial state' in the case of equilibria, for which equation (4.9) should strictly be written in terms of K and K^0.

Table 4.1. Applications of the Yukawa–Tsuno Equation[10]

Reaction	ρ	r^+	s^a	n^b
Solvolysis of t-cumyl chlorides, 90% aq. acetone, 25°C	−4.52	1.00	0.093	21
Brominolysis of benzeneboronic acids, 20% acetic acid and 0.4 M NaBr, 25°C	−3.84	2.29[c]	0.313	16
Solvolysis of benzhydryl chlorides, EtOH, 25°C	−4.14	1.20	0.095	10
Bromination of benzene derivatives by $HOBr/HClO_4$, 50% aq. dioxan, 25°C	−5.28	1.15	0.080	7
Basicities of monosubstituted azobenzenes	−2.29	0.85	0.100	11
Acid-catalyzed decomposition of ω-diazoacetophenones, acetic acid, 40°C	−0.82	0.56	0.008	12
Acetolysis of neophyl brosylates, 70°C	−3.38	0.54	0.077	7
Beckmann rearrangements:				
(1) Acetophenone oxime picryl ethers, 1,4-dichlorobutane, 70°C	−4.24	0.46	0.233	12
(2) Acetophenone oximes, 94.5% sulfuric acid, 50.9°C	−1.98	0.43	0.044	9
Diazodiphenylmethanes with benzoic acid, toluene, 25°C	−1.57	0.19	0.036	12

[a]Standard deviation of the estimate.
[b]Number of data sets.
[c]See comment in Ref. 105, p. 368.

Beckmann rearrangements in Table 4.1 differ greatly, but the r^+ values are similar, which may indicate similarity of transition states, independently of leaving group and reaction conditions. On the other hand the different r^+ value for the solvolysis of t-cumyl chlorides and of the neophyl brosylates (Table 4.1) may indicate somewhat different modes of resonance stabilization for the incipient carbonium ions in the two transition states [see (I) and (II)]. The significance of r^+ will be further discussed later.

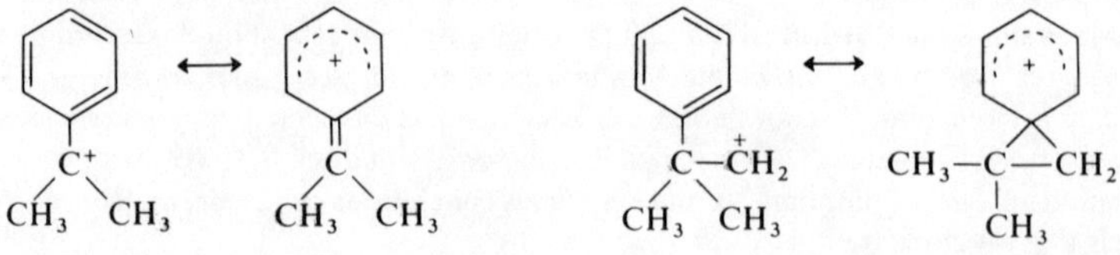

(I) (II)

Yukawa and Tsuno finally discussed the relationship of their equation to Taft's dual substituent-parameter equation (4.7) (cf. Section 4.1.4; see also Section 4.4).[13–16] We may adapt and summarize this discussion as follows.

Equation (4.7) may be rewritten as (4.11), where $q = \rho_R/\rho_I$. If σ_R is taken to be based on the ionization of *para*-substituted benzoic acids [see equation (4.6)], then the Yukawa–Tsuno equation may be rewritten as (4.12), for *para*-substituted systems. Equations (4.11) and (4.12) are

$$\log(k/k^0) = \rho_I(\sigma_I + q\sigma_R) \tag{4.11}$$

$$\log(k/k^0) = \rho(\sigma_I + \sigma_R + r^+\Delta\sigma_R^+) \tag{4.12}$$

equivalent if $\rho_I \equiv \rho$ and (4.13) holds. Thus the treatments are equivalent if $\Delta\sigma_R^+$ is linearly related to σ_R. Yukawa and Tsuno showed that there is a linear relationship of rather poor precision, i.e., the enhanced resonance effect does not accurately reflect the resonance effect in the process used to

$$q\sigma_R \equiv \sigma_R + r^+\Delta\sigma_R^+$$

or (4.13)

$$\Delta\sigma_R^+ \equiv \frac{(q-1)\sigma_R}{r^+}$$

define σ_R. The two treatments are thus not entirely equivalent and Yukawa and Tsuno argued for the superiority of their equation. A modern version of the plot of $\Delta\sigma_R^+$ vs. σ_R (values from the work of Ehrenson *et al.*[32]) is shown in Fig. 4.1.

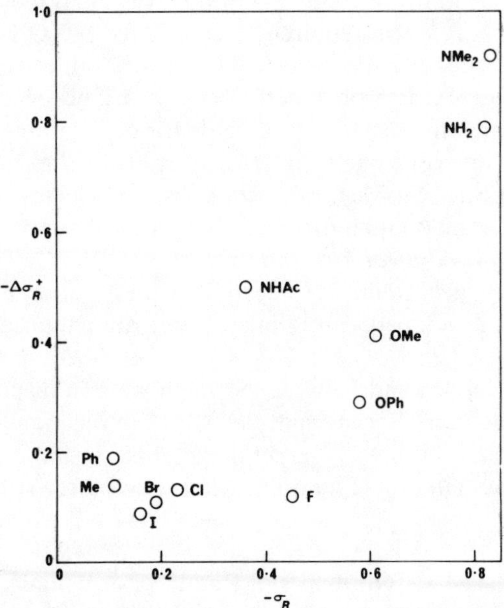

Fig. 4.1. Modern version of plot of $\Delta\sigma_R^+$ vs. σ_R. Data from Ehrenson *et al.*[32]

4.2.2. Applications of the Equation, 1959–1966

The equation received significant but not widespread attention during this period; most physical organic chemists continued with simple correlations on σ or σ^+. Important applications will now be summarized.

The Work of C. Eaborn and His Colleagues[33–40]

The Yukawa–Tsuno equation was used extensively in studies of the rates of electrophilic aromatic substitution, mainly protodemetallation of $XC_6H_4{\cdot}MR_3$. In nearly all systems σ and σ^+ plots were curved in opposite directions, while use of $(\sigma + r^+\Delta\sigma_R^+)$ with an appropriate value of r^+ straightened the plot. (Formal regression methods were not used.) This work has continued until recent years, using the 1959 form of the equation, and it is convenient to deal with all of it together.

Protodesilylation and protodegermanylation have r^+ values of 0.6–0.7, while protodestannylation and protodeplumbylation have r^+ values of about 0.4.[33–36] This was taken to indicate that the electrons of the more polarizable Ar—Sn and Ar—Pb bonds help to delocalize positive charge in the transition state, so that there is less demand on the electrons of the ring. Desilylation in the biphenyl system, 4′-$XC_6H_4{\cdot}C_6H_4SiMe_3$-4, has an r^+ value of only 0.2[37]; delocalization effects are not well transmitted between the rings. The desilylation of *o*-$XC_6H_4{\cdot}SiMe_3$ was analyzed by use of σ_p and σ_p^+ values; $r^+ = 0.3$ was required;[38] cf. 0.7 for *p*-$XC_6H_4{\cdot}SiMe_3$.[33,35] This indicates the reduced importance of the $+K$ effect relative to the inductive effect for *o*-X. For protodetritiation in trifluoroacetic acid $r^+ \approx 1.0$. Hence this reaction is especially valuable for determining σ^+ values.[39]

Eaborn and his colleagues have also studied the base-catalyzed cleavage of XC_6H_4Si(or Sn)Me_3; $+K$ *para*-substituents show *positive* deviations from the Hammett equation, even though these reactions are facilitated by electron *withdrawal* from the reaction center.[40] The Yukawa–Tsuno equation as such could not be applied, but a special equation of similar form was developed, involving σ and an enhanced resonance parameter Δ^-.

About 1962 Eaborn and R. O. C. Norman were in dispute about the Yukawa–Tsuno equation. For aromatic substitution reactions Norman and Radda[41] claimed that there was a good linear relationship between ρ and r^+. If this is so, the Yukawa–Tsuno equation may be rewritten as (4.14),

$$\log(k/k^0) = \rho\sigma + \rho^2 a\Delta\sigma_R^+ \tag{4.14}$$

where $a\rho$ has replaced r^+. Norman and Radda saw an analogy here to an equation of their own,[18] viz., (4.15), where f_p is the partial rate-factor for substitution at the *para*-position of a monosubstituted benzene, σ_G is a measure of ground-state electron density at the *para*-position, σ_P is a measure of the polarizability of the substituent, and ϕ is a measure of the

$$\log f_p = \sigma_G \phi + \sigma_P \phi^2 \tag{4.15}$$

demand of the reagent for electrons in the transition state. (σ_G and σ_P had been evaluated for a few substituents and ϕ for several reactions.†) Bott and Eaborn[42] vigorously contested the supposed relationship between ρ and r^+; for details the paper[42] should be consulted. They also criticized Norman and Radda's use of σ^n (van Bekkum *et al.*[9]) in place of σ in the Yukawa–Tsuno equation. The reasons for using σ^n were those that ultimately led Yukawa and Tsuno in 1966[12] to modify the equation in terms of σ^0 (Section 4.2.3). To Eaborn and Bott the use of σ^n instead of σ was to replace well-established experimental values from a clearcut reference reaction by values derived by various assumptions from several reactions.

Miscellaneous Applications

G. G. Smith and his colleagues found that data for the pyrolysis of 1-arylethyl acetates at 600 K could be treated well by use of σ^+.[44,45] Thus for this reaction, as for t-cumyl chloride solvolysis, r^+ is 1.0, although the ρ values differ greatly, −0.66 and −4.52 respectively, confirming that there is no general relationship between r^+ and ρ.[10,41,42] The absence of solvation in the gas phase was suggested as responsible for the higher value of $|r^+/\rho|$.[45] The pyrolysis of 1-arylethyl methyl carbonates gave $r^+ \approx 1.2$, $\rho = -0.97$[46]; the carbonate pyrolysis evidently involves a greater resonance stabilization of positive charge in the activated complex.

Crowell *et al.*[47] found large deviations from σ or σ^+ correlations for rate constants, k, of Schiff's base formation from substituted benzaldehydes and *n*-butylamine in dioxan at 25°C as shown in (4.16). The deviations were not due to changes in the relative importance of the reaction steps[4] and $k_{obs} = k_1 k_2 / k_{-1}$ throughout the reaction series. Equation (4.17) was satisfactory, but equation (4.18) was slightly better (cf. Yukawa *et al.*,

$$\mathrm{Bu}^n\mathrm{NH_2} + \mathrm{ArCHO} \underset{k_{-1}}{\overset{k_1}{\rightleftharpoons}} \mathrm{ArCH(OH)NHBu}^n \overset{k_2}{\rightarrow} \mathrm{ArCH{=}NBu}^n + \mathrm{H_2O} \tag{4.16}$$

$$\log k = -1.92\sigma + 1.36\sigma^+ - 0.957 \tag{4.17}$$

$$\log k = -1.335\sigma^0 + 0.803\sigma^+ - 0.983 \tag{4.18}$$

1966[12]). It was suggested that (k_1/k_{-1}) was correlated by σ^+, and k_2 by σ or σ^0. While the equations were regarded as equivalent to the Yukawa–Tsuno equation, the latter was not devised to deal with such a situation,

†Equation (4.15) has had little testing or further development. Its derivation has been criticized by Wells.[43]

and if an r^+ value were calculated it would not have a simple significance (see Section 4.2.5). Similar issues arise for semicarbazone formation.[48]

Other applications included those of Zwanenburg and Engberts[49] to the acid-catalyzed hydrolysis of α-diazosulfones ($r^+ \approx 0.6$), and of Slootmaekers *et al.*[50] to aryloylation of toluene with ArCOCl + $AlCl_3$ (but $r^+ = 0.94$, and simple correlation with σ^+ was said to be better; however, many useful substituents including OMe and NO_2 had to be omitted because of adduct formation with $AlCl_3$). Inukai[51] found $r^+ = 0.61$ for solvolysis of 4-biphenylyldimethylcarbinyl chlorides, but expressed some doubts as to the significance of the Yukawa–Tsuno treatment, particularly when large values of r^+ are obtained, e.g., for the brominolysis of benzeneboronic acids, $r^+ = 2.29$.[10] (See Table 4.1, footnote *c*.)

In 1963[52] Stock and Brown reviewed the quantitative treatment of directive effects in aromatic substitution. While not disputing the reality of deviations from LFER, they questioned the validity of the results obtained by applying the Yukawa–Tsuno equation to aromatic substitution. "The correlations of the data with σ^+ must be regarded as satisfactory within the framework of the precision of the data and other possible sources of the discrepancies," although they continue, "It is our view that the Yukawa–Tsuno equation provides the correct analysis of the origin of certain of the discrepancies" (see Section 4.2.5).

Towards the end of the period under review the Yukawa–Tsuno equation was first applied to additions to C=C. We discuss this in Section 4.2.4.

The nonlinearity of free energy correlations for aromatic side-chain reactions was examined by Fueno *et al.*[53] through the LCAO MO method. The concept of a sliding scale of substituent constants[9] (Section 4.1.3) was supported and the Yukawa–Tsuno equation was held to be an acceptable expression of this.

Finally in this section we must mention the first application to *nucleophilic*† reactions of an equation of the form of the Yukawa–Tsuno equation. We shall write this as (4.19). Yoshioka *et al.* (1962)[11] found that acid

$$\log (k/k^0) = \rho(\sigma + r^- \Delta\sigma_R^-) \tag{4.19}$$

dissociation constants of *N*-arylsulfanilamides ($ArNHSO_2C_6H_4NH_2$-*p*) did not conform well to σ or σ^-, but equation (4.19) with $r^- = 0.54$ gave an

†Near the start of Section 4.2.1 the terms "electrophilic reaction" and "nucleophilic reaction" were used, thereby following a usage introduced by Yukawa and Tsuno in their early papers.[10,24,25] Strictly, such terms lack clarity since all heterolytic organic reactions have both an electrophilic and a nucleophilic aspect. However, these terms are occasionally useful for referring to (*a*) processes in which $+K$ *para*-substituents show exalted resonance effects (electrophilic reactions), and (*b*) processes in which $-K$ *para*-substituents show exalted resonance effects (nucleophilic reactions). We shall make some use of these terms.

excellent correlation; similarly for *N*-arylbenzenesulfonamides, with $r^- = 0.68$. This equation has subsequently found considerable application, but mainly post-1966 (see Section 4.2.4).†

4.2.3. The Work of Y. Yukawa, Y. Tsuno, and T. Sawada, 1966

Yukawa *et al.*[12] proposed the modified relationship (4.20), where σ^0 is a "normal" substituent constant (Taft[16]) and $\Delta\bar{\sigma}_R^+ = (\sigma^+ - \sigma^0)$, the σ^+

$$\log(k/k^0) = \rho(\sigma^0 + r^+\Delta\bar{\sigma}_R^+) \tag{4.20}$$

values being those of Brown and Okamoto.[7] This equation was held to be theoretically more correct than the original form [equation (4.9)], since the σ^0 scale is a more proper origin from which to assess enhanced resonance effects. The authors did not accept Taft's σ^0 values, because of supposed limitations in their derivation. The alkaline hydrolysis of ethyl phenylacetates was selected as the defining reaction for σ^0; this has the advantage of a much larger ρ value than that of phenylacetic acid ionization. The σ^0 values for a considerable number of substituents were established essentially by the procedure of van Bekkum *et al.*[9] In most cases the values were very close to Taft's.

Values of $\Delta\bar{\sigma}_R^+$ are correlated linearly with those of $\Delta\sigma_R^+$, as in (4.21). Thus the change in the form of the Yukawa–Tsuno equation is really only a

$$\Delta\sigma_R^+ = 0.74\Delta\bar{\sigma}_R^+ \tag{4.21}$$

change in the r^+ scale: r^+ in (4.20) is zero for the ionization of phenylacetic acids, rather than the ionization of benzoic acids.

Equation (4.20) was applied to 52 "electrophilic reactivities," mainly different from those examined previously,[10] including side-chain and nuclear reactions (rates), various equilibria, and some physical data (mainly ^{19}F chemical shifts). Values of ρ, r^+, and s were tabulated as before,[10] and multiple correlation coefficients were also included. The merits of r^+ compared with ρ were reasserted, and further illustrated. Thus for the dissociation of benzoic acids in water and in various alcohols, ρ varies considerably but r^+ is almost constant (0.25), indicating that resonance stabilization scarcely varies with solvent. The authors also considered briefly the application of (4.22) to "nucleophilic reactivities"[11] but could find few systems to which to apply it.

$$\log(k/k^0) = \rho(\sigma^0 + r^-\Delta\bar{\sigma}_R^-) \tag{4.22}$$

†While due credit should be given to Yoshioka *et al.*[11] as the first to adapt the Yukawa–Tsuno equation to nucleophilic reactions, it does not seem desirable to refer to equation (4.19) by their names. It is simpler to regard the Yukawa–Tsuno equation as a collective description (cf. the similar use of the term "Schrödinger equation").

The lack of any precise relationship of r^+ to ρ was reiterated (cf. the Norman[41]–Eaborn[42] dispute, Section 4.2.2), and the relationship of the treatment to Taft's dual substituent-parameter equation (Sections 4.1.4 and 4.4) was discussed. By this time it was even clearer that the various σ_R-type scales are not collinear; each can be used with precision only in its "own" reactivity class (see Section 4.4.3). It was concluded that the Yukawa–Tsuno treatment, being more general, is to be preferred.

4.2.4. *Applications of the Equation Since 1966*

Since 1966 applications have proliferated and there are well over 200 citations of the main papers[10–12] in the principal journals.[54] Only a selection is dealt with below. Many authors show a preference for the 1959 form of the equation; presumably they distrust σ^0.

Electrophilic Aromatic Substitution

Eaborn's work[33–40] was discussed in Section 4.2.3. Olah and his colleagues have applied the equation (usually in the 1959 form) to Friedel–Crafts reactions[55,56] and to nitration.[57] Olah believes that reactions of less reactive electrophiles will have "late" transition states and will be correlated with σ^+, while those of very reactive electrophiles will have "early" transition states and will be correlated with σ. Some electrophiles may be between these extremes, and it is best not to prejudge the issue but to do correlations in a flexible way with the Yukawa–Tsuno equation (cf. the belief of Clementi *et al.*[58] in the general applicability of σ^+, which appears to conflict in some respects with an earlier view of two of these authors[59]).

Hashimoto and Morimoto[60] found that correlation with σ was satisfactory for results for the cleavage of $XC_6H_4SnEt_3$ by $Hg(OAc)_2$ in THF. This contrasts with protodestannylation[33] (Section 4.2.2), for which $r^+ \approx 0.4$. The transition state in the mercuration is perhaps closer to a π complex than to a benzenium ion.

Girault *et al.*[61] applied the Yukawa–Tsuno equation to the acetylation of α-(substituted phenyl)furans and Baas and Wepster[62] to the nitration of monoalkylbenzenes. An unusual application is to the quenching of uranyl fluorescence by aromatic molecules,[63] which is considered to involve π-complex formation and a resemblance to nitration.[64]

Additions to Multiple Carbon–Carbon Bonds

Such reactions are mainly electrophilic, and Yukawa and Tsuno's tables[10,12] contain a few examples.

Durand *et al.*[65] applied the equation to the acid-catalyzed hydration of substituted styrenes ($\rho = -4.0$ and $r^+ = 0.78$) and α-methylstyrenes ($\rho = -3.27$ and $r^+ = 0.63$). The values were interpreted in terms of a carbonium ion mechanism. Bott *et al.*[66] found $\rho = -4.3$ and $r^+ = 0.81$ for the hydration of arylacetylenes, compared with $\rho = -3.3$ and $r^+ = 0.64$ for the protodegermanylation of $XC_6H_4C{\equiv}CGeEt_3$.[67] The reactions proceed *via* analogous intermediates (III) and (IV). Probably (IV) is somewhat stabilized by electron release from $GeEt_3$, so that less demand is made on the

$$\underset{\text{(III)}}{\mathrm{Ar\overset{+}{C}{=}CH_2}} \qquad \underset{\text{(IV)}}{\mathrm{Ar\overset{+}{C}{=}CHGeEt_3}}$$

aromatic ring. Noyce and Schiavelli,[68] however, found σ^+ satisfactory for correlations involving data for the hydration of arylacetylenes.

Epoxidation of styrenes[69,70] and stilbenes[10,12] under various conditions was studied and the Yukawa–Tsuno equation applied with variable success. For the reversible acid-catalyzed addition of acetic acid to styrenes, Mollard *et al.*[71] found $r^+ \approx 0.70$ in both directions; cf. $r^+ \approx 1.0$ for the reverse reaction studied through pyrolysis of 1-arylethyl acetates.[44,45] An explanation in terms of solvation was given.

Dubois and his colleagues have measured "globale" bromination rates (Br_2/0.2 M KBr in methanol, 25°C) for C=C.[72–79] For monosubstituted 1,1-diphenylethylenes $\rho = -3.57$ and $r^+ = 0.84$.[72] (Preliminary work[73] had given a somewhat lower r^+ value through including 4- and 4,4′-substituted compounds in the same correlation; see below.) This is to be compared with −4.3 and 1.0, respectively, for styrene bromination.[74] The two rings of 1,1-diphenylethylene cannot be simultaneously coplanar with the olefinic bond, so resonance interaction with the carbonium center is reduced.

The effects of substitution in both rings are not additive and exhibit apparent saturation of resonance stabilization,[75] but most of the rate constants conformed to (4.23), where σ^+ is for the substituent with the

$$\log(k/k^0) = -3.26(\sigma^+ + \sigma) \tag{4.23}$$

larger value of $(\sigma^+ + \sigma)$ and σ is for the other. This agrees with an unsymmetrical transition state in which one of the aryl rings is coplanar with the alkenic bond and conjugated with the carbonium center (σ^+ required) and the other is sterically forced out of plane (σ required). The treatment was elaborated to deal with deviating substituents, but the Yukawa–Tsuno equation does not prove particularly useful in dealing with saturation of resonance: A sliding scale of ρ values rather than of σ values is really required.

For the bromination of related cyclic compounds (V)[76,77] the Yukawa–Tsuno equation is of value with $r^+ = 0.184$ for $n = 2$, and $r^+ = 1.383$ for

$n = 3$, values that can be interpreted in terms of differing geometries of (V).

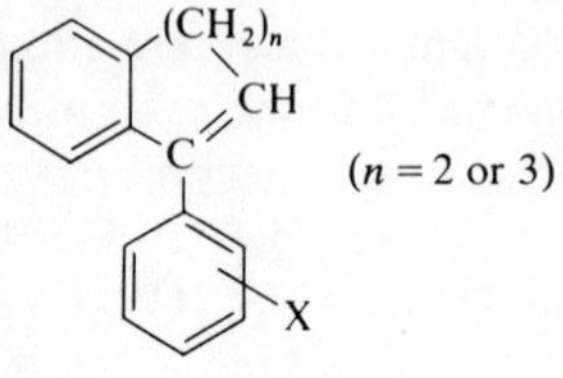

($n = 2$ or 3)

(V)

The bromination of monosubstituted *trans*-stilbenes[78] involved a very complex pattern of results, which was interpreted in terms of a substituent effect on the relative importance of two alternative activated complexes. The Yukawa–Tsuno equation expressed the data approximately, but with $\rho = -1.94$ and $r^+ = 3.05$, values which the authors considered ludicrous. "Such multiparameter correlations can be very misleading if the apparent ρ and r^+ values are not examined critically."[79]

Nucleophilic Reactions

There have now been many applications of the equation in the form first suggested by Yoshioka *et al.*[11]

Ryan and Humffray have examined several types of reaction.[80–83] In the alkaline hydrolysis of aryl acetates[80] and benzoates[81] the r^- values are 0.2 and 0.3, respectively (1959 form of the equation), indicating only a small extent of bond-breaking (release of phenoxide ion) in the transition state. On the other hand, in the reaction of picryl chloride with phenoxide ions ($r^- = 0.94$) and with anilines ($r^- = 0.68$) considerable bond formation has occurred in the relevant transition state.[82]

Štěrba, Večeřa, and their colleagues have also applied the equation (1959 form), mainly to reactions of azophenols and related compounds.[84–89] For the dissociation of azophenols (VI) in 50% ethanol–water, $\rho = 0.734$ and $r^- = 0.487$, while for (VII), $\rho = 1.223$ and $r^- = 0.286$.[84] The values for (VI) may be compared with 2.67 and 1.0, respectively, for simple phenols. These differences doubtless reflect differences in the distances and in the conjugation in the species involved. The difference between ρ values for (VI) and (VII) (0.489) indicates transmission through

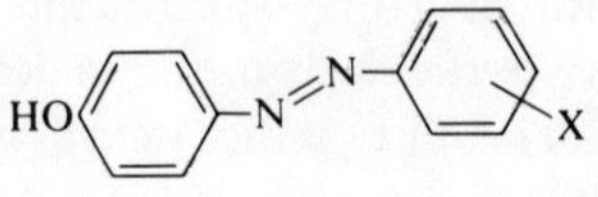

(VI) 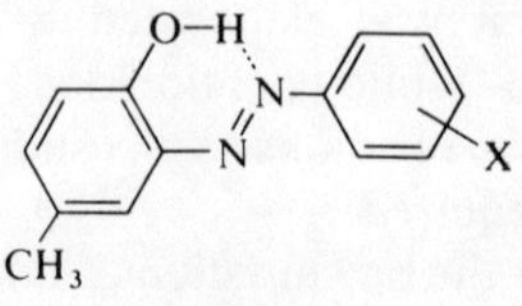 (VII)

the hydrogen bond of (VII). A correlation equation for this "two-route" transmission (4.24) is equivalent to a Yukawa–Tsuno equation with

$$\log (K/K^0) = 0.734(\sigma + 0.487\Delta\sigma^-) + 0.489\sigma \tag{4.24}$$

$\rho \approx 1.22$ and $r^- \approx 0.29$ as above. For the alkaline hydrolysis of the corresponding phenyl acetates, the ordinary Hammett equation is successful[85] (cf. above[80]). For the ionization of (VIII) $\rho = 3.59$ and $r^- = 0.61$. cf. 2.67

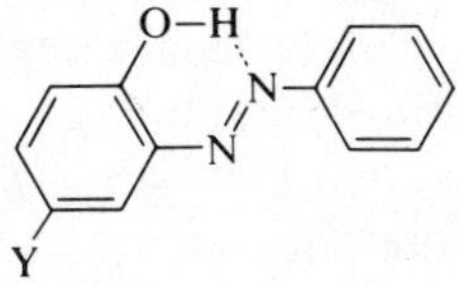

(VIII)

and 1.00, respectively, for simple phenols.[86] The larger value of ρ is attributed to transmission *via* the hydrogen bond, while the reduced r^- is attributed to competition between cross-conjugation of the reaction center with Y and with —N=N—.

Applications to nucleophilic aromatic substitution are relatively few. Dell'Erba *et al.*[90] found $\rho = 1.00$ and $r^- = 0.32$ for the piperidinodebromination of 3′- or 4′-substituted 4-bromo-3-nitrobiphenyls in methanol; cf. work with 2-halogenonitrobenzenes which found simple correlations on σ^-, with much higher ρ values (3–5).[91] Presumably the differences in ρ and r^- reflect differences in relevant distances and conjugation. Bowden and Cook[92] studied the alkaline hydrolysis of 4-substituted 1-methoxy-2-nitrobenzenes in aqueous DMSO. In 14 moles % DMSO, $\rho = 4.4$ and $r^- = 0.45$, but at higher DMSO concentrations ρ was larger and r^- was smaller. It is suggested that in protic media hydrogen-bonding assists $-K$ effects. As the protic nature of the medium is increased, resonance effects become less pronounced (r^+ is diminished), but in consequence the system becomes more dependent on substituent polar effects, i.e., ρ is increased, an example of the selectivity–reactivity principle.[93]

Applications in Spectroscopy

The equation has been applied extensively to pmr by Yukawa and Tsuno,[94–98] and by others[99] (see Chapter 8).

The O*H* chemical shifts of phenols in DMF[94] or DMSO[94,95,97] conform to (4.25), with $\rho = 1.66$ and $r^- = 0.64$ in DMF, and $\rho = 1.53$ and

$$\Delta\delta = \rho(\sigma^0 + r^- \Delta\bar{\sigma}_R) \tag{4.25}$$

$r^- = 0.67$ in DMSO (cf. Tribble and Traynham,[100] who used simple correlation with σ^- for phenols in DMSO). The ring *meta*- or *para*-protons of monosubstituted benzenes required separate correlations, e.g., equations (4.26) and (4.27) (DMA solvent). Thus there are unusually large $+K$

$$\Delta\delta_p = -39.6(\sigma_p^0 + 0.87\Delta\bar{\sigma}_R^+) + 0.24 \tag{4.26}$$

$$\Delta\delta_m = -34.5(\sigma_m^0 + 0.42\Delta\bar{\sigma}_R^+) - 0.61 \tag{4.27}$$

effects from the *meta*-position. Contributions of $r^-\Delta\bar{\sigma}_R^-$ for $-K$ substituents were sought but not found, suggesting that the benzene ring works more effectively as a π-electron acceptor than as a donor in the ground state of monosubstituted benzenes.

Laurence, Wojtkowiak, and their colleagues have applied the equation to ir data, when the vibrating group (e.g., OH, CO, or CN) is capable of conjugation with the benzene ring.[28,101] Berthelot and Laurence[102] have, however, questioned the applicability of σ^0 values based on reactions in aqueous solution to processes such as ir absorption studied in nonaqueous media.

Recently Bekárek and Pragerová[103] studied pK_a values, OH pmr, and out-of-plane OH deformation frequencies of 4- or 5-substituted 2-nitrophenols. For the three properties the Yukawa–Tsuno equation must be applied in a "double" form to allow for $-K$ substituents conjugating with 1-OH from the 4-position, and $+K$ substituents with 2-NO_2 from the 5-position. Andrianov *et al.*[104] have used the equation for correlation analysis of stretching frequencies and basicity of the aromatic amino group.

Miscellaneous Applications

Applications of the Yukawa–Tsuno equation to solvolyses continue, in the hands of Yukawa and Tsuno (chlorides)[105–107] and others (tosylates[108] and *p*-nitrobenzoates[109–111]). Yukawa and Tsuno have also used it extensively in studies of the Hofmann rearrangement of *N*-halogenobenzamides,[112] and of the basicities of ketones[113] (see also Ref. 114). Other studies involving the equation include borohydride reduction of carbonyl groups,[115] ester hydrolysis in concentrated sulfuric acid,[116] hydrolysis of sulfonyl chlorides,[117] and transmission of electronic effects through polyenes.[118]

4.2.5. Critique of the Yukawa–Tsuno Equation

The Relationship between $r^{\pm}$ and ρ

In applying the Yukawa–Tsuno equation many authors have defined the significance of $r^{\pm}$.[11,30,33,42,43,45,75,78,92] The definitions always contain the essential ideas of the original definition[10] (Section 4.2.1), i.e., $r^{\pm}$ measures the extent to which cross-conjugation of substituent with reaction center stabilizes the transition state or products relative to the initial state, on a scale established by taking $r^{\pm}$ as unity or zero for certain standard processes. As we have seen, many authors have interpreted $r^{\pm}$ values for particular reactions on this basis. Usually, however, authors dealing with the Yukawa–Tsuno equation do not define ρ, except to relate it to the Hammett equation (see, however, Refs. 33 and 92).

Definitions of ρ are often rather vague. Thus "ρ measures the susceptibility of the reaction to polar effects."[1] When a more precise definition is attempted, this usually refers to the development of charge on the side-chain reaction center. Thus Grunwald and Leffler[119] write

> . . . a positive value of ρ indicates a reaction in which a partial negative charge is developed on the side chain. Similarly a negative value of ρ should correspond to the development of a partial positive charge on the side chain. The magnitude of ρ should therefore be a measure of the magnitude of the developing charge and of the extent to which it is able to interact with the substituents. This interpretation is at least qualitatively correct when ρ is far from zero. But it is doubtful when ρ is close to zero because the sign of ρ can then be inverted by a moderate change of temperature.

We now consider further ". . . the magnitude of the developing charge and of the extent to which it is able to interact with the substituents." Various interactions can be envisaged. In the absence of conjugation of reaction center and ring, these interactions must be electrostatic, involving the partial charges developed at various points in the molecule through inductive and resonance effects. When the side chain is capable of conjugation with the ring the interactions may include cross-conjugation. It is here that ρ and $r^{\pm}$ will be influenced by the same structural factors, particularly when the reaction site is not in a side chain but at a ring carbon, as in aromatic electrophilic or nucleophilic substitution. If such resonance interactions were dominant in a limited set of selected reactions, a clear relationship between ρ and $r^{\pm}$ might well be observed. However, because factors additional to cross-conjugation influence ρ, no general relationship between ρ and $r^{\pm}$ is to be expected over any wide and varied range of reactions.[33,42,45]

Thus it seems that in principle $r^{\pm}$ should be of simpler significance than ρ and a better indicator of the structures of initial (ground) states, transition states, and products as appropriate.[10] There are, however, certain difficulties, both theoretical and practical, that beset the application of the Yukawa–Tsuno equation and that reduce confidence in the significance of $r^{\pm}$ as often measured.

The Significance of $r^{\pm}$ for Two-Stage Processes

In the above discussion of $r^{\pm}$ it is assumed that the observed rate constant is for a fixed rate-determining step throughout the entire reaction series. However, the situation may be very different for reactions involving the formation of a reaction intermediate. This was recognized by several authors some time ago[47,48,120] but was recently discussed by C. D. Johnson.[121] For the scheme (4.28) the observed rate constant, k, may have

$$\mathrm{A+B} \underset{k_{-1}}{\overset{k_1}{\rightleftharpoons}} \mathrm{I} \overset{k_2}{\rightarrow} \text{Products} \tag{4.28}$$

the significance $k = k_1$, $k = k_{-1}k_2/(k_1 + k_2)$, or $k = k_1k_2/k_{-1}$ depending on the balance between specific rates of formation and breakdown of the intermediate. If $k = k_1$, we have rate-*determining* formation of the intermediate, and if this holds throughout the reaction series, $r^{\pm}$ will have a simple significance. However, if $k = k_1k_2/k_{-1} = Kk_2$, we have rate-*limiting* breakdown of intermediate to products, and ρ and $r^{\pm}$ based on k have no simple significance.

The application of the Hammett equation to the case $k = Kk_2$ gives $\rho = \rho_1 + \rho_2$, where ρ_1 is for the pre-equilibrium (K) and ρ_2 for the rate-limiting step (k_2). It has long been recognized that these considerations apply essentially to the acid-catalyzed esterification of benzoic acids in alcohols and lead to the usual explanation of the small ρ values characteristic of such reactions.[1] In a reaction series to which the Yukawa–Tsuno equation is apparently applicable, if $k = Kk_2$ it may be shown that we have (4.29), where $r_1^{\pm}$ and $r_2^{\pm}$ are for the pre-equilibrium and rate-limiting steps, respectively. Thus in such a case $r^{\pm}$ has no clear physical significance.

$$r^{\pm} = \frac{\rho_1 r_1^{\pm} + \rho_2 r_2^{\pm}}{\rho_1 + \rho_2} \tag{4.29}$$

The situation is even more complicated for the general case $k = k_1k_2/(k_{-1} + k_2)$, particularly if there is a change in the balance of the various steps through the reaction series. In this case the application of the Yukawa–Tsuno equation would be no more than a device for trying to express gross deviations from the Hammett equation, with $r^{\pm}$ being "essentially a mathematical artifact with little theoretical significance."[121]

A special case of a meaningless r^{+} value would be one in which the pre-equilibrium actually followed, say, σ or σ^0 ($r_1^{+} = 0.0$) and the rate-limiting step followed σ^{+}($r_2^{+} = 1.00$) (cf. Ref. 47). This would lead to a value of r^{+} between 0 and 1.0 [actually $r^{+} = \rho_2 r_2^{+}/(\rho_1 + \rho_2)$].

One must therefore view with skepticism $r^{\pm}$ values for reactions known to involve a pre-equilibrium followed by a rate-limiting step. These include the acid-catalyzed decomposition of ω-diazoacetophenones, the reaction that led Yukawa and Tsuno[24,25] to the equation.[10] There are, of course, many reactions for which k is known to to be simple, and this also applies to K values for many equilibria. However, there are many reactions for which the situation is not clear. Thus while there is compelling evidence for $k = k_1$ in many electrophilic aromatic substitutions involving displacement of hydrogen (absence of a primary isotope effect), the situation is not clear for many electrophilic reactions involving the displacement of other groups.[122] This may well include some of the many displacements of MR_3 to which the Yukawa–Tsuno equation has been applied (Section 4.2.2).

Further Limitations of the Yukawa–Tsuno Treatment

The Yukawa–Tsuno equation is basically concerned with detecting and expressing deviations from the appropriate simple LFER, due to variation in cross-conjugation. This, however, is only one of the factors that can produce such deviations. Exner has discussed these in detail.[4] Change of mechanism or of rate-controlling step (cf. above) can give Hammett plots showing minima or maxima, or, less extremely, curved plots without turning points. For the latter the deviations from linearity may possibly be removed by use of the Yukawa–Tsuno equation, but $r^{\pm}$ would not be of direct physical significance. Thus various solvolyses show pronounced deviations from conformity to σ^+ in a direction corresponding to $r^+ \gg 1.0$, e.g., substituted benzyl chlorides.[1,9] However, this is considered to be due to changing proportions of S_N1 and S_N2 mechanisms, and the r^+ value, if calculated, would not give information on the resonance stabilization of the benzyl cation compared with the t-cumyl cation.

It is moreover difficult to be confident of $r^{\pm}$ values that are found to be <0.3 or 1 ± 0.3, since reactions characterized by such values will be correlated fairly well by σ (or σ^0) or $\sigma^{\pm}$ respectively.[10] This situation is aggravated by there being relatively few common substituents for which $\Delta\sigma_R^+$ is substantial and reliable; the situation is better for $\Delta\sigma_R^-$.[121]

The common *para*-substituents for which $\Delta\sigma_R^+$ is considerable are NMe_2 (−0.87), NH_2 (−0.64), OMe (−0.51), OH (−0.55), and SMe (−0.60), (from σ^+ and σ values tabulated by Exner[4]). Values for these substituents are subject to various limitations. Some of the σ^+ values have not been determined directly from the defining system[7] (NMe_2, NH_2, or OH) and must be regarded as somewhat uncertain. Further, OH is subject to interference by bases (such as Lewis basic solvents) and all the above are subject to interference by protic or Lewis acids, which may influence the inherent $+K$ effect of the substituent. Sometimes dramatic irregularities occur, e.g., OMe being apparently less electron-releasing than Me,[37] but more insidious are the slight deviations that can mimic behavior requiring the Yukawa–Tsuno equation.

Quite often applications of the Yukawa–Tsuno equation involve only one or two substituents having considerable $\Delta\sigma_R^+$ values. This may be due to the difficulty of measuring the high rate constants for the compounds with the more highly electron-releasing substituents or to problems of synthesis. Hammett's warning[123] seems timely:

> It does not always seem to be recognized that effective evidence for the Yukawa–Tsuno hypothesis is obtainable only from correlations that include more than one strong through-resonating substituent, and for which the value of r is neither close to unity nor close to zero.

The equation for nucleophilic reactivities[11] (4.19 or 4.22) seems less subject to the above difficulties.[121] A good spread of $\Delta\sigma_R^-$ values is available for common substituents, and these are probably less liable to interference by protic or Lewis acids, or Lewis bases. This equation has, on the whole, found less application than that for electrophilic reactions. It might, however, be a better subject for investigating the general utility and validity of the Yukawa–Tsuno approach.

It may reasonably be held that the Yukawa–Tsuno equation presents problems because of uncertainties in the various sigma scales involved. We drew attention above to the poor reliability of σ^+ values for various important substituents. There are, moreover, certain difficulties with σ^- values.[32,43] These may be based either on phenol or on anilinium ion ionization. The values do not agree completely, and values from the two sources should strictly never be mixed in a given correlation. While the σ values used in the 1959 form of the Yukawa–Tsuno equation are in general more secure than the values of any other polar substituent constant,[8,42] values for particular substituents may be uncertain and vary with source (Exner[4] and Chapter 10 of this volume). The σ^0 values available for the 1966 form are perhaps now more soundly based than they were ten years ago,[29,124,125] but Wepster and his colleagues[126,127] have recently cast doubts on methods used to determine σ^0 values. The σ^n values of van Bekkum *et al.*[9] are sometimes used instead of σ^0,[18,92,127–129] but Bott and Eaborn's[42] *caveat* should be recalled (Section 4.2.2). We should add that for all sigma scales, and not just σ^+ (see above) there is possible solvent dependence of sigma values for certain substituents, i.e., differences as between water and aqueous organic, protic organic, and aprotic solvents.[99,130]

The above limitations of sigma scales apply, of course, to some extent to the practice of simple correlation analysis, but multiparameter equations are more vulnerable, since at least two of the scales are involved and great significance is often attributed to relatively small deviations from simple regressions.

Difficulties for Electrophilic Aromatic Substitution

In 1963 Stock and Brown[52] discussed the Yukawa–Tsuno equation (Section 4.2.2). They appreciated its theoretical aims and the reality of at least some of the phenomena with which it had been designed to deal. Nevertheless, they wrote "It is not any easy matter to assess the real merits of the Yukawa–Tsuno correlations." (Thirteen years later this still seems to be true.) For electrophilic side-chain reactions they conceded that the equation at least yielded significant improvements in correlations. They were not convinced of this for substitution on the ring and believed that within the limits of precision of the data and other possible sources of discrepancies, simple treatment through σ^+ was successful.

Johnson[58,131] has recently raised again some of the issues with which Stock and Brown dealt. These particularly concern the "extended selectivity treatment."[52] This procedure is based on σ^+ values and deals with the behavior of a given substituent in several reactions. The logarithms of the partial rate factors for that substituent in a series of electrophilic substitutions are plotted against the corresponding ρ values. A particular example considered by Johnson is substitution at the β position of thiophen,[132] regarded as benzene in which the "substituent" is S replacing CH=CH. A fairly good straight line is obtained of slope −0.52, which is considered to be the σ^+ value for S replacing CH=CH. The straight line is taken to imply a constancy of resonance effect, irrespective of the electron demand in the transition state as measured by ρ. Johnson[58] refers also to other examples,[52] including substitution in the *para*-position of methoxybenzene. He concludes[131] that ". . . variation of r from unity is largely due to experimental error."

It should, however, be noted that there is considerable scatter of points about the extended selectivity lines and presumably this too must be attributed to experimental error. Further, the ρ values on which this and similar plots are based are presumably calculated *on the assumption that* σ^+ *correlation is applicable*. The argument is therefore somewhat circular. Thus the ρ value for bromination in aqueous acetic acid at 25°C is taken as −12.1. The Yukawa–Tsuno equation with the same data gives −9.49. For chlorination the corresponding values are −10.0 and −7.79. Examination of Stock and Brown's plots and text,[52] however, reveals that certain points, including some for *meta*-substituents, are regarded, for various reasons, as not very reliable. Less weight has apparently been given to these in drawing straight lines by eye. Least-squares treatment involving the Yukawa–Tsuno equation considers all points as equally reliable. This is a good illustration of the difficulties with which this area is fraught.

Johnson[131] also gives an example (from Stock and Brown[52]) which is considered to show a genuine variation of cross-conjugation. This is the extended selectivity treatment of biphenyl, for which the plot is a curve. The variable conjugation effect is attributed to twisting one of the phenyl rings out of plane of the other: The more negative the value of ρ, the more the rings adopt a coplanar conformation, with an increase in resonance effect across the rings. The variable resonance effect disappears when the two rings are held coplanar, as in fluorene. There are a few other examples of curvilinear extended selectivity treatment plots.[58]

Final Comments

We finally direct or redirect the attention of the reader to certain recent work bearing on the status of the Yukawa–Tsuno equation. The differing opinions of Olah[55–57] and of Johnson[58,121] on the merits of the equation are leading to a broader conflict on the significance of ρ values and the reactivity–selectivity principle.[133–135] Wepster[127,136] has made various comments on the Yukawa–Tsuno equation, which have been discussed by Wold.[137] Bancroft and Howe[138] have derived and discussed a generalized relationship that incorporates many of the usual free energy relationships, including the Yukawa–Tsuno equation.

4.3. The Swain–Lupton Treatment

4.3.1. The Work of Swain and Lupton, 1968

Swain and Lupton[8] strongly criticized the proliferation of scales of polar substituent constants and attempted a simplification by attributing to each substituent just two basic electronic characteristics: a field† constant, $\mathfrak{F}$, and a resonance constant, $\mathfrak{R}$. If this procedure is valid, the numerous sigma scales are not independent but each is a linear combination of $\mathfrak{F}$ and $\mathfrak{R}$ as in equation (4.30), where the field, f, and resonance, r, weighting

$$\sigma = f\mathfrak{F} + r\mathfrak{R} \tag{4.30}$$

factors give the blend of field and resonance effects for the system used to define the particular σ scale.§ Any one of the various σ scales should then be expressible as a linear combination of any other two, i.e., the relation (4.31) should hold. The authors illustrated this point by correlating various σ scales with σ_m and σ_p based on benzoic acid ionization. This was actually done by use of equation (4.31) modified by an intercept term, i, as in equation (4.32), to avoid giving the unsubstituted system excessive weight

†This term includes the "pure" field (direct) effect and the inductive (through-bond) effect.
§In Section 4.3 σ is often used in the general sense of polar substituent constant, and not just for the benzoic acid scale.

The treatment has undoubtedly found most application† in the correlation analysis of nmr data. This topic is dealt with in detail in Chapter 8, but the role of the Swain–Lupton equation is treated briefly below. There are also applications to optical spectroscopy. $\mathfrak{F}$ and $\mathfrak{R}$ are held by some[144] to be peculiarly suitable for the correlation analysis of biological activity data. Applications to chemical reactivity seem restricted to a few instances in some of the main areas, e.g., acidic and basic strength, electrophilic or nucleophilic substitution (ring or side chain), and electrophilic addition. There is a recent treatment of the *ortho*-effect that involves $\mathfrak{F}$ and $\mathfrak{R}$.[147]

In the space available only a few selected examples can be discussed.

The Correlation Analysis of nmr Data

Workers in the field of nmr seem more attracted to multiparameter treatments than are physical organic chemists in general, because of the undoubted difficulty of obtaining satisfactory correlations of nmr data with any one of the individual scales of σ values[148] (see Chapter 8).

Since 1971 W. F. Reynolds and his co-workers[149–150] have used the Swain–Lupton treatment extensively in correlation analysis of chemical shifts and coupling constants of ^1H or ^{13}C of various systems. The values of f and r are often discussed in relation to the various factors contributing to the effects of substituents in nmr. Gronowitz and his co-workers[151,152] (using aromatic heterocycles) and Caplin[153,154] have also used the treatment extensively. Wiley and Miller[155] applied the treatment to the CH_3 pmr of m- or p-$XC_6H_4N^+(CH_3)_3$; %$\mathfrak{R}$ values were calculated for m- and for p-X, and also for substituents in other nmr systems, but their physical meaning was questioned, e.g., sometimes %$\mathfrak{R}$ values are the same for m- as for p-substituents. Some others have been less enthusiastic as regards nmr, e.g., Dayal *et al.*[156] studied ^{19}F shielding in various systems of the type p-FC_6H_4·G·C_6H_4X-p and concluded that these data were not always well correlated by $\mathfrak{F}$ and $\mathfrak{R}$; the dual substituent-parameter equation (Section 4.4) with various σ_R scales was more satisfactory.

Applications to Optical Spectroscopy

D. W. Boykin and his colleagues[157–159] have applied the Swain–Lupton treatment extensively to ν_{CO} data of aromatic $\alpha\beta$-unsaturated ketones, with somewhat mixed success. K. Spaargaren *et al.*,[160] however, have pointed out that these workers have not always appreciated that substituents in the various ring positions must be dealt with separately in a Swain–Lupton treatment. Re-analysis of the data in the proper way is rather

†Most applications to date use the original $\mathfrak{F}$ scale[8] and not the scale as modified by Hansch *et al.*[144]

more successful.[160] Indeed, Spaargaren *et al.* regard the Swain–Lupton approach as "most elegant"[160] and apply it to their own results on ν_{CO} and rotational barriers in amides.[161] One of Boykin's papers[158] shows that when ν_{CO} is fairly well correlated by σ^+, the Swain–Lupton treatment gives worse results (cf. Section 4.3.1).

D. A. Thornton and his colleagues[162–163] have made some use of $\mathfrak{F}$ and $\mathfrak{R}$ for ir data on metal complexes with β-keto-enols, *N*-aryl-salicylaldimines, and other aromatic ligands. Laurence and Wojtkowiak[164] give numerous correlations of ir frequencies with various substituent constants. The Swain–Lupton treatment was regarded as inferior to Taft's dual substituent-parameter equation (Section 4.4).

Acid–Base Behavior

The Swain–Lupton treatment has been applied to various acid–base systems with mixed success.

Zalewski has applied the treatment to the protonation of substituted benzoic and salicylic acids[165] and $\alpha\beta$-unsaturated alicyclic ketones.[166] For the first two, correlation with σ_m and σ_p is also fairly satisfactory—cf. earlier work on the protonation of benzoic acids,[167] benzaldehydes,[168] and acetophenones,[169] for which σ_p^+ was better than σ_p.

Bergon and Calmon[170] measured the "apparent," the ketonic, and the enolic acidity constants of benzoylacetones. Swain–Lupton treatment found resonance effects to be slightly more important than in benzoic acid ionization.

Walba and Ruiz-Velasco[171] measured the pK_a values of 5-(or 6)†-substituted benzimidazoles (IX), and also the basicity constants pK^+ for the formation of the cations (X).

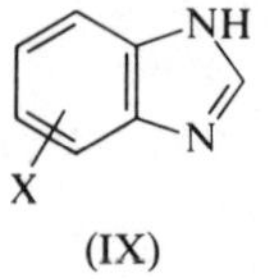

(IX) 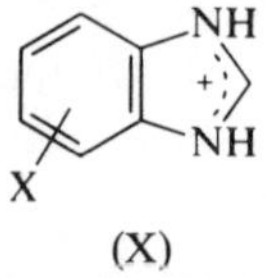 (X)

Values of pK_a and of pK^+ gave good correlations on σ_I and σ_R^0, or $\mathfrak{F}$ and $\mathfrak{R}$. %$\mathfrak{R}$ values were calculated and were about halfway between those for σ_m and σ_p^0, in accordance with the fact that in the relevant cation (X), or anion (from IX), the substituent is *meta* to one N and *para* to the other. The NH_2 substituent deviated, indicating that $\mathfrak{R}$ is not a good measure of the resonance effect of NH_2 in a system markedly different from the defining system for $\mathfrak{R}$.

†Tautomerism makes these positions equivalent.

According to Paleček and Hlavatý[172] the pK_a values of 4-substituted quinuclidinium ions (in various aqueous organic solvents) are correlated well with σ_I, but less well with $\mathfrak{F}$. The authors conclude that the "self-consistent" $\mathfrak{F}$ values based on σ_m and σ_p are less precisely applicable to saturated systems than σ_I values which are related more closely to aliphatic and alicyclic systems.

Zoltewicz and Cross[173] measured deprotonation rates (by OD^- in D_2O) at the 2- or 6-positions of 3-substituted 1-methylpyridinium ions. Taft's dual substituent-parameter equation was used with various σ_R scales; σ_R^- was best, but others were not decidedly worse. However the Swain–Lupton treatment was markedly less successful.

Electrophilic Substitution and Addition

The Swain–Lupton treatment has not been much applied to aromatic electrophilic substitution. Eaborn and Fischer[174] studied substituent effects on the detritiation of 1- and 2-tritionaphthalenes. The dual substituent-parameter equation (σ_R^0 or σ_R^+) and Swain–Lupton treatment show comparable success, but the series were not very large or the substituents particularly discriminating.

Partial rate-factors for the attack on anisole, toluene, or chlorobenzene as substrate by $p\text{-}XC_6H_4CO_2\cdot$ ($X = NO_2$, H, or Me) were measured by Kurz and Pellegrini.[175] Brown and Okamoto's[7] σ^+ values and Swain and Lupton's "revised σ^+ values" [i.e., from equation (4.42)] gave no clear difference in performance.

Various papers on side-chain reactions mention the Swain–Lupton treatment, but do not usually apply it to any great effect.

For the pyrolysis of 1,4-diaryl-1,4-dimethyl-2-tetrazenes in cumene (Nelsen and Heath[176]), σ^+ is more satisfactory than σ, but the resonance effect appeared to exceed that involved in σ^+. This was confirmed by a Swain–Lupton treatment, which gave good correlation and a $\%\mathfrak{R}$ value of 83%, cf. 66% for σ_p^+.

Trahanovsky *et al.*[177] studied the rate of oxidation of phenylacetic acids by cerium(IV) nitrate and nitric acid in aqueous acetonitrile. The reaction is complex: Benzyl radicals are formed as intermediates and σ^+ is required in correlations. Brown and Okamoto's[7] values were more satisfactory than Swain and Lupton's revised σ^+ values (cf. above).

Various addition reactions to which the Swain–Lupton treatment has been applied, usually without striking results, include bromination of cycloalkenes (Dubois and Hegarty,[77] see Section 4.2.4), addition of CCl_2 to alkenes (Sadler[178]), and formation of BF_3–aldehyde complexes (Gal *et al.*[179]).

The Swain–Lupton treatment has been used in work on solvolysis by D. S. Noyce[180–183] *et al.*, who have studied substituent effects in heterocyclic systems and their relationship to those in substituted benzenes. Modifications of the Dewar–Grisdale[22] approach are largely involved, but the Swain–Lupton treatment is also used. Thus rates of solvolysis of 5- or 6-X-1-(2-benzofuryl)ethyl *p*-nitrobenzoates[180] were separately correlated with $\mathfrak{F}$ and $\mathfrak{R}$; $\%\mathfrak{R}$ values of 44% (5-X) and 67% (6-X) were found. The latter accounts for good correlation with σ_p^+ ($\%\mathfrak{R}$, 66%), while the former shows why there is poor correlation with σ_m^+ ($\%\mathfrak{R}$, 33%). In the solvolysis of 4-, 5- or 6-X-2(2-pyridyl)-2-chloropropanes[181] correlation with σ^+ is satisfactory for 4- or 5-X systems, but 6-X required Swain–Lupton treatment; $\%\mathfrak{R} = 60\%$ for what is formally a *meta*-relationship of substituent to reaction site. Substituted 1-(4-thiazolyl)ethyl chlorides[182] and substituted l-(1-methylimidazolyl)ethyl *p*-nitrobenzoates[183] also required Swain–Lupton treatment.

Nucleophilic Reactions

There is little to say here. In the methoxide-catalyzed methanolysis of aryl acetates and carbonates in methanol (C. G. Mitton *et al.*[184]) there is no LFER correlation with the ionization constants of the phenols. A Swain–Lupton treatment finds a linear relationship to $(0.8\mathfrak{F} + \mathfrak{R})$, corresponding to $\%\mathfrak{R} = 46\%$, cf. 56% for σ^-.[8]

Swain[185] has himself used the treatment to illuminate the mechanism of displacements on ArN_2^+ in solution in the absence of strong bases, reducing agents, and light. The effect of substituents on the rate of dediazoniation showed gross deviations from any single-parameter treatment.† However equations (4.49) and (4.50) hold well and were interpreted in

$$\log (k/k_H)_m = -2.74\mathfrak{F} - 3.18\mathfrak{R} + 0.27 \tag{4.49}$$

$$\log (k/k_H)_p = -2.60\mathfrak{F} + 5.08\mathfrak{R} - 0.25 \tag{4.50}$$

terms of rate-determining cleavage to a singlet phenyl cation intermediate (for which other evidence was also adduced). The negative coefficient of $\mathfrak{R}$ in the *meta* equation signifies the stabilization of the transition state by movement of electrons from $+K$ groups to the *ortho* carbon atoms. The positive coefficient of $\mathfrak{R}$ in the *para* equation means that $+K$ groups stabilize the initial state through increasing the double bond character of the C—N bond, and have little effect on the transition state.

†For the Hammett equation, the correlation coefficient as calculated by Swain and Lupton is "imaginary" (see footnote p. 145), i.e., C^2 is negative.

Biological Activity

In the correlation analysis of a given biological activity for a series of compounds, it is often impossible to classify substituents as *ortho*, *meta*, or *para*, with respect to a certain point in the molecule believed responsible for the biological activity. The position-independent parameters $\mathfrak{F}$ and $\mathfrak{R}$ thus have an obvious advantage over the position-dependent σ_m and σ_p. Further, as pointed out by Hansch *et al.*,[144] $\mathfrak{F}$ and $\mathfrak{R}$ values for a series of substituents tend to vary in a mutually orthogonal manner, whereas σ_m and σ_p are often fairly highly correlated. $\mathfrak{F}$ and $\mathfrak{R}$ are therefore much more suitable than σ_m and σ_p for inclusion with other structural parameters in stepwise multiple regression and other regression procedures. (See Chapter 9.)

The Ortho-Effect[21,186]

Fujita and Nishioka[147] recently devised a treatment of the *ortho*-effect based on $\mathfrak{F}$ and $\mathfrak{R}$. The authors seek to include *ortho*-substituted systems in the same correlation equations as *meta*- and *para*-substituted systems. The total *ortho*-substituent effect is assumed to be composed of an "ordinary polar effect" and "proximity effects." The former is defined as equal to that of *para*-substituents and thus $\sigma_o \equiv \sigma_p$, $\sigma_o^+ \equiv \sigma_p^+$, etc. The "proximity polar effect" is considered proportional to the $\mathfrak{F}$ parameter, while the steric effect is given by the Taft E_s constant[6] for the substituent when in an *aliphatic* system.[186] The treatment cannot parameterize hydrogen-bonding and other specific intramolecular interactions. The basic equation is (4.51)

$$\log k_{o,m,p} = \rho\sigma_{o,m,p} + \delta E_s^o + f\mathfrak{F}_o + C \tag{4.51}$$

with the E_s and $\mathfrak{F}$ terms being necessarily zero for *meta*- and *para*-substituents. $\sigma_{o,m,p}$ implies a choice of sigma scale appropriate to the system.

The authors applied the equation to numerous data sets from the literature and to their own results on the alkaline hydrolysis of substituted phenyl esters.[187] The correlation equations often appear to be statistically very satisfactory, although detailed examination will be necessary to establish the validity of the treatment. The implied equality of resonance effects for *ortho*- and *para*-substituents may be adversely criticized, particularly as special σ_o values have sometimes to be used, evidently to deal with the resonance effects *not* being equal in certain cases. It should also be pointed out (as the authors do) that correlation with a σ scale and $\mathfrak{F}$ is equivalent to correlation with $\mathfrak{F}$ and $\mathfrak{R}$. Hence the treatment could be reformulated as correlation on $\mathfrak{F}$, $\mathfrak{R}$, and a steric term. The treatment is also, of course, related to Charton's[21,188–189] treatment of the *ortho*-effect in terms of σ_I, σ_R, and υ (a steric parameter). However, the attempt to include *ortho*-, *meta*-, and *para*-systems in one equation is an interesting and novel (albeit possibly restricting) feature.

4.3.3. Critique of the Swain–Lupton Treatment

The preceding section shows that the treatment has won some success as an empirical correlation procedure, although it is occasionally stated to be inferior to other types of multiparameter treatment or to accomplish little more than a one-parameter treatment. However, to the extent that σ scales are related to $\mathfrak{F}$ and $\mathfrak{R}$, the same measure of success would be achieved by any combination of σ values expressing polar effects from *meta*- and *para*-positions, e.g., σ_m and σ_p, σ_m^0 and σ_p^0, etc. The claim is made that the treatment is the most effective means of separating field (inductive) and resonance effects, and the significance of $\%\mathfrak{F}$ and $\%\mathfrak{R}$ depends on this. This aspect has been criticized.

$\mathfrak{F}$ values are basically related to Taft's σ_I values through the use of the bicyclo-octane model (Section 4.3.1). However, the $\mathfrak{F}$ values for the 42 substituents treated by Swain and Lupton[8] (and the many more included by Hansch *et al.*[144]) are immediately based on the correlation of the bicyclo-octane data with σ_m and σ_p. Equally, of course, it may be said that σ_I values emerge from a consideration of various kinds of data: aliphatic and aromatic reactivities, as well as alicyclic.

The values of $\mathfrak{R}$ are more widely criticized. Several authors have pointed out[190–193] that $\mathfrak{R}$ values depend critically on the assumption that NMe_3^+ has no resonance effect (Section 4.3.1). This group is, however, isoelectronic with Bu^t, which is known to be a $+K$ group, and there seems to be no reason for excluding *a priori* this possibility for NMe_3^+. Evidence of the $+K$ character of NMe_3^+ comes *inter alia* from the ir spectra of *para*-[194] and *meta*-disubstituted[195] benzenes, from ^{19}F nmr study of 4-fluorotrimethylanilinium ion[196] and 4-fluoro-4′-trimethylammonio-*trans*-stilbene,[193] from the σ_p^+ value for NMe_3^+,[52] and from pK_a values.[194] It may in any case be questioned whether it is wise to base a treatment of the effects of substituents, which are for the most part dipolar, on the unipole NMe_3^+. There is the further point that $\sigma_p = +0.82$ for NMe_3^+ is given an uncertainty of $\pm{>}0.2$ by McDaniel and Brown.[143] The extent to which the treatment in fact separates field (inductive) and resonance effects must be considered doubtful.[197]

There remains the question of whether the treatment really copes with enhanced resonance effects, i.e., with the reactions that require σ^+ or σ^- in a one-parameter equation. $\mathfrak{F}$ and $\mathfrak{R}$ are rather less successful in expressing these scales than they are in dealing with scales involving less extreme resonance effects [see equations (4.40) to (4.43)]. Exner[4] pointed out that the reproduction of σ^+ and σ^- with a standard deviation of as large as 0.18 cannot be considered satisfactory. Many of Swain and Lupton's "revised" σ^+ and σ^- values are unacceptably far from the experimental values based on the defining reactions, e.g., F, $\sigma_p^+(\mathrm{SL}) = -0.247$, $\sigma_p^+(\mathrm{obs}) = -0.07$; OMe, $\sigma_p^+(\mathrm{SL}) = -0.648$, $\sigma_p^+(\mathrm{obs}) = -0.778$; SMe, $\sigma_p^+(\mathrm{SL}) = -0.164$, $\sigma_p^+(\mathrm{obs}) = -0.604$. There have been few really successful applications to systems involving enhanced resonance effects. For such reactivities the Swain–Lupton treatment is no more successful than any other approach based on a fixed scale of resonance effects. This matter is referred to again in Section 4.4.4. Hansen and Hepler[198] have a treatment of substituent effects that is said to be an extension of that of Swain and Lupton.

4.4. Taft's Dual Substituent-Parameter Equation

4.4.1. Introduction

Taft and his colleagues now write the equation in the form[32]

$$P^i = \rho_I^i \sigma_I + \rho_R^i \sigma_R \tag{4.52}$$

P^i is the substituent-influenced property, expressed relative to hydrogen as substituent, e.g., for chemical reactivity $P^i = \log (k/k^0)_i$ or $\log (K/K^0)_i$, for spectroscopic frequencies $P^i = (\nu - \nu^0)_i$, etc. The right-hand side of the equation attributes the effect of substituents in ring position i to a linear combination of a polar (inductive) effect ($I^i = \rho_I^i \sigma_I$) and a resonance effect ($R^i = \rho_R^i \sigma_R$). The susceptibility constants ρ_I^i and ρ_R^i define the blend of polar and resonance effects characteristic of the influence of substituents in the position i on the property in question under given conditions, the blending constant λ being defined as ρ_R^i/ρ_I^i. The substituent constants σ_I and σ_R are position-dependent, and the Taft equation (like the Swain–Lupton treatment) is applied separately to the effects of *meta*- and *para*-substituents. The σ_I scale has long been held to be of wide general applicability, but the possibility of defining a unique, widely applicable σ_R scale is an important issue and has been much discussed.

Equation (4.52) had its origin as long ago as *ca.* 1956. For about 20 years the separation of inductive and resonance effects has been subjected to continuous scrutiny and development, largely by Taft himself and his colleagues. In this respect the equation differs in some measure from the other multiparameter equations and requires treatment in a rather different way. We shall sketch the origins and development of the equation and the critical assessment thereof, which Ehrenson *et al.* have presented.[32] We shall not attempt a survey of its applications by other authors, although we shall refer to particularly notable work.

4.4.2. Origins and Development

The origins may be seen in Taft's chapter in the book edited by M. S. Newman.[6] In one section ($V2c$) σ^* values for XCH_2 groups derived from Taft's analysis of aliphatic ester reactions[6,186] were correlated for several substituents with the σ' values of Roberts and Moreland[141] for X as a 4-substituent in the reactions of bicyclo[2.2.2.]octane-1-derivatives, viz., $\sigma'/\sigma^* = 0.45$. Taft calculated further σ' values on this basis. By assuming that the behavior of 4-X in bicyclo-octane is a good model for inductive effects in substituted benzenes and that σ' is on the same numerical scale as σ_m and σ_p, Taft evaluated resonance contributions of *m*-X and *p*-X as $(\sigma_m - \sigma')$ and $(\sigma_p - \sigma')$, respectively. There is thus the additional assumption that inductive effects operate equally from the *meta*- and *para*-positions. The resonance contributions were discussed in structural terms but not denoted by a characteristic symbol.

The symbols σ_I and σ_R first occur† in Taft's papers in 1957,[199] replacing σ' (as calculated from σ^*) and $(\sigma_p - \sigma')$. Even then σ_R^+ and σ_R^- were distinguished as enhanced resonance effects in systems requiring σ^+ and σ^-, respectively, but discussion was mainly in terms of σ_R based on benzoic acid ionization. This paper[199] was devoted mainly to correlation analysis of substituent effects on ^{19}F nmr shielding in fluorobenzenes (Gutowsky's results[200–201]): δ_m^F was found to be best correlated by σ_I, and δ_p^F by a combination of the form $\alpha\sigma_I + \beta\sigma_R$, rather than by σ_m and σ_p, respectively. For δ_p^F, $\alpha \ll \beta$, and thus it was recognized even at this early date that susceptibilities of a property to inductive and resonance effects need not necessarily be the same, even though they are defined as equal in the reference process, i.e., benzoic acid ionization, $\sigma_p = \sigma_I + \sigma_R$. The connection of σ_I and σ_R with ^{19}F nmr was thus established early on, and the correlations were regarded as proof that "σ_I and σ_R are quantitative and independent measures of the electron-withdrawing effects . . . through inductive and resonance interactions, respectively."[199]

Taft and Lewis[13] wrote equations (4.53) and (4.54) for the ionization of benzoic acid, i.e., the relayed resonance effect from the *meta*-

$$\log(K^p/K_0) = \sigma_I + \sigma_R \tag{4.53}$$

$$\log(K^m/K_0) = \sigma_I + \tfrac{1}{3}\sigma_R \tag{4.54}$$

position is $\frac{1}{3}$ of that from the *para*-position. These equations were combined to give equation (4.55), and the plot of unit slope implied was

$$\tfrac{3}{2}[\log(K^m/K_0) - \tfrac{1}{3}\log(K^p/K_0)] = \sigma_I \tag{4.55}$$

†These symbols are said[43] to have been devised in a discussion among H. C. Brown, N. C. Deno, H. H. Jaffé, R. W. Taft, and others at the Reaction Mechanisms conference, Swarthmore, Pennsylvania, September 1956.

convincingly displayed. This equation was generalized in the form (4.56), with the assumption that $\rho_I^m = \rho_I^p = \rho_I$. Equation (4.56) was shown to apply

$$\tfrac{3}{2}[\log (k^m/k_0) - \tfrac{1}{3}\log (k^p/k_0)] = \rho_I\sigma_I \tag{4.56}$$

to a considerable number of reaction series. However, in certain reaction series involving enhanced resonance effects, highly conjugating substituents required a *meta*-position resonance factor of only $\frac{1}{10}$, i.e., the equation was generalized to the form (4.57) with $\alpha = \frac{1}{3}$ when the substituent is

$$\frac{1}{1-\alpha}\left(\log\frac{k^m}{k_0} - \alpha \log\frac{k^p}{k_0}\right) = \rho_I\sigma_I \tag{4.57}$$

not capable of conjugation with the reaction center, but with $\alpha = \frac{1}{10}$ when it is.

In the above treatment ρ_R is not mentioned explicitly, but equation (4.57) implies $\rho_R^p = \rho_R^m$, otherwise the resonance effects would not disappear in the subtraction. Further ρ_I is usually very close to ρ found from the Hammett equation (with the proper selection of substituents[9]) and the implication is clearly that $\rho_I = \rho_R = \rho$. It was pointed out that this condition would not necessarily hold absolutely and indeed $\rho_I^p \neq \rho_R^p$ had already been shown for ^{19}F chemical shifts.[199]

In a later paper[14] Taft and Lewis wrote the basic equations as (4.58) and (4.59). R^m and R^p were found for a variety of substituents in various

$$\log (k^m/k_0) = I + R^m \tag{4.58}$$

$$\log (k^p/k_0) = I + R^p \tag{4.59}$$

reactions, I being taken as $\rho_I\sigma_I$ [see equation (4.56)]. Effective σ_R values, denoted $\bar{\sigma}_R$, were calculated as R/ρ_I for purposes of comparison. While $\bar{\sigma}_R^m$ values for a given substituent were reasonably constant within a selected group of substituents, corresponding $\bar{\sigma}_R^p$ values varied greatly with the reaction. It was inferred that a widely applicable and precise σ_R scale could not be devised. Thus there was the implication that various σ_R scales would be necessary for comprehensive correlation analysis.

Taft *et al.*[15] continued to examine R. For systems in which the reaction center was not conjugated with the ring, a σ_R^0 scale was defined. Values of σ_R^0 were combined with σ_I to give a σ^0 scale analogous to the σ^n scale of van Bekkum *et al.*[9] New measurements of ^{19}F nmr spectra for an extensive

$$\sigma_m^0 = \sigma_I + 0.5\sigma_R^0 \tag{4.60}$$

$$\sigma_p^0 = \sigma_I + \sigma_R^0 \tag{4.61}$$

series of *meta*- and *para*-substituted fluorobenzenes in dilute solution in carbon tetrachloride were correlated with σ_I and σ_R^0; the findings were very similar to those based on Gutowsky's results.[199–201]

The situation was reviewed by Taft in 1960.[16] A few additional matters were presented. Thus "aromatic" σ_I values were derived by considering only benzene side-chain reactivities, by using a modification (by Roberts and Jaffé[202]) of a method devised by Taft and Lewis.[14] These σ_I values agree fairly well with the "aliphatic" σ_I values.

In 1963 the correlation analysis of ^{19}F nmr data by means of σ_I and σ_R^0 was pursued in great detail.[203–204] By this time Taft *et al.* regarded the correlation of δ_m^F with σ_I and of δ_p^F with σ_I and σ_R^0 as so well established that it could be used to investigate "the effect of solvent on the inductive order"[203] and "the influence of the structure and solvent on resonance effects."[204] Thus these correlations were used to measure new σ_I and σ_R^0 values, applicable under certain conditions, and to detect solvent–substituent interactions, e.g., hydrogen bonding, or carbonyl addition.

The first suggestion that an "unconstrained" equation [i.e., (4.52)] should be used generally, was made by Ehrenson.[17,205,206] The unconstrained equation was applied (1968[207]) to substituent effects in the naphthalene series. For the various reactivities ρ_I and ρ_R were found for the different combinations of reaction center/substituent positions. Most recently Ehrenson *et al.*[32] have studied the unconstrained equation in great detail in relation to benzene reactivities. Important matters from both articles are dealt with later.

4.4.3. *Work Related to the Dual Substituent-Parameter Equation*

The correlation analysis of ir data is of particular interest.[208] Katritzky, Topsom, and their colleagues have correlated intensities of the ν_{16} ring-stretching bands of mono- and di-substituted benzenes with σ_R^0 values of the substituents. Thus data for monosubstituted benzenes[209] follow equation (4.62). This equation and analogous equations for

$$A_{\text{mono}} = 17{,}600(\sigma_R^0)^2 + 100 \tag{4.62}$$

disubstituted benzenes[194,195,210] may be used to find new σ_R^0 values and to investigate substituent interactions. Schmid[211] has shown that ν_{CH} intensities in substituted benzenes obey equations of the form (4.63), where

$$A = a\sigma_I^2 - b\sigma_I + e \tag{4.63}$$

a, *b*, and *e* are constants for a particular substituent pattern. New σ_I values can therefore be measured.

Thus σ_I and σ_R values as tabulated and used today have been derived by a variety of methods involving studies of chemical reactivity (aromatic and aliphatic), ^{19}F nmr, and ir absorption. The values have often been "averaged," made "self-consistent," or subjected to minor adjustments for various reasons, so that the values have become detached from their origins and their experimental basis is obscured.

We now mention briefly the extensive applications of the dual substituent-parameter equation by Charton,[21,188–191] who uses the equation in the form (4.64), where Q is the property to be correlated ($\log k$, ν,

$$Q = \alpha\sigma_{I,X} + \beta\sigma_{R,X} + h \tag{4.64}$$

etc.). Thus the property is not expressed relative to hydrogen as substituent [cf. equation (4.52)], and h is introduced as the appropriate intercept term (see footnote † on p. 162). Charton has applied the equation to numerous systems, both aromatic and nonaromatic[191] unsaturated systems. His work on the *ortho*-effect, in which equation (4.64) may acquire an additional steric term, merits particular attention.[21,188,189]

4.4.4. Various Pertinent Issues

The Equality of Inductive Effects from the Meta- and Para-Positions. The Scaling of σ_I

An assumption of the above kind underlay Taft's earliest discussion of resonance effects,[6] equations such as (4.53) and (4.54), and the Taft–Lewis[13] equations (4.55)–(4.57), where a unique ρ_I value is defined by assuming that $\rho_I^m = \rho_I^p = \rho_I$. It was realized, however, that this situation did not apply to ^{19}F nmr.[15,199]

This assumption, which seemed rather important in the late 1950s and in the 1960s, now seems less important and indeed to be virtually abandoned. This is because Taft and his colleagues now favor the unconstrained equation (4.52). The *meta* and *para* systems are not combined in one equation and the question of whether $\rho_I^m/\rho_I^p = 1.00$ or any other value is a matter for investigation rather than assumption. (The importance of assuming a value for the *meta*-resonance factor of $\frac{1}{3}$ or $\frac{1}{10}$ or $\frac{1}{2}$ has likewise declined.) Nevertheless the assumed equality of inductive effects has had an important influence on the development of the subject and merits discussion.

The assumption certainly seems contrary to the most simple models for the so-called inductive effect.[212] Thus whether a field model or a through-σ-bonds model is invoked, the inductive effect should be somewhat more powerful from the *meta*-position than from the *para*-position. On the other hand, if the inductive effect involves polarization of the π electrons this can be envisaged in a way that implies that the effect should operate more powerfully from the *para*-position. The resultant of all this might be that the inductive effect would in practice show only slight dependence on position.

The assumption was discussed around 1963 by several authors,[17,43,205,213,214] but received particular attention from Exner in 1966.[215] Exner measured the ionization constants (50% ethanol or 80% 2-methoxyethanol) of a large number of *meta*- or *para*-substituted benzoic acids, including many for which ordinary conjugative substituent effects could be excluded (although not necessarily hyperconjugative effects[214]), e.g., CH_2X, CF_3, CCl_3. For acids with such substituents, a plot of $\log(K^p/K^0)$ vs. $\log(K^m/K^0)$ gave a straight line of slope 1.14. Exner concluded that the inductive effect is stronger from the *para*-position by this factor. Several $-K$ substituents, e.g., NO_2, CN, SO_2Hal, surprisingly conformed to the same line, and Exner suggested that the resonance effects of such substituents (in the absence of cross-conjugation with a $+K$ substituent) were much smaller than commonly supposed. He therefore rewrote the Taft equations in the forms (4.65) and (4.66), with $\lambda = 1.14$ and $\sigma_R = 0$ for the classes of substituent referred to above. For these a plot of σ_m and $\sigma_p/1.14$ vs. Taft's σ_I values gave a line of slope 1.10. He therefore suggested that the usual σ_I values were out of scale with σ_m and σ_p by this factor, and should be appropriately multiplied. It was argued that such a correction was needed to introduce the π-inductive component, which would be absent in the saturated systems used to define the σ_I scale. Exner presented revised σ_I and σ_R values on the basis of equation (4.65) and the scale factor. The values conformed closely to equation (4.66) with

$$\sigma_p = \lambda\sigma_I + \sigma_R \tag{4.65}$$

$$\sigma_m = \sigma_I + \alpha\sigma_R \tag{4.66}$$

$\alpha = \frac{1}{3}$; cf. Taft and Lewis.[13] For $-K$ substituents, σ_R^- values were also derived. Exner's σ_I and σ_R values are tabulated in Chapter 10.

Many physical organic chemists have found $\sigma_R = 0$ for NO_2 difficult to accept, although Exner[215] mentioned several lines of evidence that suggest that in the absence of a $+K$ *para*-substituent, there is little conjugation between NO_2 and the benzene ring. Perhaps the most difficult point concerns what is usually described as the steric inhibition of the resonance of NO_2 by flanking alkyl groups. Thus 3,5-dimethyl-4-nitrobenzoic acid is weaker than *p*-nitrobenzoic acid[216] by an amount (ΔpK_a) corresponding fairly well to the removal, by twisting, of the resonance effect of the NO_2 as normally envisaged. However recent calculations by Exner[217] suggest that it may be possible to account for the observation in other ways.

After Exner had given a preliminary statement of his views,[218] Palm[214] pointed out that the linear *meta* vs. *para* relationships might imply a linear relationship between inductive effect and conjugative or hyperconjugative effect of the substituents concerned. Hammett[219] has made some admonitory comments prompted by the "Exner analysis."

Wells *et al.*[207] discussed Exner's work in relation to their own analysis of substituent effects in naphthalene. They begin with a generalized definition of the resonance parameter as given in (4.67). Taft's ordinary σ_I

$$\sigma'_R = \sigma^0_p - \gamma\sigma_I \tag{4.67}$$

scale is involved, so γ implies both matters dealt with by Exner,[215] viz. the *meta/para* difference and the scaling of σ_I. Chemical considerations limit γ to the range +0.5 to +1.3: $\gamma > 1.3$ would give negative σ'_R values for NO_2 and CN, while $\gamma < 0.5$ would give positive σ'_R values for Hal, both results being nonsensical. Wells *et al.* suggest that within these limits the correct γ value could be found by testing sets of σ_I and σ'_R values (corresponding to selected values of γ) to find which set, when applied to naphthalene reactivities, gives a minimum interdependence of ρ_I and ρ_R, and values of ρ_I which "show a positional pattern dependence which is expected by theoretical models for the polar effect." (But which model?) The authors conclude that while it is difficult to eliminate the possibility that γ could differ from unity by up to ±0.2, there is no evidence that a more rational pattern of ρ_I and ρ_R values is obtained by varying γ from unity.

The authors also examined the application of Exner's σ_I and σ_R values to naphthalene reactivities. While these are sometimes more successful than Taft's σ_I and σ^0_R for systems in which ρ_R/ρ_I is small, Exner's values proved inferior when resonance effects are relatively important, particularly for the $-K$ substituents for which, according to Exner, $\sigma_R = 0$.

Ehrenson *et al.*[32] treated only 11 *meta* reaction series (mainly ionizations) and for these ρ^m_I can be compared with the corresponding ρ^p_I. These quantities never differ by more than a few percent, and ρ^m_I is variously slightly larger or smaller than ρ^p_I. To this approximation the equivalence of inductive effects from *meta*- and *para*-positions is confirmed. We emphasize again that this equivalence is no longer a basic assumption, but the scaling problem for σ_I should not be regarded as resolved.

The Need for Several σ_R-Type Scales. Even as early as 1957–1959 Taft envisaged the need for various σ_R scales to deal with enhanced resonance effects[199] and stated that a single widely applicable and precise σ_R scale could not be devised,[14] i.e., σ_R scales defined on different bases will not show good linear interrelationships. In spite of this, much correlation analysis has been done on the assumption that a single σ_R scale is widely applicable. Thus Wells *et al.*[207] used σ^0_R values in their analysis of naphthalene reactivities, although the reaction series involved might be expected to show considerable variation in the importance of resonance effects. Further, most of Charton's work[21,188–191] has involved the σ_R scale based on benzoic acid ionization through $\sigma_R = \sigma_p - \sigma_I$. The matter was thus in a somewhat indecisive state for a long time, but considerable clarification has been provided by the work of Ehrenson *et al.*,[32] which compares thoroughly the success of various σ_R scales in the application of equation (4.52) to a wide variety of processes. The Swain–Lupton $\mathfrak{F}$ and $\mathfrak{R}$

parameters (Section 4.3) are included in the survey. We now summarize the salient features of the work.

The initial aim was to define satisfactorily four scales of σ_R values: σ_R^0, $\sigma_R(BA)$, σ_R^-, and σ_R^+. In each case this was done by applying least-squares fitting procedures to data for selected reaction series expected to exhibit resonance effects of the required type. There is great stress on the requirement that in each series data shall be available for a *minimal basis set of substituents* as follows: NMe_2, NH_2, or OMe (any two); any two $-K$ substituents (CF_3, CO_2R, CH_3CO, CN, NO_2); H† and CH_3 (both); two halogens (but not both Cl and Br). This is held to be necessary to minimize real or accidental correlations between inductive and resonance parameters of substituents. The article should be consulted for the wealth of detail involved in setting up the various scales. Twenty-four substituents (including H) were considered. A fixed σ_I scale (based on a careful reassessment of alicyclic and aliphatic reactivities) was used throughout. The $\sigma_R(BA)$ scale was based on eight reaction series (mainly ionizations of benzoic acids in various solvents). Only for the ionization of benzoic acids in water at 25°C was the condition $\rho_I^p = \rho_R^p$ imposed (to establish the scale). The σ_R^- scale was based on the ionization of anilinium ions in water at 25°C. (It is interesting that the ionization of phenols in water was found unsatisfactory for this purpose.) The σ_R^0 scale was based on 14 "insulated" reaction series (mainly $ArCH_2X$ reactivities; but see Wepster[126–127]) and also on three ^{19}F nmr sets. The σ_R^+ scale was based on a combined consideration of *t*-cumyl chloride solvolysis[7] and protodesilylation of phenyltrimethylsilanes (Eaborn[221]). (The t-cumyl chloride reaction does not meet the minimal basis set requirement.)

The application of σ_I and each of the σ_R scales in turn, as well as of $\mathfrak{F}$ and $\mathfrak{R}$,[8] was examined for a large number of series for reactivity (equilibria and rates), ir, and nmr data of *para*-substituted compounds. The series were classified under headings relating them to the various σ_R scales. It was shown very strikingly that, for any given series, the best correlation§ is

†The correlated property P_i [equation (4.52)] is always expressed relative to that for H, and no intercept term is allowed. This procedure is justified by the authors on various grounds.[220] It implicitly rejects the possibility of there being some effect connected with the replacement of H by any other substituent whatsoever (footnote ¶ on p. 127, and Ref. 31). In the present case such an effect is unlikely. Other authors, notably Charton,[21,188–191] use the equation in the form (4.64). The intercept is not always found to correspond closely to the value of the correlated property for H as substituent.

§Taft and his colleagues[32,207] use $f = SD/RMS$ as a measure of the success of correlations. SD is the root mean square of the deviations and RMS is the root mean square of the data P_i's. If the value of $f \leq 0.1$, the correlation is considered of good precision. f is not exactly the same as Exner's[4] ψ, for which the "RMS" is related to the average value of the dependent variable, and not to the value for H as substituent.[222] While ψ is a kind of disguised correlation coefficient, on an expanded numerical scale,[4] this is not true of f, although ψ and f may be approximately equivalent for a well-distributed set of data.

nearly always given by the σ_R scale based on the process which obviously bears the closest relationship to the series in question, the other σ_R scales being markedly inferior. The authors concluded that each of the σ_R scales is necessary for a satisfactory comprehensive system of correlation analysis for substituent effects in benzene, each scale having a "limited generality."

Only rarely is the Swain–Lupton treatment superior to that involving the best σ_R scale, and often it is vastly inferior. As might be expected (see Section 4.3.3) the Swain–Lupton approach is particularly poor where σ_R^+ or σ_R^- is best. The authors are very critical of Swain and Lupton's[8] exclusion of important substituents such as NMe_2,[223] and of any reactivities of the σ^- type except the ionization of phenols in water (a process shown to have its own peculiarities).[224] They conclude that "there appears little to encourage proliferation of these parameters."[225]

The Interpretation of ρ_I and ρ_R Values

The interpretation of ρ in the original Hammett equation is well developed and there is good qualitative and some quantitative understanding of how it is governed by factors such as reaction type, solvent, temperature, structural characteristics, etc. Corresponding and distinctive interpretation of ρ_I and ρ_R, and the ratio $\rho_R/\rho_I = \lambda$, is not yet so well developed, but some progress has been made, which shows the deeper insights that may be provided by the dual substituent-parameter equation compared with the original Hammett equation. We restrict ourselves to a few examples drawn from the work of Ehrenson *et al.*[32]

A comparison of λ^m and λ^p values for the same reaction clearly shows the poorer transmission of resonance effects from the *meta*-position. While λ^m/λ^p is typically *ca.* 0.4, there is considerable reaction dependence, which will no doubt receive detailed interpretation in due course.

For the ionization of benzoic acids ρ_R^p and ρ_I^p both increase with decreasing solvent polarity, but λ^p decreases at the same time. The latter is attributed to a decreasing relative importance of cross-conjugation as the polarity of the solvent is decreased. This interpretation is confirmed by absence of a solvent effect on λ^p for phenylacetic acids and other "insulated" systems.

Moreover, ρ_I and ρ_R may be of opposite sign so that λ is negative, e.g., for the rates of acid-catalyzed esterification of benzoic acids in methanol, 25°C, ρ_I^p is negative because passage from initial state to transition state involves development of positive charge on the side chain. ρ_R^p is positive because there is parallel loss of cross-conjugative stabilization of the initial state by $+K$ substituents.†

Extending a conjugated system of aromatic ring and reaction center by incorporating C=C or C≡C appropriately in the side chain has little effect on λ^P. However, steric twisting of ring and reaction center from coplanarity greatly reduces λ^P, indicating inhibition of transmission of resonance effects.

4.5. Miscellaneous Developments and Conclusion

4.5.1. Miscellaneous Developments

We now mention briefly various relevant matters that merit attention but have not found a place in the main body of this chapter.

Yukawa and Tsuno's best-known work concerns their equation[10,12] as discussed in Section 4.2. However, they have also derived inductive (σ_i) and resonance (σ_π^+ and σ_π^-) parameters to be used in equations for the effect of *meta*- and *para*-substituents separately, analogous to Taft's dual substituent-parameter equation. The origins of this work are in little-known papers of *ca.* 1966,[226,227] and accounts of it were given by Professor Yukawa to the Second Conference on LFER, Irvine, California in 1968, and to the First IUPAC Conference on Physical Organic Chemistry at Crans sur Sierre in 1972. The work appears to have had little impact (see, however, papers by Laurence and Wojtkowiak,[228] and Sekigawa[229]), although the authors have used it themselves in recent years.[94,95,97,98,105–107] For definitions of the parameters and details of the equations the reader is referred to the papers cited. Charton[197] considers that the parameters and equations lead to a rather incomplete separation of inductive and resonance effects.

S. H. Unger[230] has developed a treatment of substituent effects that is related to that of Swain and Lupton.[8] We have already mentioned (Section 4.3.2) that $\mathfrak{F}$ and $\mathfrak{R}$ tend to vary in a mutually orthogonal manner.[144] Unger seeks to develop substituent constants S and P that possess this property to an even greater extent and thereby accomplish a more complete separation of field and resonance effects. While S is based initially on the ionization of bicyclo-octane carboxylic acids (as is $\mathfrak{F}$) and P

†It may reasonably be claimed that equation (4.52) is not liable to some of the troubles of the Yukawa–Tsuno equation (Section 4.2.5) in respect of the interpretation of the reaction constants. Thus for two-stage processes [equation (4.28)] for which $k = Kk_2$, the observed ρ_I and ρ_R values will be, respectively, the sums of the ρ_I and ρ_R values of the individual steps, a simple significance compared with that of the Yukawa–Tsuno $r^{\pm}$ value in the same situation [equation (4.29)]. These considerations permit the simple interpretation for acid-catalyzed esterification given above.

values are appropriately guessed, the final values (for 42 substituents) emerge from a computerized procedure that optimizes the account given of the selected body of data. It is claimed that S and P give, on average, higher correlation coefficients than other pairs of parameters, even for data not included in the optimization. For details the thesis[230] must be consulted. A thesis by N. R. Rosenquist[231] is also relevant.

Charton has raised again[232] the question of the scaling of σ_I values and hence of σ_R values. He takes as his starting point the ionization constants of the bicyclo-octane carboxylic acids in 50% w/w ethanol–water, 25°C (like Swain and Lupton,[8] Section 4.3.1). He calculates $\rho = 1.48$ for the ionization of 4-substituted benzoic acids under the same conditions (cf. 1.65, used by Hansch *et al.*[144], Section 4.3.1), and on this basis sets up the scale for σ_I by means of equation (4.68), where pK_{aX} is for the

$$\sigma_{IX} = -0.676pK_{aX} + 4.642 \tag{4.68}$$

X-substituted bicyclo-octane acid and 4.642 is the pK_a value of the parent acid. The resulting σ_I values for about 12 substituents enable other reaction series to be analyzed and used to determine further σ_I values. Charton has calculated about 250 σ_I values.† For 23 of the most common and important substituents he has correlated his values as dependent variable with those of Ehrenson *et al.*[32] as independent variable. The regression coefficient (gradient) is 1.09 ± 0.02, with $r = 0.995$. Thus the σ_I scale of Ehrenson *et al.*[32] is found to deviate significantly from the Hammett σ_m and σ_p scale (cf. Exner,[215] Section 4.4.4, who advocates multiplying Taft's σ_I values by 1.10 to bring them onto the Hammett scale). From his σ_I values Charton goes on to calculate 145 revised σ_R values as $\sigma_p - \sigma_I$. Each σ_I or σ_R value is appraised as very reliable, reliable, approximate, or uncertain.§

Charton[197] has also recently examined the question of the relative transmission of the localized (inductive) effect from the *meta*- and *para*-positions (see Section 4.4.4). By consideration of model systems, new field-effect calculations, and the separation of inductive and resonance effects proposed above, he concludes that the localized effect is *smaller* from the *meta*-position by a factor of *ca.* 0.9. This agrees fairly well with the reciprocal of Exner's factor ($1/1.14 = 0.88$), but not with the original assumption of Taft and Lewis,[13] the findings of Ehrenson *et al.*,[32] Yukawa and Tsuno's parameters and equations mentioned above,[226,227] or certain aspects of the Swain–Lupton treatment[8] [e.g., equations (4.38) and (4.39),

†Charton wishes to use the symbol σ_L (*localized* effect) rather than σ_I, and σ_D (*delocalized* effect) rather than σ_R. Exner tabulates the values (Chapter 10), but rejects the symbolism.

§Charton's σ_R values are not the same as Exner's[215]; the calculation of the latter also involves the *meta*/*para* difference factor, $\lambda = 1.14$.

or the implications of rejecting a $+K$ effect for NMe_3^+, since σ_m and σ_p values for this group are +0.88 and +0.82, respectively[143]].

Exner[233] has also returned to this issue with a determination of the ionization constants of *meta*- and *para*-substituted benzoic acids in dimethyl sulfoxide, ethylene glycol, and methanol. The factor λ, now stated as 1.13, holds for all three systems. Other arguments are also adduced in favor of the general validity of this factor.

Finally Krygowski and Fawcett[20] have proposed an entirely different analysis of the substituent effect and a corresponding multiparameter extension of the Hammett equation. By using data for the entropies and enthalpies of ionization of substituted benzoic acids, the Hammett σ_m and σ_p values are separated into σ_S (for entropy) and σ_H (for enthalpy) for a number of common substituents. These are then used to analyze the log K_a values of about 30 other aromatic acid systems, through equation (4.69).

$$\log K_a = \log K_a^0 + \rho_H \sigma_H + \rho_S \sigma_S \tag{4.69}$$

Usually entropy effects are found to make the major contribution to the total substituent effect, the entropic reaction constant being approximately equal to the normal Hammett ρ. A comparison with σ_I and σ_R values shows that entropic and inductive effects are related, while substituents with strong resonance effects are enthalpy controlled. The initial findings of this new type of correlation analysis are of considerable interest; further applications will appear soon.[234]

4.5.2. Conclusion

It seems unwise to try to summarize a complicated situation in a few lines, and it may appear presumptuous to offer firm advice on the use of multiparameter extensions of the Hammett equation. However, there are a few points that should be made.

Many organic chemists still seem content to use the Hammett equation with σ, σ^+, or σ^-. There seems sometimes to be a rejection of multiparameter equations as physically meaningless on the grounds that "if you introduce enough parameters you can explain anything."[235] It is only too easy to misuse multiparameter equations, but the acceptance of poor correlations and inadequate explanations arising from an oversimplified approach is likewise open to criticism. Physical organic chemistry will benefit from a more widespread use of the multiparameter equations if this is linked with the accurate study of extensive reaction series, involving a discriminating choice of substituents, and a proper appreciation of statistical methods.

How far is it desirable to elaborate the parameterization and the equations in the effort to achieve a more precise explanation? This may well be a matter of subjective judgment. Those who are inclined to accept only the minimum change from the Hammett equation may be attracted to the Yukawa–Tsuno equation. This has the good features that it clearly involves the sigma scales used with the simple Hammett equation, and the enhanced resonance parameters ($\Delta\sigma_R^+$, etc.) appear to be of wide applicability. It embraces *meta* and *para* series and involves ρ like the Hammett equation; the additional reaction constant $r^{\pm}$ seems (in favorable situations) to have a simple significance. On the other hand, the equation does not make any separation of inductive and resonance effects; ρ is therefore a complex quantity, and $r^{\pm}$ is subject to various limitations (Section 4.2.5). Taft's dual substituent-parameter equation purports to make a separation of inductive and resonance effects. The minimum requirement for treating *meta* and *para* series by equation (4.52) is two equations, two substituent constants (σ_I and a σ_R type) and the generation of four reaction constants. Unless a lack of precision in dealing with certain kinds of process is acceptable, a total of four σ_R-type scales is necessary in any comprehensive system. The complexity of this system compared with the Yukawa–Tsuno equation will need ultimately to be justified in terms of the deeper insights provided and the simplicity of ρ_I^i and ρ_R^i compared with ρ and $r^{\pm}$.

The Swain–Lupton treatment may be criticized, not on the grounds that it does not work as a system of correlation analysis (it works as well as any system with a fixed scale of resonance effects), but that it claims to make a separation of substituent effects that is illusory, unless the arguments in Section 4.3.3 can be rebutted. Concern to accomplish the separation of inductive and resonance effects must not, however, be carried too far. In the final analysis these effects cannot be altogether independent. Thus while it is essential that the various substituent constants be established reliably for as many substituents as possible, it is to be hoped that undue emphasis on minor differences in scales will be avoided.

Acknowledgments

I thank Professors M. Charton, O. Exner, T. Fujita, and T. M. Krygowski for correspondence, personal discussion, and prepublication copies of articles; also Dr. S. H. Unger for a copy of his thesis.

References and Notes

1. Certain material in Section 4.1 is adapted (by kind permission of the Clarendon Press) from J. Shorter, *Correlation Analysis in Organic Chemistry: an Introduction to Linear Free-Energy Relationships*, Chap. 2 (Oxford Chemistry Series, 1973).
2. L. P. Hammett, *J. Amer. Chem. Soc.*, **59**, 96 (1937).

3. L. P. Hammett, *Physical Organic Chemistry*, Chap. 11 (McGraw-Hill, New York, 2nd edn., 1970).
4. O. Exner, in *Advances in Linear Free Energy Relationships*, Chap. 1, N. B. Chapman and J. Shorter, eds. (Plenum, London, 1972). This is a review of the Hammett equation.
5. H. H. Jaffé, *Chem. Rev.*, **53**, 191 (1953).
6. R. W. Taft, in *Steric Effects in Organic Chemistry*, Chap. 13, M. S. Newman, ed. (Wiley, New York, 1956).
7. H. C. Brown and Y. Okamoto, *J. Amer. Chem. Soc.*, **80**, 4979 (1958).
8. C. G. Swain and E. C. Lupton, *J. Amer. Chem. Soc.*, **90**, 4328 (1968).
9. H. van Bekkum, P. E. Verkade, and B. M. Wepster, *Rec. Trav. chim.*, **78**, 815 (1959).
10. Y. Yukawa and Y. Tsuno, *Bull. Chem. Soc. Japan*, **32**, 971 (1959).
11. M. Yoshioka, K. Hamamoto, and T. Kubota, *Bull. Chem. Soc. Japan*, **35**, 1723 (1962).
12. Y. Yukawa, Y. Tsuno, and M. Sawada, *Bull. Chem. Soc. Japan*, **39**, 2274 (1966).
13. R. W. Taft and I. C. Lewis, *J. Amer. Chem. Soc.*, **80**, 2436 (1958).
14. R. W. Taft and I. C. Lewis, *J. Amer. Chem. Soc.*, **81**, 5343 (1959).
15. R. W. Taft, S. Ehrenson, I. C. Lewis, and R. E. Glick, *J. Amer. Chem. Soc.*, **81**, 5352 (1959).
16. R. W. Taft, *J. Phys. Chem.*, **64**, 1805 (1960).
17. S. Ehrenson, *Progr. Phys. Org. Chem.*, **2**, 195 (1964).
18. J. R. Knowles, R. O. C. Norman, and G. K. Radda, *J. Chem. Soc.*, 4885 (1960).
19. O. Exner, *Coll. Czech. Chem. Comm.*, **31**, 65 (1966).
20. T. M. Krygowski and W. R. Fawcett, *Canad. J. Chem.*, **53**, 3622 (1975).
21. M. Charton, *Progr. Phys. Org. Chem.*, **8**, 235 (1971).
22. M. J. S. Dewar and P. J. Grisdale, *J. Amer. Chem. Soc.*, **84**, 3548 (1962).
23. M. J. S. Dewar, R. Golden, and J. M. Harris, *J. Amer. Chem. Soc.*, **93**, 4187 (1971).
24. Y. Tsuno, T. Ibata, and Y. Yukawa, *Bull. Chem. Soc. Japan*, **32**, 960 (1959).
25. Y. Yukawa and Y. Tsuno, *Bull. Chem. Soc. Japan*, **32**, 965 (1959).
26. J. E. Leffler and E. Grunwald, *Rates and Equilibria of Organic Reactions* (Wiley, New York, 1963).
27. G. R. Howe and R. R. Hiatt, *J. Org. Chem.*, **35**, 4007 (1970).
28. C. Laurence and B. Wojtkowiak, *Compt. rend.* (*C*), **264**, 1216 (1967).
29. Y. Yukawa, Y. Tsuno, and M. Sawada, *Bull. Chem. Soc. Japan*, **45**, 1198 (1972).
30. T. Imamoto, Y. Tsuno, and Y. Yukawa, *Bull. Chem. Soc. Japan*, **44**, 1632 (1971).
31. K. J. Toyne, *Tetrahedron*, **29**, 3889 (1973).
32. S. Ehrenson, R. T. C. Brownlee, and R. W. Taft, *Progr. Phys. Org. Chem.*, **10**, 1 (1973).
33. C. Eaborn and J. A. Waters, *J. Chem. Soc.*, 542 (1961).
34. C. Eaborn and K. C. Pande, *J. Chem. Soc.*, 3715, 5082 (1961).
35. C. Eaborn and P. M. Jackson, *J. Chem. Soc.* (*B*), 21 (1969).
36. C. Eaborn, Z. Lasocki, and J. A. Sperry, *J. Organometal. Chem.*, **35**, 245 (1972).
37. R. Baker, R. W. Bott, C. Eaborn, and P. M. Greaseley, *J. Chem. Soc.*, 627 (1964).
38. C. Eaborn, D. R. M. Walton, and D. J. Young, *J. Chem. Soc.* (*B*), 15 (1969).
39. R. Baker, C. Eaborn, and R. Taylor, *J.C.S. Perkin II*, 97 (1972).
40. A. R. Bassindale, C. Eaborn, R. Taylor, A. R. Thompson, D. R. M. Walton, J. Cretney, and G. J. Wright, *J. Chem. Soc.* (*B*), 1155 (1971).
41. R. O. C. Norman and G. K. Radda, *Tetrahedron Lett.*, 125 (1962).
42. R. W. Bott and C. Eaborn, *J. Chem. Soc.*, 2139 (1963).
43. P. R. Wells, *Chem. Rev.*, **63**, 171 (1963).
44. R. Taylor, G. G. Smith, and W. H. Wetzel, *J. Amer. Chem. Soc.*, **84**, 4817 (1962).
45. R. Taylor and G. G. Smith, *Tetrahedron*, **19**, 937 (1963).
46. G. G. Smith and B. L. Yates, *J. Org. Chem.*, **30**, 434 (1965).
47. T. I. Crowell, C. E. Bell, and D. H. O'Brien, *J. Amer. Chem. Soc.*, **86**, 4973 (1964).
48. W. P. Jencks, *Progr. Phys. Org. Chem.*, **2**, 63 (1964).

49. B. Zwanenberg and J. B. F. N. Engberts, *Rec. Trav. chim.*, **84**, 165 (1965).
50. P. J. Slootmaekers, A. Rasschaert, and W. Janssens, *Bull. Soc. chim. belges*, **75**, 199 (1966).
51. T. Inukai, *Bull. Chem. Soc. Japan*, **35**, 400 (1962).
52. L. M. Stock and H. C. Brown, *Adv. Phys. Org. Chem.*, **1**, 35 (1963).
53. T. Fueno, T. Okuyama, and J. Furukawa, *Bull. Chem. Soc. Japan*, **39**, 2094 (1966).
54. As indicated in *Science Citation Index* (Institute for Scientific Information, Philadelphia.
55. G. A. Olah, M. Tashiro, and S. Kobayashi, *J. Amer. Chem. Soc.*, **92**, 6369 (1970).
56. G. A. Olah, S. Kobayashi, and M. Tashiro, *J. Amer. Chem. Soc.*, **94**, 7448 (1972).
57. G. A. Olah and H. C. Lin, *J. Amer. Chem. Soc.*, **96**, 2892 (1974).
58. S. Clementi, P. Linda, and C. D. Johnson, *J.C.S. Perkin II*, 1250 (1973).
59. S. Clementi and P. Linda, *Tetrahedron*, **26**, 2869 (1970).
60. H. Hashimoto and Y. Morimoto, *J. Organometal. Chem.*, **8**, 271 (1967).
61. J. P. Girault, P. Scribe, and G. Dana, *Bull. Soc. chim. France*, 1760 (1973).
62. J. M. A. Baas and B. M. Wepster, *Rec. Trav. chim.*, **91**, 285 (1972).
63. R. Matsushima and S. Sakuraba, *J. Amer. Chem. Soc.*, **93**, 7143 (1971).
64. See, however, M. Ahmad, A. Cox, T. J. Kemp, and Q. Sultana, *J.C.S. Perkin II*, 1867 (1975).
65. J.-P. Durand, M. Davidson, M. Hellin, and F. Coussemant, *Bull. Soc. chim. France*, 52 (1966).
66. R. W. Bott, C. Eaborn, and D. R. M. Walton, *J. Chem. Soc.*, 384 (1965).
67. R. W. Bott, C. Eaborn, and D. R. M. Walton, *J. Organometal. Chem.*, **1**, 420 (1964).
68. D. S. Noyce and M. D. Schiavelli, *J. Amer. Chem. Soc.*, **90**, 1020 (1968).
69. Y. Ogata and Y. Sawaki, *Bull. Chem. Soc. Japan*, **38**, 194 (1965).
70. G. R. Howe and R. R. Hiatt, *J. Org. Chem.*, **36**, 2493 (1971).
71. M. Mollard, B. Torck, M. Hellin, and F. Coussemant, *Bull. Soc. chim. France*, 1186 (1966).
72. J.-É. Dubois, A. F. Hegarty, and E. D. Bergmann, *J. Org. Chem.*, **37**, 2218 (1972).
73. J.-É. Dubois and W. V. Wright, *Tetrahedron Lett.*, 3101 (1967).
74. J.-É. Dubois and A. Schwarcz, *Tetrahedron Lett.*, 2167 (1964).
75. A. F. Hegarty, J. S. Lomas, W. V. Wright, E. D. Bergmann, and J.-É. Dubois, *J. Org. Chem.*, **37**, 2222 (1972).
76. A. F. Hegarty and J.-É. Dubois, *Tetrahedron Lett.*, 4839 (1968).
77. J.-É. Dubois and A. F. Hegarty, *J. Chem. Soc.* (*B*), 638 (1969).
78. M.-F. Ruasse and J.-É. Dubois, *J. Org. Chem.*, **37**, 1770 (1972).
79. See further J.-É. Dubois, M. F. Ruasse, and A. Argile, *Tetrahedron*, **31**, 2921 (1975).
80. J. J. Ryan and A. A. Humffray, *J. Chem. Soc.* (*B*), 842 (1966).
81. A. A. Humffray and J. J. Ryan, *J. Chem. Soc.* (*B*), 468 (1967).
82. J. J. Ryan and A. A. Humffray, *J. Chem. Soc.* (*B*), 1300 (1967).
83. A. A. Humffray and J. J. Ryan, *J. Chem. Soc.* (*B*), 1138 (1969).
84. J. Socha, J. Horská, and M. Večeřa, *Coll. Czech. Chem. Comm.*, **34**, 2982 (1969).
85. J. Socha, J. Horská, and M. Večeřa, *Coll. Czech. Chem. Comm.*, **34**, 3178 (1969).
86. J. Socha and M. Večeřa, *Coll. Czech. Chem. Comm.*, **35**, 3072 (1970).
87. J. Jahelka, O. Macháčková, V. Štěrba, and K. Valter, *Coll. Czech. Chem. Comm.*, **38**, 3290 (1973).
88. K. Kalfus, J. Socha, and M. Večeřa, *Coll. Czech. Chem. Comm.*, **39**, 275 (1974).
89. J. Kavelek and V. Štěrba, *Coll. Czech. Chem. Comm.*, **40**, 1176 (1975).
90. C. Dell'Erba, G. Guanti, and G. Garbarino, *Tetrahedron*, **27**, 1807 (1971).
91. A. M. Porto, L. Altieri, A. J. Castro, and J. A. Brieux, *J.Chem. Soc.* (*B*), 963 (1966); and references therein.
92. K. Bowden and R. S. Cook, *J. Chem. Soc.* (*B*), 1765 (1971).

93. H. C. Brown and K. L. Nelson, *J. Amer. Chem. Soc.*, **75**, 6292 (1953).
94. Y. Tsuno, M. Fujio, Y. Takai, and Y. Yukawa, *Bull. Chem. Soc. Japan*, **45**, 1519 (1972).
95. M. Fujio, M. Mishima, Y. Tsuno, Y. Yukawa, and Y. Takai, *Bull. Chem. Soc. Japan*, **48**, 2127 (1975).
96. Y. Yukawa, Y. Tsuno, and N. Shimizu, *Bull. Chem. Soc. Japan*, **44**, 2843 (1971).
97. Y. Tsuno, M. Fujio, and Y. Yukawa, *Bull. Chem. Soc. Japan*, **48**, 3324 (1975).
98. M. Fujio, Y. Tsuno, Y. Yukawa, and Y. Takai, *Bull. Chem. Soc. Japan*, **48**, 3330 (1975).
99. J. Niwa, *Bull. Chem. Soc. Japan*, **42**, 1926 (1969).
100. M. T. Tribble and J. G. Traynham, *J. Amer. Chem. Soc.*, **91**, 379 (1969).
101. C. Laurence and B. Wojtkowiak, *Bull. Soc. chim. France*, 3833 (1971).
102. M. Berthelot and C. Laurence, *Compt. rend.* (*C*), **276**, 979 (1973).
103. V. Bekárek and K. Pragerová, *Coll. Czech. Chem. Comm.*, **40**, 1005 (1975).
104. V. F. Andrianov, A. Ya. Kaminsky, A. V. Ivanov, S. S. Ghitis, N. V. Udris, S. S. Gluzmann, and S. I. Buga, *Reakts. spos. org. Soedinenii*, **11**, 473 (1974); EE, 477.
105. Y. Tsuno, Y. Kusuyama, M. Sawada, T. Fujii, and Y. Yukawa, *Bull. Chem. Soc. Japan*, **48**, 3337 (1975).
106. Y. Tsuno, M. Sawada, T. Fujii, and Y. Yukawa, *Bull. Chem. Soc. Japan*, **48**, 3347 (1975).
107. Y. Tsuno, M. Sawada, T. Fujii, Y. Tairaka, and Y. Yukawa, *Bull. Chem. Soc. Japan*, **48**, 3356 (1975).
108. D. S. Noyce and R. L. Castenson, *J. Amer. Chem. Soc.*, **95**, 1247 (1973).
109. H. Tanida and H. Matsumura, *J. Amer. Chem. Soc.*, **95**, 1586 (1973).
110. J. S. Lomas and J.-É. Dubois, *J. Org. Chem.*, **40**, 3303 (1975).
111. J. S. Lomas and J.-É. Dubois, *Tetrahedron Lett.*, 407 (1976).
112. T. Imamoto, Y. Tsuno, and Y. Yukawa, *Bull. Chem. Soc. Japan*, **44**, 1632, 1639, 1644 (1971).
113. Y. Yukawa, Y. Tsuno, and M. Sawada, *Bull. Chem. Soc. Japan*, **45**, 1210 (1972).
114. J. Mindl and M. Večeřa, *Coll. Czech. Chem. Comm.*, **35**, 950 (1970).
115. G. T. Bruce, A. R. Cooksey, and K. J. Morgan, *J.C.S. Perkin II*, 551 (1975).
116. H. van Bekkum, H. M. A. Buurmans, B. M. Wepster, and A. M. van Wijk, *Rec. Trav. chim.*, **88**, 301 (1969).
117. A. R. Haughton, R. M. Laird, and M. J. Spence, *J.C.S. Perkin II*, 637 (1975).
118. K. A. Tskhadadze, V. A. Bren', and V. I. Minkin, *Reakts. spos, org. Soedinenii*, **11**, 495 (1974): EE 499.
119. Ref. 26, p. 177.
120. A. Williams and R. A. Naylor, *J. Chem. Soc.* (*B*), 1967 (1971).
121. C. D. Johnson, *The Hammett Equation*, Chap. 3 (Cambridge University Press, 1973).
122. R. Taylor, in *Comprehensive Chemical Kinetics*, Vol. 13, *Reactions of Aromatic compounds*, Chap. 1, C. H. Bamford and C. F. H. Tipper, eds. (Elsevier, Amsterdam, 1972).
123. Ref. 3, p. 365.
124. M. Sawada, Y. Tsuno, and Y. Yukawa, *Bull. Chem. Soc. Japan*, **45**, 1206 (1972).
125. M. Berthelot, C. Laurence, and B. Wojtkowiak, *J. Chim. phys.*, **76**, 1629 (1973).
126. A. J. Hoefnagel, J. C. Monshouwer, E. C. G. Snorn, and B. M. Wepster, *J. Amer. Chem. Soc.*, **95**, 5350 (1973).
127. A. J. Hoefnagel and B. M. Wepster, *J. Amer. Chem. Soc.*, **95**, 5357 (1973).
128. J. Basters, H. van Bekkum, and L. L. van Reijen, *Rec. Trav. chim.*, **89**, 491 (1970).
129. A. P. G. Kieboom, J. F. de Kreuk, and H. van Bekkum, *J. Catalysis*, **20**, 58 (1971).
130. Ref. 121, Chap. 1.
131. Ref. 121, Chap. 4.
132. S. Clementi, P. Linda, and G. Marino, *J. Chem. Soc.* (*B*), 1153 (1970).
133. C. D. Johnson and K. Schofield, *J. Amer. Chem. Soc.*, **95**, 270 (1973).

134. C. D. Johnson, *Chem. Rev.*, **75**, 755 (1975).
135. R. N. McDonald and J. M. Richmond, *J.C.S. Chem. Comm.*, 333 (1974).
136. B. M. Wepster, *J. Amer. Chem. Soc.*, **95**, 102 (1973).
137. S. Wold, *Acta Chem. Scand.*, **27**, 3602 (1973).
138. K. C. C. Bancroft and G. R. Howe, *J. Chem. Soc.* (*B*), 1221 (1971).
139. For a simple account see the appendix in Ref. 1.
140. See Ref. 8 for sources.
141. J. D. Roberts and W. T. Moreland, *J. Amer. Chem. Soc.*, **75**, 2167 (1953).
142. L. Daub and J. M. Vandenbelt, *J. Amer. Chem. Soc.*, **71**, 2414 (1949).
143. D. H. McDaniel and H. C. Brown, *J. Org. Chem.*, **23**, 420 (1958).
144. C. Hansch, A. Leo, S. H. Unger, K. H. Kim, D. Nikaitani, and E. J. Lien, *J. Med. Chem.*, **16**, 1207 (1973).
145. H. D. Holtz and L. M. Stock, *J. Amer. Chem. Soc.*, **86**, 5188 (1964).
146. F. W. Baker, R. C. Parish, and L. M. Stock, *J. Amer. Chem. Soc.*, **89**, 5677 (1967).
147. T. Fujita and T. Nishioka, *Progr. Phys. Org. Chem.*, **12**, 49 (1976).
148. T. Yokoyama, G. R. Wiley, and S. I. Miller, *J. Org. Chem.*, **34**, 1859 (1969.
149. D. A. Dawson, G. K. Hamer, and W. F. Reynolds, *Canad. J. Chem.*, **52**, 39 (1974); and earlier papers.
150. D. A. Dawson, and W. F. Reynolds, *Canad. J. Chem.*, **53**, 373 (1975).
151. S. Rodmar, S. Gronowitz, and U. Rosén, *Acta Chem. Scand.*, **25**, 3841 (1971).
152. S. Gronowitz, I. Johnson, and S. Rodmar, *Acta Chem. Scand.*, **26**, 1726 (1972).
153. G. A. Caplin, *Org. Magn. Reson.*, **5**, 169 (1973).
154. G. A. Caplin, *Org. Magn. Reson.*, **6**, 99 (1974).
155. G. R. Wiley and S. I. Miller, *J. Org. Chem.*, **37**, 767 (1972).
156. S. K. Dayal, S. Ehrenson, and R. W. Taft, *J. Amer. Chem. Soc.*, **94**, 9113 (1972).
157. N. L. Silver and D. W. Boykin, *J. Org. Chem.*, **35**, 759 (1970).
158. W. F. Winecoff III and D. W. Boykin, *J. Org. Chem.*, **37**, 674 (1972).
159. A. Perjéssy, D. W. Boykin, L. Fišera, A. Krutošíková, and J. Kováč, *J. Org. Chem.*, **38**, 1807 (1973).
160. K. Spaargaren, C. Kruk, T. A. Molenaar-Langefeld, P. K. Korver, P. J. van der Haak, and Th.J. de Boer, *Spectrochim. Acta*, **28A**, 965 (1972).
161. K. Spaargaren, P. K. Korver, P. J. van der Haak, and Th.J. de Boer, *Org. Magn. Reson.*, **3**, 605 (1971).
162. P. R. Johnson and D. A. Thornton, *J. Inorg. Nuclear Chem.*, **37**, 461 (1975).
163. G. C. Percy and D. A. Thornton, *J. Inorg. Nuclear Chem.*, **35**, 2319 (1973); and earlier papers.
164. C. Laurence and B. Wojtkowiak, *Bull. Soc. chim. France*, 3874 (1971).
165. R. I. Zalewski, *Bull. Acad. polon. Sci., Sér. Sci chim.*, **20**, 853 (1972).
166. R. I. Zalewski, *Bull. Acad. polon. Sci., Sér. Sci. chim.*, **19**, 351 (1971).
167. R. Stewart and K. Yates, *J. Amer. Chem. Soc.*, **82**, 4059 (1960).
168. K. Yates and R. Stewart, *Canad. J. Chem.*, **37**, 664 (1959).
169. R. Stewart and K. Yates, *J. Amer. Chem. Soc.*, **80**, 6355 (1958).
170. M. Bergon and J.-P. Calmon, *Bull. Soc. chim. France*, 1020 (1972).
171. H. Walba and R. Ruiz-Velasco, *J. Org. Chem.*, **34**, 3315 (1969).
172. J. Paleček and J. Hlavatý, *Coll. Czech. Chem. Comm.*, **38**, 1985 (1973).
173. J. A. Zoltewicz and R. E. Cross, *J.C.S. Perkin II*, 1363 (1974).
174. C. Eaborn and A. Fischer, *J. Chem. Soc.* (*B*), 152 (1969).
175. M. E. Kurz and M. Pellegrini, *J. Org. Chem.*, **35**, 990 (1970).
176. S. F. Nelsen and D. H. Heath, *J. Amer. Chem. Soc.*, **91**, 6452 (1969).
177. W. S. Trahanovsky, J. Cramer, and D. W. Brixius, *J. Amer. Chem. Soc.*, **96**, 1077 (1974).
178. I. H. Sadler, *J. Chem. Soc.* (*B*), 1024 (1969).

179. J.-F. Gal, L. Elegant, and M. Azzaro, *Bull. Soc. chim. France*, 1150 (1973).
180. D. S. Noyce and R. W. Nichols, *J. Org. Chem.*, **37**, 4306 (1972).
181. D. S. Noyce and J. A. Virgilio, *J. Org. Chem.*, **38**, 2660 (1973).
182. D. S. Noyce and S. A. Fike, *J. Org. Chem.*, **38**, 3321 (1973).
183. D. S. Noyce and G. T. Stowe, *J. Org. Chem.*, **38**, 3762 (1973).
184. C. G. Mitton, R. L. Schowen, M. Gresser, and J. Shapley, *J. Amer. Chem. Soc.*, **91**, 2036 (1969).
185. C. G. Swain, J. E. Sheats, and K. G. Harbison, *J. Amer. Chem. Soc.*, **97**, 783 (1975).
186. J. Shorter, in *Advances in Linear Free Energy Relationships*, Chap. 2, N. B. Chapman and J. Shorter, eds. (Plenum, London, 1972).
187. T. Nishioka, T. Fujita, K. Kitamura, and M. Nakajima, *J. Org. Chem.*, **40**, 2520 (1975).
188. M. Charton, *J. Org. Chem.*, **40**, 407 (1975); and earlier papers.
189. M. Charton, *J. Amer. Chem. Soc.*, **97**, 1552, 3691 (1975); and other papers in press or in course of preparation.
190. M. Charton, *J. Org. Chem.*, **36**, 266 (1971).
191. M. Charton, *Progr. Phys. Org. Chem.*, **10**, 81 (1973).
192. D. Holtz, *Chem. Rev.*, **71**, 139 (1971).
193. I. R. Ager, L. Phillips, T. J. Tewson, and V. Wray, *J.C.S. Perkin II*, 1979 (1972).
194. P. J. Q. English, A. R. Katritzky, T. T. Tidwell, and R. D. Topsom, *J. Amer. Chem. Soc.*, **90**, 1767 (1968).
195. N. C. Cutress, T. B. Grindley, A. R. Katritzky, M. V. Sinnott, and R. D. Topsom, *J.C.S. Perkin II*, 2255 (1972).
196. R. W. Taft, personal communication in Ref. 194.
197. M. Charton, personal communications and forthcoming publications.
198. L. D. Hansen and L. G. Hepler, *Canad. J. Chem.*, **50**, 1030 (1972).
199. R. W. Taft, *J. Amer. Chem. Soc.*, **79**, 1045 (1957).
200. H. S. Gutowsky, D. W. McCall, B. R. McGarvey, and L. H. Meyer, *J. Amer. Chem. Soc.*, **74**, 4809 (1952).
201. L. H. Meyer and H. S. Gutowsky, *J. Phys. Chem.*, **57**, 481 (1953).
202. J. L. Roberts and H. H. Jaffé, *J. Amer. Chem. Soc.*, **81**, 1635 (1959).
203. R. W. Taft, E. Price, I. R. Fox, I. C. Lewis, K. K. Andersen, and G. T. Davis, *J. Amer. Chem. Soc.*, **85**, 709 (1963).
204. R. W. Taft, E. Price, I. R. Fox, I. C. Lewis, K. K. Andersen, and G. T. Davis, *J. Amer. Chem. Soc.*, **85**, 3146 (1963).
205. S. Ehrenson, *Tetrahedron Lett.*, 351 (1964).
206. S. Ehrenson, *Abstracts of Papers at Symposium on LFER, Durham, North Carolina*, 1964, reprinted in *Progr. Phys. Org. Chem.*, **6**, 222 (1968).
207. P. R. Wells, S. Ehrenson, and R. W. Taft, *Progr. Phys. Org. Chem.*, **6**, 147 (1968).
208. A. R. Katritzky and R. D. Topsom, in *Advances in Linear Free Energy Relationships*, Chap. 3, N. B. Chapman and J. Shorter, eds. (Plenum, London, 1972).
209. R. T. C. Brownlee, R. E. J. Hutchinson, A. R. Katritzky, T. T. Tidwell, and R. D. Topsom, *J. Amer. Chem. Soc.*, **90**, 1757 (1968).
210. A. R. Katritzky, M. V. Sinnott, T. T. Tidwell, and R. D. Topsom, *J. Amer. Chem. Soc.*, **91**, 628 (1969).
211. E. D. Schmid, *Spectrochim. Acta*, **22**, 1659 (1966); and previous papers referred to therein.
212. A. R. Katritzky and R. D. Topsom, *J. Chem. Educ.*, **48**, 427 (1971).
213. O. Exner, *Tetrahedron Lett.*, **815** (1963).
214. V. A. Palm and A. V. Tuulmets, *Reakts. spos. org. Soedinenii*, **1**(1), 33 (1964).
215. O. Exner, *Coll. Czech. Chem. Comm.*, **31**, 65 (1966).
216. J. P. Schaefer and T. J. Miraglia, *J. Amer. Chem. Soc.*, **86**, 64 (1964).
217. V. Všetečka and O. Exner, *Coll. Czech. Chem. Comm.*, **39**, 1140 (1974).

218. *Trudy konferentsii po problemam primeneniya korrelattsionnykh uravnenii v organicheskoi khimii* (*Proceedings of the Conference on Applications of Correlation Equations in Organic Chemistry*), Tartu, 1962 (in Russian, with English summaries); O. Exner's lecture is on pp. 67–86, with summary on pp. 86–87.
219. Ref. 3, p. 382.
220. Ref. 32, p. 38.
221. Collected data, Ref. 122, p. 324.
222. Ref. 139 is not quite correct on this point.
223. Ref. 32, p. 12.
224. Ref. 32, pp. 13–20, 50–53.
225. Ref. 32, p. 35.
226. Y. Yukawa and Y. Tsuno, *Nippon Kagaku Zasshi*, **86**, 873 (1965).
227. Y. Yukawa and Y. Tsuno, *Mem. Inst. Sci. Ind. Res., Osaka University*, **23**, 71 (1966); *Chem. Abs.*, **67**, 73057 (1967).
228. C. Laurence and B. Wojtkowiak, *Ann. Chim.* (*France*), **5**, 163 (1970). (This is an excellent review, in French, of the Hammett equation and its extensions.)
229. K. Sekigawa, *Tetrahedron*, **28**, 505, 515 (1972).
230. S. H. Unger, Ph.D. Thesis, M.I.T., 1970.
231. N. R. Rosenquist, Ph.D. Thesis, M.I.T., 1973.
232. M. Charton, personal communications and forthcoming publications.
233. O. Exner and K. Kalfus, *Coll. Czech. Chem. Comm.*, **41**, 569 (1976).
234. T. M. Krygowski, personal communication.
235. Various organic chemists in the hearing of the present author.

5

Applications of Linear Free Energy Relationships to Polycyclic Arenes and to Heterocyclic Compounds

M. Charton

M. Charton • Department of Chemistry, Pratt Institute, Brooklyn, New York 11205, U.S.A.

5.1. Introduction

5.1.1. Objectives

The purpose of this chapter is briefly to review the application of the Hammett equation, the extended Hammett equation,† and related linear free energy relationships to polycyclic arenes and to heterocyclic compounds. No comprehensive review of the application of linear free energy relationships to polynuclear aromatic systems exists. Wells *et al.*[1] have described the application of the extended Hammett equation to substituted naphthalene series, and Zuman[2] has reviewed the application of the Hammett equation to polarographic data for condensed polycyclic aromatic compounds. No other specialized reviews of this area are available. For heterocyclic compounds, an excellent comprehensive review by Jaffé and Jones[3] appeared in 1964. Since then, Tomasik has published an extensive review (1971)[4] and a monograph (1974)[5] on the application of linear free energy relationships to heterocyclic compounds. Unfortunately, both of these works are in Polish, and thus are not readily accessible to most English-speaking chemists. Zuman[2] has reviewed the application of the Hammett equation to polarographic data for mono- and polycyclic heterocyclic compounds. It is clear then, that there is a real need for a review dealing with polycyclic arenes and with heterocyclic compounds.

5.1.2. The Hammett, the Extended Hammett, and the Dewar–Grisdale Equations

The Hammett equation is too well known to require discussion here. Recent reviews by Shorter,[6] Johnson,[7] and Exner[8] may be consulted as well as an introductory account by Charton.[9] The extended Hammett equation has been reviewed briefly by Exner,[8] Charton,[9,10] Wells *et al.*,[1] and by Ehrenson *et al.*[11] A more detailed review by Shorter appears in this volume.[12] The Dewar–Grisdale method[13] has been treated by Exner[8] and by Charton.[9] As it has received less attention than the extended Hammett and the Hammett approach, it will be considered in more detail here.

† This term is used in this chapter to refer to the following equation:

$$Q_X = \alpha\sigma_{I,X} + \beta\sigma_{R,X} + h$$

This relationship is another form of the equation referred to by Taft as the DSP (dual substituent-parameter) equation (see Chapter 4). The two equations are not equivalent from the viewpoint of multiple linear regression analysis. The equation below is referred to by the present author as the LDS (localized, delocalized, steric) equation:

$$Q_X = \alpha\sigma_{I,X} + \beta\sigma_{R,X} + \psi\upsilon_X + h$$

It is obtained from the extended Hammett equation by the inclusion of a term for steric effects.

The Hammett equation was originally proposed to correlate data for *meta*-.and *para*-substituted benzene sets.† The greatest structural variation allowed other than that of the substituent was the introduction of a side chain between the reaction site and the benzene ring. It is desirable, however, to be able to apply structure–reactivity correlations to any kind of organic compound and potentially to covalent inorganic compounds as well. It is for this purpose that the extended Hammett equation and the Dewar–Grisdale treatment were designed. The advantage of these methods over the simple Hammett equation is that they are potentially capable of handling any combination of localized (field and/or inductive) and delocalized (resonance) effects. They achieve this versatility in different ways, however. In order to account for these different approaches, let us describe a set of data that is to be studied, as derived from compounds having the form XGY, where X is the variable substituent, Y is the reaction site at which the measured phenomena occur, and G is the skeletal group to which X and Y are attached. For the extended equation, σ is a function only of X, while the reaction constants α and β (or ρ_I and ρ_R as they are also designated) are functions of both Y and G. In this method a substituent may be characterized by a single localized-effect substituent constant and a few delocalized-effect substituent constants. By contrast, in the Dewar–Grisdale treatment the substituent constant is a function of X and G, and the reaction constant ρ is a function only of Y. Thus, in this method, new substituent constants are required for every possible location in every conceivable group, G. The Dewar–Grisdale treatment requires only a single value of ρ (in theory at least) for any given Y. The extended Hammett equation requires different values of α and β for each different G, and skeletal groups that differ in the locations of X and Y require different values of α and β. All other considerations aside, it must be pointed out that the extended Hammett equation requires fewer values of σ, α, and β combined, than the Dewar–Grisdale treatment requires values of σ and ρ for application to any given system, such as substituted naphthalene sets.

According to the original Dewar–Grisdale (DG) treatment,[13] the localized and delocalized electrical effects are represented by the quantities F and M, respectively. Substituent constants are calculated from equation (5.1), where r_{ij} represents the distance between the atom i of the skeletal

$$\sigma_{\mathrm{X,DG}} = F_{\mathrm{X}}/r_{ij} + M_{\mathrm{X}} q_{ij} \tag{5.1}$$

†There is no agreement at present as to what a group of data points obtained for a collection of structurally related compounds should be called. Traditionally, they have been referred to as a reaction series. Since, from the point of view of correlation analysis, such data points must be treated statistically, and from a mathematical viewpoint "set" is a better term than "series," and finally, since the data may involve physical as well as chemical properties, they will be referred to as *sets*.

group G to which the substituent, X, is bonded, and the atom j of the group G to which the reaction site, Y, is bonded. The distance is given in units of benzene bond length (1.40 Å). The quantity q_{ij} is the charge on the atom bearing Y in the carbanion $ArCH_2^-$; the charge is calculated by the method of Longuet-Higgins.[14] The F and M values can now be calculated from σ_m and σ_p constants by substituting appropriate values for r_{ij} and q_{ij}. Thus we have equations (5.2) and (5.3). Once F_X and M_X are known, σ constants

$$\sigma_{m-X} = F_X/1.73 \tag{5.2}$$

$$\sigma_{p-X} = F_X/2 + M_X/7 \tag{5.3}$$

can be calculated for any structure. The most useful test of the validity of the DG treatment is based on the fact that, in order for it to hold, the ρ values obtained from the Hammett equation (5.4), by using the calculated

$$Q_X = \rho\sigma_{X,DG} + h \tag{5.4}$$

$\sigma_{X,DG}$ values, must be independent of G, and in fact should be constant for a given reaction site. As an example of a test of the validity of the DG treatment, Charton[15] has examined the correlation of pK_a data for several sets of quinolines and isoquinolines by use of equation (5.4). The ρ values obtained were not constant—in fact they varied with G. Furthermore, Snyder[16] reported that a plot of σ_{obs} for 6- or 7-substituted quinolines against σ_{DG}, although linear, had a slope of 0.7, whereas the Dewar-Grisdale treatment requires that the slope be 1.0. Thus, this treatment is probably in error. It must also be noted that if the Taft analysis of the electrical effect is correct,[11] the DG F values must include a resonance effect.

Dewar, Golden, and Harris[17] (DGH) have modified the DG treatment. They proposed that substituent constants for carboxylic acids be calculated from equation (5.5). The model used for carboxylic acids is

$$\sigma_{imX} = F_X R_{im} + M_X q_{im} + M_{FX} \sum_{k \neq m} \frac{q_{ik}}{r_{kn}} \tag{5.5}$$

shown in Fig. 5.1. The quantities F and M retain their previous meaning, while M_F is the mesomeric field effect, an additional resonance effect. The quantity q_{ik} represents the charge at the position k in the skeletal group. The quantity R_{im} is given by the expression (5.6), where r_{in} is the distance

$$R_{im} = 1/r_{in} - 0.9/r_{jn} \tag{5.6}$$

from i to n and r_{jn} the distance from j to n, with both distances in units of the benzene bond length. The F parameters were calculated from data for 4-substituted bicyclo[2.2.2]octane-1-carboxylic acids. The M and M_F parameters were then calculated from the σ_m and σ_p constants by means of

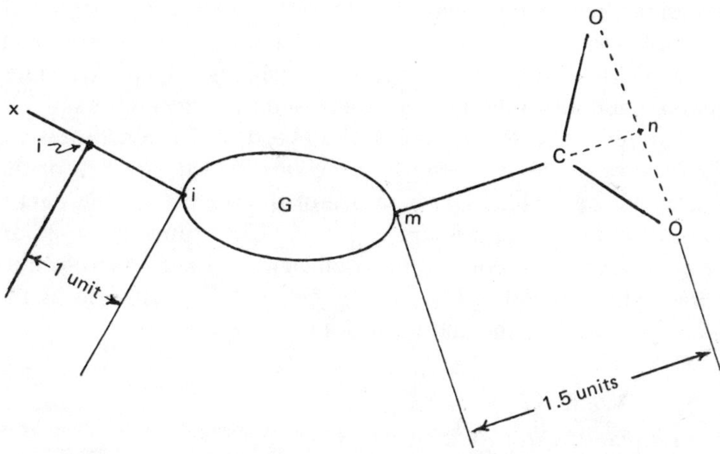

Fig. 5.1. Model used by Dewar *et al.*[17] for substituted carboxylic acids.

equation (5.5). Dewar *et al.* provided a test of their method by correlating observed σ constants with their σ_{im} constants. Excluding values for which steric factors may be significant, they obtained equation (5.7).

$$\sigma_{\mathrm{obs,X}} = 0.892\sigma_{im,\mathrm{X}} + 0.0087 \tag{5.7}$$

This result shows that the DGH treatment contains some error, as the slope of the correlation line should be 1.00, not 0.892. It would be useful to examine the correlation of data for various skeletal groups, G, to determine whether ρ remains constant. This has not yet been done. At present, the use of the extended Hammett equation seems preferable to that of the DG or the DGH treatment.

Data sets for polycyclic aromatic and heterocyclic compounds have been correlated by the simple Hammett equation by using substituent constants such as σ_m, σ_p, $\overset{0}{\sigma_p}$, $\overset{-}{\sigma_p}$, $\overset{+}{\sigma_p}$, and σ_I. Such a treatment requires that the composition of the electrical effect of the substituents in the set studied be comparable to that of the substituent constant used. The DG and the DGH method and multiparameter equations such as the extended Hammett and the Yukawa–Tsuno relation have also been applied to data sets for polycyclic aromatic and heterocyclic compounds. Finally, some authors have proposed sets of substituent constants applicable to a particular polycyclic aromatic or heterocyclic skeletal group. It is of considerable interest to compare these different relationships, in the hope of determining which correlation equations are more useful. For best results such a study should involve the application of the three methods to well-characterized data sets[10,11] and a comparison of the statistics obtained by least-mean-squares regression analysis. Regrettably, well-characterized data sets are rather rare, as are correlations of data sets by all three methods. We shall make comparisons whenever such results are available. The problem of goodness of fit has been discussed by many authors.[6,8,10,11] Our personal preference is for the use of the F test, which constitutes a test of the significance of the regression coefficients, and the square of the correlation coefficient, which is a measure of the fraction of the data that is accounted for by the regression equation. Unfortunately, most authors have reported only the correlation coefficient. We shall therefore use the square and the confidence level of the correlation coefficient as the best *available* indicators of the goodness of fit.

5.2. Polynuclear Aromatic Systems

The polycyclic aromatic systems studied may conveniently be divided into three categories: those in which the polycyclic aromatic system is the skeletal group G in XGY; those in which the polycyclic aromatic system is XG; and those in which it is the substituent, X. Since the latter two categories have received little attention, they will be discussed under the same heading. We have limited our coverage to those papers in which the polycyclic system is fully aromatic. Thus, applications of the Hammett and the extended Hammett equation to naphtho- and anthra-quinones will not be considered. We have also excluded series in which the polynuclear aromatic system is present as a constant substituent.

In view of the chaotic situation in the literature with regard to correlations of ionization constants, we shall always report values of α, β, or ρ for such correlations *without* sign whether the quantity correlated was $\log K_a$, pK_a, pK_{BH^+}, or half-neutralization potential. Actually, for correlations with $\log K_a$ and pK_a, the regression coefficients should be positive and negative, respectively. A similar situation exists for ^1H, ^{13}C, and ^{19}F chemical shifts. Again, in all correlations of chemical shifts reported in this work, the regression coefficients α, β, and ρ will be reported without sign. One would normally expect a positive sign for chemical shift correlations, as electron acceptor groups should cause deshielding.

5.2.1. The Polynuclear Aromatic System as a Skeletal Group

Applications of the Hammett, the extended Hammett, and the DG method to substituted polynuclear aromatic systems are now considered.

Biphenyls, Terphenyls, and Fluorenes

Only those biphenyl derivatives with the substituent in one ring and the reaction site in the other will be treated, because when substituent and reaction site are in the same ring, the series is equivalent to substituted benzenes with a constant phenyl substituent. Of the nine possible combinations of substituent and reaction-site location, the 4′-substituted 4-Y-biphenyls have received the most attention. Berliner and Blommers[18] first reported the application of the Hammett equation to the ionization constants of 3′- or 4′-substituted biphenyl-4-carboxylic acids, using σ_m and σ_p constants. This method is generally applicable to biphenyls with the substituent in one ring and the reaction site in the other, as the ring to which the reaction site is attached can be considered a side chain, Z, and the system then falls into the category XC_6H_4ZY, for which the simple Hammett equation has long been used. Data sets for substituted biphenyls, which have been correlated by use of the simple Hammett equation, and for which a correlation coefficient is available, are given in Table 5.1. The results obtained are fairly good. Application of the DGH method to ^{19}F chemical shifts of substituted biphenyls gave acceptable results.[26] ^{13}C chemical shifts of 4-substituted biphenyls were successfully correlated[25] by use of equation (5.8), where *F*, *M*, and M_F are the electrical-effect

$$\Delta\delta = aAF + bBM + cCM_F \tag{5.8}$$

parameters of the DGH method, and *A*, *B*, and *C* are the corresponding weighting factors. Correlations by use of the Yukawa–Tsuno equation of rate constants for electrophilic substitution reactions of 4′-substituted biphenyls at the 4-position have been carried out but no statistics were reported.[31] Correlations by use of the extended Hammett equation are given in Table 5.2.[32] ^{19}F chemical shifts for 4′-substituted 4-fluoro-biphenyls have been correlated with the aid of the DSP equation[33]; best results were obtained with σ_R(BA). Values of the *f* statistic originally proposed by Wells *et al.*[1] are given and indicate good results. According to Shorter,[6] the *f* statistic is related to the correlation coefficient (see, however, footnote on p. 162).

The results obtained for the correlation of data for substituted fluorenes by use of the Hammett equation are acceptable, but too sparse to permit the drawing of any conclusions. It is interesting to note that the correlation of proton chemical shifts for 2-substituted fluorenes was invariably best with σ_p, regardless of the position of the proton. Several reports have appeared regarding the correlation of rate constants for the reactions of 2-, 3-, or 4-substituted fluorenones with $NaBH_4$, and for the solvolysis of 2-, 3-, or 4-substituted 9-diazofluorenes by the DSP equation.[34–37] Unfortunately, no statistics were reported and therefore no

Table 5.1. Correlations of Data for Biphenyls, Terphenyls, and Fluorenes by the Hammett Equation

i,j[a]	Y[b]	G[c]	Q[d]	Subst.[e] constant	Reagent	Reaction conditions	ρ	$100r^2$[f]	n[g]	Ref.
4',4 3',4	CO_2H	Biphenyl	$\log K_a$	σ_m, σ_p		50% Aq. butylcellosolve, 25°C	0.482	96.2[h]	9	19
4',4 3',4	CO_2Et	Biphenyl	$\log k$	σ_m, σ_p	OH^-	88% EtOH–H_2O, 25°C	0.608	99.8[h]	7	19
4',4	NH_3^+	Biphenyl	pK_a	σ_p		10% v/v EtOH–H_2O, 20°C	0.67	86.5[i]	6	20
3',4	NH_3^+	Biphenyl	pK_a	σ_m		10% v/v EtOH–H_2O, 20°C	0.69	88.2[i]	5	20
4',4	NH_3^+	Biphenyl	pK_a	σ_p^o		Aq. dioxan, 25°C	1.06	96.8[j]	4	21
4',4 3',4	CO_2H	Biphenyl	$\log k$	σ_m, σ_p	Ph_2CN_2	EtOH, 30°C	0.224	99.0[k]	4	22
4',4	NH_3^+	Biphenyl	pK_a	σ_p		50% EtOH–H_2O, 20°C	0.574	99.6[h]	5	23
3',4 4',4	Br	3-Nitrobiphenyl	$\log k$	σ_m, σ_p	Piperidine	MeOH	1.00	99.4[h]	7	24
3',4	—	Biphenyl	$\delta\,^{13}C$	σ_m		10% C_6H_6–$COMe_2$	0.509	72.0[h]	12	25
3',4	—	Biphenyl	$\delta\,^{13}C$	σ_m^+		10% C_6H_6–$COMe_2$	0.534	75.3[h]	12	25
3',4	—	Biphenyl	$\delta\,^{13}C$	σ_p^+		10% C_6H_6–$COMe_2$	0.366	91.1[h]	12	25
2',4	—	Biphenyl	$\delta\,^{13}C$	σ_p^+		10% C_6H_6–$COMe_2$	0.640	90.6[h]	12	25
4',4	—	Biphenyl	$\delta\,^{13}C$	σ_p^+		10% C_6H_6–$COMe_2$	1.57	98.2[h]	12	25
4',4	—	Biphenyl	$\delta\,^{13}C$	σ_p		10% C_6H_6–$COMe_2$	2.18	99.4[h]	12	25
4',4	F	2,6,2',6'-Tetramethyl-biphenyl	$\delta\,^{19}F$	$\sigma_{4',4^o}$			16.9	87.8[i]	6	26
4',4 4'',4	F	Biphenyl, terphenyl	$\delta\,^{19}F$	$\sigma_{4',4}$; $\sigma_{4'',4^o}$			18.0	89.5[h]	32	26
4',4	NH_3^+	2-Phenylbiphenyl	pK_a	σ_p, σ_p^-		50% EtOH–H_2O, 20°C	0.202	98.4[h]	5	23
2,9		Fluorene	$\log k$	σ_m	OBr^-	75% Dioxan–H_2O, 25°C	4.40	97.2[h]	6	27
2,9	CH_2	Fluorene	pK_a	σ_m		25°C	6.30	95.3[k]	6	28

2,9; (2,7),9	C:CHPh	Fluorene	$E_{1/2}$	σ_m	0.23	96.4[k]	5	29
2,5	H	Fluorene	δ^1H	σ_p	30.5	96.0[h]	9	30
2,6	H	Fluorene	δ^1H	σ_p	14.5	96.0[h]	8	30
2,7	H	Fluorene	δ^1H	σ_p	23.8	96.0[h]	9	30
2,8	H	Fluorene	δ^1H	σ_p	15.9	96.0[h]	7	30
2,9	H	Fluorene	δ^1H	σ_p	28.6	96.0[h]	9	30

[a] i and j are the positions of the substituent and reaction site respectively. [b] Reaction site.
[c] Skeletal group. [d] Quantity correlated. [e] Substituent constant used in the correlation.
[f] Percentage of the data accounted for by the regression equation. Superscripts indicate the confidence level (CL) of the correlation coefficient.
[g] Number of points in the set. [h] 99.9%. [i] 99.0%. [j] 98.0%. [k] 99.5%.
[l] 95.0%. [m] 90.0%. [n] <90.0%. [o] DGH substituent constants.

Table 5.2. Correlations of Biphenyl Sets by the Extended Hammett Equation

i,j[a]	Y[b]	Q[c]	Reagent	Reaction conditions	α	β	n[d]	$100R^2$[e]	Ref.
4′,4	CO_2Et	log k	OH^-	88.7% w/w EtOH–H_2O, 25°C	0.590	0.633	6	99.8[f]	32
4′,4	CO_2Et	log k	OH^-	88.7% w/w EtOH–H_2O, 40°C	0.570	0.608	6	99.9[f]	32
2′,4	NH_3^+	pK_a		50% EtOH–H_2O, 20°C	0.619	0.217	6	97.6[g]	10
2′,4	CO_2H	pK_a		50% Butylcellosolve–H_2O, 20°C	0.299	0.197	9	87.2[f]	10

[a] Location of the substituent and the reaction site, respectively. [b] Reaction site.
[c] Quantity correlated. [d] Number of points in the set.
[e] Percentage of the data accounted for by the regression equation. Superscript indicates the confidence level of the correlation coefficient.
[f] 99.9%. [g] 99.5%. [h] 99.0%. [i] 98.0%.
[j] 95.0%. [k] 90.0%. [l] <90.0%.

comparison between the use of the simple Hammett equation and the multiple-parameter relations can be made for fluorenes. Correlations of data for substituted biphenyls by use of the extended Hammett equation or the DSP equation are certainly effective. Regrettably, the number of sets for which a comparison between the extended Hammett and the Hammett equation can be made is too small to allow any conclusions to be drawn.

Mention should also be made of the use of the Jaffé equation (5.9) for correlating pK_a values of 2-substituted fluorenes.[28] The Jaffé equation was

$$\log (k/k^0) = \rho_1\sigma_m + \rho_2\sigma_p \tag{5.9}$$

proposed for the purpose of expressing the transmission of the substituent effect by each of two possible paths, assuming that the localized effect is purely inductive in character (i.e., transmission of the localized effect is through bonds). We may write for σ_m and σ_p, equations (5.10) and (5.11),

$$\sigma_m = \lambda_m\sigma_L + \delta_m\sigma_D \tag{5.10}$$

$$\sigma_p = \lambda_p\sigma_L + \delta_p\sigma_D \tag{5.11}$$

where σ_L is a measure of the localized (field and/or inductive) effect, σ_D is a measure of the delocalized (resonance) effect, and λ and δ are constants. Substituting from (5.10) and (5.11) into (5.9) gives (5.12), (5.13), and (5.14); (5.14) is a form of the extended Hammett equation. Thus, correlation with the Jaffé equation is completely equivalent to correlation with

$$\log (k/k^0) = \rho_1(\lambda_m\sigma_L + \delta_m\sigma_D) + \rho_2(\lambda_p\sigma_L + \delta_p\sigma_D) \tag{5.12}$$

$$= (\rho_1\lambda_m + \rho_2\lambda_p)\sigma_L + (\rho_1\delta_m + \rho_2\delta_p)\sigma_D \tag{5.13}$$

$$\log k = \alpha\sigma_L + \beta\sigma_D + \log k^0 \tag{5.14}$$

the extended Hammett equation. Furthermore, unless and until it can be conclusively established that transmission of the localized electrical effect is almost entirely by the inductive mechanism, no conclusions should be drawn from the magnitude of ρ_1 and of ρ_2.

Naphthalenes

There are several approaches to the correlation of data for substituted naphthalenes. One method is the definition of substituent constants for use with substituted naphthalenes. This method requires a separate substituent constant for each possible combination of reaction site and substituent location. Since separate σ, σ^+, and σ^- values could conceivably be required, the resulting proliferation of substituent constants would be enormous. There are two variations of this method extant. In both of these, the ρ value required for the definition of σ constants is obtained from the corresponding substituted benzene set undergoing the same reaction under the same conditions. In the first variation the reference compound is the appropriate unsubstituted benzene derivative, whereas in the second variation (σ_{ij} constants), the reference compound is the appropriate unsubstituted naphthalene derivative. The first variation was adopted by Price and Michel,[38] who calculated σ values from rates of alkaline hydrolysis of substituted ethyl naphthalene-2-carboxylates. This work represents the first application of correlation analysis to naphthalene derivatives. Ionization constants of substituted 1-naphthoic acids and rate constants for hydrolysis of substituted ethyl naphthalene-1-carboxylates were used to calculate σ constants by both variations of the above method.[39] Values of σ for the NO_2 group calculated by the second variation were obtained from rates of alkaline hydrolysis of ethyl nitronaphthalene-1-carboxylates.[40] Wells and Ward[41] have tabulated a number of σ_{ij} constants from various reactions. Bryson has also obtained σ_{ij} constants from the ionization constants of substituted naphthylamines.[42] He reported that the σ_{31} and σ_{42} constants can be correlated by equation (5.15). Values of σ_{ij} for the

$$\sigma = \lambda\sigma_I + \delta\sigma_R \tag{5.15}$$

methyl group have been calculated from the ionization constants of x-methyl-1-naphthoic acids.[43] A review of σ_{ij} constants has appeared.[44] The values of σ_{81} were obtained from rate constants for the reactions of 8-substituted 1-naphthoic acids with diazodiphenylmethane in butylcellosolve.[45] Ionization constants and rate constants for reactions with diazodiphenylmethane in other solvents were correlated with these σ constants. Ionization constants of methylsulfonylnaphthalene carboxylic acids have been treated by use of the Hammett equation and σ_{ij} constants.[46] Interaction moments of nitronaphthylamines also were correlated[47] with the σ_{ij} values of Bryson. It has been suggested that the linearity of the plot of pK_{ij} values for cyanonaphthalene carboxylic acids against pK_{ij} values for nitronaphthalene carboxylic acids indicates that universal values of substituent constants independent of substituent position can be obtained.[48] This is incorrect. Linear plots such as the one referred to above will occur only when the substituents involved have similar electrical effects. The report that the ratio $\sigma_{ij,SO_2Me}/\sigma_{ij,NO_2}$ is independent of i and j[49] is again an example of this special case. In a significant paper, Wells and Adcock[50] correlated rate constants for alkaline hydrolysis of methyl 6- and 7-substituted 2-naphthoates by use of (5.16), with $\Delta = \Delta pK_a$ for the related acids. Equation (5.16) is equivalent to the

$$\log k = m\Delta + c \tag{5.16}$$

Hammett equation with Δ proportional to σ_{ij}. The results of the correlations showed that the Hammett equation applies to naphthalene data sets with excellent precision. It is noteworthy, however, that ρ values for the 6,2 and 7,2 data sets differ. Other examples of different ρ values were cited. Furthermore, ρ values for naphthalene data sets are frequently different from those for the corresponding benzene data sets.

An alternative method for the application of the Hammett equation to naphthalene derivates is by the use of the σ_m and σ_p constants for various types of naphthalene data sets. The obvious analogy between 4,1-naphthalene and *para*-substituted benzene derivatives, and between 3,1- and 4,2-naphthalene and *meta*-substituted benzene derivatives, has led to a number of studies using σ_p and σ_m, respectively. Sometimes the simple Hammett equation has been applied to naphthlene derivatives[57] in which the substituent was in positions 5, 6, 7, or 8. Correlations by the Hammett equation are reported in Table 5.3. The sets studied cover a reasonably wide range of types of data, and the results show that the use of the simple Hammett equation is indeed justified for 3,1-, 4,1-, and 6,2-naphthalenes.†

Naphthalene derivatives, from the initial reports of Dewar and Grisdale to the present, have been used as a test of the DG and the DGH method. Thus, pK_a values for substituted naphthylammonium ions calculated by means of the DG treatment often show significant deviations from the experimental values.[59] The DG method was reportedly unsuccessful when applied to 8,2- and 7,1-naphthoic acids.[60] Adcock and Dewar[61] proposed a modified DG method by the introduction of a cos θ term to improve the estimate of the field effect. The agreement between observed and calculated ^{19}F chemical shifts of substituted fluoronaphthalenes is generally good.[61,62] Values of pK_a for 8,2- and 5,1-naphtholsulfonates calculated by the DG method deviate significantly from the observed values.[63] Correlations with the DG and the DGH constants for which statistics are reported are relatively rare. Some examples are given in Table 5.3: The correlations are generally satisfactory. There are still flaws in the method, however. Excluding the ρ values for *meta*-benzene and 4,2-naphthalene sets which are not well correlated, the variation in ρ is from -7.61 for 3,1-naphthalenes to -30.6 for 6,2-naphthalenes when ^{19}F chemical shifts are correlated with the DGH constants. Furthermore, a correlation of observed σ_{ij} values with those calculated by the DGH method has a slope significantly different from unity. It seems then, that even the DGH method retains some degree of error, although further examination is necessary to prove this point.

Bancroft and Howe first proposed the application of the DG method to the calculation of σ^+ constants,[65] by simply using σ_m^+ and σ_p^+ in place of σ_m and σ_p for the calculation of F^+ and M^+. Eaborn and Fischer[66] have

† We shall frequently find it convenient to refer to a skeletal group as *i,j*-arene, where *i* is the location of the substituent and *j* that of the reaction site.

*Table 5.3. Correlations of Naphthalene Sets by the Hammett Equation**

i,j[a]	Y[b]	Q[d]	Subst. constant	Reagent	Reaction conditions	ρ	$100\,r^{2}$[f]	n[g]	Ref.
3,1	CO_2Et	log k	σ_m, σ_p	OH^-	85% w/w EtOH–H_2O	2.21	99.6[h]	8	51
4,1									
4,1; 3,1	CO_2Et	log k	σ_m, σ_p	OH^-	85% w/w EtOH–H_2O	2.21	99.4[h]	9	52
4,1: 3,1	O^-	log k	σ_m, σ_p	MeI	EtOH	−1.06	99.0[h]	5	53
3,1	OH	τ_{OH}	σ_m, σ_p^0		DMSO	1.55	99.2		54
4,1									
4,1	q	log k	σ_p	MeO^-	MeOH	−2.19	98.6		55
4,1	q	log k	σ_p^0	MeO^-	MeOH	−2.31	98.2		55
4,1	r	ν_{CO}	σ_p			15.1	95.6[h]	7	56
4,1	s	ν_{CO}	σ_p			11.1	99.4[h]	8	56
8,1	CH_2Cl	log k	σ_m		80% $COMe_2$–H_2O	−3.01	99.0[h]	5	57
4,1	OH	pK_a	σ_p		50% v/v EtOH–H_2O	3.30	96.0[h]	6	58
4,1	OH	pK_a	σ_p^-		50% v/v EtOH–H_2O	3.16	97.6[h]	6	58
6,2	OH	pK_a	σ_p		50% v/v EtOH–H_2O	1.53	98.0[h]	6	58
6,2	OH	pK_a	σ_p^-		50% v/v EtOH–H_2O	1.45	98.6[h]	6	58
5,1	OH	pK_a	σ_p		50% v/v EtOH–H_2O	1.07	85.7[i]	6	58
5,1	OH	pK_a	σ_p^-		50% v/v EtOH–H_2O	0.99	81.9[j]	6	58
4,1	OH	δ^1H	σ_p		DMSO	1.61	99.6[h]	6	58
4,1	OH	δ^1H	σ_p^-		DMSO	1.51	97.6[h]	6	58
6,2	OH	δ^1H	σ_p		DMSO	0.76	99.2[h]	6	58
6,2	OH	δ^1H	σ_p^-		DMSO	0.71	96.8[h]	6	58
5,1	OH	δ^1H	σ_p		DMSO	0.52	80.3[j]	6	58
5,1	OH	δ^1H	σ_p^-		DMSO	0.47	74.1[l]	6	58
5,1	H	δ^1H	$\sigma_{51}{}^p$		CCl_4	—	82.8[j]	6	64
5,2	H	δ^1H	$\sigma_{52}{}^p$		CCl_4	—	94.1[k]	6	64
8,2	H	δ^1H	$\sigma_{82}{}^p$		CCl_4	—	98.0[h]	6	64
3,1	F	δ^{19}F	$\sigma_{31}{}^o$			7.61	94.9[k]	7	17

continued overleaf

Table 5.3—continued

i,j^a	Y^b	Q^d	Subst. constant	Reagent	Reaction conditions	ρ	$100r^{2\,f}$	n^g	Ref.
4,1	F	$\delta^{19}F$	$\sigma_{41}{}^o$			23.9	97.4^h	7	17
6,1	F	$\delta^{19}F$	$\sigma_{61}{}^o$			10.6	93.7^i	5	17
7,1	F	$\delta^{19}F$	$\sigma_{71}{}^o$			14.1	97.0^h	9	17
4,2	F	$\delta^{19}F$	$\sigma_{42}{}^o$			0.400	3.4^n	6	17
6,2	F	$\delta^{19}F$	$\sigma_{62}{}^o$			30.6	99.0^h	9	17
7,2	F	$\delta^{19}F$	$\sigma_{72}{}^o$			17.8	93.1^h	9	17
8,2	F	$\delta^{19}F$	$\sigma_{82}{}^o$			17.3	97.4^h	9	17
3,2; 4,2 5,2; 6,2 7,2	NH_3^+	pK_a	σ_m, σ_p			2.2	93.3^h	18	69
3,2; 4,2 5,2; 6,2 7,2	NH_3^+	pK_a	σ_{32}, σ_{42}, σ_{52}, σ_{62}, $\sigma_{72}{}^p$			2.9	90.1^h	17	69

*For footnotes other than those below, see Table 5.1.
[p]Dewar–Grisdale constants. [q]3-(4-Substituted 1-naphthylmethylene)phthalides.
[r]*Trans*-3-(4-substituted 1-naphthylmethylene)-4,7-dithia-4,5,6,7-tetrahydrophthalides.
[s]*Trans*-3-(4-substituted 1-naphthylmethylene)phthalides.

expanded on this work and examined the Wheland intermediate as an alternative model for the calculation of σ_{ij}^+ constants. The Wheland intermediate is said to be the better model. Regrettably, no statistics for a correlation with the σ_{ij}^+ constants were reported. Plots of log k_{rel} vs. σ_{ij}^+ for detritiation of substituted 2-tritionaphthalenes show considerable scatter. A report has appeared[67] that examines various modifications of the DG treatment applied to the calculation of σ_{ij}^+ constants. The field effect is said to give about the same results with the use of either $1/r$ or $\cos\theta/r$. Forsyth[68] has proposed modifications of the DG method, which have given very good results, allowing data for all the substituted 1-tritionaphthalenes considered to lie on one line, and for all the substituted 2-tritionaphthalenes to lie on another. It should be noted, however, that the range of substituent types under consideration was very small, as only methyl, halogeno, methoxy, and hydrogen have been examined. A plot of σ_{obs}^+ vs σ_{calc}^+ for a wide range of substituent types again showed considerable scatter.

A number of authors have reported the application of the extended Hammett and the DSP equation to naphthalene sets.[1,32,70,71] Correlations by the extended Hammett equation are in Table 5.4. Results are generally good. Comparisons with the Hammett equation are possible for pK_a values and OH chemical shifts of 4,1-, 6,2-, and 5,1-substituted naphthols. The

*Table 5.4. Correlation of Naphthalene Sets by the Extended Hammett Equation and the DSP Equation**

i,j^a	Y^b	Q^c	Reagent	Reaction cond.	α	β	n^d	$100R^{2e}$	Ref.
3,1	CO_2H	pK_a		50% v/v EtOH–H_2O, 25°C	1.37	0.426	5	99.0[f]	32
4,1	CO_2H	pK_a		50% v/v EtOH–H_2O, 25°C	1.64	1.85	9	99.4[f]	32
5,1	CO_2H	pK_a		50% v/v EtOH–H_2O, 25°C	1.02	0.569	9	97.8[f]	32
6,1	CO_2H	pK_a		50% v/v EtOH–H_2O, 25°C	0.793	0.547	8	99.6[f]	32
7,1	CO_2H	pK_a		50% v/v EtOH–H_2O, 25°C	0.702	0.568	7	98.4[f]	32
4,2	CO_2H	pK_a		50% v/v EtOH–H_2O, 25°C	1.25	0.497	10	96.0[f]	32
5,2	CO_2H	pK_a		50% v/v EtOH–H_2O, 25°C	0.851	0.429	6	99.6[f]	32
6,2	CO_2H	pK_a		50% v/v EtOH–H_2O, 25°C	0.812	0.633	10	99.0[f]	32
7,2	CO_2H	pK_a		50% v/v EtOH–H_2O, 25°C	0.754	0.411	12	95.8[f]	32

continued overleaf

Table 5.4.—continued

i,j^a	Y^b	Q^c	Reagent	Reaction cond.	α	β	n^d	$100R^{2\,e}$	Ref.
8,2	CO_2H	pK_a		50% v/v EtOH–H_2O, 25°C	0.588	0.548	10	95.5^f	32
4,1	CO_2Et	log k	OH^-	85% w/w EtOH–H_2O	2.37	2.23	6	99.4^f	32
2,1	CO_2H	ν_{CO}		Dioxan	10.7	11.4	6	84.6^j	32
1,2	H	δ^1H		c-C_6H_{12}	0.794	1.53	10	91.0^f	72
3,2	H	δ^1H		c-C_6H_{12}	0.348	1.14	10	87.8^f	72
$1,2^m$	H	δ^1H		CCl_4	1.08	1.52	7	93.5^g	72
$3,2^m$	H	δ^1H		CCl_4	0.828	1.08	7	90.8^g	72
1,–	$-^n$	log K_e	$PkOH^o$	$CHCl_3$, 27°C	−0.684	−0.850	10	71.9^g	73
2,–	$-^n$	log K_e	$PkOH^o$	$CHCl_3$, 27°C	−0.747	−0.524	11	86.1^f	73
1,–	$-^n$	log K_e	$PkOH^o$	$CHCl_3$, 28.5°C	−1.59	−1.68	9	91.0^f	73
2,–	$-^n$	log K_e	$PkOH^o$	$CHCl_3$, 28.5°C	−1.63	−1.15	10	91.0^f	73
4,1	OH	pK_a		50% v/v EtOH–H_2O	3.01	3.98	6	94.5^h	58
$4,1^p$	OH	pK_a		50% v/v EtOH–H_2O	2.88	3.87	6	98.2^g	58
$4,1^q$	OH	pK_a		50% v/v EtOH–H_2O	3.76	3.03	6	90.1^i	58
6,2	OH	pK_a		50% v/v EtOH–H_2O	1.41	1.81	6	97.6^g	58
$6,2^p$	OH	pK_a		50% v/v EtOH–H_2O	1.32	1.78	6	99.0^f	58
$6,2^q$	OH	pK_a		50% v/v EtOH–H_2O	1.82	1.36	6	94.9^g	58
5,1	OH	pK_a		50% v/v EtOH–H_2O	0.67	2.01	6	93.9^h	58
$5,1^p$	OH	pK_a		50% v/v EtOH–H_2O	0.59	2.03	6	91.0^i	58
$5,1^q$	OH	pK_a		50% v/v EtOH–H_2O	0.91	1.29	6	94.3^h	58
4,1	OH	δ^1H		DMSO	1.60	1.64	6	99.4^f	58
$4,1^p$	OH	δ^1H		DMSO	1.46	1.64	6	96.4^g	58
$4,1^q$	OH	δ^1H		DMSO	2.16	1.27	6	98.8^f	58
6,2	OH	δ^1H		DMSO	0.75	0.78	6	98.6^f	58
$6,2^p$	OH	δ^1H		DMSO	0.68	0.78	6	95.1^g	58
$6,2^q$	OH	δ^1H		DMSO	1.02	0.59	6	98.8^f	58
5,1	OH	δ^1H		DMSO	0.31	1.01	6	86.7^i	58
$5,1^p$	OH	δ^1H		DMSO	0.25	1.04	6	81.9^j	58
$5,1^q$	OH	δ^1H		DMSO	0.45	0.64	6	87.6^i	58

*For footnotes other than those below, see Table 5.2.
[m] 6-Methoxynapthalenes. [n] The entire molecule is the reaction site.
[o] Picric acid. [p] $\sigma_{\bar{R}}$ defined as $(\sigma_p^- - \sigma_I)$ was used in place of σ_R.
[q] The Swain–Lupton $\mathfrak{F}$ and $\mathfrak{R}$ values (see Chapter 4) were used in place of σ_I and σ_R.

correlations obtained do not seem to be significantly better than those obtained by the application of the simple Hammett equation.

Only one paper has appeared in which the Yukawa–Tsuno equation is applied to naphthalene sets.[74] The Jaffé equation was proposed for use with naphthalene sets[75] but does not seem to have been much applied. Several papers have appeared in which the Thirot equation[76] (5.17) has

$$Q = \rho(\sigma_X - r\Delta\sigma_{1X} - s\Delta\sigma_{2X}) \tag{5.17}$$

been applied to the correlation of data for substituted naphthalenes.[77–80] In this relation, σ_X is a localized electrical-effect parameter based on the Taft

σ^* values, $\Delta\sigma_1$ and $\Delta\sigma_2$ are resonance parameters in effect, the former loosely related to σ_R^- and the latter to σ_R^+. The interpretation of $\Delta\sigma_1$ and $\Delta\sigma_2$ is not clear and the results reported to date do not seem to be an improvement upon those obtained with simpler relationships.

Square roots of the integral intensities of bands for skeletal vibrations of substituted naphthalenes were correlated by σ_R^0 constants. Better results were obtained for 1-substituted than for 2-substituted naphthalenes. The difference in the ρ values is ascribed to a *peri*-interaction in the 1-substituted naphthalenes.[81] Polarographic half-wave potentials of 1- and 2-alkylnaphthalenes have been correlated with the Taft σ^* values.[82] The Hammett and DSP equations have been used to obtain estimates of the microconstants in the ionization of 3-amino-1- and 4-amino-2-naphthoic acids.[83]

Application of the Brown selectivity relationship to naphthalene reacting at positions 1 and 2 resulted in curved plots,[84] although a plot of $\log f_1$ vs. $\log f_2$ is linear[85] (f_1 and f_2 are the partial rate factors for reaction at positions 1 and 2).

Other Polynuclear Aromatic Systems

The only systems that have been studied are anthracenes, phenanthrenes, and azulenes. One of the first papers in this area was that of Norman and Ralph,[86] who examined the correlation of pK_a values of 10-substituted anthracene-9-carboxylic acids and the carbonyl stretching frequencies of their methyl esters by use of the simple Hammett equation. The σ_I value for NO_2 was required although the other groups required σ_p^n. This was ascribed to *peri*-interactions with the hydrogen atoms at positions 1 and 8. Thus, a constant steric effect is exerted upon the substituent at position 10 and the reaction site at position 9, which results in steric inhibition of resonance. The quite good results of correlations with the simple Hammett equation are reported in Table 5.5.

The DGH method has been applied to the ^{19}F chemical shifts of 10-substituted 9-fluoroanthracenes,[26] and to the pK_a values of 8-substituted anthracene-1-carboxylic acids.[17,90] Although statistics are not reported, a plot of the pK_a values vs. the DGH σ constants for 2,8-naphthoic, 1,7-naphthoic, and 8,1-anthroic acids shows a considerable amount of scatter. Bowden and his co-workers have defined two sets of substituent constants for 10,9-anthracenes, from the ionization constants of the acids,[91] and from the rate constants for the reactions of the acids with diazodiphenylmethane in ethanol at 30°C.[45] A correlation of proton chemical shifts of 9-substituted anthracenes was carried out, with the substituent constants defined from pK_a values, but with poor results.

*Table 5.5. Correlations by the Hammett Equation for Polycyclic Arenes**

i,j[a]	Y[b]	G[c]	Q[d]	Subst.[e] constant	Reagent	Reaction conditions	ρ	$100r^2$ [f]	n[g]	Ref.
10,9	CO_2H	Anthracene	$\log k$	$\sigma_p{}^n$	Ph_2CN_2	EtOH, 30°C	0.697	90.6[l]	4	22
10,9	9–C	Anthracene	$\log k_{rel}$	σ_p^+	$\cdot CCl_3$	C_6H_6, 70°C	−0.83	99.6[h]	9	87
10,9	F	Anthracene	$\delta^{19}F$	$\sigma_{10,9}$		$CHCl_3$	15.7	94.1[h]	9	26
10,9	T	Phenanthrene	$\log k$	σ_p^+		CF_3CO_2H		100.0[h]	6	68
3,1	CH_2CH_2–OTs	Azulene	$\log k$	σ_p^0		AcOH	−3.60	99.0[h]	5	88
3,1	CO_2H	Azulene	pK_a	σ_m		50% v/v EtOH–H_2O, 25°C	1.93	95.1[h]	7	89
3,1	CO_2H	Azulene	pK_a	σ_p^0		50% v/v EtOH–H_2O, 25°C	1.52	98.2[h]	7	89

*For footnotes, see Table 5.1.

*Table 5.6. Correlations by the Extended Hammett and the Yukawa–Tsuno Equation for Polycyclic Aromatic Hydrocarbons(G)**

i,j[a]	Y[b]	Q[c]	G	Reagent	Reaction conditions	α	β	n[d]	$100R^2$[e]	Ref.
9, –	Ring	log k	Anthracene	Maleic anhydride	Dioxan, 130°C	−3.84	−4.23	5	97.2[i]	93
9, 10, –	Ring	log k	Anthracene	Maleic anhydride	Dioxan, 130°C	−4.93	−6.91	4	98.6[k]	93
2, –	Ring	log (% syn/% anti)	Anthracene	Maleic anhydride	Benzene	−0.262	−0.123	4	90.6[l]	93
2,1[m]	CO_2H	pK_a	Azulene		50% v/v EtOH–H_2O, 25°C	1.23	1.61	5	99.9[f]	89
3,1[m]	CO_2H	pK_a	Azulene		50% v/v EtOH–H_2O, 25°C	1.01	0.97	7	98.4[f]	89
6,1[m]	CO_2H	pK_a	Azulene		50% v/v EtOH–H_2O, 25°C	0.71	0.93	4	99.4[j]	89
2,1[n]	CO_2H	pK_a	Azulene		50% v/v EtOH–H_2O, 25°C	2.15[o]	0.4[p]	5	100.0[f]	89
3,1[n]	CO_2H	pK_a	Azulene		50% v/v EtOH–H_2O, 25°C	1.68[o]	−0.18[p]	7	99.6[f]	89
6,1[n]	CO_2H	pK_a	Azulene		50% v/v EtOH–H_2O, 25°C	1.22[o]	0.05[p]	4	99.6[j]	89

*For footnotes other than those below, see Table 5.2.
[m] Using the Swain–Lupton constants. [n] Using the Yukawa–Tsuno equation. [o] ρ. [p] r.

There are only two papers that report the application of the extended Hammett equation or DSP equation to anthracenes.[92,93] The results for which statistics were reported are in Table 5.6. Polarographic half-wave potentials of 1-, 2-, and 9-alkylanthracenes were correlated with the σ^* values.[94]

One set of experimental results for the phenanthrenes has been reported: the rate constants for detritiation of 10-substituted 9-tritiophenanthrenes.[95a] Successful correlations by the simple Hammett equation, the Yukawa–Tsuno, and the DSP equation, and various modifications of the DG method have appeared.[95a,68]

Azulenes have received some attention. pK_a values and rate constants have been correlated by the simple Hammett, the extended Hammett, and the Yukawa–Tsuno equation.[88,89,95b,96,97]

Transmission of Substituent Effects through Polynuclear Aromatic Systems

The available evidence indicates that the localized effect is predominantly a field effect,[9] as is shown by the calculation of α values by means of the Kirkwood–Westheimer equation. The α values are coefficients of the localized-effect substituent constants in the extended Hammett equation and β values are coefficients for the delocalized effect. The correlation of β values has been carried out by means of equation (5.18), where q_{G} is the formal negative charge at the carbon atom that

$$\beta_{\mathrm{G}} = mq_{\mathrm{G}} + c \tag{5.18}$$

bears the reaction site in the model compound, $^{-}CH_2GH$.[32] The extent of transmission of the localized effect can be described by γ_L, given by (5.19).

$$\gamma_{L,\mathrm{G}} = \alpha_{\mathrm{G}}/\alpha_{\mathrm{G}^0} \tag{5.19}$$

Similarly, for the delocalized effect we have (5.20). Values of γ_L and γ_D are

$$\gamma_{D,\mathrm{G}} = \beta_{\mathrm{G}}/\beta_{\mathrm{G}^0} \tag{5.20}$$

presented in Table 5.7. The values given were obtained from correlations of the ionization constants of the appropriate carboxylic acids by use of the extended Hammett equation.[32] As G^0, the 4,1-benzene group was chosen.

It is also of interest to consider the variation of the composition of the electrical effect. For this purpose the quantity P_R is useful. It is given by the expression (5.21). Values of P_R are also given in Table 5.7. These values

$$P_R = 100\beta/(\alpha + \beta) \tag{5.21}$$

clearly show the variation in composition of the electrical effect with change in the skeletal group G. They also show why correlation by the simple Hammett equation is successful. The substituent constants, σ_m, σ_p^0,

Table 5.7. Values of γ_L, γ_D, and P_R

i,j Skeletal Group	γ_L	γ_D	P_R
3,1-Benzene	1.00	0.33	25.0
4,1-Benzene	1.00	1.00	50.0
4′,4-Biphenyl[a]	0.341	0.341	50.0
3,1-Naphthalene	0.945	0.294	23.7
4,1-Naphthalene	1.13	1.28	53.0
5,1-Naphthalene	0.703	0.392	35.8
6,1-Naphthalene	0.547	0.377	40.8
7,1-Naphthalene	0.484	0.392	44.7
4,2-Naphthalene	0.862	0.343	28.4
5,2-Naphthalene	0.587	0.296	33.5
6,2-Naphthalene	0.560	0.437	43.8
7,2-Naphthalene	0.520	0.283	35.3
8,2-Naphthalene	0.406	0.378	48.2
1,8-Anthracenyl[b]	0.119	0.0784	39.7

[a] Estimated values from ρ for correlation by use of the Hammett equation.
[b] $X = CO_2^-$ was excluded from the correlation.

and σ_p have P_R values of 25.0, 40.0, and 50.0, respectively. Obviously, 3,1- and 4,2-naphthalene derivatives can be correlated by use of σ_m; 5,1-, 6,1-, 7,1-, 7,2-, and 6,2-naphthalenes and 8,1-anthracene derivatives with σ_p^0; and 4,4′-biphenyl, and 4,1- and 8,2-naphthalene derivatives with σ_p constants. The 5,2-naphthalene derivatives have a P_R value that is intermediate between the values corresponding to σ_m and σ_p^0.

As expected, γ_D values for 6,2- and 8,2-naphthalene derivatives in which substituent and reaction site are conjugated are greater than γ_D values for 5,2- and 7,2-naphthalene derivatives in which substituent and reaction site are not conjugated. It is noteworthy that the 5,1-, 6,1-, and 7,1-naphthalene derivatives all have essentially the same value of γ_D, a value comparable to that of the 6,2- and 8,2-naphthalene derivatives, although with the 6,1-naphthalene derivatives the substituent and reaction site are not conjugated. One of these observations can be accounted for qualitatively in the following manner: The q values for 5,1- and 7,1-naphthalene are greater than those for 6,2- and 8,2-naphthalene. On the basis of equation (5.18), γ_D should be larger for the former than for the latter. The *peri*-interaction between the carboxy group in position 1 and the hydrogen in position 8 should prevent molecular planarity and therefore reduce the value of γ_D. At present, the magnitude of γ_D for the 6,1-naphthalene derivatives is inexplicable.

The double *peri*-interaction in 10,9-anthracene derivatives results in steric inhibition of resonance in both the reaction site and the substituent.

It would be interesting to examine the γ_D value for the 10,9-anthracene derivatives. Unfortunately a sufficiently well-characterized set of pK_a values has not been measured.

5.2.2. Polynuclear Aromatic Systems as Substituents

Substituent constants of polynuclear aromatic systems may be divided into three types:

(a) *Normal* substituent constants. In the definition of these the aryl group is X in the defining set, XGY, and the reference compound used in defining the substituent constants is HGY.

(b) *Replacement* (positional) substituent constants. In the definition of these the aryl group is HG in HGY, and the reference compound is PhY.

(c) *Special* substituent constants. In the definition of these the aryl group is X in the set XY, and the reference compound is X^0Y, where X^0 is also an aryl group.

Normal Substituent Constants

Values of σ_I for the 1- and the 2-naphthyl group have been reported.[98] Rate constants of H–D exchange of deuterioarenes catalyzed by lithium cyclohexylamide were correlated with σ_I constants and with F_i, which is given by equation (5.22), where r_{ij} is the distance between the

$$F_i = \sum 1/r_{ij} \tag{5.22}$$

reacting carbon atom, C_i, and a remaining carbon atom, C_j. Thus, F_i values can be used to calculate σ_I values for aryl groups.[99] It must be noted, however, that this correlation of rate constants for H–D exchange with σ_I involved only three points in the data set.

Replacement Substituent Constants

The first example extant of a replacement substituent constant is the calculation by Hammett of a substituent constant for the fusion of CH=CH—CH=CH on to a benzene ring at the 3,4-positions.[100] This substituent constant is simply the replacement constant for the 2-naphthyl group. Later, values of σ^* and E_s were calculated for the 2,3-CH=CH—CH=CH group, which are equivalent to replacement substituent constants for the 1-naphthyl group.[101] Replacement substituent constants were calculated in the usual manner from the ionization constants of 2-arenecarboxylic acids and from those of 1-arenecarboxylic acids by means of (5.23), where 0.41 is a "quantum-mechanically derived

$$\sigma_{ar} = [\log(k/k^0) - 0.41]\rho^{-1} \tag{5.23}$$

correction for the steric effect".[102] Rate constants for the alkaline hydrolysis of arylmethyl benzoates were used to determine σ^0-type replacement constants.[103] The σ^0 constants are given by expression (5.24),

$$\sigma^0 = 0.06\Delta F_i + 1.51\pi_{rr} + 0.06 \qquad (5.24)$$

where F_i is given by equation (5.22) and π_{rr} is the self-atom polarizability. By using $\Delta\sigma_R^+$ values obtained from the solvolysis of 1-aryl-1-chloroethanes, and the σ^0 constants defined above, ionization constants of aryl methyl ketones, rates of alkaline hydrolysis of ethyl arenecarboxylates, rates of protodesilylation of trimethylsilylarenes, and of solvolysis of 1-aryl-1-chloroethanes and of arylmethyl tosylates were correlated by use of the Yukawa–Tsuno equation.[104] Brown and Inukai have determined σ^+-type replacement constants for aryl groups from the rates of solvolysis of 2-aryl-2-chloropropanes.[105] Rate constants for the solvolysis of arylchlorophenylmethanes were correlated with these σ^+-type constants.[106] Values of σ^+(replacement) for the 1- and the 2-biphenylene group were determined from rates of pyrolysis of 1-arylethyl acetates.[107,108] From a correlation of rates of acetolysis of 2-arylethylmercuric perchlorates with those of arylmethylmercuric perchlorates, ρ^+ was evaluated,[109] and σ^+ (replacement) constants were also obtained from the acetolysis of arylmethyl tosylates.[110] These σ^+ constants were then correlated with the results of SCF–MO calculations of various types; CNDO/2 calculations gave the best results. Relative rates of protodetritiation were also treated by means of these constants,[111] and σ^+ (replacement) constants were also calculated.[112] Streitwieser proposed an equation[113] of the form (5.25),

$$\log(k_i/k_{\text{Ph}}) = \rho_{\text{ar}}\sigma_{i,\text{ar}} \qquad (5.25)$$

where $\rho_{\text{ar}} \equiv 1.000$ for the solvolysis of arylmethyl chlorides in 80% aqueous ethanol at 50°C, and $\sigma_{i,\text{ar}} \equiv 0.000$ for the phenyl group. Thus, the $\sigma_{i,\text{ar}}$ constants are actually σ^+(replacement) constants. Rates of alkaline hydrolysis of methyl arylacetates have been correlated by use of $\sigma_{i,\text{ar}}$ constants. The data for hindered and unhindered compounds lie on separate lines.[114] Rates of H–D exchange of (deuteriomethyl)arenes with lithium cyclohexylamide and rates of acetolysis of arylmethylmercuric perchlorates[116] have also been correlated by use of $\sigma_{i,\text{ar}}$ constants.[115] Values of $\sigma_{i,\text{ar}}$ are linearly related to ΔE_π values obtained from the acetolysis rates of arylmethylmercuric perchlorates.[116] Also, σ^+ (replacement) constants have been used to correlate the SC ratio in the resonance-theory treatment of electrophilic substitution.[117] The SC ratio is the ratio of the number of Kekulé structures for the postulated cation to that for the neutral reactant. The equation (5.26) has been derived, where f_1 and f_2 are partial-rate factors

$$\log f_1 = [\sigma^+_{(\text{repl,1-naphth})}/\sigma^+_{(\text{repl,2-naphth})}] \log f_2 \qquad (5.26)$$

for electrophilic aromatic substitution at the 1- and the 2-position of naphthalene, and has been applied to f values for various electrophilic substitutions. Deviations from linearity are believed to be due to steric effects encountered in substitution at the 1-position.[118]

Special Substituent Constants

Streitwieser has suggested the relation (5.27), where ρ^* is defined as 1.000 for the protonation of arenes and σ_r is defined as 0.000 for the

$$\log(k_r/k_{1\text{-naph}}) = \rho^*\sigma_r \tag{5.27}$$

1-position of naphthalene.[112] The σ_r constants have been used to correlate uv spectral properties,[119,120] polarographic half-wave potentials,[119–121] ionization potentials,[119] and proton chemical shifts[119,122,123] of arenes. They have also been applied to relative rates of the reactions of arenes with N_2F_4, with poor results,[124] to the acetolysis of arylmethylmercuric perchlorates,[116] and to rates of light-initiated decomposition of arenes.[125] The σ_r constants can be qualitatively related to indices of aromaticity.[126] Polarographic half-wave potentials of diaryl ketones and of quinones can be correlated by use of equation (5.28).[127] The Streitwieser σ_r constants

$$E_{1/2} = \rho^*\sigma_r + c \tag{5.28}$$

are linearly related to the σ^+ (replacement) constants of Eaborn and his co-workers.[128,129]

5.3. Heterocyclic Systems

Because of the large number of heterocyclic systems that have been studied, the organization of the heterocyclic portion of this chapter differs from that of the previous section on polynuclear aromatic systems.

5.3.1. Transmission of Substituent Effects through the Ring

As for the polynuclear aromatic systems, a major topic is that of the heterocyclic system functioning as the skeletal group, G, in XGY. It is convenient to divide this topic into two parts; the first is that in which the reaction site, Y, is a group attached to the heterocyclic ring, the second is that in which the reaction site is a heteroatom in the ring. The first part will now be considered.

5-Membered Rings

The first recorded example of the application of the Hammett equation to heteroarenes was the correlation by Hammett of ionization constants of 5-substituted 2-furoic acids with the σ_p constants.[100] Many authors have carried out similar correlations by the simple Hammett equation for XGY, where G is a furan, thiophen, selenophen, tellurophen, or pyrrole ring. In such correlations there are several possible locations of substituent and reaction site [(I)–(VI)] Z = O, S, Se, Te, or NH. The *ortho*-substituted systems have received little attention, unfortunately. The

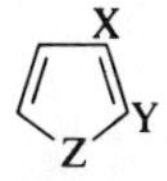

(I) 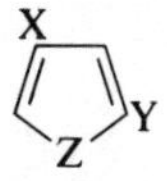 (II) 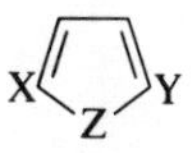 (III)

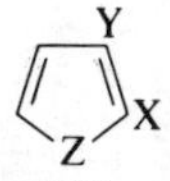

(IV)  (V) 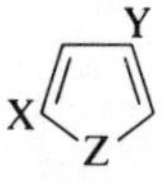 (VI)

systems (II) and (III) are considered to bear a formal resemblance to *meta*-substituted benzenes in which CH=CH is replaced by a heteroatom. Data for many such sets have been correlated by use of σ_m constants. The 5,2-system (III) is considered formally analogous to *para*-substituted benzene derivatives; such sets have generally been correlated with σ_p, σ_p^0, σ_p^-, or σ_p^+ constants depending on the nature of the reaction site.[149] Correlations by the simple Hammett equation of the 4,2-, 5,2-, and 5,3-systems are in Table 5.8, and 4,2- and 5,2-compounds are often included in the same set. Some reactions of thiazoles have also been studied.[177] Examination of the results in Table 5.8 indicates that correlation by the Hammett equation using the appropriate substituent constants is generally effective. Significant correlation was obtained in all of the sets in Table 5.8, with 26 of the 32 sets having confidence levels for the correlation coefficient >99.0%, while 11 sets had $100r^2$ values ≥99.0, and 13 more had values <99.0 but ≥96.0. Correlations using the simple Hammett equation are, however, not always highly successful. Thus, the $E_{1/2}$ values of 5-substituted 2-nitrothiophens in water at pH 3.10 and the rate constants for esterification of 5-substituted thiophen-2-carboxylic acids gave very poor values of $100r^2$ (see Table 5.8). Rate constants for the reactions of phenylhydrazine with 5-substituted furfurals were not correlated with either σ or σ^+.[141]

Of further interest are the observations that ionization constants of 5-substituted furan-3-carboxylic acids give the best correlation with σ_p, rather than, as expected,[142] with σ_m, and rates of permanganate oxidation

Table 5.8. 5-Membered Ring Sets Correlated by the Hammett Equation

i,j	Y	Z[a]	Subst const.	Q[b]	Reagent	Reaction conditions	ρ	n[c]	100r[2 d]	Ref.
5,2	CO_2H	O	σ_p	pK_a		H_2O, 25°C	1.40	6	97.6[e]	3
5,2	CO_2H	O	σ_p	log k	Ph_2CN_2	EtOH, 15.5°C	1.00	7	98.2[e]	130
5,2	CO_2H	O	σ_p	log k	Ph_2CN_2	EtOH, 25°C	0.975	7	98.8[e]	130
5,2	CHO	O	σ_p^+	log k	5-Nitro-2-phenylsulfonylmethylfuran	AcOH, 118°C	1.815	5	94.1[f]	131
5,2	CHO	O	σ_p^+	log k	5-Nitro-2-phenylsulfonylmethylfuran	MeOH, 41°C	1.10	6	90.3[g]	131
5,2	CHO	O	σ_m	log k	5-Nitro-2-phenylsulfonylmethylfuran	H_2O, 25°C pH 11.5–13.3	−1.30	6	99.5[e]	132
5,2	CO_2H	S	σ_p	pK_a		H_2O, 25°C	1.10	5	97.6[g]	3
5,3	CO_2H	S	σ_m	pK_a		H_2O, 25°C	1.31	4	99.0[g]	3
4,2	CO_2H	S	σ_m	pK_a		H_2O, 25°C	0.97	5	98.0[g]	3
5,2	CO_2Et	S	σ_p	log k	OH^-	85% EtOH–H_2O, 30°C	1.87	4	96.6[h]	3
5,2	CO_2Et	S	σ_p	log k	OH^-	85% EtOH–H_2O, 40°C	1.86	4	95.1[i]	3
5,3	CO_2Et	S	σ_m	log k	OH^-	85% EtOH–H_2O, 25°C	1.63	4	99.6[g]	3
4,2	CO_2Et	S	σ_m	log k	OH^-	85% EtOH–H_2O, 25°C	2.88	4	99.2[g]	3
4,2	CO_2Et	S	σ_m	log k	OH^-	85% EtOH–H_2O, 35°C	2.88	4	99.4[g]	3
5,2	CO_2Et	O	σ_p	log k	OH^-	85% EtOH–H_2O, 25°C	3.07	4	96.6[g]	3
5,2	CO_2H	S	σ_p	log k	MeOH	MeOH, 45°C	−0.34	5	78.1[i]	3
4,2 5,2	SO_2Cl	S	σ_m, σ_p	log k	$PhNH_2$	MeOH, 25°C	1.70	7	94.1[e]	133
5,2	COCl	S	σ_p	log k	$PhNH_2$	C_6H_6, 25°C	1.79	5	98.2[g]	134
4,2 5,2	CO_2H	Se	σ_m, σ_p	pK_a		H_2O, 25°C	1.23	7	99.6[e]	135
5,2	CO_2H	Te	σ_p	pK_a		H_2O, 25°C	1.20	5	99.8[e]	136
4,2 5,2	CO_2H	NH	σ_m, σ_p	pK_a		H_2O, 25°C	1.65	9	97.6[e]	137

4,3	CO_2H	NH[j]	σ_m	pK_a	50% EtOH–H_2O, 20°C	2.18	6	98.0[e]	178
4,2	CO_2H	NH[k]	σ_m, σ_p	pK_a	78.5% MCS–H_2O, 20°C	2.38	4	99.2[g]	178
5,2									
5,3	CO_2H	NH[l]	σ_m	pK_a	50% EtOH–H_2O	1.78	4	94.3[i]	178
5,2	NO_2	O	σ_p	$E_{1/2}$	H_2O, pH 12	0.25	6	98.0[e]	138
5,2	NO_2	O	σ_p	$E_{1/2}$	DMF	0.50	6	94.1[g]	138
5,2	SCN	S	σ_p	$E_{1/2}$	60% EtOH–H_2O	0.48	7	96.6[e]	139
5,2	NO_2	S	σ_p	$E_{1/2}$	H_2O, pH 3.10	0.20	12	51.6[f]	3
5,2	NO_2	S	σ_p	$E_{1/2}$	H_2O, pH 2.9, 25°C	0.30	4	99.8[e]	3
5,2	NO_2	S	σ_p	$E_{1/2}$	H_2O, pH 5.0, 25°C	0.40	4	99.4[g]	3
5,2	NO_2	S	σ_p	$E_{1/2}$	H_2O, pH 7.6, 25°C	0.46	4	99.6[g]	3
4,2	MeCHO	S	σ_m^+, σ_p^+	log k	80% EtOH–H_2O, 25°C	−6.79	9	98.6[e]	140
5,2	PNB								

[a]Heteroatom. [b]Quantity correlated. [c]Number of points in the set.
[d]Percentage of the data accounted for by the correlation. Superscripts indicate the confidence level of the correlation coefficient.
[e]99.9%. [f]99.0%. [g]99.5%. [h]98%. [i]95.0%.
[j]4-Substituted 3,5-dimethylpyrrole-2-carboxylic acids.
[k]4,5-Disubstituted 3-methylpyrrole-2-carboxylic acids.
[l]5-Substituted 2,4-dimethylpyrrole-3-carboxylic acids.

of 5-substituted furfurals are best correlated with σ_m, rather than, as expected,[132] with σ_p. It is, of course, of great interest to compare correlations obtained with the Hammett equation with those obtained by application of a multiparameter equation. For this purpose, correlations obtained with the extended Hammett equation[143] are reported in Table 5.9. As the number of degrees of freedom is not the same when a data set is correlated by the extended Hammett equation as when it is correlated by the simple Hammett equation, the most useful available statistic for comparison is $100r^2$. Of the eight sets that have been correlated with the aid of both equations, four show no significant difference, and the remainder show a significant improvement when the extended Hammett equation is used.

Noyce and his co-workers[179–181,140] studied the application of the DG treatment to rates of solvolysis of compounds of the type XGCHMe–L, where X is a substituent, G a heterocyclic skeletal group, and L a leaving group. In order to make use of the DG method the q_{ij} term in equation (5.1) was replaced by Δq_{ij}, which is defined as the difference in regional charge (the sum of the charges on a carbon atom and on any hydrogen atoms bonded to it) at atom i in the un-ionized molecule and in the heteroarylmethyl cation. Thus we have (5.29), which is analogous to (5.1).

$$\sigma_{ij}^{+} = F_{\mathrm{X}}^{+}/r_{ij} + \Delta q_{ij} M_{\mathrm{X}}^{+} \tag{5.29}$$

F^+ and M^+ are determined from (5.29) by means of the σ_p^+ and σ_m^+ constants, with r_{ij} and Δq_{ij} obtained for the benzyl cation and toluene. Values of Δq_{ij} were obtained from INDO quantum chemical calculations. Unfortunately, the authors do not report statistics for these correlations that permit a comparison of the DG method with the extended Hammett equation or the simple Hammett equation.

Substituent constants applicable to groups in the 5-position of furans and thiophens bearing the reaction site in the 3-position have been defined from the ionization constants of the corresponding carboxylic acids.[142,148] The thiophen constants have been successfully applied to rates of elimination reactions of XGCHMe–L, with G = 5,2-thiophen, L = Br or OTs.[152a]

6-Membered Rings

Correlations of data for XGY, where G is a 6-membered heteroarene, are almost entirely limited to pyridine derivatives. Sets that were correlated by the simple Hammett equation are given in Table 5.10. There are four possible pyridine derivatives with reaction site and substituent *ortho* (2,3; 3,2; 3,4; 4,3), four *meta* (2,4; 4,2; 5,3), and two *para* (5,2; 2,5). In all these systems both substituent and reaction site are bound to carbon.

Table 5.9. Correlation of 5-Membered Ring Sets by the Extended Hammett Equation

i,j	Y	Z[a]	*Q*[b]	Reagent	Reaction conditions	α	β	*n*[c]	100R^2[d]	Ref.[e]
5,2	CO_2H	O	log K_a		H_2O, 24°C	1.37	1.50	6	97.8[f]	144
5,2	CO_2H	O	pK_a		80% MCS–H_2O, 25°C	1.86	1.86	8	97.6[g]	147, 145
5,2	CO_2Et	O	log *k*	OH^-	85% EtOH–H_2O, 35°C	2.83	3.44	4	100[g]	146
5,2	CO_2Et	O	log *k*	OH^-	85% EtOH–H_2O, 25°C	2.95	3.57	4	100[h]	146
5,2	CO_2Et	O	log *k*	OH^-	85% EtOH–H_2O, 45°C	2.71	3.54	4	100[i]	146
5,2	CO_2H	O	log *k*	Ph_2CN_2	EtOH, 25°C	0.987	0.849	7	99.2[g]	130
5,2	CO_2H	O	log *k*	Ph_2CN_2	EtOH, 15.5°C	1.01	0.878	7	98.5[g]	130
5,2	CO_2H	S	pK_a		78.1% EtOH–H_2O, 25°C	1.86	0.789	4	99.7[j]	39
5,2	CO_2H	S	pK_a		80% MCS–H_2O, 25°C	2.10	1.19	8	99.5[g]	147, 145
5,2	CO_2H	S	pK_a		H_2O, 25°C	1.24	1.13	6	99.7[g]	148
5,2	CO_2Et	S	log *k*	H_3O^+	62% w/w $COMe_2$–H_2O, 100°C	0.161	0.694	5	94.3[k]	150
5,2	CO_2Et	S	log *k*	H_3O^+	62% w/w $COMe_2$–H_2O, 50°C	0.176	0.879	5	98.4[h]	150
5,2	CO_2Et	S	log *k*	OH^-	62% w/w $COMe_2$–H_2O, 25°C	2.63	2.21	5	100[g]	150
5,2	CO_2Et	S	log *k*	OH^-	85% w/w EtOH–H_2O, 30°C	2.10	0.883	4	99.7[k]	151
5,2	CO_2Et	S	log *k*	OH^-	85% w/w EtOH–H_2O, 40°C	2.14	0.661	4	99.1[k]	151
5,2	CO_2H	Se	pK_a		H_2O, 25°C	1.27	1.15	5	99.9[g]	135
5,2	CO_2H	Te	pK_a		H_2O, 25°C	1.18	1.24	5	99.9[g]	136
5,2	CO_2H	NH	pK_a		H_2O, 25°C	1.62	2.08	5	99.3[i]	137
5,2	CO_2H	NH	pK_a		H_2O, 25°C	1.40	1.09	5	99.4[i]	137

[a] Heteroatom. [b] Quantity correlated. [c] Number of points in the set.
[d] Percentage of the data accounted for by correlation. Superscripts indicate confidence level of the *F* test for significance of the correlation.
[e] Reference to source of data. [f] 99.5%. [g] 99.9%.
[h] 97.5%. [i] 99.0%. [j] 95.0%. [k] 90.0%. [l] 98.0%.

Table 5.10. *Pyridine Sets Correlated by the Hammett Equation**

i,j	Y	Subst. constant	Q^b	Reagent	Reaction conditions	ρ	n^c	$100r^{2\ d}$	Ref.
5,3	CO_2Et	σ_m	$\log k$	OH^-	85% EtOH–H_2O, 0°C	2.32	4	99.8^e	152b
5,3	CO_2Et	σ_m	$\log k$	OH^-	85% EtOH–H_2O, 15°C	2.13	5	98.8^e	152b
5,3	CO_2Et	σ_m	$\log k$	OH^-	85% EtOH–H_2O, 25°C	2.26	6	97.6^e	152b
5,3	CO_2H	σ_m	$\log K_a$		50% EtOH–H_2O, 25°C	1.73	5	98.8^e	152b
5,2	CO_2Me	σ_m, σ_p	$\log k$	OH^-	85% MeOH–H_2O, 25°C	1.00	9	99.8^e	153
4,2									
5,3	NO_2	σ_m	$E_{1/2}$		DMF, 20°C	0.458	8	98.2^e	154
2,4	NO_2	σ_m	$E_{1/2}$		DMF, 20°C	0.424	7	92.9^e	155
5,2	NO_2	σ_p	$E_{1/2}$		DMF, 20°C	0.408	6	99.9^e	156
2,5	NO_2	σ_p	$E_{1/2}$		DMF, 20°C	0.358	6	97.8^e	156
6,2	NO_2	σ_m	$E_{1/2}$		DMF, 20°C	0.440	6	98.6^e	157
4,2	NO_2	σ_m	$E_{1/2}$		DMF, 20°C	0.295	6	99.6^e	157

*For footnotes, see Table 5.8.

*Table 5.11. Correlations of Pyridine Sets by the Extended Hammett Equation**

i,j	Y	Q^b	Reagent	Reaction conditions	α	β	n^c	$100R^{2\,d}$	Ref.
5,3	CO_2Et	log *k*	OH^-	85% EtOH–H_2O, 15°C	2.15	1.10	6	97.8^f	153
5,3	CO_2Et	log *k*	OH^-	85% EtOH–H_2O, 0°C	2.51	1.01	5	99.7^f	153
5,3	CO_2H	pK_a		50% EtOH–H_2O, 25°C	1.72	0.489	5	99.5^i	153
2,5	NO_2	$E_{1/2}$		DMF, 20°C	0.506	0.405	5	99.6	156
5,2	NO_2	$E_{1/2}$		DMF, 20°C	0.457	0.335	5	99.8	156
5,3	NO_2	$E_{1/2}$		DMF, 20°C	0.406	0.180	8	99.4	154
6,2	NO_2	$E_{1/2}$		DMF, 20°C	0.534	−0.015	6	99.0	157
2,4	NO_2	$E_{1/2}$		DMF, 20°C	0.377	0.213	7	96.8	155
4,2	NO_2	$E_{1/2}$		DMF 20°C	0.304	0.123	6	99.8	157

*For footnotes, see Table 5.9.

There are also three possible derivatives with the reaction site bound to nitrogen (2,1; 3,1; 4,1) and three systems with the substituent bound to nitrogen (1,2; 1,3; 1,4). Of these systems, the 3,5; 2,4; 4,2; 6,2; 5,2; and 2,5 have been examined. Excellent results, as determined by the confidence level of the correlation coefficient, were obtained with all of the 11 sets reported in Table 5.10. The simple Hammett equation is obviously effective in correlating data for sets in which the skeletal group is a pyridine ring.

The few results obtained for the correlation of pyridine derivatives by the extended Hammett equation are in Table 5.11. The results reported for correlations of $E_{1/2}$ were obtained by Tomasik by using the corrected σ_I and σ_R values of Exner.[8] The other correlations are from Ref. 143. The results obtained by use of the extended Hammett equation are an improvement on those for the simple Hammett equation in three of the five sets for which comparison is possible.

Polynuclear Systems

It is useful to divide these systems into two groups: fused-ring type and separate-ring type, analogous to biphenyl. The latter can be divided into three subcategories: XArHarY, XHarArY, and XHarHarY (where Ar and Har are aromatic and heteroaromatic rings, respectively). Correlations of data for the first two subcategories are in Table 5.12. Values of pK_a for 5-substituted 2,2′-bithienyl-5′-carboxylic acids and 5-substituted 2,2′-biselenyl-5′-carboxylic acids have been determined and ρ values were reported.[158] Regrettably, no statistics of any kind were given.

No attempts at correlations of data for systems of these types by use of multiparameter equations or the DG treatment have appeared. The results

*Table 5.12. Data for Polynuclear Systems Correlated by the Hammett Equation**

i,j	Y	Skeletal group	Subst. constant	Q^b	Reagent	Reaction conditions	ρ	n^c	$100r^{2\,d}$	Ref.
4′,5	CO_2H	1′,2-Phenylfuryl†	σ_p	pK_a		50% v/v EtOH–H_2O	0.481	8	96.2[e]	160
4′,5	CHO	1′,2-Phenylfuryl†	σ_p^+	log *k*	5-Nitrofuryl phenyl sulfone	AcOH, 118°C	0.346	7	96.0[e]	131
4′,5	CHO	1′,2-Phenylfuryl†	σ_p^+	log *k*	5-Nitrofuryl phenyl sulfone	MeOH, 41°C	0.403	9	92.2[e]	131
5′,4	CO_2H	2,1′-Furylphenyl	σ_p	pK_a		50% v/v EtOH–H_2O, 25°C	0.555	10	97.4[e]	161
5′,4	CO_2Me	2,1′-Furylphenyl	σ_p	log *k*	OH^-	60% w/w $COMe_2$–H_2O, 25°C	1.056	9	98.6[e]	161
3′,4 4′,4	CO_2H	1′,2-Phenylthiazyl	σ_m, σ_p	pK_a		30% EtOH–H_2O, 25°C	0.263	5	97.8[g]	163
3′,4 4′,4	CO_2H	1′,2-Phenylthiazyl	σ_m, σ_p	pK_a		50% EtOH–H_2O, 25°C	0.356	5	97.4[g]	163
3′,4 4′,4	CO_2H	1′,2-Phenylthiazyl	σ_m, σ_p	pK_a		70% EtOH–H_2O, 25°C	0.394	5	97.6[g]	163
4′,6	NCS	1′,2-Phenylbenzothiazolyl	σ_p	log *k*	OH^-	H_2O, 25°C	0.50	6	95.1[e]	164
4′,6	NCS	1′,2-Phenylbenzothiazolyl	σ_p	log *k*	$H_2NCH_2CO_2H$	H_2O, 25°C	0.51	6	89.5[g]	164
4′,2″ 3′,2″ 2′,2″	CO_2H	1′,5-Phenylfuryl-2-(1″-vinyl)¶	σ_m, σ_p	pK_a		80% MCS–H_2O, 25°C	0.417	12	85.4[e]	157
4′,2″ 3′,2″	CO_2Me	1′,5-Phenylfuryl-2-(1″-vinyl)¶	σ_m, σ_p	log *k*	OH^-	60% v/v $COMe_2$–H_2O, 25°C	0.479	9	87.8[e]	157
6,3 5,3	CO_2H	Indole	σ_m, σ_p	pK_a		50% EtOH–H_2O, 26°C	0.670	6	90.6[g]	165
6,3 5,3	CO_2H	Indole	σ_m, σ_p	pK_a		95% EtOH–H_2O, 26°C	1.277	5	93.3[g]	165

6,3 5,3	CO_2H	Benzisoxazole	σ_m, σ_p^-	pK_a		Not stated	0.29	5	99.4[e]	166
4′,2	CHO	1′,5-(Phenylthio)furyl	σ_p	log k	$CH_2(CN)_2$	MeOH–$C_5H_{11}N$, 25°C	0.85	7	96.0[e]	167*a*
4′,2	CHO	1′,5-(Phenylthio)furyl	σ_p	log k	$NCCH_2CO_2Me$	MeOH–$C_5H_{11}N$, 25°C	0.77	7	96.0[e]	167*a*
4′,2	CHO	1′,5-(Phenylthio)furyl	σ_p	log k	4-$C_2NC_6H_4CH_2CN$	MeOH–$C_5H_{11}N$, 25°C	0.33	7	90.3[g]	167*a*
4′,4	NCO_2Me	3-Phenylsydnoneimines	σ_m^0, σ_p^0	pK_a		H_2O	0.69	6	97.2[g]	167*b*

*For footnotes, see Table 5.8.

†Structure:

¶Structure:

*Table 5.13. Correlations of Data for Polycyclic Systems by the Extended Hammett Equation**

i,j	Y	Skeletal group	Q^b	Reagent	Reaction conditions	α	β	n^c	$100R^{2\,d}$	Ref.
5,2	CO_2H	Benzofuran	pK_a		50% EtOH–H_2O, 25°C	1.02	1.10	5	99.5[f]	168
5,2	CO_2H	Benzofuran	log k	Ph_2CN_2	EtOH, 30°C	0.599	0.608	6	99.2[g]	159
5,2	CO_2Et	Benzofuran	log k	OH^-	85% EtOH–H_2O, 35°C	1.52	1.64	5	99.3[f]	168
5,2	CO_2H	Indole	pK_a		50% EtOH–H_2O, 25°C	1.42	0.787	5	97.2[l]	168
5,2	CO_2H	Indole	log k	Ph_2CN_2	EtOH, 30°C	0.455	0.211	5	99.4[f]	159
5,2	CO_2Et	Indole	log k	OH^-	85% EtOH–H_2O, 35°C	1.08	0.907	5	99.8[f]	168
5,3	CO_2H	Indole	log k	Ph_2CN_2	EtOH, 30°C	1.30	1.05	5	99.7[f]	159

*For footnotes, see Table 5.9

obtained with the simple Hammett equation seem to be quite satisfactory.

Data for fused-ring systems that have been correlated by the simple Hammett equation are also given in Table 5.12. Again, the statistics suggest that the simple Hammett equation generally gives adequate correlations. Some results obtained for correlations with the extended Hammett equation are in Table 5.13. Unfortunately, the sets are not sufficiently large to make the results conclusive. They do, however, suggest that application of the extended Hammett equation to these ring types would be useful.

Jaffé and Jones[3] and Bowden and Parkin[159] have reported correlations for fused ring systems by the Jaffé equation. As was noted previously, this equation is equivalent in terms of correlation analysis to the extended Hammett equation, except for the fact that it can accommodate substituents in more than one location in a single set—something that the extended Hammett equation cannot do. Interpretation of the regression coefficients is difficult, however. Judging by the values of the correlation coefficients, very good results were obtained. Bowden and Parkin[159] have also correlated data for fused-ring systems with the aid of a modification of the DG method. Substituting from equation (5.1) into the simple Hammett equation gives (5.30).

$$Q_X = \rho_D(F_X/r_{ij}) + \rho_D M_X q_{ij} + h \tag{5.30}$$

The data were correlated by multiple regression analysis with the parameters F/r_{ij} and M as substituent constants, ρ_D and $\rho_D q_{ij}$ being the reaction constants. This method provides another test of the validity of the theory of the DG method, which requires that ρ_D be constant for a given reaction and set of reaction conditions. Although very good correlations were always obtained, values of ρ_D for the reaction of heteroaryl carboxylic acids with diazodiphenylmethane in ethanol at 30°C are as follows: 5,2-benzofuran, 0.99; 6,2- and 5,2-indole, 0.87; 6,3- and 5,3-indole, 1.85. Thus, the DG treatment again seems to be inadequate.

Reaction Site Bonded to a Heteroatom

The only compounds of this type that have been studied are pyridine derivatives. Those that have been most frequently examined are the *N*-oxides. Correlations by the simple Hammett equation were best when σ^+ and σ^- constants were used for electron-donor and -acceptor substituents (by resonance), respectively. This is in accord with Jaffé's suggestion.[169] The few sets that have been correlated by the simple Hammett equation and for which statistics have been reported are in Table 5.14. A set of substituent constants for use with pyridine-*N*-oxides was defined from pK_a values of the protonated compounds in water[170] for substituents in the 3- and the

*Table 5.14. Correlation of Data for Pyridines by the Hammett Equation**†*

i, j	Y	Subst. const.	Q^b	Reagent	Reaction conditions	ρ	n^c	$100r^{2\,d}$	Ref.
3,1 4,1	O	σ_m, σ_p	log K_e	Br_2	CCl_4, 30°C	−1.23	4	97.6[h]	171
3,1 4,1	O	σ_m, σ_p	log K_e	I_2	CCl_4, 30°C	−2.66	5	99.2[e]	171
3,1 4,1	O	σ_m, σ_p	log K_e	IBr	CCl_4, 30°C	−3.02	5	99.0[e]	171
3,1 4,1	O	σ_m, σ_p	log K_e	ICl	CCl_4, 30°C	−3.75	5	99.0[e]	171
2,1	OH	σ_p	pK_a		H_2O, 20°C	4.13	6	92.2[g]	172
3,1 4,1	O	$\sigma_{m\mathrm{PyNO}}$ $\sigma_{p\mathrm{PyNO}}$	log K_e	Br_2	CCl_4, 30°C	−1.08	4	99.8[e]	171
3,1 4,1	O	$\sigma_{m\mathrm{PyNO}}$ $\sigma_{p\mathrm{PyNO}}$	log K_e	I_2	CCl_4, 30°C	−1.75	5	99.8[e]	171
3,1 4,1	O	$\sigma_{m\mathrm{PyNO}}$ $\sigma_{p\mathrm{PyNO}}$	log K_e	IBr	CCl_4, 30°C	−1.99	5	99.8[e]	171
3,1 4,1	O	$\sigma_{m\mathrm{PyNO}}$ $\sigma_{p\mathrm{PyNO}}$	log K_e	ICl	CCl_4, 30°C	−2.46	5	99.8[e]	171
3,1 4,1	Me	σ_m, σ_p^0	log k	OH^-	H_2O, 65°C	6.5	8	93.1[e]	173

*For footnotes, see Table 5.8.
†Reaction site on heteroatom. See also Ref. 3.

4-position. The *para*-substituent constants generally bear a strong resemblance to the σ^+ constants for groups that are electron donors by resonance and to the σ^- constants for groups that are electron acceptors by resonance, in agreement with Jaffé's proposal. Although a number of correlations by the simple Hammett equation by using the σ_{PyNO} constants have been reported, usually no statistics are given. The sets for which statistics are available are in Table 5.14. Based on the sparse results available it seems that better results are indeed obtained with the σ_{PyNO} constants. Further investigation is certainly necessary.

Only one instance of correlation by the extended Hammett equation has been reported,[174] namely, that of the pK_a values of 2-substituted 1-hydroxypyridinium ions. Both the extended Hammett and the LDS equations were used. No steric effect was observed, and very good correlations were obtained by use of the extended Hammett equation.

Extent of Transmission and Composition of Substituent Effects

Comparisons of the ease of transmission of substituent effects have been made by many authors.[147,160,175,185] It has been suggested that transmission of substituent effects in XGY is related to the aromaticity of

Table 5.15. Values of γ_L, γ_D, and P_R

i,j-Skeletal group[a]	γ_L	γ_D	P_R
3,1-Benzene	1.00	0.33	25.0
4,1-Benzene	1.00	1.00	50.0
5,2-Furan	1.37	1.50	52.0
4,2-Pyrrole	1.40	1.09	43.8
5,2-Pyrrole	1.62	2.08	56.2
5,2-Thiophen	1.24	1.13	47.7
5,2-Selenophen	1.27	1.15	47.5
5,2-Tellurophen	1.18	1.24	51.2
5,3-Pyridine	1.19	0.337	22.1
5,2-Indole	0.979	0.543	35.7
5,2-Benzofuran	0.703	0.759	51.9
5,2-Furan[b]	1.17[c]	0.986[c]	46.2[c]
5,2-Benzofuran[b]	0.711	0.706	50.4
5,2-Indole[b]	0.540	0.245	31.7
5,3-Indole[b]	1.54	1.22	44.7
5,2-Furan[d]	1.11	1.47	54.9
5,2-Thiophen[e]	1.04	0.944	45.7
5,2-Benzofuran[d]	0.598	0.701	51.9
5,2-Indole[d]	0.425	0.388	45.6

[a] *i* and *j* refer to positions of substituent and reaction site respectively.
[b] From reactions of the carboxylic acids with diazodiphenylmethane in ethanol at 30°C.
[c] At 25°C.
[d] From hydrolysis of the ethyl esters in 85% ethanol–water at 35°C (the reference set was in 87.83% ethanol–water at 30°C).
[e] From alkaline hydrolysis of the ethyl esters in 62% acetone–water at 25°C.

G and lies in the sequence furan < selenophen < thiophen < benzene.[135] Values of γ_L, γ_D, and P_R calculated from α and β values obtained by correlation of ionization constants of $XGCO_2H$ by the extended Hammett equation are in Table 5.15.[143] The γ_L values for the 5-membered 5,2-heteroarenes, with the exception of the value for pyrrole, do seem to be roughly related to aromaticity. The γ_D values for 5,2-furan, 5,2-pyrrole, and 5,2-thiophen are in the order of decreasing σ_R values of the corresponding MeZ groups, where Z is the heteroatom.

The values of γ_L and γ_D obtained from different reactions vary considerably. For the γ_L values, one factor is the change in the geometry of the system with change in reaction site. For the γ_D values the extent of resonance interaction between reaction site and substituent may be a factor. It might be argued that a major factor is the uncertainty in the α and β values. That this is not of major importance, at least in most cases, is shown by the relative constancy of the P_R values. Thus, P_R for 5,2-furan is 51 ± 5; for 5,2-benzofuran, 51 ± 2; for 5,2-thiophen, 47 ± 1. The only exception is 5,2-indole, for which the P_R values obtained from ionization of the acids and reactions of the acids with diazodiphenylmethane are in good agreement; $P_R = 34 \pm 2$, while the value obtained from ester hydrolysis is significantly different. Undoubtedly, the interpretation of γ_L, γ_D, and P_R values would be much easier if the data sets used to obtain the necessary α and β values were well characterized.

Another significant observation is that P_R for 5,2-indole is very much less for two of the three sets for which data are available than is the value for 5,2-benzofuran. At present, this seems inexplicable. The P_R values of 5,2-furan, -pyrrole, -thiophen, -selenophen, and -tellurophen are all about 50.0, indicating that these systems will give good correlations with σ_p. The P_R value for 4,2-pyrrole seems rather high, as it is formally analogous to 3,1-benzene.

Values of σ^+ for solvolysis of 1-heteroaryl-1-L-ethanes, where L is a leaving group, have been correlated with Δq, which is the difference in regional charge at the substituent site for the unsubstituted neutral molecule and the cationic transition state.[176]

5.3.2. The Heteroatom as the Reaction Site

5-Membered Rings

A number of reports have appeared on the correlation of pK_a values by the simple Hammett equation. Systems studied include 3(5)- and 4-substituted pyrazoles,[182] 1-, 2-, and 4(5)-substituted imidazoles,[183–185] 5-

Table 5.16. Results of Correlations of pK_a values for 5-Membered Ring Sets by the Hammett Equation[a]

i,j[b]	G[c]	Subst. constant	Reaction conditions	ρ	n[d]	$100r^2$[e]	Ref.
2,1	Imidazole	σ_m	H_2O, 25°C	10.9	6	99.7[f]	184
2,3	1-Methylimidazole	σ_m	H_2O, 25°C	10.6	3	99.8[g]	184
4,3	1-Methylimidazole	σ_m	H_2O, 25°C	10.5	4	99.0[h]	184
5,3	1-Methylimidazole	σ_m	H_2O, 25°C	6.31	3	99.8[g]	184
1,3	Imidazole	σ_I	H_2O, 25°C	10.7	6	98.6[f]	183
4(5) 1(3)	Imidazole	σ_m	H_2O, 25°C	10.0	7	99.4[f]	183
4,3	1-Methylimidazole	σ_m	H_2O, 25°C	10.4	5	99.2[f]	185
5,3	1-Methylimidazole	σ_m	H_2O, 25°C	7.0	4	98.4[i]	185
2,1	Imidazole	σ_m	H_2O, 25°C	10.4	8	98.2[f]	185
5,	3-Nitro-1,2,4-triazole	σ_m	H_2O, 20°C	8.92	7	98.6[f]	188
5,	Tetrazole	σ_m	H_2O, 25°C	6.90	9	98.6[f]	184

[a] All data are pK_{BH^+} values except for the 3-nitro-1,2,4-imidazole and tetrazole sets.
[b] Locations of substituent and reacting heteroatom, respectively.
[c] Skeletal group. [d] Number of points in the set.
[e] Percentage of data accounted for by the correlation. Superscripts indicate confidence levels of the correlation coefficient.
[f] 99.9%. [g] 95.0%. [h] 99.5%. [i] 99.0%.
[j] 98.0%. [k] 97.5%. [l] 90.0%. [m] <90.0%.

Table 5.17. Results of Correlations of pK_a Values for 5-Membered Heteroarenes by the Extended Hammett Equation[a]

i,j[b]	G[c]	Reaction conditions	α	β	n[d]	$100R^2$[e]	Ref.
2,1	Imidazole	H_2O, 25°C	11.0	3.32	5	99.9[f]	174
4,3	1-Methylimidazole	H_2O, 25°C	10.4	3.70	4	98.8[l]	191
5,3	1-Methylimidazole	H_2O, 25°C	7.24	3.73	4	100.0[h]	191
4(5) 1(3)	Imidazole	H_2O, 25°C	6.72	2.77	5	97.6[j]	191
5,1(2)	Tetrazole	H_2O, 25°C	7.22	2.20	9	99.0[f]	191

[a] All data are pK_{BH^+} values except for 4(5)-substituted imidazoles and 5-substituted tetrazoles. For other footnotes see Table 5.16.

substituted-1,2,4-triazoles,[186–188] 5-substituted tetrazoles,[189] and oxazoles.[190] Results of the correlations, which are generally very good, are in Table 5.16. It is somewhat surprising that such good results were obtained for the 4(5)-substituted imidazoles, 5-substituted 3-nitro-1,2,4-triazoles, and 5-substituted tetrazoles, all of which are capable of annular tautomerism.

Only two papers have appeared in which the extended Hammett equation has been applied to pK_a values for 5-membered rings. These results are given in Table 5.17. The correlations so obtained are very good, but the size of the sets and the number of sets studied are both too small to permit definitive conclusions to be drawn. The use of the extended Hammett equation has not resulted in significantly better correlation than that obtained by the simple Hammett equation for the four sets for which comparisons can be made. The advantage in the use of the extended Hammett equation is based on the belief that a better estimate of the composition of the electrical effect can be made.

6-Membered Rings

Since the pioneering paper of Jaffé and Doak,[192] considerable work has been done on the application of the simple Hammett equation to pyridines, and to a less extent to other 6-membered aza-arenes. Correlations for which statistics were reported are in Table 5.18. Generally, good results were obtained. Much work has been done on the correlation of data for pyridines by the extended Hammett and the DSP equation. Results of the correlations are in Table 5.19. Ehrenson, *et al.* have reported that the best results are obtained by the use of σ_R^+ constants.[11] A correlation of gas-phase basicities of 3- and 4-substituted pyridines by the DSP equation has been carried out.[202] It is interesting that correlations of rate

Table 5.18. Results of Correlations of Data for 6-Membered Rings by the Hammett Equation[a]

i,j[b]	G[c]	Q	Subst. constant	Reagent	Reaction conditions	ρ	n[d]	$100r^2$[e]	Ref.
3,1 4,1	Pyridine	pK_{BH^+}	σ_m, σ_p		H_2O, 25°C	5.77	28	96.4[f]	192
3,1 4,1	Pyridine	log k	σ_m, σ_p	MeI	$PhNO_2$, 30°C	−1.27	8	44.0[l]	3
3,1 4,1	Pyridine	log k	σ_m, σ_p	EtI	$PhNO_2$, 60°C	−1.28	8	54.2[g]	3
3,1 4,1	Pyridine	log k	σ_m, σ_p	Pr^iI	$PhNO_2$, 80°C	−1.07	8	59.6[g]	3
3,1	Pyridine	log k	σ_m	$CH_2{=}CHCH_2Br$	$MeNO_2$, 60°C	−2.53	8	90.4[f]	3
4,1	Pyridine	log k	σ_p	$CH_2{=}CHCH_2Br$	$MeNO_2$, 60°C	−1.42	5	98.6[f]	3
3,1 4,1	Pyridine	log k	σ_m, σ_p	Methyloxiran	H_2O, 30°C	−0.56	4	88.7[l]	3
3,1 4,1	Pyridine	log K_e	σ_m, σ_p	Ag^+	H_2O, 25°C	−2.07	11	81.2[f]	3
3,1 4,1[n]	Pyridine	pK_{BH^+}	σ_m, σ_p		H_2O, 25°C	5.13	2	93.7[f]	193
3,1 4,1	Pyridine	pK_{BH^+}	σ_m, σ_p		H_2O, 20°C	5.70	20	97.0[f]	172
3,1 4,1	Pyridine	log k	σ_m, σ_p	MeI	$PhNO_2$, 25°C	−2.57	7	93.9[f]	172
3,1 4,1	Pyridine	log k	σ_m, σ_p	BzO_2H	$PhNO_2$, 25°C	−2.23	5	96.2[h]	172
2,1	Pyridine	pK_{BH^+}	σ_I		H_2O, 25°C	10.3	13	98.6[f]	172
2,1	Pyridine	pK_{BH^+}	σ_m		H_2O, 20°C	9.04	10	94.5[f]	172
2,1	Pyridine	pK_{BH^+}	σ_I		50% EtOH–H_2O, 25°C	8.47	6	98.4[f]	172
2,1	Pyridine	log k	σ_I	MeI	$PhNO_2$, 25°C	−4.45	4	95.6[g]	172

continued overleaf

Table 5.18—continued

i,j[b]	G[c]	Q	Subst. constant	Reagent	Reaction conditions	ρ	n[d]	$100r^2$ [e]	Ref.
2,1	Pyridine	$\log k$	σ_I	BzO_2H	$PhNO_2$, 25°C	−4.84	4	100.0[f]	172
3,1 4,1[n]	Pyridine	$\log k$	σ_m, σ_p	n-$C_{12}H_{25}Br$	MeOH, 75°C	−0.847	16	96.8[f]	194
3,1 4,1[n]	Pyridine	$\log k$	σ_m, σ_p	n-$C_{12}H_{25}Br$	DMF, 50°C	−1.06	15	97.2[f]	194
3,1 4,1[n]	Pyridine	$\log K_f$	σ_m, σ_p	I_2	CCl_4, 25°C	−2.25	10	96.0[f]	195
3,1 4,1[n]	Pyridine	$\log K_f$	σ_m, σ_p	I_2	CCl_4, 30°C	−2.15	9	99.0[f]	196
3,1 4,1[n]	Pyridine	$\log K_f$	σ_m, σ_p	Br_2	CCl_4, 30°C	−0.92	5	96.4[h]	196
3,1 4,1[n]	Pyridine	$\log K_f$	σ_m, σ_p	IBr	CCl_4, 30°C	−2.80	10	99.4[f]	196
3,1 4,1[n]	Pyridine	$\log K_f$	σ_m, σ_p	ICl	CCl_4, 30°C	−3.33	7	99.4[f]	196
3,1	Pyridine	$\log K_f$	σ_m^0	TsBr, 3-$ClC_6H_4NH_2$	1 : 1, $PhNO_2$–c–C_6H_{12}	−3.33	5	97.6[h]	197
3,1; 4,1	Pyridine	pK_{BH^+}	σ_m, σ_p, σ_p^-		H_2O	5.0	19	20.3[l]	198
4,1	Pyridine	pK_{BH^+}	σ_p, σ_p^-		H_2O	18.3	7	74.0[j]	198
3,1; 4,1	Pyridine	pK_{BH^+}	σ_m, σ_p^-		H_2O	9.7	5	96.0[h]	198
3,1	Pyridine	pK_{BH^+}	σ_m		H_sO, 25°C	6.36	12	94.1[f]	199
6,2	3-Dimethyl-aminopyridazine	pK_{BH^+}	σ_m		H_2O, 25°C	6.14	6	89.7[h]	203
6,1	3-Chloro-pyridazine	pK_{BH^+}	σ_p		H_2O, 25°C	6.79	5	99.6[f]	203
6,1	Pyridazone	pK_{BH^+}	σ_p		H_2O, 25°C	2.92	7	98.2[f]	203

5,3	Orotic acid	pK_a	σ_m	H_2O, 25°C	5.78	7	96.0[f]	204
5,1	2-Chloro-pyrimidine	HNP[o]	σ_m	Ac_2O	273	5	99.0[f]	205
4(6),1(3)	2-Chloro-pyrimidine	HNP[o]	σ_p^+	Ac_2O	64	6	99.0[f]	205
5,1	2,4-Diamino-pyrimidine	pK_a	σ_m	H_2O, 20°C	6.28	10	95.6[f]	206
6,1	2,4-Diamino-pyrimidine	pK_a	σ_m	H_2O, 20°C	7.95	8	87.8[f]	206
2,1	Pyrimidine	pK_a	σ_p	H_2O, 20°C	3.66	6	98.8[f]	206
4(6),1(3)	Pyrimidine	pK_a	σ_p	H_2O, 20°C	5.85	7	93.7[f]	206
4(6),1(3)	2-Methylamino-pyrimidine	pK_a	σ_p	H_2O, 20°C	5.72	4	99.0[i]	206
6,1(3)	4-Aminopyrimidine	pK_a	σ_m	H_2O, 20°C	9.38	4	97.8[i]	206
2,1	4,6-Diamino-1,3,5-triazine	pK_a	σ_m	H_2O, 25°C	4.28	13	93.3[f]	207
5,1	4-Amino-2-methyl-pyrimidine	pK_a	σ_m	H_2O, 25°C	5.399	5	100.0[f]	208
5,1	4-Dimethylamino-2-methylpyrimidine	pK_a	σ_m	H_2O, 25°C	5.53	4	99.8[f]	208
5,1; 4,1	2-Methylpyrimidine	pK_a	$(\sigma_m + \sigma_o)$	H_2O, 25°C	5.014	15	98.6[f]	208

[a] For footnotes, other than those below, see Table 5.16.
[n] Includes disubstituted compounds.
[o] Half-neutralization potential.

Table 5.19. *Results of Correlations of Data for 6-Membered Rings by the Extended Hammett Equation*[a]

i,j[b]	G[c]	Q	Reaction conditions	α	β	n[d]	$100R^2$[e]	Ref.
2,1	Pyridine	pK_{BH^+}	H_2O, 25°C	11.3	2.31	12	95.8[f]	174
2,1	Pyridine	pK_{BH^+}	H_2O, 20°C	9.18	2.64	10	95.3[d]	174
2,1[n]	Pyridine	$\log K_e$	CCl_4, 20°C	−1.43	1.19	4	99.0[l]	10
3,1	Pyridine	pK_{BH^+}	H_2O, 20°C	6.22	1.84	16	98.8[f]	200
4,1	Pyridine	pK_{BH^+}	H_2O, 20°C	5.38	1.51	20	98.6[f]	200
2,1[o]	Pyridine	pK_{BH^+}	H_2O, 20°C	6.16	2.19	5	98.4[i]	201
3,1[o]	Pyridine	pK_{BH^+}	H_2O, 20°C	3.77	1.88	5	98.2[i]	201
4,1[o]	Pyridine	pK_{BH^+}	H_2O, 20°C	2.55	5.70	5	97.8[j]	201
2,1	Pyrimidine	pK_{BH^+}	H_2O, 20°C	6.30	3.29	10	97.8[f]	209
6,1	2,4-Diaminopyrimidine	pK_{BH^+}	H_2O, 20°C	7.21	0.063	10	95.3[f]	209
5,1	2,4-Diaminopyrimidine	pK_{BH^+}	H_2O, 20°C	5.52	2.07	14	93.3[f]	209
6,1	4-Aminopyrimidine	pK_{BH^+}	H_2O, 20°C	8.21	1.01	6	99.8[f]	209
6,1	4-Dimethylaminopyrimidine	pK_{BH^+}	H_2O, 20°C	8.64	5.52	5	99.2[l]	209

[a] All correlations with σ_I, σ_R unless otherwise noted. For footnotes other than those below, see Table 5.16.
[n] Complexation with phenol. [o] The Swain–Lupton $\mathfrak{F}$ and $\mathfrak{R}$ constants were used.

constants for the reactions of allyl bromide with substituted pyridines by the Hammett equation required separate lines for 3- and 4-substituted compounds.[3] The pK_{BH^+} values of 3- and 4-substituted pyridines have different α values (−6.22 and −5.38, respectively). This result agrees with field-effect calculations for 3- and 4-substituted pyridines, which suggest that the ratio, α_m/α_p, is about 1.1.

Pyridazines, pyrazines, 1,3,5-triazines, and in particular, pyrimidines have also been examined. The results of correlations by the Hammett and the extended Hammett equation are in Tables 5.18 and 5.19, respectively. The results obtained for 6-substituted 3-chloropyridazines are quite striking. Protonation should occur at N^1, which is *ortho* to the substituent. For 2-substituted pyridines, quinolines, and isoquinolines, the best correlation by the Hammett equation is invariably obtained with σ_I or σ_m. The results of Cookson and Cheeseman[203] show a very good correlation with σ_p. It would be useful to examine a well-characterized set of pK_{BH^+} measurements for these compounds. Although Roth and Strelitz[206] found the best correlation for 2-substituted pyrimidines to be with σ_p, Charton, by using a larger set, and the extended Hammett equation, obtained a P_R value of 34.3, which is between the values for σ_m and σ_p^0, in terms of the composition of the electrical effect. The data for 6-substituted 2,4-diaminopyrimidines, which probably undergo protonation at N^1, depend only on $\sigma_I (P_R = 0)$, although they too involve protonation at an *ortho* nitrogen atom. These results point up the use of two-electrical-parameter equations in determining the composition of the electrical effect. The correlations obtained by use of the extended Hammett equation were quite successful.

Polynuclear Systems

Many reports on the application of the Hammett equation to polycyclic heteroarenes have appeared. Correlations for which statistics are available are in Table 5.20. Generally, good results were obtained. Correlations by the extended Hammett and the Yukawa–Tsuno equation are reported in Table 5.21. Again, results are very good. This is particularly interesting in the case of the 5(6)-substituted imidazoles where annular tautomerism is possible.

All the systems considered above involved nitrogen as the heteroatom. The Hammett equation has been applied to the peroxybenzoic acid oxidation at the sulfur atom in 3-substituted benzo[*b*]thiophens and dibenzothiophens.[225] Statistics, however, were not reported.

Nonaromatic Heterocycles

A number of authors have applied the Hammett equation to nonaromatic heterocycles. Results of the correlations are in Table 5.22.

Table 5.20. Results of Correlations of Data for Polynuclear Heteroarenes by the Hammett Equation[a]

i,j[b]	G[c]	Q	Subst. constant	Reaction conditions	ρ	n[d]	$100r^2$[e]	Ref.
3′,3 4′,3	1′, 1-Phenylimidazole	pK_{BH^+}	σ_m, σ_p^-	50% $EtOH–H_2O$, 20°C	0.753	8	91.8[f]	210
3′,3 4′,3[q]	1′, 1-Phenylimidazole	$\log k$	σ_m	$COMe_2$, 50°C	−0.395	8	90.1[f]	210
3′,3 4′,3[r]	1′, 1-Phenylpyrazole	$\log k$	σ_m, σ_p	Sulfolan, 60°C	−1.10	7	99.8[f]	211
4′,1[s]	3-Styrylpyridine	$\log K_e$	σ_p	CH_2Cl_2, 30°C	−0.391	7	98.2[f]	212
4′,1[s]	4-Styrylpyridine	$\log K_e$	σ_p	CH_2Cl_2, 30°C	−0.389	5	99.6[f]	212
2,1	Benzimidazole	pK_{BH^+}	σ_I	H_2O, 25°C	6.73	6	96.6[f]	184
2,1	Benzimidazole	pK_{BH^+}	σ_m	H_2O, 25°C	7.92	10	98.4[f]	184
2,1	Benzimidazole	pK_{BH^+}	σ_m	5% $EtOH—H_2O$, 30°C	3.08	8	93.7[f]	184
2,1	Benzimidazole	pK_{BH^+}	σ_p	50% $EtOH–H_2O$, 25°C	3.85	8	91.6[f]	184
5(6),1(3) 4(7),1(3)	2-Chlorobenzimidazole	pK_{BH^+}	$(\sigma_m + \sigma_p)$	2% $MeOH–H_2O$, 25°C	2.07	7	97.8[f]	213
4,1; 5,1 6,1; 7,1	Indole	pK_{BH^+}	$\sigma_m(4,7)$ $\sigma_p(5,6)$	H_2O	2.56	9	99.2[f]	214
(4,5,6,7),1	Benzotriazole	pK_{BH^+}	$\sum\sigma$	H_2O	3.09	20	96.8[f]	214
(4,5,6,7),1	2-Trifluoromethylbenzotriazole	pK_{BH^+}	$\sum\sigma$	H_2O	3.39	29	98.6[f]	214
2,1	Benzimidazole	pK_{BH^+}	σ_m	H_2O	7.70	5	94.9[h]	214
2,1	5-Chlorobenzimidazole	pK_{BH^+}	σ_m	H_2O	7.03	4	38.4[f]	214
2,1	5-Nitrobenzimidazole	pK_{BH^+}	σ_m	H_2O	9.52	10	92.5[f]	214
2,1	4,5,6,7-Tetrachlorobenzimidazole	pK_{BH^+}	σ_m	H_2O	8.95	7	97.0[f]	214
4′,1	(1′)2-Phenyl-4,5-diphenylimi- dazole	pK_{BH^+}	σ_p	H_2O	1.88	6	99.8[f]	215
5,4	4-Azaindole	pK_{BH^+}	σ_I	H_2O	5.37		95.1	216
6,5	5-Azaindole	pK_{BH^+}	σ_I	H_2O	6.35		77.8	216
2,1	Quinoline	pK_{BH^+}	σ_m	H_2O, 25°C	10.4	7	90.3[h]	172

1,2	Isoquinoline	pK_{BH^+}	σ_m	H_2O, 20°C	13.7	4	92.3[g]	172
3,1; 4,1	Quinoline	pK_{BH^+}	σ_m, σ_p	H_2O, 20°C	5.70	20	97.0[f]	172
3,1; 4,1	Quinoline	pK_{BH^+}	σ_m, σ_p	H_2O, 25°C	5.72	9	96.8[f]	172
3,1; 4,1	Quinoline	pK_{BH^+}	σ_m, σ_p	10% Et OH–H_2O, 25°C	5.15	7	94.9[f]	172
4,1	Quinoline	pK_{BH^+}	σ_p	50% EtOH–H_2O, 21–22°C	6.15	4	94.5[g]	172
4,	Cinnoline	pK_{BH^+}	σ_p	50% EtOH–H_2O, 21–25°C	4.61	4	82.8[l]	217
4,	Cinnoline	pK_{BH^+}	σ_p	H_2O, 20°C	6.01	4	88.7[l]	217
5,1	Quinoline	pK_{BH^+}	σ_m	H_2O, 20°C	2.10	7	91.8[f]	15
5,1	Quinoline	pK_{BH^+}	σ_{51}[n]	H_2O, 20°C	3.87	7	95.3[f]	15
7,1	Quinoline	pK_{BH^+}	σ_p	H_2O, 20°C	2.96	5	99.6[f]	15
7,1	Quinoline	pK_{BH^+}	σ_{71}[n]	H_2O, 20°C	4.96	5	99.8[f]	15
8,1	Quinoline	pK_{BH^+}	σ_I	H_2O, 20°C	3.75	4	95.6[g]	15
4,2	Isoquinoline	pK_{BH^+}	σ_m	H_2O, 20°C	5.63	4	99.8[f]	15
5,2	Isoquinoline	pK_{BH^+}	σ_m	H_2O, 20°C	2.49	4	94.5[g]	15
5,2	Isoquinoline	pK_{BH^+}	σ_{52}[n]	H_2O, 20°C	4.33	4	95.1[g]	15
6,2	Isoquinoline	pK_{BH^+}	σ_p	H_2O, 20°C	2.57	3	99.9[i]	15
6,2	Isoquinoline	pK_{BH^+}	σ_{62}[n]	H_2O, 20°C	5.25	3	99.4[g]	15
7,2	Isoquinoline	pK_{BH^+}	σ_p	H_2O, 20°C	1.79	4	97.2[i]	15
7,2	Isoquinoline	pK_{BH^+}	σ_{72}[n]	H_2O, 20°C	5.71	4	94.5[g]	15
8,2	Isoquinoline	pK_{BH^+}	σ_m	H_2O, 20°C	2.75	3	99.2[l]	15
8,2	Isoquinoline	pK_{BH^+}	σ_{82}[n]	H_2O, 20°C	3.10	3	99.0[l]	15
5,1	Quinoline	$\log K_B$	σ_m	10% EtOH–H_2O, 25°C	−3.35	4	98.6[i]	15
5,1	Quinoline	$\log K_B$	σ_{51}[n]	10% EtOH–H_2O, 25°C	−6.08	4	98.0[i]	15
6,1	Quinoline	$\log K_B$	σ_m	10% EtOH–H_2O, 25°C	−2.51	5	97.6[h]	15
6,1	Quinoline	$\log K_B$	σ_{61}[n]	10% EtOH–H_2O, 25°C	−4.32	5	98.4[f]	15
7,1	Quinoline	$\log K_B$	σ_m	10% EtOH–H_2O, 25°C	−2.84	5	95.3[h]	15
7,1	Quinoline	$\log K_B$	σ_{71}[n]	10% EtOH–H_2O, 25°C	−4.95	5	99.4[f]	15
8,1	Quinoline	$\log K_B$	σ_m	10% EtOH–H_2O, 25°C	−4.25	5	98.2[h]	15
4′,1 3′,1	2-Phenylperimidine	pK_{BH^+}	σ_m, σ_p	MeCN, 25°C	1.95	6	98.8[f]	218
5,1	Benzoxazoline-2-one	pK_a	σ_m^0, σ_p^0	$COMe_2$–H_2O, 25°C	2.56	10	96.0[f]	219
5,1	Benzoxazoline-2-thione	pK_a	σ_m^0, σ_p^0	$COMe_2$–H_2O, 25°C	2.63	8	98.2[f]	219

continued overleaf

Table 5.20—continued

i,j[b]	G[c]	Q	Subst. constant	Reaction conditions	ρ	n[d]	$100r^2$[e]	Ref.
3′,1 4′,1	2-Phenylbenzoxazole	pK_{BH^+}	σ_m^+, σ_p^+	MeCN, 25°C	1.87	10	96.8[f]	215
3′,1 4′,1	2-Phenylbenzothiazole	pK_{BH^+}	σ_m^+, σ_p^+	MeCN, 25°C	1.60	9	97.6[f]	215
3′,1 4′,1	2-Phenylbenzimidazole	pK_{BH^+}	σ_m, σ_p	MeCN, 25°C	2.12	9	99.2[f]	215
3′,1 4′,1	2-Subst. phenyl-4-phenyl-10H-indeno[1,2,g]quinoline	pK_{BH^+}	σ_m, σ_p	4:1 $CHCl_3$–MeCN, 20°C	2.00	8	98.6[f]	220
3′,1 4′,1	2-Subst. phenyl-4-styryl-10H-indeno[1,2,*g*]quinoline	pK_{BH^+}	σ_m^0, σ_p^0	4:1 $CHCl_3$–MeCN, 20°C	1.70	8	98.2[f]	220
2,9; 3,9	Carbazole	HNP	σ_m, σ_p	20% EtOH–H_2SO_4	5.0	6	98.6[f]	223
3′,1[t] 4′,1	Bis-4(5)phenyl-5(4)-2-naphthyl-imidazole	$\log k$	σ_m, σ_p	Toluene, 60°C	−0.402	5	99.6[f]	224
5,1; 5′,1	1,11-Diethyl-3,3-diphenylimido-carbocyanines	pK_{BH^+}	σ_p	57% EtOH–H_2O	3.16	10	95.1[f]	232
5′,1; 5,1[p]	Y = CMe_2, R = Me	pK_{BH^+}	σ_p	57% EtOH–H_2O	1.04	4	97.4[i]	233
5′,1; 5,1[p]	Y = S, R = Et	pK_{BH^+}	σ_p	57% EtOH–H_2O	1.19	19	99.2[f]	233
6′,1; 6,1[p]	Y = CH = CH, R = Et	pK_{BH^+}	σ_p	57% EtOH–H_2O	3.90	6	98.6[f]	233
5′,1; 5,1[p]	Y = NEt, R = Et	pK_{BH^+}	σ_p	57% EtOH–H_2O	3.23	18	94.5[f]	233

[a] For footnotes other than those below, see Table 5.16.
[n] DG Constants.
[o] Half-neutralization potential. [p] Carbocyanines of the structure shown to the right.
[q] Reagent = EtI. [r] Reagent = $(MeO)_2SO_2$. [s] Reagent = I_2.
[t] Dissociation into radicals.

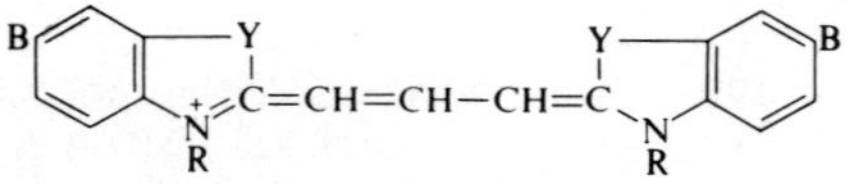

Table 5.21. Results of Correlations of Data for Polynuclear Heteroarenes by the Extended Hammett Equation[a]

i,j	G[c]	Q	Reaction conditions	α	β	n[d]	$100R^2$[e]	Ref.
4,1	Quinoline	pK_{BH^+}	H_2O, 25°C	11.2	2.77	6	88.2[j]	174
2,1	Benzimidazole	pK_{BH^+}	H_2O, 25°C	12.4	4.54	6	95.8[h]	174
2,1	Benzimidazole	pK_B	H_2O, 25°C	7.15	2.95	9	95.8[f]	174
2,1	5,6,7,8-Tetrahydronaphth-[2,3]imidazole	pK_{BH^+}	H_2O, 20°C	8.43	2.96	5	99.0[h]	174
5(6)[n] 1(3)	Benzimidazole	ΔpK_{BH^+}	H_2O, 25°C	2.61	2.10	7	99.8[f]	221
5,3[n]	1-Methylbenzimidazole	ΔpK_{BH^+}	50% v/v EtOH–H_2O, 25°C	2.94	2.43	6	99.2[f]	221
5(6) 1(3)[o]	Benzimidazole	ΔpK_{BH^+}	H_2O, 25°C	1.62	1.66	8	99.6[f]	222
5(6) 1(3)[o]	Benzimidazole	ΔpK_a	H_2O, 25°C	1.43	1.52	8	98.4[f]	222
5(6) 1(3)[n]	Benzimidazole	ΔpK_a	H_2O, 25°C	2.29	1.96	8	97.8[f]	222
3′,1 4′,1[p]	2-Phenylperimidine	pK_{BH^+}	MeCN, 25°C	2.00	0.991	6	98.2[h]	218
3′,1 4′,1[q]	2-Phenylperimidine	pK_{BH^+}	MeCN, 25°C	1.93	0.155	6	99.0[f]	218
5,1	Quinoline	pK_{BH^+}	H_2O, 20°C	3.17	1.99	7	97.6[f]	15
6,1	Quinoline	pK_{BH^+}	H_2O, 20°C	2.96	1.49	8	98.8[f]	15
7,1	Quinoline	pK_{BH^+}	H_2O, 20°C	3.28	2.73	5	99.4[h]	15
3,1	Quinoline	pK_{BH^+}	H_2O, 20°C	5.40	0.838	6	99.8[f]	32
4,1	Quinoline	pK_{BH^+}	H_2O, 20°C	5.45	6.29	6	99.0[f]	32
5,2	Isoquinoline	pK_{BH^+}	H_2O, 20°C	2.64	0.754	4	98.2[l]	32
7,2	Isoquinoline	pK_{BH^+}	H_2O, 20°C	2.50	1.38	4	100.0[i]	32
4,2	Isoquinoline	pK_{BH^+}	H_2O, 20°C	5.68	1.87	4	100.0[j]	32
4′,4	1′, 1-Styrylpyridine	pK_{BH^+}	H_2O, 20°C	0.897	0.768	5	98.2[i]	32
4′,4	1′, 1-Phenylethynylpyridine	pK_{BH^+}	H_2O, 20°C	0.393	0.515	5	99.6[h]	32

[a] All correlations with σ_I and σ_R unless otherwise stated. For footnotes other than those below see Table 5.16.
[n] σ_R^0 in place of σ_R. [o] Swain–Lupton $\mathfrak{F}$, $\mathfrak{R}$ in place of σ_I, σ_R. [p] σ_R^+ in place of σ_R.
[q] Correlation by the use of the Yukawa-Tsuno equation, $\alpha = \rho$; $\beta = r$.

Table 5.22. Results of Correlations of Data for Nonaromatic Heterocyclic Sets by the Hammett Equation[a]

i,j[b]	G[c]	Q	Subst. constant	Reaction conditions	ρ	n[d]	$100r^2$ [e]	Ref.
2,1[m]	Tetrahydrofuran	$\Delta\nu_{OH}$	σ^*	CCl_4	294	8	91.6[f]	226
2,1[m]	Tetrahydrofuran	$\log K_e$	σ^*	CCl_4	8.04	8	95.1[f]	226
2,1[m]	Tetrahydrofuran	A_{rel}	σ^*	CCl_4	29.5	8	98.2[f]	226
4,1	*Trans*-1-methylpiperidine	pK_{BH^+}	σ^*	MeOH	1.23	9	98.8[f]	227
4,1	*Trans*-(*cis*-4-hydroxy-4-methyl)piperidine	pK_{BH}	σ^*	MeOH	1.1	5	97.6[h]	227
4,1	*Trans*-(*cis*-4-hydroxy-1,2-dimethyl)piperidine	pK_{BH^+}	σ^*	MeOH	1.19	5	98.8[f]	228
4,1	*Trans*-(*cis*-4-hydroxy-1,3-dimethyl)piperidine	pK_{BH^+}	σ^*	MeOH	1.2	5	92.2[i]	228
4,1	*Trans*-(*cis*-4-hydroxy-2-methyl)decahydroquinoline	pK_{BH^+}	σ^*	MeOH	0.9	5	95.6[h]	228
4,1	*a,a*-2-Methyldecahydroquinoline	pK_{BH^+}	σ^*	H_2O, 25°C	0.11	5	99.6[f]	229
4,1	*e,a*-2-Methyldecahydroquinoline	pK_{BH^+}	σ^*	H_2O, 25°C	0.43	5	99.6[f]	229
4,1	*a,e*-2-Methyldecahydroquinoline	pK_{BH^+}	σ^*	H_2O, 25°C	0.39	5	99.4[f]	229
4,1	*e,e*-2-Methyldecahydroquinoline	pK_{BH^+}	σ^*	H_2O, 25°C	0.38	5	99.8[f]	229
4,1	Quinuclidine	$\log (K_{BH^+}/K^0_{BH^+})$	σ_I	5% v/v EtOH–H_2O, 25°C	5.11	17	98.0[f]	230
4,1	Quinuclidine	$\log (K_{BH^+}/K^0_{BH^+})$	σ_I	50% w/w EtOH–H_2O, 25°C	5.09	17	98.2[f]	230
4,1	Quinuclidine	$\log (K_{BH^+}/K^0_{BH^+})$	σ_I	80% MCS–H_2O, 25°C	4.41	7	97.8[f]	230
4,1[n]	Quinuclidine	$\log (k/k^0)$	σ_I	MeOH, 25°C	−0.91	14	92.7[f]	130
4,1	Quinuclidine	pK_{BH^+}	σ^*	H_2O, 25°C	—	7	99.6[f]	231

[a] For footnotes other than those below, see Table 5.16.
[m] Reagent = PhOH. [n] Reagent = MeI.

Extent of Transmission of Substituent Effects

As γ values cannot be calculated for the ionization of aza-arenes (unless the reference group is changed from 4,1-benzene to some heteroaryl group) it will be convenient to describe the extent of transmission of substituent effects by comparison of α and β values for proton transfer at nitrogen. Such values are given in Table 5.23, together with values of P_R. The α values for those sets in which the substituent is *ortho* to the protonated nitrogen all lie between 8.4 and 12.4; β values lie between 2.31 and 4.54; and P_R values between 17.0 and 26.8, with the exception of data for the 2-substituted pyrimidines, which show an unusually low α value of 6.30 and an unusually high P_R value of 34.3. A comparison of P_R values for naphthalenes (Table 5.7) with those for quinolines and isoquinolines is given in Table 5.24. There is considerable similarity in the majority of the systems.

Values of β for aza-arenes are correlated by equation (5.18),[32] thus making possible the prediction of these values from the regression equation. The utility of the extended Hammett equation is demonstrated by a consideration of the P_R values in Table 5.23. Half of the values in the table lie between the P_R values of the simple substituent constants. It is also pertinent that the P_R values for 5,1-naphthalene and -quinoline derivatives

Table 5.23. Values of α, β, and P_R for Aza-arene Ionization at 20°C[32,209]

Aza-arene	α	β	P_R
2-Substituted pyridines	11.3[a]	2.31[a]	17.0[a]
	9.18	2.64	22.3
3-Substituted pyridines	6.22	1.84	22.8
4-Substituted pyridines	5.38	5.11	48.7
2-Substituted pyridines	11.2	2.77	19.8
3-Substituted quinolines	5.40	0.838	13.4
4-Substituted quinolines	5.45	6.29	53.6
5-Substituted quinolines	3.17	1.49	32.0
6-Substituted quinolines	2.96	1.49	33.5
7-Substituted quinolines	3.28	2.73	45.4
4-Substituted isoquinolines	5.68	1.87	24.8
5-Substituted isoquinolines	2.64	0.754	22.2
7-Substituted isoquinolines	2.50	1.38	35.6
4-(4-Substituted styryl)-pyridines	0.897	0.768	46.1
4-(4-Substituted phenylethynyl)pyridines	0.393	0.515	56.7
2-Substituted imidazoles	11.0	3.32	23.2
2-Substituted benzimidazoles	12.4[a]	4.54[a]	26.8[a]
2-Substituted 5,6,7,8-Tetrahydronaphth[2,3]imidazoles	8.43	2.96	26.0
2-Substituted pyrimidines	6.30	3.29	34.3

[a] At 25°C.

Table 5.24. P_R Values of Naphthalenes, Quinolines, and Isoquinolines

i,j	Naphthalene	Quinoline or isoquinoline
3,1	23.7	13.4
4,1	53.0	53.6
5,1	35.8	32.0
6,1	40.8	33.5
7,1	44.7	45.4
4,2	28.4	24.8
5,2	33.5	22.2
7,2	48.2	35.6

are between σ_m and σ_p^0, although this *i,j* combination is conjugated. The 7,1 combination, which is also conjugated, has P_R values between those of σ_p^0 and σ_p, as might have been expected.

5.3.3. The Heterocyclic Group as the Reaction Site

5-Membered Rings

Many studies have appeared on the application of the Hammett equation to electrophilic aromatic substitution in five-membered heteroarenes. The subject has been reviewed.[234–236] Nucleophilic aromatic substitution has also received considerable attention. Results of correlations by the Hammett equation for these and other reactions are in Table 5.25. The correlations obtained are very good. The values obtained for electrophilic aromatic substitution of thiophen and benzene are in Table 5.26. Values are generally comparable except for those for acetylation.

The one correlation by a multiparameter equation for which results are available is given in Table 5.27. The results are about the same as those obtained with the simple Hammett equation. There are sufficient data available for electrophilic and nucleophilic aromatic substitution to make a study of the application of the extended Hammett equation worthwhile. It is certainly possible that the P_R values for the heteroarene skeletal groups are significantly different frbm those of the σ_m, σ_p, σ_p^0, and σ_p^+ constants.

A report has appeared that describes the application of the Taft–Pavelich equation (5.31) to rate constants for the decomposition of 1-substituted sydnonimines.[248]

$$\log (k/k^0) = \rho\sigma^* + \delta E_S \tag{5.31}$$

Table 5.25. Results of Correlations of Data for 5-Membered Heteroarenes by the Hammett Equation[a]

i,j[b]	L[c]	G[d]	Subst. constant	Reagent	Reaction conditions	ρ	n[e]	$100r^2$ [f]	Ref.
2,5	H	Thiophen	σ_p^+	Cl_2	AcOH, 25°C	−9.39	5	94.5[g]	237
5,2	Br	3-NO_2-thiophen	σ_p^-	Piperidine	MeOH, 20°C	3.18	7	99.0[h]	238
5,2	Br	3-NO_2-4-Me-thiophen	σ_p^-	Piperidine	MeOH, 20°C	3.24	7	99.0[h]	238
4,3	Br	2-NO_2-thiophen	σ_p	PhS^-	MeOH, 20°C	3.95	8	98.6[h]	239
2,3	Br	4-NO_2-thiophen	σ_p^-	PhS^-	MeOH, 20°C	8.18	8	99.6[h]	239
5,2	Cl	3-NO_2-thiophen	σ_p^-	Piperidine	MeOH, 20°C	3.42	7	100.0[h]	240
5,2	Br	3-NO_2-thiophen	σ_p^-	Piperidine	MeOH, 20°C	3.31	8	100.0[h]	240
5,2	I	3-NO_2-thiophen	σ_p^-	Piperidine	MeOH, 20°C	3.26	7	99.8[h]	240
5,2	$OC_6H_4NO_2$-4	3-NO_2-thiophen	σ_p^-	Piperidine	MeOH, 20°C	2.65	7	99.6[h]	240
5,2	SO_2Ph	3-NO_2-thiophen	σ_p^-	Piperidine	MeOH, 20°C	4.02	8	99.9[h]	240
3,2	Br	5-NO_2-thiophen	σ_p^-	Piperidine	MeOH, 20°C	4.02	8	99.6[h]	241
5,2	Br	3-NO_2-selenophen	σ_p^-	Piperidine	EtOH, 20°C	3.15	8	99.0[h]	242
4,2 5,2	Cl	Thiazole	σ_m, σ_p	PhS^-	MeOH, 50°C	5.30	7	97.6[h]	243
4,2 5,2	Cl	Thiazole	σ_m, σ_p	PhSH	MeOH, 50°C	*ca.*−3	6	89.5[l]	244
4,2	Cl	Thiazole	σ_m	MeO^-	MeOH, 50°C	5.88	4	99.6[i]	245
4,2	Cl	Thiazole	σ_m	PhS^-	MeOH, 50°C	5.14	4	99.0[i]	245
2,5 3,5	H	Thiophen	σ_m, σ_p^0	MeO^-	MeOD, 50°C	4.55	10	96.4[h]	146
2,5 3,5	H	Thiophen	σ_m, σ_p^0	MeO^-	MeOD, 140°C	4.97	8	96.8[h]	246
3,2	H	Thiophen	σ_I	MeO^-	MeOD, 140°C	8.12	5	98.6[h]	246
2,[o]	H	Thiophen	σ_p	TCNE[n]	CCl_4, 20°C	1.6	7	98.0[h]	247

[a] All correlations with log k unless otherwise noted. [b] Locations of substituent and leaving group if any. [c] Leaving group. [d] Skeletal group.
[e] Number of points in the set. [f] Percentage of the data accounted for by the correlation. Superscripts indicate confidence levels of the correlation coefficient.
[g] 99.0%. [h] 99.9%. [i] 98.0%. [j] 95.0%. [k] 90.0%. [l] <90.0%. [m] 99.5%. [n] Correlated with K_e. [o] Charge-transfer complex formation

Table 5.26. Values of ρ for Electrophilic Substitution in Thiophen and Benzene[a]

Reaction	Thiophen	Benzene
Bromination	−10.0	−12.1
Chlorination	−9.4[b]	−9.6[b]
Protodedeuteriation	−7.6	—
Protodetritiation	−7.2	−8.2
Acetylation	−5.6	−9.1
Trifluoroacetylation	−7.4	—
Mercuriation	−5.3	−4.0

[a]From Ref. 235 unless otherwise noted. [b]Reference 237.

6-Membered Rings

Partial-rate factors for the electrophilic aromatic substitution of polysubstituted aromatic compounds, including pyridinium ions, can be correlated by the Miller equation[8] (5.32), where $\sum \pi$ is the sum of the inter-

$$\log f = \rho \sum \sigma^{+} + q \sum \pi \qquad (5.32)$$

action terms.[249] Some work has appeared involving the application of the Hammett equation to nucleophilic aromatic substitution and to base-catalyzed H–D exchange. Results of the correlations for these and other reactions are in Table 5.28. They are quite satisfactory. The results obtained for H–D exchange at an *ortho*-position in pyridines, pyridine-*N*-oxides, pyridinium ions, and thiophens agree with the previously observed results for benzene,[10] in their almost complete dependence on σ_I, and in the one case for which a significant resonance effect was detected, it depended on σ_R^-. By contrast, ionization constants of 2-substituted pyridinium ions are a function largely of σ_I, but with a significant dependence on σ_R^+. The reason for the lack of dependence on σ_R or σ_R^- in most of the sets studied is probably due to almost all of these sets being composed of substituents that are electron donors by resonance. Had the sets included several substituents that are electron acceptors by resonance it is quite likely that a significant dependence on σ_R^- would have been found for the H–D exchange reaction.

The few available correlations by the extended Hammett equation are given in Table 5.27. The results agree with the above remarks concerning the H–D exchange reaction.

Polynuclear Systems

Many papers have appeared on the application of the Hammett equation to reactions of the heterocyclic ring in polynuclear systems. The results

Table 5.27. Results of Correlations of Data for Heteroarenes by the Yukawa–Tsuno and the Extended Hammett Equation[a]

i,j[b]	G[d]	Reagent	Reaction conditions	α	β	$100R^2$[f]	n[e]	Ref.
2,5[m]	Thiophen	Cl_2	AcOH, 25°C	−8.31	1.56	97.8[i]	5	237
3,2[o]	1-Me-pyridinium	OD^-	D_2O, 75°C	7.9	1.4	99.9[h]	6	253
1,2[p]	Pyridinium	OD^-	D_2O, 75°C	14.5	−0.428[q]	91.0[i]	6	252
3,2[p]	Pyridine	MeO^-	MeOD, 140°C	4.74	0.770[q]	89.1[i]	6	254
4,[m]	5-Substituted-phenyl-5-phenyl-2-phenylimino-Δ^3-1,3,4-oxadiazoline	[n]	PhCl, 104.4°C	−1.39	0.55	98.0[k]	4	265

[a] All sets are correlations of log k and for all sets, the leaving group is H. Correlations by the extended Hammett equation using σ_I and σ_R unless otherwise noted. This table follows the format of Table 5.25.
[m] Correlated by the Yukawa–Tsuno equation, $\alpha = \rho$, $\beta = r$. [n] Thermolysis. [o] σ_R^- was used.
[p] Data from reference, correlation from Ref. 209. [q] Not significant.

Table 5.28. Results of Correlations of Data for 6-Membered Heteroarenes by the Hammett Equation[a]

i,j[b]	L[c]	G[d]	Subst. constant	Reagent	Reaction conditions	ρ	n[e]	$100r^{2}$[f]	Ref.
6,3	Cl	Pyridazine	σ_p	MeO^-	MeOH	6.82	8	84.6[h]	250
4,2	Cl	1,3,5-Triazine	σ_m	OH^-	H_2O, 0°C	8.25	11	97.6[h]	251
6,2									
4,2	Cl	1,3,5-Triazine	σ_m	OH^-	H_2O, 100°C	5.87	11	94.7[h]	251
6,2									
1,2	H	Pyridine	σ_I	OD^-	D_2O, 25°C	14.5	6	91.2[i]	252
3,2	H	1-Methylpyridine	σ_I	OD^-	D_2O, 25°C	8.9	8	96.0[h]	253
3,6	H	1-Methylpyridine	σ_p^0	OD^-	D_2O, 25°C	3.7	6	90.3[m]	253
3,4	H	1-Methylpyridine	σ_I	MeO^-	MeOD, 140°C	4.41	6	85.4[g]	254[p]
4,2	H	Pyridine-*N*-oxide	σ_I	MeO^-	MeOD, 50°C	8.81	5	95.5[m]	255[p]
4,2	H	Pyridine-*N*-oxide	σ_I	MeO^-	MeOD, 190°C	9.66	5	96.4[m]	255[p]
4,2	H	Pyridine-*N*-oxide	σ_I	ND_2^-	ND_3, 120°C	7.05	4	99.8[h]	256[p]

[a] All sets are correlations of rate constants. For footnotes other than that below, see Table 5.25.
[p] Data from reference, correlation from Ref. 209.

of the correlations are in Table 5.29. Very good correlations were generally obtained. Results for the only set that was correlated by a multiparameter equation are given in Table 5.27.

Correlations with σ_p of pK_R^+ of 6- or 7-substituted chromylium and thiachromylium perchlorates by the Hammett equation have been reported. The data for 6- and 7-substituted compounds lie on different lines.[275,276] Correlation with σ_p is not unreasonable in view of the fact that the reaction occurs at the 2-position, and the P_R values for 6,2- and 7,2-naphthalene, and 7,2-isoquinoline are roughly equivalent to that for σ_p^0.

The ρ values obtained for the reaction of substituted styrenes with 9-acridizinium ions bearing a constant substituent are a linear function of the σ_p values of the constant substituents.[277]

No conclusions can be drawn concerning the relative merits of the Hammett and multiparameter equations as regards the subject of this section. The latter are surely deserving of more attention that they have received. Their use could provide a much better estimate of the extent of the transmission and the composition of electrical effects.

5.3.4. Substituent Effects of Heterocyclic Groups

Substituent constants of heterocyclic systems can be divided into four types, three of which are the same as those for polynuclear aromatic systems (see Section 5.2.2). Thus there are (*a*) normal substituent constants, (*b*) replacement substituent constants, (*c*) special substituent constants, defined as on p. 196, but substituting heteroaryl for aryl. The fourth type (*d*) consists of σ_I and σ_R values for a heteroatom that may or may not bear an attached group.

The Heterocyclic Ring as a Substituent: Normal Substituent Constants

Values of σ_I, σ_R, σ_m, σ_p, σ_p^+, σ_p^-, and other substituent constants have been reported for various heteroaryl groups. The available values are collected in Table 5.30. As can be seen from the table, there are many gaps. Where two or more values for a given σ constant are listed, they were obtained by different methods or under different conditions. Some authors have reported σ constants for a series of substituted heteroarenes.[150,283,289] In such cases, only the constants for the parent heteroarene are given in the table. In some cases, the type of substituent constant was not given in the original work and has been assigned on the basis of the method of determination.

Table 5.29. Results of Correlations of Data for Polynuclear Heterocyclic Systems by the Hammett Equation[a]

i,j[b]	G[d]	Q	Subst. constant	Reaction conditions	ρ	n[e]	$100r^2$[f]	Ref.
3′,2[n] 4′,2	*N*-Phenylpyridinium	log k	σ_m^0, σ_p^0	1:1MeOH–DMSO, 25°C	1.90	6	99.8[h]	257*b*
3′,2[n] 4′,2	*N*-Phenylpyridinium	log K_e	σ_m^0, σ_p^0	MeOH, 25°C	3.27	5	99.6[h]	257*a*
4′,-	2-(Substituted phenyl)-3,4,5-triphenylpyrrole	$E_{1/2}$	σ_p^+	DMF, 25°C	0.17	6	97.4[m]	258
4′,- 4″,-	2,5-Di(substituted phenyl)-3,4-diphenylpyrrole	$E_{1/2}$	σ_p^+	DMF, 25°C	0.15	7	97.8[h]	258
4′,-	2-(Substituted phenyl)-3,4-diphenyl-5-(4-biphenylyl)pyrrole	$E_{1/2}$	σ_p^+	DMF, 25°C	0.14	5	96.4[m]	258
4′,-	2-(Substituted phenyl)-3-4-diphenyl-5-(2-naphthyl)pyrrole	$E_{1/2}$	σ_p^+	DMF, 25°C	0.13	5	96.0[m]	258
3′,-[o] 4′,-	4-Phenyl-5-(substituted phenyl)-1,3,4-thiadiazole-2-thione	log k	σ_m^0, σ_p^0	50% v/v EtOH–H_2O, 20°C	1.10	12	89.5[h]	259
3′,-[o] 4′,-	4-Phenyl-5-(substituted phenyl)-1,3,4-thiadiazole-2-thione	log k	σ_m^0, σ_p^0	50% v/v EtOH–H_2O, 35°C	1.27	12	89.3[h]	259
4′,- 4″,-	2-(4-Quinolyl)-4,5-di-(substituted phenyl)imidazole	$E_{1/2}$	σ_p	Dioxan–H_2O, 25°C	0.141	5	89.3[i]	260
4′,- 4″,	2-(6-Quinolyl)-4,5-di-(substituted phenyl)imidazole	$E_{1/2}$	σ_p	Dioxan–H_2O, 25°C	0.135	5	98.2[m]	260
4′,- 4″,-	2-(7-Quinolyl)-4,5-di-(substituted phenyl)imidazole	$E_{1/2}$	σ_p	Dioxan–H_2O, 25°C	0.122	5	98.4[h]	260
4′,- 4″,-	2-(9-Acridinyl)-4,5,-di-(substituted phenyl)imidazole	$E_{1/2}$	σ_p	Dioxan–H_2O, 25°C	0.105	4	99.2[m]	260
4′,-[p]	3-(Substituted phenyl)-1,5-diphenylverdazyl	$E_{1/2}$	σ_p	DMF, 25°C	0.117	6	98.0[h]	261
4′,-[p]	3-(Substituted phenyl)-1,5-diphenylverdazyl	$E_{1/2}$	σ_p	Propylene carbonate, 25°C	0.131	6	98.8[h]	263
4′,-[p]	3-(Substituted phenyl)-1,5-diphenylverdazyl	$E_{1/2}$	σ_p	MeOH, 25°C	0.119	6	98.4[h]	261

4',-	3-(Substituted phenyl)-1,5-diphenylverdazyl	$E_{1/2}$	σ_p	DMF, 25°C	0.078	6	98.6	261
4',-[q]	3-(Substituted phenyl)-1,5-diphenylverdazyl	$E_{1/2}$	σ_p	Propylene carbonate, 25°C	0.071	6	95.8[h]	263
4',-[q]	3-(Substituted phenyl)-1,5-diphenylverdazyl	$E_{1/2}$	σ_p	MeOH, 25°C	0.042	6	87.6[g]	261
4',-[r]	3-(Substituted phenyl)-1,5-diphenylverdazyl	log k	σ_p^+	EtOH	−0.31	5	99.0[h]	262
4',-[r]	3-(Substituted phenyl)-1,5-diphenylverdazyl	log k	σ_p^+	Pyridine	−0.23	5	99.6[h]	262
4',-[r]	3-(Substituted phenyl)-1,5-diphenylverdazyl	log k	σ_p^+	DMF	−0.23	5	98.4[h]	262
3',5[s] 4',5	2-(Substituted phenyl)-5-methoxyoxazole-*N*-methyl-*N*-phenyl-2-carboxamide	log k	σ_m^+, σ_p^+	$PhNO_2$, 95.3°C	−1.16	8	95.3[h]	264
5,2[t]	1,3-Dimethylbenzimidazole	log k	$\frac{1}{2}(\sigma_m + \sigma_p)$	80% v/v MeOH–H_2O, 40°C	1.92	4	98.0[g]	266
5,2	1-Methyl-2-methylenebenzothiazole dimer	log K_e	σ_m	MeCN, 25°C	−1.53	5	86.4[l]	267
5,2[u] 6,2	1-Ethyl-2-methylenebenzimidazole	log k	σ_m, σ_p	MeCN, 25°C	1.34	13	98.0[h]	267
4,-; 5,-	Benzthiadiazole	$E_{1/2}$	σ_m^0, σ_p^0	DMF	0.610	11	98.4[h]	268
4,-; 5,-	Benzselendiazole	$E_{1/2}$	σ_m^0, σ_p^0	DMF	0.549	11	99.6[h]	268
6,2	2-Azido-1-ethylquinolinium	$E_{1/2}$	σ_p	H_2O, 25°C	0.122	8	98.2[h]	269
6,2	1-Cyanoquinolinium	pK_{ROH}	σ_p	H_2O	−6.16	7	98.0[h]	270
1,2	5-Nitroisoquinolinium	pK_{ROH}	σ^*	H_2O	−3.7	9	96.4[h]	271
1,2	1,8-Naphthyridinium	pK_{ROH}	σ^*	H_2O	−4.9	7	95.5[h]	271
	Porphin	$E_{1/2}$	σ_p	DMF	0.27	7	97.6[h]	272
9,-[v]	Acridizinium	log k	σ_p	Sulfolan, 130°C	1.13	5	90.3[i]	273
9,-[w]	Acridizinium	log k	σ_p	DMSO, 65°C	1.69	10	98.8[h]	274

[a] For footnotes *b–m*, see Table 5.25.
[n] Reagent, MeO^-. [o] Reagent, OH^-.
[p] Oxidation potentials. [q] Reduction potentials. [r] Reagent, 1,2-diphenylhydrazine. [s] Leaving group, OMe.
[t] Leaving group, Cl; reagent, PhSH.
[u] Reagent, 1,3-diethyl-5-(*NN*-dimethylaminomethylene)-2-thiobarbituric acid.
[v] Reagent, acrylonitrile. [w] Reagent, styrene.

Table 5.30. Substituent Constants for Heteroaryl Groups

Group	σ_m	σ_p	σ_I	σ_R	σ_m^+	σ_p^+
2-Furyl	0.06[a]	0.02[a]	0.04[a]		0.10[a]	−0.39[a]
3-Furyl					0.10[c]	−0.45[c]
2-Thienyl	0.09[e]	0.05[e]	0.21[d]		0.16[b]	−0.43[b]
						−0.23[c]
					0.15[c]	−0.33[c]
					0.15[c]	−0.38[c]
3-Thienyl	0.03[e]	−0.02[e]			0.08[b]	−0.38[b]
1-Pyrryl		0.10[f]				
		0.21[f]				
1-Pyrazolyl		0.19[f]				
		0.23[f]				
1-Imidazolyl		0.24[f]				
		0.45[f]				
1-(1H-1,2,4-triazolyl)		0.37[f]				
		0.44[f]				
1-(4H-1,2,4-triazolyl)		0.33[f]				
1-(1H-1,2,3-triazolyl)		0.40[f]				
		0.48[f]				
1-(2H-1,2,3-triazolyl)		0.36[f]				
		0.36[f]				
1-(1H-tetrazolyl)	0.52[g]	0.50[g]	0.54[g]			
		0.71[f]				
1-(2H-tetrazolyl)		0.59[f]				
		0.62[f]				
2-Pyridyl	0.49[i]	0.69[i]	0.40[i]			
	0.28[i]	0.32[i]	0.28[i]			
3-Pyridyl	0.70[j]	0.83[i]	0.73[j]			
	0.15[j]	0.31[i]	0.20[j]			
4-Pyridyl	0.49[i]	0.68[i]	0.27[i]			
	0.27[i]	0.35[i]	0.24[i]			
2-(4,6-Dimethyl-1,3,5-triazinyl)	0.25[k]	0.39[k]	0.15[k]	0.24[k]		
1-(1-Methylindazolyl)		0.27[f]				
		0.25[f]				
1-Benzimidazolyl		0.38[f]				
		0.50[f]				
1-Carbazolyl		0.39[f]				
		0.43[f]				
2-Benzoxazolyl	0.30[o]	0.33[o]	0.28[o]	0.05[o]		
2-Benzothiazolyl	0.27[o]	0.29[o]	0.26[o]	0.03[o]		
2-(1-Phenylbenzimidazolyl)	0.17[o]	0.21[o]	0.14[o]	0.08[o]		
2-(1-Methylbenzimidazolyl)			0.07[o]	0.04[o]		
1-Indolyl			−0.01[d]			

Table 5.30—continued

	σ_m^-	σ_p^-	σ^*	σ_m^0	σ_p^0	σ_R^0
2-Furyl	0.11[a]	0.21[a]	1.08[h]			
3-Furyl			0.65[h]			
2-Thienyl	0.11[e]	0.19[e]	0.93[h]			
3-Thienyl	0.07[e]	0.13[e]	0.65[h]			
1-(1H-tetrazolyl)	0.60[g]	0.57[g]		0.52[g]	0.502[g]	−0.04[g]
3-Sydnonyl		0.71[r]	4.81[l]			
3-Benzyl-4-sydnonyl			3.31[l]			
2-Pyridinyl						0.23[i]
						0.00[i]
3-Pyridinyl						0.08[i]
						−0.01[i]
4-Pyridinyl						0.57[i]
						0.08[i]
2-Pyridine-*N*-oxide	0.23[m]	0.27[m]				
4-Pyridine-*N*-oxide		0.33[m]				
2-Pyridinium	–	0.75[m]				
4-Pyridinium		0.65[m]				
2-(1,3,5-Triazinyl)		0.61[n]				
2-(4,6-Dimethyl-1,3,5-triazinyl)						0.18[k]
2-Quinolyl			1.15[p]			
1-Isoquinolyl			1.19[p]			
4-Quinolyl			1.22[q]			

[a]Ref. 278. [b]Ref. 279. [c]Ref. 280. [d]Ref. 98. [e]Ref. 281. [f]Ref. 282. [g]Ref. 283.
[h]Ref. 150. [i]Ref. 287. [j]Ref. 288. [k]Ref. 290. [l]Ref. 284. [m]Ref. 286.
[n]Ref. 289. [o]Ref. 291. [p]Ref. 292. [q]Ref. 293. [r]Ref. 285.

The Heteroatom as a Substituent: Replacement Substituent Constants

Replacement constants for heteroarenes were first proposed by Jaffé,[19] who suggested that a substituent constant could be defined for the replacement of —CH— or —CH=CH— in benzene by a heteroatom. These constants are defined, then, from a correlation of data for a set of benzene derivatives by the Hammett equation. The reference compound is the unsubstituted benzene derivative. In early publications in this area, little attention was paid to the point that more than one type of substituent constant could be obtained. Values of σ^+, σ^-, σ, σ^0, σ_I, σ_R, and σ_R^0 have all been reported for various heteroarenes. These values are in Table 5.31. When the substituent type was not indicated in the original work, it has been assigned on the basis of the method of determination.

There is a tendency to report replacement substituent constants in the form $\geqslant N^+H$ for pyridinium or $\geqslant N$ for pyridine. As the constants for nitrogen atoms, for example, can vary from one heteroarene to another, they will be listed in Table 5.31 by the heteroarene from which they were derived.

Inspection of the σ (replacement) values tabulated shows that there is a considerable degree of variation even within a given type of substituent constant, such as σ^+ or σ^- for any given heteroaryl group. If substituent constants are to be useful, it is of course necessary for them to be reasonably constant for any given group and type of substituent effect. It seems then, that the application of the Hammett equation to sets that include heteroaryl groups, by using σ (replacement) constants is at best of limited utility.

The effect of solvent upon σ (replacement) constants for pyridyl groups determined from rate constants for the saponification of substituted

Table 5.31. Replacement Substituent Constants for Heteroaryl Groups

Group	σ	σ^+	σ^-	σ^0	σ_I
2-Furyl	0.61[a]	−0.37[a]	0.1[cc]	0.57[r]	0.4[ii]
	1.08[a]	−0.98[a]		0.6[cc]	
		−0.89[l]			
	0.24[a]	−0.85[m]			
		−0.13[n]			
	0.28[a]	−0.86[o]			
		−0.75[q]			
	0.29[a]	−1.32[r]			
	0.80[a]	−0.95[s]			
	0.66[a]				
	0.10[b]				
	0.23[b]				
	0.32[m]				
3-Furyl	0.25[a]	−0.45[j]		0.42[ii]	
		0.51[s]			
	0.04[m]	−0.42[l]			
		−0.44[m]			
		0.00[n]			
		−0.74[r]			
2-Thienyl	0.36[a]	−0.27[a]	0.58[a]	0.50[r]	0.6[ii]
			0.4[cc]	0.4[cc]	
	0.71[a]	−0.85[a]	0.3[cc]	0.7[cc]	
		−0.80[ll]			
		−0.80[a]			
	−0.03[a]	−0.79[k]			
		−0.80[l]			
	0.05[a]	−0.76[m]			
		−0.44[n]			
	0.01[a]	−0.83[o]			
	−0.06[b]	−0.68[q]			
	−0.15[b]	−1.22[r]			
	0.03[m]	−0.85[s]			

Table 5.31—continued

Group	σ	σ^+	σ^-	σ^0	σ_I
3-Thienyl	0.12[a]	−0.45[a]		0.26[r]	0.6[jj]
		−0.47[ll]			
		−0.52[k]		0.2[cc]	
	0.00[a]	−0.38[l]		0.2[cc]	
	−0.01[a]	−0.44[m]			
	0.04[m]	−0.10[n]			
		−0.45[o]			
		−0.24[q]			
		−0.62[r]			
		−0.49[s]			
2-Selenyl	0.28[a]	−0.95[o]		0.55[r]	
		−1.28[r]			
2-Pyrrolyl	−0.25[a]	−1.61[m]			
	−0.58[m]	−1.33[n]			
	−0.15[kk]	−1.7[o]			
		−2.0[kk]			
3-Pyrrolyl	−0.94[m]	−1.20[m]			
	−0.75[kk]	−0.54[n]			
		−1.71[o]			
		−1.5[kk]			
2-Telluryl		−1.0[p]			
		−0.90[p]			
		−0.90[p]			
		−0.90[p]			
3-Isothiazolyl		0.65[t]			
4-Isothiazolyl		−0.04[t]			
5-Isothazolyl		0.67[t]			
2-Thiazolyl		0.26[u]			
4-Thiazolyl		−0.01[u]			
5-Thiazolyl		−0.18[u]			
2-Pyridyl	0.81[a]	0.45[a]	0.28[a]	1.0[hh]	
	1.1[a]	0.56[h]	0.41[a]	0.8[hh]	
	0.73[f]			1.0[ii]	
	0.37[a]	0.73[v]	0.58[a]	0.9[ii]	
	0.61[f]	0.72[v]			
	0.92[a]	0.72[v]	1.00[f]		
	0.81[f]	0.75[w]			
	1.02[a]		1.00[gg]		
	1.0[d]		1.1[hh]		
	0.75[c]		1.00[hh]		
	0.88[e]				
	0.68[g]				
3-Pyridyl	0.62[a]	0.62[a]	0.76[a]	0.5[hh]	
	0.27[a]	0.57[v]		0.4[hh]	
	1.3[a]	0.60[h]	0.59[f]	0.4[ii]	
	0.55[a]	0.47[v]	0.67[f]		
	0.40[a]	0.45[v]	0.58[gg]		
	0.10[a]	0.54[w]	0.59[hh]		

continued overleaf

Table 5.31—continued

Group	σ	σ^+	σ^-	σ^0	σ_I
3-Pyridyl	0.34[a]	0.78[mm]	0.6[hh]		
(*contd.*)	0.62[a]				
	0.65[c]				
	0.6[d]				
	0.74[e]				
	0.45[f]				
	0.53[f]				
	0.51[f]				
	0.55[f]				
	0.47[f]				
	0.62[f]				
	0.33[g]				
	0.42[i]				
4-Pyridyl	0.93[a]	0.67[a]	1.26[a]	0.6[hh]	
	0.95[a]	0.83[h]	1.17[f]	0.7[hh]	
	1.6[a]	0.68[a]	1.00[f]	0.5[ii]	
	0.99[a]	1.13[v]	1.16[gg]		
	0.93[a]	1.2[v]	1.17[hh]		
	0.96[c]	1.14[v]	1.1[hh]		
	0.8[d]				
	1.11[e]				
	0.6[g]				
	0.76[f]				
	0.85[f]				
	0.99[f]				
	0.91[f]				
	0.93[f]				
	0.93[f]				
2-(1-Methylpyridylium)			4.58[ff]		
			2.99[gg]		
3-(1-Methylpyridylium)			1.58[gg]		
4-(1-Methylpyridylium)			3.16[ff]		
			2.32[gg]		
2-Pyridylium	2.2[d]				
	3.16[f]	3.21[h]	3.18[f]	2.1[ii]	
	3.41[f]		2.49[f]	2.2[ii]	
	3.45[f]		4.38[ff]		
	2.98[f]				
3-Pyridylium	2.1[a]	2.07[a]	2.1[a]		
	2.09[f]	1.58[f]	2.22[f]		
	1.9[d]	2.18[h]	2.12[f]		
	2.32[f]	1.82[y]	2.02[dd]		
	2.06[f]				
	2.04[f]				
	2.18[f]				
4-Pyridylium	2.3[a]	1.97[a]	4.0[a]	1.3[ii]	
	1.3[d]	2.42[h]	2.32[f]	1.7[ii]	
	2.34[f]	1.88[mm]	3.4[z]		

Table 5.31—continued

Group	σ	σ^+	σ^-	σ^0	σ_I
4-Pyridylium	2.63[f]	1.89[mm]	4.06[dd]		
(*contd.*)	2.86[f]		3.19[ff]		
	2.80[f]				
2-Pyridyl-*N*-oxide	−0.40[e]	0.68[w]	1.5[hh]	1.0[hh]	
	1.48[a]		1.6[hh]	1.0[hh]	
				1.0[ii]	
				1.4[ii]	
3-Pyridyl-*N*-oxide	0.7[d]	1.18[a]	1.59[a]	0.7[hh]	
	1.31[e]	0.81[x]	1.18[hh]	0.8[hh]	
	1.48[h]	1.99[y]	1.1[hh]	0.6[ii]	
		0.8[z]		0.7[ii]	
4-Pyridyl-*N*-oxide	1.35[a]	0.23[a]	1.88[a]	0.5[hh]	
	0.4[d]	0.45[w]	2.0[z]	0.7[hh]	
	1.14[e]	0.02[x]	1.58[hh]	0.6[ii]	
	1.35[h]		1.2[hh]	0.4[ii]	
3-(1-Hydroxypyridylium)	2.1[z]	2.3[a]			
	2.3[z]	2.25[dd]			
4-(1-Hydroxypyridylium)			3.9[a]		
			3.9[z]		
			3.49[dd]		
3-Pyranyl		3.0[z]			
4-Pyranyl			5.8[z]		
3-Thiapyranyl		3.2[z]			
4-Thiapyranyl			5.4[z]		
2-Benzo[*b*]furyl		−0.49[s]	0.8[cc]	0.7[cc]	
		−0.65[q]		0.7[cc]	
3-Benzo[*b*]furyl		−0.48[s]			
2-Benzo[*b*]thienyl		−0.53[q]	0.6[cc]	0.5[cc]	
		−0.46[s]		1.0[cc]	
		−0.49[aa]			
3-Benzo[*b*]thienyl		−0.46[q]		0.5[cc]	
		−0.54[s]		0.5[cc]	
		−0.56[aa]			
4-Benzo[*b*]thienyl		−0.25[aa]			
5-Benzo[*b*]thienyl		−0.34[aa]			
6-Benzo[*b*]thienyl		−0.42[aa]			
7-Benzo[*b*]thienyl		−0.11[aa]			
2-(1-Methylindolyl)			0.5[cc]	0.2[cc]	
2-Benzoxazolyl	1.08[ee]		1.7[cc]	1.4[cc]	
				1.2[cc]	
2-Benzothiazolyl	0.72[ee]		1.7[cc]	1.3[cc]	
			1.6[cc]	1.6[cc]	
			1.80[ee]		
2-Benzoselenazolyl			1.8[cc]	1.2[cc]	
				1.9[cc]	
2-(1-Methylbenzimidazolyl)	−0.24[ee]		1.27[ee]	0.7[cc]	
			1.4[cc]	0.5[cc]	
			1.5[cc]		

continued overleaf

Table 5.31—continued

Group	σ	σ^+	σ^-	σ^0	σ_I
2-Quinolyl	0.00[ee]	0.73[bb]	1.33[ee]	1.3[ii]	
3-Quinolyl		0.08[bb]			
4-Quinolyl		0.75[bb]	1.38[ii]	0.4[ii]	
5-Quinolyl		−0.11[bb]			
6-Quinolyl		0.07[bb]	0.45[ii]	0.5[ii]	
7-Quinolyl		0.15[bb]			
8-Quinolyl		0.07[bb]			

[a]Ref. 3. [b]Ref. 294. [c]Ref. 295. [d]Ref. 296. [e]Ref. 297. [f]Ref. 298. [g]Ref. 299.
[h]Ref. 300. [i]Ref. 301. [j]Ref. 302. [k]Ref. 303. [l]Ref. 304. [m]Ref. 305.
[n]Ref. 306. [o]Ref. 307. [p]Ref. 308. [q]Ref. 309*a*. [r]Ref. 309*b*. [s]Ref. 310.
[t]Ref. 311. [u]Ref. 312. [v]Ref. 313. [w]Ref. 314. [x]Ref. 315. [y]Ref. 316.
[z]Ref. 317. [aa]Ref. 318. [bb]Ref. 319. [cc]Ref. 320. [dd]Ref. 321. [ee]Ref. 322.
[ff]Ref. 323. [gg]Ref. 324. [hh]Ref. 325. [ii]Ref. 326. [jj]Ref. 327. [kk]Ref. 328.
[ll]Ref. 331. [mm]Ref. 337.

alkyl benzoates and pyridinecarboxylates in aqueous methanol, acetone, dioxan, and DMSO has been reported.[329] Only the 2-pyridyl group seems to show a significant solvent effect.

The σ, σ^-, and σ^+ (replacement) values obtained for pyrimidine, pyrazine, and pyridazine show that in a given reaction the effect of the heteroatoms is generally additive,[324,330,337] as can be seen from Table 5.32. The degree of additivity is greater for σ than for σ^-, the greatest deviation is for 2-pyrazyl. Examples of correlations with σ (replacement) constants giving excellent results are reported by Blanch.[332,333a]

Special Substituent Constants

Ionization constants of *N*-methyl- or *N*-ethyl-2-aminoheteroarenes were used to define σ_{het} constants by means of equation (5.33).[333b] Rates of reaction of 2-azido-*N*-ethylheteroarenes with 1-(4-sulfonatophenyl)-3-methylpyrazolin-5-one,[334] of reactions of 2-azido-*N*-ethylheteroarenes with cyanide ion,[335] and of hydrolysis of the resulting adducts[335] were

Table 5.32. Additivity of σ(Replacement) Values

Heteroaryl Group	σ	σ^-	σ^+	Equivalent	$\sum\sigma$	$\sum\sigma^-$	$\sum\sigma^+$
5-Pyrimidyl	1.30	1.54	1.10	3-Py+3-Py	1.30	1.16	1.10
2-Pyrazinyl	1.40	1.72		2-Py+4-Py	1.71	2.16	
3-Pyridazyl	1.40	1.66		2-Py+3-Py	1.40	1.58	
2-Pyrimidyl	1.50	1.98		2-Py+2-Py	1.50	2.00	
4-Pyridazyl	1.61	1.69		3-Py+4-Py	1.61	1.74	
4-Pyrimidyl	1.71	1.78		2-Py+4-Py	1.71	2.16	
5-Pyrimidylium			3.90				3.76

correlated with the σ_{het} values. Polarographic half-wave potentials of 2-azido-*N*-ethylheteroarenes were also studied.[336]

$$\log (K_{a,het}/K_{a,py}) \equiv \sigma_{het} \tag{5.33}$$

Localized and Delocalized Electrical Effect Constants

It is theoretically possible to determine σ_I and $\sigma_R^{\cdot}$ values for a heteroatom which may or may not bear an attached group. A number of such constants have been reported for nitrogen in 6-membered rings and they are collected in Table 5.33. They may be used in correlations by the extended Hammett and the DSP equation. Data for ^{35}Cl nuclear quadrupole resonance spectra have been correlated by the equation (5.34).[340,341] Alternatively, the Yukawa–Tsuno equation may be used for the correlation of data for heteroaryl groups. Thus, rates of solvolysis of 2-arylethyl tosylates were employed to obtain Fk_Δ values.† These values were best correlated by a modification of the Yukawa–Tsuno equation,[342] viz., (5.35), with $100R^2 = 94.0$, significant at the 99.9% confidence level.

$$\nu\,^{35}\mathrm{Cl} = \alpha_o \sum \sigma_{Io} + \beta_o \sum \sigma^0_{Ro} + \alpha_m \sum \sigma_{Im} + \beta_m \sum \sigma^0_{Rm} + \alpha_p \sum \sigma_{Ip} + \beta_p \sum \sigma^0_{Rp} + h \tag{5.34}$$

$$\log (k/k^0) = \rho[\sigma_{repl} + r(\sigma^+_{repl} - \sigma_{repl})] \tag{5.35}$$

5.3.5. Tautomerism

The application of the Hammett and the extended Hammett equation to tautomerism in heterocyclic compounds has been recently reviewed in an excellent comprehensive monograph.[343]

Table 5.33. Values of σ_I and σ^0_R for Nitrogen Atoms with Substituents Attached in 6-Membered Rings

Substituent	σ_I	σ^0_{Rm}
O^-	0.90[b,c]	−0.21[a] −0.20[b]
Me		0.28[a]
—	0.45[c]	0.24[a] 0.25[b]
OMe		0.17[a]
OBu^t		0.14[a]
OBF_3^-		0.12[a]
BH_3^-		0.21[a]
H	1.0[c]	0.3[c]

[a]Reference 338. [b]Reference 339. [c]Reference 340.

† "Rates of the participating pathway" (see Ref. 342).

Annular Tautomerism

The Hammett equation has been used to derive equation (5.36) for annular tautomerism of 4(5)-substituted imidazoles, where K_T is given by

$$\log K_T = (\rho_1 - \rho_3)\sigma_{m-X} \tag{5.36}$$

(5.37), and where ρ_1 and ρ_3 are the reaction constants for the ionization of the 5-substituted and 4-substituted imidazoles (B_1 and B_3, respectively).

$$K_T = C_{B_3}/C_{B_1} \tag{5.37}$$

Evaluation of ρ_1 and ρ_3 led to the conclusion that B_1 predominates with strong electron-donor groups and B_3 with strong electron-acceptor groups.[184,185] The σ_{imid} constants defined from substituted 1-methylimidazole pK_a values have been used to predict constants for tautomeric equilibria of 5-substituted histamines.[185] Values of pK_T and f_1 for 5-substituted tetrazoles (f_1 is the fraction of the substance present as tautomer 1) were calculated from the extended Hammett equation by using α and β values obtained from appropriate model systems.[191] It has been proposed that successful correlation of ionization constants for 4-substituted pyrimidines by use of σ_p constants indicates[208] protonation at N^1. On the basis of correlations by the extended Hammett equation it was concluded that the 5- and the 6-tautomers of 5(6)-substituted benzimidazoles occur to nearly the same extent.[222] The successful correlation of pK_a values of 4-substituted cinnolinium ions with σ_p constants was believed[217] to indicate protonation at N^1. The Jaffé equation (5.9) and the Hammett equation have been used to investigate annular tautomerism in 6-substituted 2,3-dimethylquinoxalines.[344–345] Values of pK_T and f_1 for 2-, 3-, 4-, and 5-substituted 1,10-phenanthrolines were obtained from the extended Hammett equation by using appropriate model systems for the estimation of α and β values.[346]

Ionization constants of 5-substituted 4-hydroxy-2-methylpyrimidines, 5-substituted 1,2-dimethyl-4(1H)-pyrimidones, and 5-substituted 2,3-dimethyl-4(3H)-pyrimidones were correlated with σ_m constants. The success of the correlations is believed to imply that pK_T is constant.[347]

Other Types of Tautomerism

Many heterocyclic compounds have two or more nonequivalent acidic and/or basic sites. Such compounds exist as a mixture of tautomers. The problem was discussed in detail by Jaffé and Jones,[3] who have shown that K_T can be estimated by means of the Hammett equation.

Ionization constants of 4-, 6-, and 8-substituted quinoline-2-carboxylic acids were correlated with the aid of the DG method, by using substituent constants calculated by assuming protonation at $-CO_2^-$ and others calculated by assuming protonation at nitrogen. Best results were obtained for protonation at $-CO_2^-$. Comparison of the ρ values obtained with the ρ value for 4-substituted quinolines confirms the results obtained.[348]

Equilibrium constants for ring–chain tautomerism of 5-(4-substituted phenyl)-5-hydroxy-1,2-isoxazolines are well correlated by the Hammett equation, using σ^+ constants, and by the DSP equation using σ_R^+ constants. An attempt to correlate equilibrium constants for ring–chain tautomerism of 5-substituted 5-hydroxy-2-isoxazolines, in which the substituent is alkyl, phenyl, or diphenylmethyl, with the Taft–Pavelich equation was unsuccessful.[349]

5.3.6. Physical Properties of Heterocyclic Compounds

In this section, we consider correlations of nmr chemical shifts and coupling constants, ir frequencies and intensities, ionization potentials, dipole moments, and other physical properties.

5-Membered Rings

Ir intensities of ring-stretching bands for 2-substituted thiophens and furans,[350] and 2- and 3-substituted selenophens,[351] are correlated by (5.38),

$$A = a(\sigma_R^0 + \lambda + b)^2 + c \tag{5.38}$$

where λ is a constant that accounts for direct interactions between the heteroatom and the substituent. The λ constants for substituents with empty orbitals and for π-acceptors are given by equations (5.39) and (5.40), or (5.41), where K_X and K_A are constants characteristic of the two types of

$$\lambda = K_X \sigma_{R,\text{repl}}^0 \tag{5.39}$$

$$\lambda = 6.5\sigma_{R,\text{acptor}}^0(\sigma_{R,\text{repl}}^0)^2 \tag{5.40}$$

$$\lambda = K_A(\sigma_{\text{repl}}^+ - \sigma_{\text{repl}}^0) \tag{5.41}$$

delocalized electron-acceptor groups, and the σ replacement substituent constants are for the parent heteroaryl group. Other ir bands of thiophens and furans have also been studied.[352,353] Thus, the CH intensities were said to be governed by the Schmid[354] equation (5.42).

$$I = (a\sigma_I - b)^2 + c^2 \tag{5.42}$$

Correlations by the Hammett equation of ir frequencies of substituted heteroarenes in which the absorption involves a functional group attached to the ring have received some attention. Examples are given in Table 5.34.

Table 5.34. *Results of Correlations of Spectroscopic Data by the Hammett Equation*

i,j[a]	G[b]	Y[c]	*Q*	Subst. constant	ρ	n[d]	$100r^2$[e]	Ref.
2,	Thiophen		IP	σ_p^+	0.93	6	94.9[g]	360
5,2[o]	Thiophen	Ac	ν_{CO}	σ_p	0.075	6	90.4[g]	3
3,1:4,1[p]	Pyridine-*N*-oxide		ν_{NO}	σ_m, σ_p	0.029	6	95.6[f]	3
3,1:4,1[q]	Pyridine-*N*-oxide		ν_{NO}	σ_m, σ_p	0.033	4	99.6[g]	3
3,1:4,1[r]	Pyridine-*N*-oxide		ν_{NO}	σ_m, σ_p	0.033	4	99.4[g]	3
3,1:4,1[p]	*Trans*-pyridine-*N*-oxide-		$\nu_{CH:CH}$	σ_m, σ_p	0.018	6	99.6[f]	3
3,1:4,1[p]	ethylene dichloroplatinum(II)		ν_{NO}	σ_m, σ_p	0.030	6	88.4[h]	3
3,1:4,1	1-Methylpyridinium	CH_3	δ^1H_4 δ^1H_3	σ_R^0	0.38	14	91.6[f]	365
3,1:4,1	1-Methylpyridinium		$A_{NC}^{1/2}$		114	8	94.1[f]	365
2,3	Pyrazine		δ^1H	σ_p^+	0.64	12	97.0[f]	366
2,5	Pyrazine		δ^1H	σ_p^+	0.57	12	96.8[f]	366
2,6	Pyrazine		δ^1H	σ_m^+	0.75	6	95.3[f]	366
2,6[m]	Pyrazine		δ^1H	σ_p^0	1.21	6	99.6[f]	366
3′,2; 4′,2[o]	Phenylfuran	CHO	ν_{CO}	σ_m, σ_p	8.97	10	95.5[f]	367
3′,2; 4′,2[s]	Phenylfuran	CHO	ν_{CO}	σ_m, σ_p	11.8	10	99.2[f]	367
3′,2; 4′,2[s]	Phenylthiophen	CHO	ν_{CO}	σ_m, σ_p	9.69	9	96.0[f]	368
3′,2; 4′,2[o]	Phenylthiophen	CHO	ν_{CO}	σ_m, σ_p	5.55	8	83.7[g]	368
3′,[s] 4′,	2-(5-Substituted-phenyl-2-thenylidene)-1,3-indandione	C=O	$\bar{\nu}$	σ_m^+, σ_p^+	3.54	9	98.2[f]	368
3′,[s] 4′,	2-(5-Substituted-phenyl-2-thenylidene)-1,3-indandione	C=O	ν_{as}	σ_m^+, σ_p^+	3.05	9	93.1[f]	368
3′,[s] 4′,	2-(5-Substituted-phenyl-2-thenylidene)-1,3-indandione	C=O	ν_s	σ_m^+, σ_p^+	3.39	9	90.3[f]	368
3′,[o] 4′,	2-(5-Substituted-phenyl-2-thenylidene)-1,3-indandione	C=O	ν_{as}	σ_m^+, σ_p^+	2.89	9	97.0[f]	368
3′,[o] 4′,	2-(5-Substituted-phenyl-2-thenylidene)-1,3-indandione	C=O	ν_{as}	σ_m^+, σ_p^+	2.26	9	95.6[f]	368

3′,-[o] 4′,-	2-(5-Substituted-phenyl-2-thenylidene)-1,3-indandione	C=O	ν_s	σ_m^+, σ_p^+	3.52	9	93.1[f]	368
3′,1; 4′,1[o]	*Trans*-1-phenyl-3-(5-substituted-phenyl-2-furyl)propenone	C=O	ν_{CO}, *s-cis*	σ_m^+, σ_p^+	3.32	9	97.8[f]	369
3′,1; 4′,1[o]	*Trans*-1-phenyl-3-(5-substituted-phenyl-2-furyl)propenone	C=O	ν_{CO}, *s-trans*	σ_m^+, σ_p^+	4.74	8	98.0[f]	369
3′,1; 4′,1[s]	*Trans*-1-phenyl-3-(5-substituted-phenyl-2-furyl)propenone	C=O	ν_{CO}, *s-cis*	σ_m^+, σ_p^+	3.13	9	98.0[f]	369
3′,1; 4′,1[o]	*Trans*-1-phenyl-3-(5-substituted-phenyl-2-thienyl)propenone	C=O	ν_{CO}, *s-cis*	σ_m^+, σ_p^+	2.49	8	94.1[f]	369
3′,1; 4′,1[o]	*Trans*-1-phenyl-3-(5-substituted-phenyl-2-thienyl)propenone	C=O	ν_{CO}, *s-trans*	σ_m^+, σ_p^+	2.82	8	97.8[f]	369
3′,1; 4′,1[s]	*Trans*-1-phenyl-3-(5-substituted-phenyl-2-thienyl)propenone	C=O	ν_{CO}, *s-trans*	σ_m^+, σ_p^+	3.99	8	98.6[f]	369
3′,1; 4′,1[s]	*Trans*-1-phenyl-3-(5-substituted-phenyl-2-thienyl)propenone	C=O	ν_{CO}, *s-cis*	σ_m^+, σ_p^+	2.72	8	96.0[f]	369
3′,[s] 4′,	2-(5-Substituted-phenyl-2-furfurylidene)-1,3-indandione	C=O	$\bar{\nu}$	σ_m^+, σ_p^+	4.13	11	96.2[f]	370
3′,[s] 4′,	2-(5-Substituted-phenyl-2-furfurylidene)-1,3-indandione	C=O	ν_{as}	σ_m, σ_p	7.21	11	93.7[f]	370
3′,[s] 4′,	2-(5-Substituted-phenyl-2-furfurylidene)-1,3-indandione	C=O	ν_s	σ_m^+, σ_p^+	3.39	11	96.6[f]	370
3′,[o] 4′,	2-(5-Substituted-phenyl-2-furfuryldiene)-1,3-indandione	C=O	$\bar{\nu}$	σ_m^+, σ_p^+	4.74	10	94.3[f]	370
3′,[o] 4′,	2-(5-Substituted-phenyl-2-furfuryldiene)-1,3-indandione	C=O	ν_{as}	σ_m, σ_p	5.84	10	96.4[f]	370
3′,[o] 4′,	2-(5-Substituted-phenyl-2-furfurylidene)-1,3-indandione	C=O	ν_s	σ_m, σ_p	5.31	10	95.3[f]	370
4,	3-Substituted-phenyl-5-phenyl-1,2,4-oxadiazole		IP	σ_p^+		5	98.0[g]	371
4,	3-Phenyl-5-substituted-phenyl-1,2,4-oxadiazole		IP	σ_p^+		5	98.0[g]	371
3′,5 4′,5	1-(Substituted-phenyl)-3-hydroxy-4-phenyl-1,2,4-triazolium		δ^1H	σ_m^+, σ_p^+	23.1	6	96.8[f]	372

continued overleaf

Table 5.34—continued

i,j[a]	G[b]	Y[c]	Q	Subst. constant	ρ	n[d]	$100r^2$[e]	Ref.
3′,5 4′,5	1-Phenyl-4-(substituted-phenyl)-3-hydroxy-1,2,4-triazolium		δ^1H	σ_m^+, σ_p^+	14.5	13	92.5[f]	372
5,2; 6,2	Benzoxazole	CH_3	δ^1H	σ_m, σ_p	6.82	7	87.6[g]	373
5,2; 6,2	Benzothiazole	CH_3	δ^1H	σ_m, σ_p	7.31	7	91.4[f]	373
5,2; 6,2	Benzoselenazole	CH_3	δ^1H	σ_m, σ_p	6.72	7	91.8[f]	373
5,2[s]	1,3-Dimethylbenzimidazolone	C=O	ν_{CO}	F^v	16.4	8	96.6[f]	374
2,8; 6,8[t]	Purine		δ^1H	σ_p^+	0.342	17	96.8[f]	375
6,8[t]	Purine		δ^1H	σ_p^+	0.359	9	98.2[f]	376
6,2[t]	Purine		δ^1H	σ_R	0.899	9	93.5[f]	376
2,5[u]	Thieno[2,3-*b*][1]benzothiophen		δ^1H	σ_p^0	0.15	7	96.0[f]	377
2,6[u]	Thieno[2,3-*b*][1]benzothiophen		δ^1H	σ_p^0	0.17	7	98.4[f]	377
2,5[u]	Thieno[3,2-*b*][1]benzothiophen		δ^1H	σ_p^0	0.15	7	92.0[f]	377
2,6[u]	Thieno[3,2-*b*][1]benzothiophen		δ^1H	σ_p^0	0.21	7	93.3[f]	377
3′,9 4′,9	9-Substituted-phenyl-3,3,6-tetramethyl-1,8-diketo-octahydroxanthen		δ^1H	σ_m, σ_p	9.29	8	92.2[f]	378

[a]Locations of substituent and reaction site. [b]Skeletal group.
[c]Probe site. [d]Number of points in the set.
[e]Percentage of the data accounted for by the regression equation. Superscripts indicate the confidence levels of the correlation coefficients.
[f]99.9%. [g]99.5%. [h]99.0%. [i]90.0%. [j]95.0%. [k]90.0%. [l]<90.0%.
[m]Delocalized-effect electron-acceptor groups only. [n]Delocalized-effect electron-donor groups only.
[o]Solvent, CCl_4. [p]Nujol mull. [q]Solid. [r]Solvent, CS_2. [s]Solvent, $CHCl_3$.
[t]Solvent, DMSO. [u]Solvent, $CDCl_3$. [v]Dewar–Grisdale F values (see p.177).

A number of correlations of ^{13}C, ^{19}F and ^{1}H chemical shifts by multiparameter equations have appeared.[355–359] The results are in Table 5.35. Correlations are somewhat better for $\delta^{1}H$ than for $\delta^{13}C$. The 2,3-, 3,2-, 2,5-, and to a less extent, the 2,4-derivatives all show a large preponderance of the delocalized effect. The 3,4- and 3,5-derivatives are also somewhat more dependent on the delocalized effect. The β values for the thiophens and selenophens are quite similar; those for the furans are different. It has been suggested that α and β for the thiophens, selenophens, and furans are a linear function of the electronegativity of the heteroatom. Proton chemical shifts of H_4 in 1,3,5-substituted pyrazoles have been correlated by equation (5.43), where δ_S^4 is a constant characteristic of the solvent and α_1, α_3, and α_5 are constants that represent the

$$\delta H_4 = \delta_S^4 + \alpha_1 + \alpha_3 + \alpha_5 \tag{5.43}$$

effect of replacing a methyl group by some substituent at positions 1, 3, and 5, respectively. The constants are said to be a linear function of σ_p.[361] Ionization potentials of 2-substituted thiophens, furans, and pyrroles are correlated by σ_p^+ constants.[360,362] The quantity log (a_{NX}/a_{NH}), where a_N is the hyperfine splitting constant for 5-substituted 2-thienyl nitroxides, is linear in σ_p^-.

6-Membered Rings

Intensities of the CH stretching vibrations of 2-, 3-, and 4-substituted pyridines are described by equations derived from the Schmid equation[364] (5.42). Correlations of data by the Hammett equation are reported in Table 5.34. No reports involving the use of multiparameter equations have appeared.

Polynuclear Systems

Correlations of ir and nmr spectral data by the Hammett equation and by multiparameter equations are in Table 5.34 and 5.35, respectively. The results in Table 5.34 show that ir and nmr spectra can give good correlations by means of the simple Hammett equation. Perjéssy and his coworkers have made attempts to use ir spectra as a means of determining the extent of transmission of electrical effects.[367–370,379] It would be well to examine correlations of carbonyl stretching frequencies for naphthalene derivatives, and to compare the γ_L, and γ_D, and P_R values obtained with those known for carboxylic acid ionization in order to establish whether or not ir spectra are useful for the measurement of the transmission of electrical effects.

*Table 5.35. Results of Correlations of Nmr Data by the Extended Hammett Equation**

i,j^a	Skeletal group	Q	Conditions	α	β	n^d	$100R^{2\,e}$	Ref.
2,3	Thiophen	$\delta^{13}C$	$COMe_2$–d^6	5.8	46.7	8	86.5[g]	357
2,4	Thiophen	$\delta^{13}C$	$COMe_2$–d^6	0.9	5.9	8	86.5[g]	357
2,5	Thiophen	$\delta^{13}C$	$COMe_2$–d^6	0.4	33.5	8	98.0[f]	357
3,2	Thiophen	$\delta^{13}C$	$COMe_2$–d^6	5.2	55.5	8	94.1[f]	357
3,4	Thiophen	$\delta^{13}C$	$COMe_2$–d^6	8.2	15.1	8	60.8[i]	357
3,5	Thiophen	$\delta^{13}C$	$COMe_2$–d^6	1.9	2.9	8	86.5[g]	357
2,3	Thiophen	$\delta^{1}H$	$COMe_2$–d^6	0.34	2.15	9	96.0[f]	357
2,4	Thiophen	$\delta^{1}H$	$COMe_2$–d^6	0.08	0.73	9	92.2[f]	357
2,5	Thiophen	$\delta^{1}H$	$COMe_2$–d^6	0.05	1.65	9	94.1[f]	357
3,2	Thiophen	$\delta^{1}H$	$COMe_2$–d^6	0.34	2.42	8	98.0[f]	357
3,4	Thiophen	$\delta^{1}H$	$COMe_2$–d^6	0.12	1.19	8	77.4[i]	357
3,5	Thiophen	$\delta^{1}H$	$COMe_2$–d^6	0.06	0.40	8	51.8[k]	357
2,5	Thiophen	$\delta^{19}F$	$COMe_2$–d^6	7.5	25.0	12	92.2[f]	357
2,3	Selenophen	$\delta^{1}H$	$COMe_2$–d^6	0.34	2.67	8	98.0[f]	358
2,4	Selenophen	$\delta^{1}H$	$COMe_2$–d^6	0.04	0.93	8	96.0[f]	358
2,5	Selenophen	$\delta^{1}H$	$COMe_2$–d^6	0.18	1.99	8	98.0[f]	358
3,2	Selenophen	$\delta^{1}H$	$COMe_2$–d^6	0.57	3.23	7	98.0[f]	358
3,4	Selenophen	$\delta^{1}H$	$COMe_2$–d^6	0.33	1.04	7	81.0[i]	358
3,5	Selenophen	$\delta^{1}H$	$COMe_2$–d^6	0.30	0.38	7	88.4[h]	358
2,3	Selenophen	$\delta^{13}C$	$COMe_2$–d^6	5.7	46.6	8	86.5[g]	358
2,4	Selenophen	$\delta^{13}C$	$COMe_2$–d^6	1.6	6.5	8	75.6[i]	358
2,5	Selenophen	$\delta^{13}C$	$COMe_2$–d^6	2.4	34.0	8	98.0[f]	358
3,2	Selenophen	$\delta^{13}C$	$COMe_2$–d^6	3.7	61.7	7	96.0[f]	358
3,4	Selenophen	$\delta^{13}C$	$COMe_2$–d^6	7.2	9.1	7	41.0[l]	358
3,5	Selenophen	$\delta^{13}C$	$COMe_2$–d^6	2.6	4.1	7	79.2[i]	358
2,3	Furan	$\delta^{13}C$	$COMe_2$–d^6	11.2	58.7	7	86.5[h]	359
2,4	Furan	$\delta^{13}C$	$COMe_2$–d^6	1.1	0.9	7	32.5[l]	359
2,5	Furan	$\delta^{13}C$	$COMe_2$–d^6	0.3	20.4	7	94.1[g]	359
3,2	Furan	$\delta^{13}C$	$COMe_2$–d^6	2.6	36.9	7	94.1[f]	359
3,5	Furan	$\delta^{13}C$	$COMe_2$–d^6	0.3	20.4	7	72.3[i]	359
2,3	Furan	$\delta^{1}H$	$COMe_2$–d^6	0.63	2.89	7	98.0[f]	359
2,4	Furan	$\delta^{1}H$	$COMe_2$–d^6	0.34	0.62	7	81.0[i]	359
2,5	Furan	$\delta^{1}H$	$COMe_2$–d^6	0.23	1.24	7	98.0[f]	359
3,2	Furan	$\delta^{1}H$	$COMe_2$–d^6	0.78	1.81	7	82.8[i]	359
3,4	Furan	$\delta^{1}H$	$COMe_2$–d^6	0.45	0.76	7	92.2[g]	359
3,5	Furan	$\delta^{1}H$	$COMe_2$–d^6	0.17	0.57	7	81.0[i]	359
6,8[m]	Purine	$\delta^{1}H$	DMSO	0.346	0.363	9	98.0[f]	376
5,4[n]	Benzo[*b*]thiophen	$\delta^{1}H$	$CDCl_3$	0.809	1.33	17	75.5[f]	72
5,6[n]	Benzo[*b*]thiophen	$\delta^{1}H$	$CDCl_3$	0.701	1.12	17	73.8[f]	72

*All correlations with the Swain–Lupton constants unless otherwise noted. Footnotes other than those below are as for Table 5.34.
[m]Correlated with σ_I, σ_R^+. [n]Correlated with σ_I, σ_R.

5.3.7. Miscellaneous Applications of the Hammett Equation

Values of ΔR_M, where R_M is defined by (5.44) for paper chromatography of 9-substituted acridine-*N*-oxides, are linear in σ_p.[380]

$$R_M = \log (R_F^{-1} - 1) \tag{5.44}$$

Chemical shifts of H_4 in substituted borazines have been correlated by σ_R^0 constants.[381] Ionization constants of hexasubstituted cyclotriphosphatriazines and octasubstituted cyclotetraphosphatetrazines have been examined with the aid of the extended Hammett equation.[382] The square roots of the intensity of the bands for CH vibrations in ir spectra of *C*-substituted carboranes and neocarboranes are a linear function of σ_I; $100r^2 = 95.8$, $n = 6$.[383]

5.4. Conclusions

The results described above show the wide applicability of correlation analysis to polynuclear aromatic and heterocyclic systems. These studies have been carried out by using the Hammett, the extended Hammett, the Jaffé, the Thirot, and the Yukawa–Tsuno equation, and the Dewar–Grisdale and the Dewar–Grisdale–Harris treatment. The question then arises, which of these methods, if any, is preferable.

Correlations with the simple Hammett equation suffer from the fact that the composition of the electrical effect in the systems under study frequently deviates from that of the available simple substituent constants. This problem, can, of course, be avoided by the use of sets of substituent constants that are defined so that P_R increases in 5% increments from 0 to 100. Such sets would be required for σ_R, σ_R^+, and σ_R^- at least. They have the advantage that data can be plotted, and the right range of σ constants can be chosen by inspection. It would nevertheless be simpler and more convenient to use a multiparameter equation such as the extended Hammett equation. Such a relation can accommodate any composition of the electrical effect.

The available results suggest that multiparameter equations can give better correlations than does the simple Hammett equation. This conclusion is supported by the comparisons made in this work and by the reports of others. It is probably due to the fact that multiparameter equations can be adjusted to any electrical effect composition.

The most important advantages of the multiparameter equations are as follows.

(*a*) They permit a more precise determination of the composition of the electrical effect.

(*b*) They provide values of the coefficients of the localized and delocalized effects that are more useful as a measure of transmission of these electrical effects than are the ρ values. Such values should make possible a better understanding of the transmission of electrical effects.

There are, however, inherent disadvantages in the use of multiparameter equations, e.g., data cannot be plotted before the correlation is attempted. A major disadvantage is that multiparameter equations require more data points to ensure reliability in the results.

The choice of correlation equation depends on the purpose of the correlation and the number and type of data points available. If the only function of the correlation is roughly to establish the type, composition, and magnitude of the electrical effect, the simple Hammett equation is sufficient. For the best possible correlation, or a much more precise determination of the magnitude, type, and composition of the electrical effect, it is necessary to resort to either the DG or the DGH treatment and their modifications, or to multiparameter equations. The DG and the DGH treatment are fundamentally in error. There are a number of reactions in which correlation with the DG σ constants results in ρ values for various skeletal groups that are significantly different from one another. The DG treatment requires that for a given phenomenon (reaction or physical property) under study, and a given set of conditions (medium, temperature, etc.) ρ should have the same value for all skeletal groups. The DG method may therefore be excluded as a useful treatment for precise purposes. The utility of the DGH method is still in doubt. There are indications that it is not entirely satisfactory. The correlation between σ_{calc} and σ_{obs} for naphthalene derivatives is not very good. This is attributed to a *peri*-interaction in 1-naphthalene derivatives, and, in fact, exclusion of such values from the correlation results in some improvement. This points up a great disadvantage of the DGH treatment, however. If skeletal groups in which a constant steric effect occurs cannot be treated, the method is obviously severely limited in its usefulness. The most successful of the modified DG and DGH treatments are the SUMCHIN 1 and 2 method of Forsyth.[68] These methods are presumably also subject to error due to the presence of constant steric effects. It seems, then, that the DG and the DGH method and their modifications do not provide the method of choice. If they did, it would, of course, be simple to calculate σ^0 and σ^- constants by this approach, but in view of its limited applicability this is not useful.

The most generally preferred equations are of the multiparameter type. These equations include the extended Hammett, the DSP, the Yukawa–Tsuno, the Jaffé, and the Thirot relations (assuming steric effects are either constant or absent). The Thirot equation does not seem to give better results than the two-electrical-parameter equations. As was pointed out previously, the Jaffé equation, which has the distinction of being the first of the multiparameter equations, is limited by the difficulty of interpreting the regression coefficients, ρ_1 and ρ_2. The Yukawa–Tsuno equation has the advantage over the extended Hammett and the DSP equation in that *meta*- and *para*-substituted benzene derivatives and analogs can be treated as members of the same set. This advantage is of limited utility in heterocyclic chemistry, however. The Yukawa–Tsuno equation can be related to the extended Hammett and the DSP equation only if σ_p^+ is related to σ_p or σ_p^0. The approximate relationships (5.45)–(5.47) hold. It

$$\sigma_p^0 = \sigma_I + 0.67\sigma_R \tag{5.45}$$

$$\sigma_p = \sigma_I + \sigma_R \tag{5.46}$$

$$\sigma_p^+ = \sigma_I + 1.60\sigma_R \tag{5.47}$$

has been claimed, however,[11] that (5.48)–(5.50) hold, and σ_R^+ or $\sigma_R^0 \neq$ const. $\times\, \sigma_R$. If this claim is valid, as it seems to be, the Yukawa–Tsuno

$$\sigma_p^+ = \sigma_I + \sigma_R^+ \tag{5.48}$$

$$\sigma_p = \sigma_I + \sigma_R \tag{5.49}$$

$$\sigma_p^0 = \sigma_I + \sigma_R^0 \tag{5.50}$$

equation as now written is incorrect. Corrected versions of the Yukawa–Tsuno equation are given in (5.51)–(5.53).

$$Q_X = \rho\sigma_X^+ + r\sigma_R^+ + h \tag{5.51}$$

$$Q_X = \rho\sigma_X^- + r\sigma_R^- + h \tag{5.52}$$

$$Q_X = \rho\sigma_X + r\sigma_R + h \tag{5.53}$$

The extended Hammett and the DSP equation have the advantage that the regression coefficients are easily understood and useful. They provide measures of the transmission of the localized and delocalized electrical effects, and of the composition of the overall electrical effect, information that is not as readily obtainable from the Yukawa–Tsuno equation.

The difference between the extended Hammett and the DSP equation is that the latter is required to pass through the origin. In terms of the multiple linear regression analysis method, which is used to correlate data by the extended Hammett and DSP equations, this introduces a constraint on the correlation by assigning a special weight to the effect of the hydrogen atom as a substituent. There is no reason for doing this other than the fact that it is the type of model originally proposed by Hammett. Against it is the argument that it assigns an unjustifiable and unnecessary importance to the hydrogen substituent. In view of this point, and the above remarks concerning the other multiparameter equations, the extended Hammett equation seems to be the best choice, at present.

In order to apply the extended Hammett equation it is necessary to have available substituent parameters that are measures of the localized and delocalized electrical effects. There are now available two different sets of such constants: those proposed by Taft and his co-workers, and those proposed by Swain and Lupton and modified by Hansch *et al.* Arguments against the Swain–Lupton–Hansch constants seem overwhelming.[9,384] The Taft σ_I constants also seem to be in error in some cases. Thus, the values for alkyl groups are probably too large. It is also not certain that they are properly scaled. In view of the obvious deficiencies of the Swain–Lupton–Hansch constants, however, the σ_I, σ_R, σ_R^+, σ_R^-, and σ_R^0 constants are definitely preferable.

In order to correlate data for systems in which steric effects occur, it is necessary to use a three-parameter equation in which the third parameter represents the steric effect. The LDS equation, which is obtained from the extended Hammett equation by the addition of a term involving the steric parameter υ is best for this purpose.

It is impossible to overemphasize the need for well-characterized data sets if the best results of correlation analysis are to be obtained. Such sets must have a sufficient number of and a wide variety of substituents, and a sufficient range of σ (or υ) values. It is true that useful information can be extracted from sets that are not well characterized, but the results obtained from such sets are frequently misleading.

5.5. Recent Developments

Since the body of this review was completed, a number of papers of interest have appeared. Rate constants for the reactions of 4′-substituted 3-bromo-4-nitrobiphenyls with piperidine were correlated by the Hammett equation by using σ_p constants for H and halogen substituents, and $\sigma_p + 0.32(\sigma_p^- - \sigma_p)$ for electron-acceptor groups (Ac, NO_2, $MeSO_2$).[385] The rate constants for the alkaline hydrolysis of methyl 3-substituted thiophen-2-carboxylates are very much better correlated by the Yukawa–Tsuno equation than by the simple Hammett equation, another example of the advantage in the use of multiparameter equations (see Tables 5.37 and 5.38).[398] Bolton and Burley[386] have used a correlation between rate constants for solvolysis of 7-substituted 2-α-chlorobenzyl-fluorenes (F) and those of 4′-substituted 4-α-chlorobenzylbiphenyls (B) to determine the interplanar angle η in the biphenyls. Correlating $\log (k_X/k_H)_F$ with $\log (k_X/k_H)_B$ gives (5.54). It is assumed that the only effect on relative sensitivity to substituents as between the fluorenes and the biphenyls is that of the interplanar angle, i.e., it is assumed that (5.55) holds. Taking η_F as 0, $1/\cos^2 \eta_B = 1.45$, and $\eta_B = 34°$. In fact, sensitivity to substituent effects

$$\log (k_X/k_H)_F = 1.45 \log (k_X/k_H)_B \tag{5.54}$$

$$\log \left(\frac{k_X}{k_H}\right)F = \frac{\cos^2 \eta_F}{\cos^2 \eta_B} \log \left(\frac{k_X}{k_H}\right)B \tag{5.55}$$

depends upon molecular geometry as regards the localized effect, and upon constant substituents. The geometry of the 7,2-fluorene system differs significantly from that of the biphenyl system. Furthermore, the effect of the C-9 methylene group, which acts as a constant substituent, on the reactivity of the fluorene system must be considered. Thus, the conclusions of Bolton and Burley regarding the interplanar angle are open to question.

Additional examples of the correlation of data for substituted naphthalenes with Hammett σ constants are assembled in Table 5.36. The Dewar–Grisdale method has been applied unsuccessfully to pK_a values in water at 25°C of 5- and 7-substituted naphthalene-1-sulfonamides and 8-substituted naphthalene-2-sulfonamides. The results are improved by the use of σ^+_{DG} constants calculated by the DG method from σ^+_m and σ^+_p values.[389]

Values of σ_{ij} calculated from pK_a values of substituted 1- and 2-naphthoic acids in 50% v/v aqueous ethanol appear to agree fairly well with the corresponding DG constants.[390]

Some very useful new work has appeared on *ortho* effects in heterocyclic systems.[392,393] Rate constants for the reactions of 3-substituted 5-nitro-2-thienyl phenyl sulfones with piperidine were correlated with σ_p^- constants, while rate constants for alkaline hydrolysis of methyl 3-substituted thiophen-2-carboxylates were correlated by use of the Yukawa–Tsuno equation. Results are given in Table 5.37 and 5.38, respectively. Other examples of heteroaryl skeletal group sets are given in Table 5.37.

Ionization constants of 1-substituted imidazolinium ions were correlated[396] by use of a form of the Hammett equation (5.56). Polarographic

$$\log (k_X/k_H) = \rho\bar{\sigma} + c \tag{5.56}$$

$$[\bar{\sigma} = \sigma_I + (\rho_R/\rho_I)\sigma_R^-]$$

half-wave potentials and d–d transition energies of nickel (II) complexes of (VII) (p. 257) have been correlated with the σ_p^- constants.[398]

Rate and equilibrium constants for the addition reactions of heteroarenes have been correlated by use of the Hammett equation; results are in Table 5.37.[399–401] New σ^+ (replacement) values for 2- and 3-furyl[404]; 2- and 3-thienyl[404]; 2-, 3-, and 4-pyrazolyl[402]; and for 4-isoxazolyl[402] were determined, as were σ^- (replacement) values for 2-, 3-, and

Table 5.36. *Correlation of Data for Biphenyl and Naphthalene Sets by the Hammett Equation*[a]

ij	Y	G	*Q*	Reagent	Reaction conditions	Subst. constant	ρ	$100r^2$	*n*	Ref.
4′,3	Br	Biphenyl	log *k*	$C_5H_{11}N$	MeOH, 40°C	σ_p^b	0.35	99.6^d	7	385
6,2	$BrCH_2CO$	Naphthalene	log *k*	—	80% EtOH–H_2O, 80°C	*c*	0.33	99.2^d	7	387
4,1	Ac	Naphthalene	log *k*	$[Fe(CN)_6]^{3-}$	15% v/v EtOH–H_2O, 35°C	σ_p	1.3	88.7^e	6	388
i,1	CO_2Me	Naphthalene	log *k*	OH^-	80% v/v MeOH–H_2O, 30°C	σ_{ij}	0.997	87.2^f	6	391
i,1	CO_2Me	Naphthalene	log *k*	H_3O^+	50% v/v dioxan–H_2O, °C	σ_{ij}	0.595	88.2^d	8	391

[a] For meaning of column headings, see Table 5.1. [b] See text.
[c] Not reported. [d] 99.9% CL. [e] 99.5% CL. [f] 99.0% CL.

Table 5.37. Correlation of Heterocyclic Sets by the Hammett Equation[a]

ij	Y	G	Q	Reagent	Reaction conditions	Subst. constant	ρ	$100r^2$	n	Ref.
3,2	SO_2Ph	Thiophen	log k	$C_5H_{11}N$	MeOH, 20°C	σ_p^-	4.60	99.0[i]	5[b]	392
4,4′	O^-	[c]	log k	MeI	$CHCl_3$, 25°C	σ	−0.40	98.6[i]	18	395
4,4′	O^-	[c]	log k	MeI	MeCN, 25°C	σ	−0.15	89.9[i]	18	395
3,2	CO_2Me	Thiophen	log k	OH^-	80% v/v MeOH–H_2O, 20°C	σ^n	3.46	91.2[i]	8	393
—	—	[d]	$E_{1/2}$	—	DMF	σ_p^-	0.645	98.8[i]	7	398
—	—	[d]	$E_{d\text{–}d}$	—	$CHCl_3$	σ_p^-	2.21	98.2[i]	7	398
3′ or 4′,4	—	1-Phenyl-pyridinium	log k^e	Cyclohexanone anion	MeOH, 25°C	σ^o	−1.48	99.7[i]	7	399
3′ or 4′,4	—	1-Phenyl-pyridinium	log k^f	Cyclohexanone anion	MeOH, 25°C	σ^o	0.97	98.4[i]	6	399
3′ or 4′,2	—	1-Benzyl-5-nitro-quinolinium	$pK_{R^+}^e$	—	H_2O, 24.6°C	σ	−1.32	99.6[i]	7	400
3′ or 4′,2	—	1-Benzyl-5-nitro-quinolinium	log k^e	—	H_2O, 24.6°C	σ	−0.74	98.6[i]	6	400
3′ or 4′,2	—	1-Benzyl-5-nitro-quinolinium	log k^f	OH^-	H_2O, 24.6°C	σ	0.45	90.3[i]	6	400
3′ or 4′,1	—	2-Benzyl-5-nitroiso-quinolinium	$pK_{R^+}^e$	—	H_2O, 24.6°C	σ	−1.14	99.4[i]	9	400
3′ or 4′,1	—	2-Benzyl-5-nitroiso-quinolinium	log k^e	—	H_2O, 24.6°C	σ	−0.61	99.6[i]	9	400
3′ or 4′,1	—	2-Benzyl-5-nitroiso-quinolinium	log k^f	OH^-	H_2O, 24.6°C	σ	0.51	98.4[i]	9	400

continued overleaf

Table 5.37—continued

ij	Y	G	*Q*	Reagent	Reaction conditions	Subst. constant	ρ	$100r^2$	*n*	Ref.
3,1 and 4	—	2-Methyl-isoquinolinium	log *k*	CH_2:CHOEt	DMSO, 25°C	σ	3.47	95.8[i]	6[g]	401
—	—	CH_2[h]	$\Delta\delta^1H$	—	$CDCl_3$	σ_p	−1.21	98.4[k]	3	394
—	—	$SiMe_2$[h]	$\Delta\delta^1H$	—	$CDCl_3$	σ_p	−1.11	92.2[l]	3	394
—	—	BPh[h]	$\Delta\delta^1H$	—	$CDCl_3$	σ_p	−1.21	93.3[l]	3	394
—	—	PPh[h]	$\Delta\delta^1H$	—	$CDCl_3$	σ_p	−1.28	98.0[m]	4	394
—	—	P(O)Ph[h]	$\Delta\delta^1H$	—	$CDCl_3$	σ_p	−1.34	98.2[k]	3	394
4′,5	—	2-Phenylfuran	$\delta^{13}C$	—	$CDCl_3$	σ_p	2.24	96.3[i]	8	407
4′,2	—	2-Phenylfuran	$\delta^{13}C$	—	$CDCl_3$	σ_p	−2.20	85.2[j]	8	407
4′,3	—	2-Phenylfuran	$\delta^{13}C$	—	$CDCl_3$	σ_p	4.76	95.8[i]	8	407
4′,4	—	2-Phenylfuran	$\delta^{13}C$	—	$CDCl_3$	σ_p	0.598	84.3[j]	8	407
4′,5	—	5-Methyl-2-phenylfuran	$\delta^{13}C$	—	$CDCl_3$	σ_p	2.83	94[m]	6	407
4′,2	—	5-Methyl-2-phenylfuran	$\delta^{13}C$	—	$CDCl_3$	σ_p	−2.10	91[n]	6	407
4′,3	—	5-Methyl-2-phenylfuran	$\delta^{13}C$	—	$CDCl_3$	σ_p	4.92	97[j]	6	407
4′,4	—	5-Methyl-2-phenylfuran	$\delta^{13}C$	—	$CDCl_3$	σ_p	1.01	77[k]	6	407

4′,5	—	5-Methyl-2-phenylpyrrole	$\delta^{13}C$	—	$CDCl_3$	σ_p	4.14	96[n]	5	407
4′,2	—	5-Methyl-2-phenylpyrrole	$\delta^{13}C$	—	$CDCl_3$	σ_p	−2.5	97.2[n]	5	407
4′,3	—	5-Methyl-2-phenylpyrrole	$\delta^{13}C$	—	$CDCl_3$	σ_p	5.03	96.8[n]	5	407
4′,4	—	5-Methyl-2-phenylpyrrole	$\delta^{13}C$	—	$CDCl_3$	σ_p	1.72	91[o]	5	407

[a] For meaning of column headings see Table 5.1.
[b] X = Br, Me, SO_2Me were excluded from the correlation.
[c] The skeletal groups 4-(substituted-phenyl)-1-(4-oxy-2,6-dipheny phenyl)-2,6-diphenylpyridinium.
[d] Skeletal group is the nickel(II) complex of (VII).
[e] Rate or equilibrium constant for dissociation of the adduct.
[f] Rate constant for formation of the adduct.
[q] X = Bu^t deviates.
[h] Group Z in (VIII), chemical shift of Me protons.
[i] 99.9% Cl. [j] 99.5% CL. [k] 90.0% CL. [l] <90.0% CL.
[m] 99.0% CL. [n] 98.0% CL. [o] 95.0% CL.

Table 5.38. Results of Correlations of Heteroarene Sets with the Extended Hammett and the Dual Substituent-Parameter Equation[a]

ij	Y	G	Q	Reaction conditions	α	β	$100R^2$	n	Ref.
3,2	CO_2Me	Thiophen	$\log k$[b]	80% v/v $MeOH\text{-}H_2O$, 20°C	2.77	0.44	99.2[e]	8	393
2,1	—	Pyridine	$\nu_{OH\cdots N}$	MeOH in CCl_4	230	−4		8[c]	397
3,1	—	Pyridine	$\nu_{OH\cdots N}$	MeOH in CCl_4	129	37		8[c]	397
4,1	—	Pyridine	$\nu_{OH\cdots N}$	MeOH in CCl_4	111	46		9[c]	397
5,2	—	Thieno-[3,2-*b*]-thiophen	$\delta^{13}C$	$(CD_3)_2CO$	3.9	18.0	98[f]	6[d]	408
6,2	—	Thieno-[3,2-*b*]-thiophen	$\delta^{13}C$	$(CD_3)_2CO$	0.2	2.8	90[g]	6[d]	408
7,2	—	Thieno-[3,2-*b*]-thiophen	$\delta^{13}C$	$(CD_3)_2CO$	0.9	17.2	90[h]	5[d]	408
5,2	—	Thieno-[2,3-*b*]-thiophen	$\delta^{13}C$	$(CD_3)_2CO$	1.7	7.4	92[h]	6[d]	408
7,2	—	Thieno-[2,3-*b*]-thiophen	$\delta^{13}C$	$(CD_3)_2CO$	0.1	27.5	98[i]	5[d]	408
8,2	—	Thieno-[2,3-*b*]-thiophen	$\delta^{13}C$	$(CD_3)_2CO$	−2.7	5.0	83[j]	5[d]	408

[a] For meaning of column headings see Tables 5.1 and 5.2.
[b] Alkaline hydrolysis, correlated by use of the Yukawa–Tsuno equation, $\alpha = \rho$, $\beta = r^-$.
[c] Correlated by use of the extended Hammett equation using σ_R^+ for 2,1 and 4,1; σ_R^0 for 3,1; f values are reported as a measure of goodness of fit.
[d] Correlated by use of the extended Hammett equation using the Swain–Lupton constants.
[e] 99.9% CL. [f] 95.0% CL. [g] 98.0% CL.
[h] 95.0% CL. [i] 99.0% CL. [j] 90.0% CL.

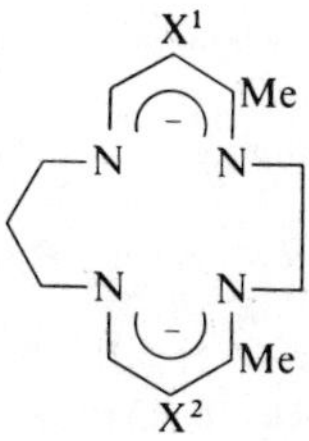

(VII)

4-pyridyl.[403] A correlation of $\Delta G^{\ddagger}$ values for rotational barriers in heteroarene *NN*-dimethylcarboxamides with σ (replacement) constants has been reported.[405]

Carbonyl-stretching frequencies of the *cis* and *trans* conformers of methyl 6-substituted picolinates and nicotinates have been correlated by use of the DSP equation, σ_R^0 being used as the delocalized-effect parameter. For the picolinates the electrical effect shows a larger resonance contribution than would ordinarily be expected for a *meta*-substituted system.[406] Further correlations of proton and ^{13}C chemical shifts by use of the Hammett equation have appeared.[394,395] The results obtained for the chemical shifts of the methyl protons in (VIII) are interesting as regards the lack of dependence of ρ on the nature of Z. Although the number of data points in each set is very small, the constancy of ρ seems fairly certain. Correlations of ^{13}C chemical shifts with Swain–Lupton constants have also been carried out.[408]

Finally, attention must be directed to the appearance of a major review of the applications of the Hammett equation to heterocyclic compounds.[409]

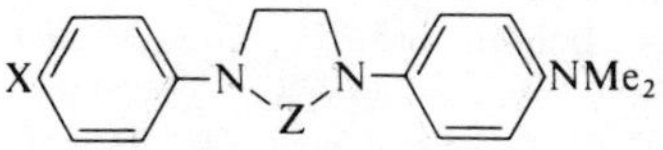

(VIII)

Z = CH_2, $SiMe_2$, BPh, PPh, or P(O)Ph

References

1. P. R. Wells, S. Ehrenson, and R. W. Taft, *Progr. Phys. Org. Chem.*, **6**, 147 (1968).
2. P. Zuman, *Substituent Effects in Organic Polarography*, pp. 131, 219, 247 (Plenum, New York, 1967).
3. H. H. Jaffé and H. L. Jones, *Adv. Heterocyclic Chem.*, **3**, 209 (1964).
4. P. Tomasik, *Prace Naukowe Inst. Chem. i. Technol. Nafty i Wegla, Politech. Wroclawskiej* No. 4, 3 (1973).
5. P. Tomasik, *Chemia polaczen heterocyklicznych w ujeciu hammettowskim* (Wroclaw Technical University, Wroclaw, 1974).

6. J. Shorter, *Correlation Analysis in Organic Chemistry*, pp. 8, 17, 45 (Clarendon Press, Oxford, 1973).
7. C. D. Johnson, *The Hammett Equation*, pp. 1, 83, 86, (Cambridge University Press, London, 1973).
8. O. Exner, in *Advances in Linear Free Energy Relationships*, Chap. 1, N. B. Chapman and J. Shorter, eds. (Plenum, London, 1972).
9. M. Charton, *Chem Tech.*, 502 (1974); 245 (1975).
10. M. Charton, *Progr. Phys. Org. Chem.*, **8**, 235 (1971).
11. S. Ehrenson, R. T. C. Brownlee, and R. W. Taft, *Progr. Phys. Org. Chem.*, **10**, 1 (1973).
12. J. Shorter, this volume, Chapter 4.
13. M. J. S. Dewar and P. J. Grisdale, *J. Amer. Chem Soc.*, **84**, 3539, 3546, 3548 (1962).
14. See A. Streitwieser, *Molecular Orbital Theory for Organic Chemists*, p. 54 (Wiley, New York, 1961).
15. M. Charton, *J. Org. Chem.*, **30**, 3341 (1965).
16. L. R. Snyder, *J. Chromatog.*, **17**, 73 (1965).
17. M. J. S. Dewar, R. Golden, and J. M. Harris, *J. Amer. Chem. Soc.*, **93**, 4187 (1971).
18. E. Berliner and E. A. Blommers, *J. Amer. Chem. Soc.*, **73**, 2479 (1951).
19. H. H. Jaffé, *Chem. Rev.*, **53**, 191 (1953).
20. J. P. Idoux, V. S. Cantwell, J. Hinton, S. O. Nelson, P. Hollier, and R. Zarrillo, *J. Org. Chem.*, **39**, 3946 (1974).
21. E. V. Titov, N. G. Korzhenevskaya, R. S. Popova, and L. M. Litvinenko, *Ukrain. khim. Zhur.*, **37**, 790 (1971).
22. K. Bowden, N. B. Chapman, and J. Shorter, *Canad. J. Chem.*, **42**, 1979 (1964).
23. E. Czerwinska-Fejgin and W. Polaczkowa, *Roczniki Chem.*, **41**, 1759 (1967).
24. C. Dell'Erba, G. Guanti, and G. Garbarino, *Tetrahedron*, **27**, 1807 (1971).
25. E. M. Schulman, K. A. Christensen, D. M. Grant, and C. Walling, *J. Org. Chem.*, **39**, 2686 (1974).
26. W. Adcock, M. J. S. Dewar, R. Golden, and M. A. Zeb, *J. Amer. Chem. Soc.*, **97**, 2198 (1975).
27. Y. Ogata, K. Nagura, and M. Kozuka, *J. Org. Chem.*, **40**, 615 (1975).
28. K. Bowden, A. F. Cockerill, and J. R. Gilbert, *J. Chem. Soc. (B)*, 179 (1970).
29. J. F. Archer and J. Grimshaw, *J. Chem. Soc. (B)*, 266 (1969).
30. A. Mathieu, J-C. Milano, and J. Douris, *Bull. Soc. chim. France*, 299 (1974).
31. R. Baker, R. W. Bott, C. Eaborn, and P. M. Greasley, *J. Chem. Soc.*, 627 (1974).
32. M. Charton, *J. Org. Chem.*, **39**, 2797 (1974).
33. S. K. Dayal and R. W. Taft, *J. Amer. Chem. Soc.*, **95**, 5595 (1973).
34. A. J. Harget, K. D. Warren, and J. R. Yandle, *J. Chem. Soc. (B)*, 214 (1968).
35. K. D. Warren and J. R. Yandle, *J. Chem. Soc.*, 4221 (1965).
36. J. A. Parry and K. D. Warren, *J. Chem. Soc.*, 4049 (1965).
37. K. D. Warren and J. R. Yandle, *J. Chem. Soc.*, 5518 (1965).
38. C. C. Price and R. H. Michel, *J. Amer. Chem. Soc.*, **74**, 3651 (1952).
39. C. C. Price, E. C. Mertz, and J. Wilson, *J. Amer. Chem. Soc.* **76**, 5131 (1954).
40. A. Fischer, J. D. Murdoch, J. Packer, R. D. Topsom, and J. Vaughan, *J. Chem. Soc.*, 4358 (1957).
41. P. R. Wells and E. R. Ward, *Chem. Ind.*, 528 (1958).
42. A. Bryson, *J. Amer. Chem. Soc.*, **82**, 4862 (1960).
43. J.-M. Bonnier and J. Rinaudo, *Bull. Soc. chim. France*, 3901 (1966).
44. B. Demian, *Stud. Cercet. Chim.*, **16**, 911 (1968).
45. K. Bowden and D. C. Parkin, *Canad. J. Chem.*, **47**, 185 (1969).
46. V. A. Koptyug, R. N. Berezina, and V. P. Petrov, *Zhur. org. Khim.*, **2**, 1696 (1966); EE 1672.
47. J. H. Richards and S. Walker, *Tetrahedron*, **20**, 841 (1964).

48. V. A. Koptyug and V. P. Petrov, *Reakts. spos. org. Soedinenii*, **1**(2), 48 (1964).
49. V. A. Koptyug, V. P. Petrov, and T. N. Gerasimova, *Reakts. spos. org. Soedinenii*, **1**(2), 43 (1964).
50. P. R. Wells and W. Adcock, *Austral. J. Chem.*, **19**, 221 (1966).
51. A. Fischer, W. J. Mitchell, G. S. Ogilvie, J. Packer, J. E. Packer, and J. Vaughan, *J. Chem. Soc.*, 1426 (1958).
52. A. Fischer, H. M. Fountain, and J. Vaughan, *J. Chem. Soc.*, 1310 (1959).
53. A. Fischer, M. A. Riddolls, and J. Vaughan, *J. Chem. Soc. (B)*, 106 (1966).
54. A. Fischer, M. C. A. Opie, J. Vaughan, and G. J. Wright, *J.C.S. Perkin II*, 319 (1972).
55. M. Livař, P. Hrnčiar, and A. Kopecka, *Chem. Zvesti*, **28**, 402 (1974).
56. A. Perjéssy and P. Hrnčiar, *Coll. Czech. Chem. Comm.*, **37**, 1706 (1972).
57. D. C. Kleinfelter and P. H. Chen, *J. Org. Chem.*, **34**, 1741 (1969).
58. T. Bisanz and M. Bukowska, *Roczniki Chem.*, **48**, 777 (1974).
59. D. D. Perrin, *J. Chem. Soc.*, 5590 (1965).
60. P. R. Wells and W. Adcock, *Austral. J. Chem.*, **18**, 1365 (1965).
61. W. Adcock and M. J. S. Dewar, *J. Amer. Chem. Soc.*, **89**, 379 (1967).
62. W. Adcock, P. D. Bettress, and S. Q. A. Rizvi, *Austral. J. Chem.*, **23**, 1921 (1970).
63. J. T. van Gemert, *Austral. J. Chem.*, **22**, 1883 (1969).
64. W. B. Smith, D. L. Deavenport, and A. M. Ihrig, *J. Amer. Chem. Soc.*, **94**, 1959 (1972).
65. K. C. C. Bancroft and G. R. Howe, *Tetrahedron Lett.* 4207 (1967).
66. C. Eaborn and A. Fischer, *J. Chem. Soc. (B)*, 152 (1969).
67. K. C. C. Bancroft and G. R. Howe *J. Chem. Soc. (B)*, 1221 (1971).
68. D. A. Forsyth, *J. Amer. Chem. Soc.*, **95**, 3594 (1973).
69. N. N. Zatsepina, I. F. Tupitsyn, V. P. Dustina, Yu. M. Kapustin, and Yu. L. Kaminskii, *Reakts spos. org. Soedinenii*, **9**, 745 (1972).
70. A. Bryson and R. W. Matthews. *Austral. J. Chem.*, **16**, 401 (1963).
71. P. R. Wells, D. P. Arnold, and D. Doddrell, *J,S,C, Perkin II* 1745 (1974).
72. M. Charton, *J. Org. Chem.*, **36**, 266 (1971).
73. M. Charton, *J. Org. Chem.*, **31**, 2991 (1966).
74. Y. Yukawa, Y. Tsuno, and N. Shimizu, *Bull. Chem. Soc. Japan*, **44**, 3175 (1971).
75. H. H. Jaffé, *J. Amer. Chem. Soc.*, **76**, 4261 (1954).
76. G. Thirot, *Bull. Soc. chim. Fr.*, 739, 3559 (1967).
77. C. Kirsche and C. Caullet, *Compt. rend. C*, **272**, 1331 (1971).
78. J. Mayer, G. Thirot, and C. Caullet, *Bull. Soc. chim. France*, 4129 (1971).
79. C. Kirsche, G. Thirot, C. Caullet, and J. M. Juillerat, *Bull. Soc. chim. France*, 765 (1974).
80. C. Caullet, J. Mayer, I. Mentre, and J. Virey, *Compt. rend. B.*, **268**, 1237 (1969).
81. N. N. Zatsepina, A. A. Kane, N. S. Kolodina, and I. F. Tupitsyn, *Reakts. spos. org. Soedinenii*, **10**, 353 (1973).
82. L. H. Klemm and A. J. Kohlik, *J. Org. Chem.*, **28**, 2044 (1963).
83. A. Bryson and R. W. Matthews, *Austral. J. Chem.*, **14**, 237 (1961).
84. S. Clementi, P. Linda, and C. D. Johnson, *J.C.S. Perkin II*, 1250 (1973).
85. H. Cerfontain and A. Telder, *Rec. Trav. chim.*, **86**, 527 (1967).
86. R. O. C. Norman and P. D. Ralph, *J. Chem. Soc.*, 2221 (1961).
87. J. C. Arnold, G. J. Gleicher, and J. D. Unruh, *J. Amer. Chem. Soc.*, **96**, 787 (1974).
88. R. N. McDonald and J. R. Curtis, *J. Amer. Chem. Soc.*, **93**, 2530 (1971).
89. R. N. McDonald, R. R. Reitz, and J. M. Richmond, *J. Org. Chem.*, **41**, 1822 (1976).
90. R. Golden and L. M. Stock, *J. Amer. Chem. Soc.*, **94**, 3080 (1972).
91. K. Bowden, J. G. Irving, and M. J. Price, *Canad. J. Chem.*, **46**, 3903 (1968).
92. G. L. Anderson, R. C. Parish, and L. M. Stock, *J. Amer. Chem. Soc.*, **93**, 6984 (1971).
93. M. Charton, *J. Org. Chem.*, **31**, 3745 (1966).
94. L. H. Klemm, A. J. Kohlik, and K. B. Desai, *J. Org. Chem.*, **28**, 625 (1963).

95*a*. C. Eaborn, A. Fischer, and D. R. Killpack, *J. Chem. Soc.* (*B*), 2142 (1971).
95*b*. C. Weiss, W. Engewald, and H. Müller, *Z. Chem.*, **3**, 307 (1963).
96. J. Schulze and F. A. Long, *J. Amer. Chem. Soc.*, **86**, 331 (1964).
97. F. A. Long and J. Schulze, *J. Amer. Chem. Soc.*, **86**, 327 (1964).
98. M. Charton, *J. Org. Chem.*, **29**, 1222 (1964).
99. A. Streitwieser and R. G. Lawler, *J. Amer. Chem. Soc.*, **87**, 5388 (1965).
100. L. P. Hammett, *Physical Organic Chemistry*, p. 188 (McGraw-Hill, New York, 1940).
101. J. Packer, J. Vaughan, and E. Wong, *J. Org. Chem.*, **23**, 1373 (1958).
102. Y. Otsuji, M. Kubo, and E. Imoto, *Bull. Univ. Osaka Prefecture, Ser. A*, **7**, 61(1960); *Chem. Abs.*, **54**, 24796 (1960).
103. M. Sawada, Y. Tsuno, and Y. Yukawa, *Bull. Chem. Soc. Japan*, **45**, 1206 (1972).
104. Y. Yukawa, Y. Tsuno, and M. Sawada, *Bull. Chem. Soc. Japan*, **45**, 1210 (1972).
105. H. C. Brown and T. Inukai, *J. Amer. Chem. Soc.*, **83**, 4825 (1961).
106. L. Verbit and E. Berliner, *J. Amer. Chem. Soc.*, **86**, 3307 (1964).
107. J. Blatchly and R. Taylor, *J. Chem. Soc.* (*B*), 1402 (1968)
108. R. Taylor, M. P. David, and J. F. W. McOmie, *J.C.S. Perkin II*, 162 (1972).
109. R. J. Ouellette and B. G. van Leuwen, *J. Amer. Chem. Soc.*, **90**, 7061 (1968).
110. A. Streitwieser, H. A. Hammond, R. H. Jagow, R. M. Williams, R. G. Jesaitis, C. J. Chang, and R. Wolf, *J. Amer. Chem. Soc.*, **92**, 5141 (1970).
111. A. Streitwieser, A. Lewis, I. Schwager, R. W. Fish, and S. Labana, *J. Amer. Chem. Soc.*, **92**, 6525 (1970).
112. R. Baker, C. Eaborn, and R. Taylor, *J.C.S. Perkin II*, 97 (1972).
113. A. Streitwieser, *Molecular Orbital Theory for Organic Chemists*, pp. 326, 369 (Wiley, New York, 1961).
114. N. Acton and E. Berliner, *J. Amer. Chem. Soc.*, **86**, 3312 (1964).
115. A. Streitwieser and W. C. Langworthy, *J. Amer. Chem. Soc.*, **85**, 1757 (1963).
116. B. G. van Leuwen and R. J. Ouellette, *J. Amer. Chem. Soc.*, **90**, 7056 (1968).
117. W. C. Herndon, *J. Org. Chem.*, **40**, 3583 (1975).
118. H. Cerfontain and A. Telder, *Rec. Trav. chim.*, **86**, 527 (1967).
119. T. M. Krygowski, *Bull. Acad. polon. Sci., Sér. Sci. chim.*, **19**, 49 (1971).
120. T. M. Krygowski, *Bull. Acad. polon. Sci., Sér. Sci. chim.*, **19**, 61 (1971).
121. T. M. Krygowski, M. Stencel, and Z. Galus, *J. Electroanalyt. Chem. Interfacial Electrochem.*, **39**, 395 (1972).
122. J. Kuthan, *Coll. Czech. Chem. Comm.*, **35**, 714 (1970).
123. B. Kamieński and T. M. Krygowski, *Tetrahedron Lett.* 681 (1972).
124. H. Cerfontain, *J. Chem. Soc.*, 6602 (1965).
125. L. Paalme, A. Tuulmets, U. Kirso, and M. Gubergrits, *Reakts. spos. org. Soedinenii*, **11**, 313 (1974); EE, 315.
126. T. M. Krygowski and J. Kruszewski, *Bull. Acad. polon. Sci., Sér. Sci. chim.*, **20**, 993 (1972).
127. T. M. Krygowski, *Bull. Acad. polon. Sci., Sér. Sci. chim.*, **19**, 433 (1971).
128. T. M. Krygowski, *Tetrahedron*, **28**, 4981 (1972).
129. T. M. Krygowski, *Wiadomosci Chem.*, **28**, 37 (1974).
130. W. K. Kwok, R. A. M. O'Ferrall, and S. I. Miller, *Tetrahedron*, **20**, 1913 (1964).
131. V. Knoppová, A. Jurášek, J. Kováč, and M. Guttman, *Coll. Czech. Chem. Comm.*, **40**, 399 (1975).
132. F. Freeman, J. B. Brant, N. B. Hester, A. A. Kamego, M. L. Kasner, T. G. McLaughlin, and E. W. Paull, *J. Org. Chem.*, **35**, 982 (1970).
133. E. Maccarone, G. Musumarra, and G. A. Tomaselli, *Ann. Chim.* (*Italy*), **63**, 861 (1973).
134. G. Alberghina, A. Arcoria, S. Fisichella, and G. Scarlata, *Gazz. Chim. Ital.*, **103**, 319 (1973).
135. D. Spinelli, G. Guanti, and C. Dell'Erba, *Ricera sci.* **38**, 1048 (1968).

136. F. Fringuelli, G. Marino, and A. Taticchi, *J.C.S. Perkin II* 1738 (1972).
137. F. Fringuelli, G. Marino, and G. Savelli, *Tetrahedron*, **25**, 5815 (1969).
138. J. Stradins, I. Kravis, G. Reihmanis, S. Hillers, *Khim. Heterotsikl. Soedinenii*, 1309 (1972).
139. F. M. Stoyanovich, S. G. Mairanovskii, Yu. L. Gol'dfarb and I. A. D'yachenko, *Izvest. Akad. Nauk. SSR, Ser. Khim.*, 1439 (1971); EE, 1342.
140. D. S. Noyce, C. A. Lipinski, and R. W. Nichols, *J. Org. Chem.*, **37**, 2615 (1972).
141. L. do Amaral, *J. Org. Chem.*, **37**, 1433 (1972).
142. F. Freeman, *J. Chem. Educ.*., **47**, 140 (1970).
143. M. Charton and P. Friedman, unpubished results.
144. W. E. Catlin, *Iowa State Coll. J. Sci.*, **10**, 65 (1935).
145. H. Hopff and A. Krieger, *Helv. Chim. Acta*, **44**, 1058 (1961).
146. E. Imoto, Y. Otsuji, and H. Inoue, *Nippon Kagaku Zasshi*, **77**, 809 (1956).
147. O. Exner and W. Simon, *Coll. Czech. Chem. Comm.*, **29**, 2016 (1964).
148. A. R. Butler, *J. Chem. Soc.* (*B*), 867 (1970).
149. E. Imoto and R. Motoyama, *Bull. Naniwa Univ.*, **2A**, 127 (1954); *Chem. Abs.*, **49**, 9614 (1955).
150. P. A. Ten Thije and M. J. Janssen, *Rec. Trav. chim.*, **84**, 1169 (1965).
151. E. Imoto, Y. Otsuji, and J. Hirai, *Nippon Kagaku Zasshi*, **77**, 804 (1956); *Chem. Abs.*, **52**, 9066 (1958).
152*a*. E. Baciocchi, V. Mancini, and P. Perucci, *J.C.S. Perkin II*, 821 (1975).
152*b*. Y. Ueno and E. Imoto, *Nippon Kagaku Zasshi*, **88**, 1210 (1967); *Chem. Abs.*, **69**, 66782 (1968).
153. A. D. Campbell, S. Y. Chooi, L. W. Deady, and R. A. Shanks, *J. Chem. Soc.* (*B*), 1063 (1970).
154. T. Batkowski, M. K. Kalinowski, and P. Tomasik, *Ann. Chim.* (*Italy*), **63**, 121 (1973).
155. M. K. Kalinowski, J. Skarzewski, Z. Skrowaczewska, and P. Tomasik *Ann. Chim.* (*Italy*), **63**, 129 (1973).
156. W. Drzeniek and P. Tomasik, *Ann. Chim.* (*Italy*), **63**, 135 (1973).
157. A. Krutošíková, J. Surá, J. Kováč, and S. Juhás, *Coll. Czech. Chem. Comm.*, **40**, 3362 (1975).
158. C. Dell'Erba, D. Spinelli, and G. Garbarino, *Gazz. Chim. Ital.*, **100**. 777 (1970).
159. K. Bowden and D. C. Parkin, *Canad. J. Chem.*, **44**, 1493 (1966).
160. A. Krutošíková, J. Kováč, J. Rentka, and M. Čakrt, *Coll. Czech. Chem. Comm.*, **39**, 767 (1974).
161. L. Fišera, J. Surá, J. Kováč, and M. Lucký, *Coll. Czech Chem. Comm.*, **39**, 1711 (1974).
162. L. Fišera, J. Kováč, M. Lucký, and J. Surá, *Chem. Zvesti* **28**, 386 (1974).
163. A. Benkó, J. Zsakó, and P. Nagy, *Chem. Ber.*, **100**, 2178 (1967).
164. A. Martvoň, J. Surá, and M. Černayová, *Coll. Czech. Chem. Comm.*, **39**, 1356 (1974).
165. M. S. Melzer, *J. Org. Chem.*, **27**, 496 (1962).
166. M. L. Casey, D. S. Kemp, K. G. Paul, and D. D. Cox, *J. Org. Chem.*, **38**, 2294 (1973).
167*a*. R. Kada, V. Knoppová and J. Kováč, *Coll. Czech. Chem. Comm.*, **40**, 1563 (1975).
167*b*. L. E. Kholodov and V. G. Yashunskii, *Reakts. sposob. org. Soedinenii*, **1**(1), 77 (1964).
168. Y. Otsuji and H. H. Jaffé, *Abstr.* 137*th Mtg. Amer. Chem. Soc., Cleveland*, 1960, p. 76-O.
169. H. H. Jaffé, *J. Org. Chem.*, **23**, 1790 (1958).
170. J. H. Nelson, R. G. Garvey, and R. O. Ragsdale, *J. Heterocyclic Chem.*, **4**, 591 (1967).
171. G. Beggiato, G. G. Aloisi, and U. Mazzucato, *J.C.S. Faraday I*, **70**, 628 (1974).
172. M. Charton, *J. Amer. Chem. Soc.*, **86**, 2033 (1964).
173. I. F. Tupitsyn, N. N. Zatsepina, and A. V. Kirova, *Reakts. sposob. org. Soedinenii*, **9**, 223 (1972).

174. M. Charton, *J. Org. Chem.*, **36**, 882 (1971).
175. L. A. Kutulya, L. P. Pivovarevich, V. G. Gordienko, S. V. Tsukerman, and V. F. Lavrushin, *Reakts. spos. org. Soedinenii*, **9**, 1043 (1972).
176. D. A. Forsyth and D. S. Noyce, *Tetrahedron Lett.*, 3893 (1972).
177. E. Imoto and Y. Otsuji, *Bull. Univ. Osaka Prefecture, Ser. A.* **6**, 115 (1958); *Chem. Abs.*, **53**, 3027 (1959).
178. T. A. Melent'eva, L. V. Kazanskaya, and V. M. Berezovskii, *Doklady Akad. Nauk. SSR.*, **175**, 354 (1967); EE, 624.
179. D. S. Noyce and R. W. Nichols, *Tetrahedron Lett.*, 3889 (1972).
180. D. S. Noyce and H. J. Pavez, *J. Org. Chem.*, **37**, 2620 (1972).
181. D. S. Noyce and H. J. Pavez, *J. Org. Chem.*, **37**, 2623 (1972).
182. J. Elguero, E. Gonzalez, and R. Jaguier, *Bull. Soc. chim. France*, 5009 (1968).
183. M. J. Collis and G. R. Edwards, *Chem. Ind.*, 1097 (1971).
184. M. Charton, *J. Org. Chem.*, **30**, 3346 (1965).
185. C. R. Ganellin, *Jerusalem Symp. Quantum Chem. Biochem.*, No. 7 (*Mol. and Quantum Pharmacol.*), **43** (1974).
186. C. F. Kröger and W. Freiberg, *Z. Chem.*, **5**, 381 (1965).
187. W. Freiberg and C. F. Kröger, *Chimia*, **21**, 159 (1967).
188. L. I. Bagal and M. S. Pevzner, *Khim. Heterotsikl. Soedinenii*, 558 (1970).
189. J. A. Caruso, P. G. Sears, and A. I. Popov, *J. Phys. Chem.*, **71**, 1756 (1967).
190. D. J. Brown and P. B. Ghosh, *J. Chem. Soc.* (*B*), 270 (1969).
191. M. Charton, *J. Chem. Soc.* (*B*), 1240 (1969).
192. H. H. Jaffé and G. O. Doak, *J. Amer. Chem. Soc.*, **77**, 4441 (1955).
193. M. R. Chakrabarty, C. S. Handloser, and M. W. Mosher, *J.C.S. Perkin II*, 938 (1973).
194. K. Murai, S. Takeuchi, and C. Kimura, *Nippon Kagaku Kaishi*, 95 (1973).
195. W. J. McKinney, M. K. Wong, and A. I. Popov, *Inorg. Chem.*, **7**, 1001 (1968).
196. G. G. Aloisi, G. Beggiato, and U. Mazzucato, *Trans. Faraday Soc.*, **66**, 3075 (1970).
197. V. A. Savyolova, T. N. Solomoichenko, and L. M. Litvinenko, *Reakts. spos. org. Soedinenii*, **9**, 665 (1972).
198. H. H. Jaffé and H. L. Jones, *J. Org. Chem.*, **30**, 964 (1965).
199. A. Bryson, *J. Amer. Chem. Soc.*, **82**, 4871 (1960).
200. M. Charton, *Abstr. 154th Mtg. Amer. Chem. Soc., Chicago*, 1967, S-137.
201. T. Matsui and M. Nagano, *Chem. Pharm. Bull.*, **22**, 2123 (1974).
202. M. Taagepera, W. G. Henderson, R. T. C. Brownlee, J. L. Beauchamp, D. Holtz, and R. W. Taft, *J. Amer. Chem. Soc.*, **94**, 1369 (1972).
203. R. F. Cookson and G. W. H. Cheeseman, *J.C.S. Perkin II*, 392 (1972).
204. E. R. Tucci, C. H. Ke, and N. C. Li, *J. Inorg. Nuclear. Chem.*, **29**, 1657 (1967).
205. O. A. Zagulayaeva, E. G. Saikovich, and V. P. Mamaev, *Reakts. spos. org. Soedinenii*, **7**, 112, (1970); EE, 49.
206. B. Roth and J. Z. Strelitz, *J. Org. Chem.*, **34**, 821 (1969).
207. T. Tashiro and M. Yasuda, *Chem. High Polymers* (*Japan*), 853 (1969); *Chem. Abs.*, **72**, 120880 (1970).
208. S. Mizukami and E. Hirai, *J. Org. Chem.*, **31**, 1199 (1966).
209. M. Charton, unpublished results.
210. A. F. Pozharskii, L. M. Sitkina, A. M. Simonov, and T. N. Chegolya, *Khim. Heterotsikl. Soedinenii*, 209 (1970).
211. L. W. Deady, R. G. McLoughlin, and M. R. Grimmett, *Austral. J. Chem.*, **28**, 1861 (1975).
212. U. Mazzucato, G. Aloisi, and G. Cauzzo, *Trans. Faraday Soc.*, **62**, 2685 (1966).
213. A. Ricci and P. Vivarelli, *Boll. sci. Fac. Chim. ind.* (*Bologna*), **24**, 249 (1966); *Chem. Abs.*, **66**, 119452 (1967).
214. W. C. Aten and K. H. Büchel, *Z. Naturforsch.*, **25b**, 961 (1970).

215. V. A. Bren', V. I. Minkin, A. D. Garnovskii, E. V. Botkina, and B. S. Tenaisechuk, *Reakts. spos. org. Soedinenii*, **5**, 651 (1968); EE, 264.
216. L. N. Yakhontov, M. A. Portnov, V. A. Azimov, and E. I. Lapan, *Zhur. org. Khim.*, **5**, 956 (1969); EE, 942.
217. M. Charton, *J. Chem. Soc.*, 5884 (1964).
218. L. L. Popova, I. D. Sadekov, and V. I. Minkin, *Reakts. spos. org. Soedinenii*, **6**, 47 (1969); EE, 17.
219. N. A. Vorontsova, N. L. Poznanskaya, O. N. Vlasov, and N. I. Shvetsov-Shilovskii, *Reakts. spos. org. Soedinenii*, **5**, 665 (1968); EE 270.
220. N. S. Kozlov, V. I. Letunov, I. S. Berdinskii and V. V. Misenzhnikov, *Reakts. spos. org. Soedinenii*, **7**, 1058 (1970); EE, 479.
221. H. Walba, D. L. Stiggall, and S. M. Coutts, *J. Org. Chem.*, **32**, 1954 (1967).
222. H. Walba and R. Ruiz-Velasco, Jr., *J. Org. Chem.*, **34**, 3315 (1969).
223. H. J. Chen, L. E. Hakka, R. L. Hinman, A. J. Kresge, and E. B. Whipple, *J. Amer. Chem. Soc.*, **93**, 5102 (1971).
224. B. S. Tanaseichuk, A. A. Bardina, and A. A. Khomenko, *Khim. Heterotsikl. Soedinenii*, 1255 (1971).
225. R. Ponec and M. Procházka, *Coll. Czech. Chem. Comm.*, **39**, 2088 (1974).
226. M. Kratochvil, M. Sedlackova, A. Zahradnicova, and Z. Babak, *Chem. Zvesti*, **25**, 137 (1971).
227. T. D. Sokolova, S. V. Bogatkov, Yu. F. Malina, B. V. Unkovsky, and E. M. Cherkasova, *Reakts. spos. org. Soedinenii*, **5**, 160 (1968); EE, 65.
228. T. D. Sokolova, S. V. Bogatkov, Yu. F. Malina, and B. V. Unkovsky, *Reakts. spos. org. Soedinenii*, **7**, 626 (1970); EE, 278.
229. G. S. Litvinenko, V. I. Artyukhin, A. A. Andrusenko, D. V. Sokolov, V. V. Sosnova, and M. N. Akimova, *Reakts. spos. org. Soedinenii*, **7**, 960 (1970); EE, 435.
230. J. Paleček and J. Hlavatý, *Z. Chem.*, **9**, 428 (1969).
231. C. A. Grob, W. Simon, and D. Treffert, *Angew. Chem. Intern. Edn.*, **12**, 319 (1973); GE, **85**, 310 (1973).
232. E. B. Lifshits, L. M. Yagupol'skii, D. Ya. Naroditskaya, and E. S. Kozlova, *Reakts. spos. org. Soedinenii*, **6**, 317 (1969); EE, 133.
233. E. B. Lifshits, N. S. Spasokukotskii, L. M. Yagupol'skii, E. S. Kozlova, D. Ya Naroditskaya, and I. I. Levkoev, *Zhur. obschchei. Khim.*, **38**, 2025 (1968); EE, 1965.
234. G. Marino, *Adv. Heterocyclic Chem.*, **13**, 235 (1971).
235. G. Marino, *Pr. Nauk. Inst. Chem. Tech. Nafty Wegla Politech. Wroclaw*, No. 15, 89 (1973) (in English).
236. G. Marino, *Chim. Ind.*, (*Milan*) **55**, 349 (1973).
237. R. N. McDonald and J. M. Richmond, *J.C.S. Chem. Comm.*, 333 (1974).
238. D. Spinelli, G. Consiglio, and A. Corrao, *J.C.S. Perkin, II*, 1866 (1972).
239. D. Spinelli, G. Guanti, and C. Dell'Erba, *J.C.S. Perkin II*, 441 (1972).
240. D. Spinelli and G. Consiglio, *J.C.S. Perkin II*, 989 (1975).
241. D. Spinelli, G. Consiglio, R. Noto, and A. Corrao, *J.C.S. Perkin II*, 620 (1975).
242. C. Dell'Erba, A. Guareschi, and D. Spinelli, *J. Heterocyclic Chem.*, **4**, 438 (1967).
243. M. Bosco, L. Forlani, V. Liturri, P. Riccio, and P. E. Todesco, *J. Chem. Soc.* (*B*), 1373 (1971).
244. M. Bosco, V. Liturri, L. Troisi, L. Forlani, and P. E. Todesco, *J.C.S. Perkin II*, 508 (1974).
245. G. Bartoli, O. Sciacovelli, M. Bosco, L. Forlani, and P. E. Todesco, *J. Org. Chem.*, **40**, 1275 (1975).
246. N. N. Zatsepina, Yu. L. Kaminskii and I. F. Tupitsyn, *Reakts. spos. org. Soedinenii*, **6**, 448 (1969); EE, 187.
247. G. G. Aloisi, S. Santini, and G. Savelli, *J.C.S. Faraday I*, 2045 (1975).

248. L. E. Kholodov, A. M. Khelem, and V. G. Yashunskii, *Zhur. org. Khim.* **3**, 1870 (1967); EE, 1825.
249. S. Clementi, C. D. Johnson, and A. R. Katritzky, *J.C.S. Perkin II* 1294 (1974).
250. J. H. M. Hill and J. G. Krause, *J. Org. Chem.*, **29**, 1642 (1964).
251. T. N. Bykhovskaya and O. N. Vlasov, *Reakts. spos. org. Soedinenii*, **4**, 510 (1967); EE, 210.
252. J. A. Zoltewicz and L. S. Helmick, *J. Amer. Chem. Soc.*, **92**, 7547 (1970).
253. J. A. Zoltewicz and R. E. Cross, *J.C.S. Perkin II*, 1363 (1974).
254. I. F. Tupitsyn, N. N. Zatsepina, A. V. Kirova, and Yu. M. Kapustin *Reakts. spos. org. Soedinenii*, **5**, 601 (1968); EE, 243.
255. I. F. Tupitsyn, N. N. Zatsepina, Yu. M. Kapustin, and A. V. Kirova, *Reakts. spos. org. Soedinenii*, **5**, 613 (1968); EE, 268.
256. N. N. Zatsepina, I. F. Tupitsyn, A. V. Kirova, and A. I. Belyashova, *Reakts. spos. org. Soedinenii*, **6**, 257 (1969); EE, 109.
257*a*. J. Kaválek, A. Lyčka, V. Macháček, and V. Štěrba *Coll. Czech. Chem. Comm.*, **40**, 1166 (1975).
257*b*. K. Beránek, A. Lyčka, and V. Štěrba, *Coll. Czech. Chem. Comm.*, **40**, 1919 (1975).
258. V. N. Shishkin, S. L. Vlasova, and B. S. Tanaseichuk, *Khim. Heterosikl. Soedinenii*, 1520 (1972).
259. P. B. Talukdar, S. Banerjee, and A. C. Chakraborty, *Indian J. Chem.*, **9**, 827 (1971).
260. Yu.A. Rozin, V. E. Blokhin, Z. V. Pushkareva, and V. I. Elin, *Khim. Heterotsikl. Soedinenii*, 990 (1974).
261. O. M. Polumbrik, G. F. Dvorko, N. G. Vasil'kevich, and V. A. Kuznetsov, *Reakts. spos. org. Soedinenii*, **9**, 357 (1972).
262. O. M. Polumbrik, N. G. Vasil'kevich, and G. F. Dvorko, *Zhur. org. Khim.*, **11**, 770 (1975); EE, 763.
263. O. M. Polumbrik, G. F. Dvorko, N. G. Vasil'kevich, and V. A. Kuznetsov, *Teor. Eksp, Khim.*, **9**, 375 (1973); EE, 294.
264. M. J. S. Dewar and I. J. Turchi, *J. Amer. Chem. Soc.*, **96**, 6148 (1974).
265. P. R. West and J. Warkentin, *J. Org. Chem.*, **34**, 3233 (1969).
266. G. Seconi, P. Vivarelli, and A. Ricci, *J. Chem. Soc.* (*B*), 254 (1970).
267. J. R. Owen, *Eastman Org. Chem. Bull.*, **43**, 3 (1971).
268. Z. V. Todres, S. I. Zhdanov, and V. S. Tsveniashvili, *Izv. Akad. Nauk. SSR Ser. Khim.*, 975 (1968); EE, 934.
269. H. Balli and D. Schelz, *Helv. Chim. Acta*, **58**, 448 (1975).
270. C. J. Cooksey and M. D. Johnson, *J. Chem. Soc.* (*B*), 1191 (1968).
271. J. W. Bunting and W. G. Meathrel, *Canad. J. Chem.*, **52**, 962 (1974).
272. V. G. Mairanovskii, V. M. Mamaev, G. V. Ponomarev, R. I. Marinova, and R. P. Evestigneeva, *Zhur. obschei Khim.* **44**, 2508 (1974); EE, 2468.
273. C. K. Bradsher, C. R. Miles, N. A. Porter, and I. J. Westerman, *Tetrahedron Lett.*, 4969 (1972).
274. I. J. Westerman and C. K. Bradsher, *J. Org. Chem.*, **36**, 969 (1971).
275. I. Degani, R. Fochi, and G. Spunta, *Gazz. Chim. Ital.*, **97**, 388 (1967).
276. G. Canalini, I. Degani, R. Rochi, and G. Spunta, *Ann. Chim.* (*Italy*) **57**, 1045 (1967).
277. N. A. Porter, I. J. Westerman, T. G. Wallis, and C. K. Bradsher, *J. Amer. Chem. Soc.*, **96**, 5104 (1974).
278. F. Fringuelli, G. Marino, and A. Taticchi, *J. Chem. Soc.* (*B*)., 2304 (1971).
279. F. Fringuelli, G. Marino, and A. Taticchi, *J. Chem. Soc.* (*B*)., 2302 (1971).
280. F. Fringuelli, G. Marino, and A. Taticchi, *J.C.S. Perkin II* 158 (1972).
281. F. Fringuelli, G. Marino, and A. Taticchi, *J. Chem. Soc.* (*B*) 1595 (1970).
282. P. Bouchet, C. Coquelet, and J. Elguero, *J.C.S. Perkin II* 449 (1974).
283. J. C. Kauer and W. A. Sheppard, *J. Org. Chem.*, **32**, 3580 (1967).

284. E. V. Borisov, L. E. Kholodov, and V. G. Yashunskii, *Reakts. spos. org. Soedinenii*, **7**, 704 (1970); EE, 314.
285. T. L. Chan, J. Miller, and F. Stansfield, *J. Chem. Soc.*, 1213 (1964).
286. A. R. Katritzky and P. Simmons, *J. Chem. Soc.*, 1511 (1960).
287. E. E. Pasternak and P. Tomasik, *Bull. Acad. polon Sci., Sér. Sci. chim.*, **23**, 57 (1975).
288. E. E. Pasternak and P. Tomasik, *Bull. Acad. polon Sci., Sér. Sci. chim.*, **23**, 797 (1975).
289. Y. Ohto, Y. Hashida, S. Sekiguchi, and K. Matsui, *Bull. Chem. Soc. Japan*, **47**, 1301 (1974).
290. H. L. Nyquist and B. Wolfe, *J. Org. Chem.*, **39**, 2591 (1974).
291. V. F. Bystrov, Zh. N. Belaya, B. E. Gruz, G. P. Syrova A. I. Tomachev, L. M. Shulezhko, and L. M. Yagupol'skii, *Zhur. obschei Khim.*, **38**, 1001 (1968); EE, 963.
292. J. Chodkowski and T. Giovanoli-Jakubczak, *Roczniki Chem.*, **43**, 1037 (1969).
293. J. Chodkowski and T. Giovanoli-Jakubczak, *Roczniki Chem.*, **44**, 1289 (1970).
294. Š. Toma, *Coll. Czech. Chem. Comm.*, **34**, 2771 (1969).
295. A. D. Campbell, S. Y. Chooi, L. W. Deady, and R. A. Shanks, *Austral. J. Chem.*, **23**, 203 (1970).
296. A. R. Katritzky and F. J. Swinbourne, *J. Chem. Soc.*, 6707 (1965).
297. D. M. Dimitrijević, Z. D. Tadić, M. M. Mišić-Vuković, and M. Muškatirović, *J.C.S. Perkin II*, 1051 (1974).
298. J. H. Blanch, *J. Chem. Soc.* (*B*), 937 (1966).
299. R. Joeckle, E. D. Schmid, and R. Mecke, *Z. Naturforsch.*, **21a**, 1906 (1966).
300. D. D. Perrin, *J. Chem. Soc.*, 5590 (1965).
301. D. R. Eaton, R. E. Benson, C. G. Bottomley, and A. D. Josey, *J. Amer. Chem. Soc.*, **94**, 5996 (1972).
302. R. Taylor, *J. Chem. Soc.* (*B*)., 1364 (1970).
303. S. Clementi, P. Linda, and G. Marino, *J. Chem. Soc.* (*B*)., 1153 (1970).
304. R. Taylor, *J. Chem. Soc.* (*B*), 1397 (1968).
305. G. T. Bruce, A. R. Cooksey, and K. J. Morgan, *J.C.S. Perkin II*, 551 (1975).
306. L. W. Deady, R. A. Shanks, and R. D. Topsom, *Tetrahedron Lett.*, 1881 (1973).
307. K. Schwetlick and K. Unverferth, *J. prakt. Chem.*, **314**, 603 (1972).
308. F. Fringuelli, G. Marino, and A. Taticchi, *Gazz. Chim. Ital.*, **102**, 534 (1972).
309*a*. G. G. Smith and J. A. Kirby, *J. Heterocyclic Chem.*, **8**, 1101 (1971).
309*b*. N. N. Zatsepina, Yu. L. Kaminskii, and I. F. Tupitsyn, *Reakts. spos. org. Soedinenii*, **6**, 778 (1969); EE, 333.
310. E. A. Hill, M. L. Gross. M. Stasiewicz, and M. Manion, *J. Amer. Chem. Soc.*, **91**, 7381 (1969).
311. D. S. Noyce and B. B. Sandel, *J. Org. Chem.*, **40**, 3381 (1975).
312. D. S. Noyce and S. A. Fike, *J. Org. Chem.*, **38**, 3316 (1973).
313. T. J. Broxton, G. L. Butt, L. W. Deady, S. H. Toh, R. D. Topsom, A. Fischer, and M. W. Morgan, *Canad. J. Chem.*, **51**, 1620 (1973).
314. D. S. Noyce, J. A. Virgilio, and B. Bartman, *J. Org. Chem.*, **38**, 2657 (1973).
315. R. Taylor, *J.C.S. Perkin II*, 277 (1975).
316. G. P. Bean and A. R. Katritzky, *J. Chem. Soc.* (*B*), 864 (1968).
317. P. Bellingham, C. D. Johnson, and A. R. Katritzky, *J. Chem. Soc.* (*B*), 866 (1968).
318. D. S. Noyce and D. A. Forsyth, *J. Org. Chem.*, **39**, 2828 (1974).
319. R. Taylor, *J. Chem. Soc.* (*B*), 2382 (1971).
320. N. N. Zatsepina, I. F. Tupitsyn, Yu. L. Kaminskii, and N. S. Kolodina, *Reakts. spos. org. Soedinenii*, **6**, 766 (1969); EE, 327.
321. P. Forsythe, R. Frampton, C. D. Johnson, and A. R. Katritzky, *J.C.S. Perkin II*, 671 (1972).
322. N. N. Zatsepina, Yu. L. Kaminskii, and I. F. Tupitsyn, *Reakts. spos. org. Soedinenii*, **4**, 433 (1967); EE, 177.

323. A. Gordon, A. R. Katritzky, and S. K. Roy *J. Chem. Soc.* (*B*), 556 (1968).
324. T. L. Chan and J. Miller, *Austral. J. Chem.*, **20**, 1595 (1967).
325. I. F. Tupitsyn, N. N. Zatsepina, N. S. Kolodina, and A. A. Kane, *Reakts. spos. org. Soedinenii*, **5**, 931 (1968); EE 387.
326. N. N. Zatsepina, A. V. Kirova, and I. F. Tupitsyn, *Reakts. spos. org. Soedinenii*, **5**, 70 (1968); EE, 27.
327. I. F. Tupitsyn, N. N. Zatsepina, N. Kolodina, and Yu. L. Kaminskii, *Reakts. spos. org. Soedinenii*, **6**, 458 (1969); EE, 192.
328. M. K. A. Khan and K. J. Morgan, *Tetrahedron*, **21**, 2197 (1965).
329. L. W. Deady and R. A. Shanks, *Austral. J. Chem.*, **25**, 2363 (1972).
330. L. W. Deady, D. J. Foskey, and R. A. Shanks, *J. Chem. Soc.* (*B*), 1962 (1971).
331. D. S. Noyce, C. A. Lipinski, and G. M. Loudon, *J. Org. Chem.*, **35**, 1718 (1970).
332. J. H. Blanch, *J. Chem. Soc.* (*B*)., 167 (1968).
333*a*. J. H. Blanch and J. Andersen, *J. Chem. Soc.* (*B*), 169 (1968).
333*b*. H. Balli, B. Hellrung, and H. Hinsken, *Helv. Chim. Acta*, **57**, 1174 (1974).
344. P. Vetešńik, J. Kaválek, V. Beránek, and O. Exner, *Coll. Czech. Chem. Comm.*, **33**, 566 (1968).
335. B. Hellrung and M. Balli, *Helv. Chim. Acta*, **58**, 605 (1975).
336. B. Hellrung and H. Balli, *Helv. Chim. Acta*, **57**, 1185 (1974).
337. A. R. Katritzky, M. Kingsland, and O. S. Tee, *J. Chem. Soc.* (*B*)., 1484 (1968).
338. A. R. Katritzky, C. R. Palmer, F. J. Swinbourne, T. T. Tidwell, and R. D. Topsom, *J. Amer. Chem. Soc.*, **91**, 636 (1969).
339. I. F. Tupitsyn, N. N. Zatsepina, and N. S. Kolodina, *Reakts. spos. org. Soedinenii*, **6**, 11 (1969); EE, 1.
340. I. F. Tupitsyn, N. N. Zatsepina, N. S. Kolodina, and A. V. Kirova *Reakts. spos. org. Soedinenii*, **9**, 1075 (1972).
341. C. J. Turner, *J.C.S. Perkin II*, 1250 (1975).
342. D. S. Noyce and R. L. Castenson, *J. Amer. Chem. Soc.*, **95**, 1247 (1973).
343. J. Elguero, C. Marzin, A. R. Katritzky, and P. Linda, *The Tautomerism of Heterocycles* (Academic Press, New York, 1976).
344. P. Vetešńik, J. Kaválek, V. Beránek, and O. Exner, *Coll. Czech. Chem. Comm.*, **33**, 566 (1968).
345. P. Vetešńik, J. Kaválek, and V. Beránek, *Coll. Czech. Chem. Comm.*, **36**, 2486 (1971).
346. M. Charton, *J. Org. Chem.*, **31**, 3739 (1966).
347. T. Kitagawa, S. Mizukami, and E. Hirai, *Chem. and Pharm. Bull.* (*Japan*) **22**, 1239 (1974)
348. C. W. Donaldson and M. M. Joullié, *J. Org. Chem.*, **33**, 1504 (1968).
349. R. Escale, R. Jacquier, B. Ly, F. Petrus, and J. Verducci, *Tetrahedron*, **32**, 1369 (1976).
350. J. M. Angelelli, A. R. Katritzky, R. F. Pinzelli, and R. D. Topsom, *Tetrahedron*, **28**, 2037 (1972).
351. G. P. Ford, T. B. Grindley, A. R. Katritzky, M. Shome, J. Morel, C. Paulmier, and R. D. Topsom, *J. Mol. Structure*, **27**, 195 (1975).
352. J-J. Péron, J-M Lebas, and P. Saumagne, *Compt. rend. B*, **267**, 586 (1968).
353. M. Senechal and P. Saumagne, *Comp. rend. B*, **276**, 79 (1973).
354. E. D. Schmid and J. Bellanato, *Z. Elektrochem.*, **65**, 362 (1961).
355. R. A. Gavars and J. P. Stradins, *Reakts. spos. org. Soedinenii*, **2**(1), 22 (1965).
356. J.-P. Morizur and Y. Pascal, *Bull. Soc. chim. France*, 2296 (1966).
357. S. Gronowitz, I. Johnson, and A. B. Hornfeldt, *Chem. Scripta*, **7**, 76 (1975).
358. S. Gronowitz, I. Johnson, and A. B. Hornfeldt, *Chem. Scripta*, **7**, 111 (1975).
359. S. Gronowitz, I. Johnson, and A. B. Hornfeldt, *Chem. Scripta*, **7**, 211 (1975).
360. S. Pignataro, P. Linda, and G. Marino, *Ricera sci.*, **39**, 668 (1969).
361. L. G. Tensmeyer and C. Ainsworth, *J. Org. Chem.*, **31**, 1878 (1966).
362. P. Linda, G. Marino, and S. Pignataro, *J. Chem. Soc.* (*B*)., 1585 (1971).

363. C. M. Camaggi, R. Leardini, and G. Placucci, *J.C.S. Perkin II*, 1195 (1974).
364. E. D. Schmid and R. Joeckle, *Spectrochim. Acta*, **22**, 1645 (1966).
365. N. N. Zatsepina, I. F. Tupitsyn, N. S. Kolodina, and A. V. Kirova, *Reakts. spos. org. Soedinenii*, **8**, 805 (1971).
366. G. S. Marx and P. E. Spoerri, *J. Org. Chem.*, **37**, 111 (1972).
367. A. Perjéssy, R. Frimm, and P. Hrnčiar, *Coll. Czech. Chem. Comm.*, **37**, 3302 (1972).
368. A. Perjéssy, P. Hrnčiar, R. Frimm, and L. Fišera, *Tetrahedron*, **28**, 3781 (1972).
369. A. Perjéssy, D. W. Boykin, Jr., L. Fišera, A. Krutošíkova, and J. Kováč, *J. Org. Chem.*, **38**, 1807 (1973).
370. A. Perjéssy, P. Hrnčiar, and A. Krutošíkova, *Tetrahedron*, **28**, 1025 (1972).
371. A. Selva, P. Traldi, L. F. Zerilli, B. Cavalleri, and G. G. Gallo, *Ann. Chim.* (*Italy*), **64**, 229 (1974).
372. R. F. Smith, J. L. Deutsch, P. A. Almeter, D. S. Johnson, S. M. Roblyer, and T. C. Rosenthal, *J. Heterocyclic Chem.*, **7**, 671 (1970).
373. G. Di Modica, E. Barni, and A. Gasco, *J. Heterocyclic Chem.*, **2**, 457 (1965).
374. D. Simov, B. S. Gulubov, and V. B. Kalcheva, *Compt. rend. Acad. bul. Sci.*, **27**, 663 (1974).
375. W. C. Coburn, Jr., M. C. Thorpe, J. A. Montgomery, and K. Hewson, *J. Org. Chem.*, **30**, 1110 (1965).
376. W. C. Coburn, Jr., M. C. Thorpe, J. A. Montgomery, and K. Hewson, *J. Org. Chem.*, **30**, 1114 (1965).
377. D. F. Ewing, D. N. Gregory, and R. M. Scrowston, *Org. Magn. Reson.*, **6**, 293 (1974).
378. M. Salmón, A. Jiménez, and R. Zawadzi, *Org. Magn. Reson.*, **5**, 5 (1973).
379. A. Perjéssy, A. Krutošíkova, and A. F. Olejnik, *Tetrahedron*, **31**, 2936 (1975).
380. M. Ionescu, I. Goia, and H. Mantsch, *Rev. Roumaine Chim.*, **11**, 243 (1966).
381. O. T. Beachley, Jr., *J. Amer. Chem. Soc.*, **92**, 5372 (1970).
382. M. Charton, *J. Org. Chem.*, **34**, 1877 (1969).
383. L. A. Leites, L. E. Vinogradova, V. N. Kalinin, and L. I. Zakharkin, *Izvest. Akad. Nauk. SSSR.*, *Ser. Khim.*, 1016 (1968); EE, 970.
384. M. Charton, *Progr. Phys. Org. Chem.*, **10**, 81 (1973).
385. G. Guanti, M. Novi, G. Garbarino, and C. Dell'Erba, *J.C.S. Perkin II*, 137 (1977).
386. R. Bolton and R. E. M. Burley, *J.C.S. Perkin II*, 426 (1977).
387. P. Ananthakrishna Nadar and C. Gnanasekaran, *J.C.S. Perkin II*, 1893 (1976).
388. P. Ananthakrishna Nadar, A. Shanmugasundaram, and R. Murugesan, *Ind. J. Chem.*, **14a**, 146 (1976).
389. C. Kirsche, G. Thirot, C. Caullet, and J.-M. Juillerat, *Bull. Soc. chim. France*, 101 (1975).
390. R. Sivaprakasam, *J. Mississippi Acad. Sci.*, **20**, 7 (1975).
391. R. Sivaprakasam, *J. Mississippi Acad. Sci.*, **20**, 14 (1975).
392. D. Spinelli, G. Consiglio, and R. Noto, *J.C.S. Perkim II*, 1495 (1976).
393. D. Spinelli, R. Noto, G. Consiglio, and A. Storace, *J.C.S. Perkin II*, 1805 (1976).
394. M. K. Das and J. J. Zuckerman, *J. Amer. Chem. Soc.*, **99**, 1354 (1977).
395. C. Reichardt and R. Müller, *Justus Liebigs Ann. Chem.*, 1953 (1976).
396. A. C. M. Paiva, L. Juliano, and P. Boschcov, *J. Amer. Chem. Soc.*, **98**, 7645 (1976).
397. C. Laurence and M. Luçon, *Canad. J. Chem.*, **54**, 2021 (1976).
398. D. G. Pillsbury and D. H. Busch, *J. Amer. Chem. Soc.*, **98**, 7836 (1976).
399. J. Kaválek, A. Lyčka, V. Macháček, and V. Štěrba, *Coll. Czech. Chem. Comm.*, **41**, 67 (1976).
400. J. W. Bunting and D. J. Norris, *J. Amer. Chem. Soc.*, **99**, 1189 (1977).
401. C. K. Bradsher, T. G. Wallis, I. J. Westerman, and N. A. Porter, *J. Amer. Chem. Soc.*, **99**, 2588 (1977).
402. D. S. Noyce and B. B. Sandel, *J. Org. Chem.*, **41**, 3640 (1976).

403. T. J. Broxton, L. W. Deady, and Y. T. Pang, *J. Amer. Chem. Soc.*, **99**, 2268 (1977).
404. E. Maccarone, G. Musumarra, and G. A. Tomaselli, *J.C.S. Perkin II*, 906 (1976).
405. M. Davis, R. Lakhan, and B. Ternai *J. Org. Chem.*, **41**, 3591 (1976).
406. L. W. Deady, P. M. Harrison, and R. D. Topsom, *Spectrochim. Acta*, **31A**, 1671 (1975).
407. G. Dana, O. Convert, J.-P. Girault, and E. Mulliez, *Canad. J. Chem.*, **54**, 1827 (1976).
408. S. Gronowitz, I. Johnson, and A. Bugge, *Acta. Chem. Scand., B*, **30**, 417 (1976).
409. P. Tomasik and C. D. Johnson, *Adv. Heterocyclic Chem.*, **20**, 1 (1976).

6

Substituent Effects in Olefinic Systems

G. P. Ford, A. R. Katritzky, and R. D. Topsom

6.1. Introduction

Most of the early work in the field of linear free energy relationships was concerned with benzene derivatives. Interest in saturated organic

G. P. Ford • University of Texas at Austin, Austin, Texas 78712, U.S.A.
A. R. Katritzky • University of East Anglia, Norwich, NR4 7TJ, England.
R. D. Topsom • La Trobe University, Bundoora, Australia, 3083.

molecules has recently led to an understanding[1–3] of the differing modes of transmission of electronic effects of substituents. By contrast, relatively little attention has been paid to similar effects in simple olefins. Substituent effects on the reactions of the unsaturated linkage in olefins were well discussed by Charton[4] as part of a general review on nonaromatic unsaturated systems, but little attention was paid there to physical properties, apart from some tabulated correlations. Recently, theoretical, nuclear magnetic resonance, and infrared studies have each helped towards the attainment of a considerably better understanding of interactions in simple olefins, and it is thus timely to present a more detailed review.

The first systematic study of substituent effects in substituted ethylenes did not appear until 1958, when pK_a values of a series of *trans*-3-substituted acrylic acids were successfully correlated[5] with σ_p values. The preference for studies with benzene derivatives rather than the conceptually simpler alkene derivatives arises for several reasons. Thus, many simple substituted benzenes are readily synthesized and purified, and they are usually stable compounds of convenient volatility and convenient solubility in organic solvents. The benzene ring also provides a rigid framework with at least two sites for a substituent relative to a probe (such as a carboxy group) free of steric interaction. Such factors are generally less favorable for alkene derivatives.

Since substituent interactions are of such significance in benzene derivatives it is timely to consider what complementary or additional information can be gained from results from ethylenes. We concentrate in this review on systems in which the substituent, denoted by Y throughout, is joined directly to a carbon atom of the double bond. There has been a considerable number of studies on substituted phenylethylenes, such as cinnamic acids, and substituted styrenes and stilbenes, but in these systems the main resonance interactions involve the substituent and the phenyl group, the double bond acting as a transmission system as in cinnamic acids, or as a probe, as in nmr studies on substituted styrenes. In such compounds the field effect of the substituent may cause additional polarization effects[6] which complicate the analysis. We discuss such systems briefly in Section 6.10.

In this chapter, we give particular attention to the following three points. (*a*) How far do the interactions of substituents with a double bond parallel those with benzene? (*b*) To what extent are the interactions additive in each of the various classes of disubstituted ethylenes; how important is through-conjugation? (*c*) How far can resonance parameters be used as a guide to rotational barriers about the substituent–vinyl bonds?

Correlations. Early correlations of data were made either with general substituent properties such as electronegativity, or with simple Hammett σ parameters (or σ^* for substituents attached to saturated carbon). More recent correlations also involve substituent parameters such as σ^+ or σ^-, which are applicable to conditions of differing electron demand. However, all such single-parameter correlations suffer from the built-in requirement of a constant relationship between inductive and resonance contributions as shown in equation (6.1), where $\bar{\sigma}$ and $\bar{\sigma}_R$ can be σ^- and σ_R^-, σ^0 and σ_R^0,

$$\bar{\sigma} = \sigma_I + \bar{\sigma}_R \tag{6.1}$$

σ and σ_R(BA), or σ^+ and σ_R^+. The σ values were established for benzene systems and it has been amply demonstrated[1,7,8] that the ratio of resonance to inductive contributions for substituted benzenes covers a very wide range, depending on the property or reaction considered. Use of a single-parameter approach can thus lead to erroneous conclusions,[1,9] and any proper correlation for ethylenes requires the use of the dual substituent-parameter (DSP) method. Charton[4] used a DSP correlation of the form shown in equation (6.2). Here P_Y is the datum, σ_I and σ_R are the Hammett

$$P_Y = \alpha\sigma_I + \beta\sigma_R + h \tag{6.2}$$

σ_I and σ_R(BA) values for the substituent Y, and h is a constant to allow the parent compound (Y = H) to be treated as just one data point. We believe that it is better to avoid this additional parameter and constrain the data by using the parent compound figures as the origin, by analogy with the constraint on σ_I and σ_R values.

Moreover, it seems to us essential to incorporate the various values of $\bar{\sigma}_R$ as alternatives in fitting the data. Experience, mainly in the benzene series, has shown that when one particular $\bar{\sigma}_R$ scale shows a markedly better fit than the others, then the result is invariably meaningful in chemical terms.[1,7] We therefore use equation (6.3) in the analyses reported

$$P - P_0 = \rho_I\sigma_I + \rho_R\bar{\sigma}_R \tag{6.3}$$

below. However, it is necessary[1,7] to process data for an adequate and representative set of substituents if we wish to establish whether a correlation does or does not exist; unfortunately many of the investigations on alkenes are deficient in this respect. We used the following substituents in our program: NMe_2, OMe, F, Cl, Me, Ph, CF_3, COMe, CO_2R, CN, NO_2; substituents with heavier first atoms such as S, Si, or Br are omitted to avoid anisotropic, d-orbital, or mass effects, and OH and NH_2 are omitted because of possible hydrogen bonding. In some analyses of ethylenes, OR or R groups, where R is a higher homologue of methyl, have to be treated as OMe or Me. "Goodness of fit" (f) is given in terms of the standard deviation of the estimates divided by the root mean square of the data, expressed as a percentage. Experience[1,7] shows that a fit of better than 10% indicates a good correlation while one of better than 20% implies a fair one. Correlations having f higher than 30% are regarded as having no significance and are omitted.

6.2. Molecular Orbital Calculations

Ab initio calculations at the STO-3G level have proved useful in examining substituent interactions in monosubstituted benzenes[10,11] but surprisingly few results are available[12] for substituted ethylenes. However, a series of CNDO/2 calculations has been published[13] and these should be informative, especially since such calculations for benzene derivatives[13] seem to parallel the *ab initio* calculations.[10,11] The CNDO/2 method reproduces well the dipole moments of a wide range of organic molecules,[14] and has been used widely in the aromatic series to study ^{13}C and ^{19}F chemical shifts,[13,15] and in the calculation of infrared intensities (see Section 6.6).

In Table 6.1 we list the calculated[13] charge densities for a series of monosubstituted ethylenes. The representation used throughout this review is shown in structure (I). A DSP correlation of the total π-electron transfer between the substituent and the vinyl group gives equation (6.4)

$$\sum \Delta q_\pi = -0.05\sigma_I + 0.157\sigma_R^0 \tag{6.4}$$

with a goodness of fit (f, see Section 6.1) of 4.5%. Figure 6.1 shows a least-squares plot against σ_R^0, constrained to pass through the origin; the slope is 0.16 electron per unit of σ_R^0. Clearly therefore we can interpret

Table 6.1. Charge Densities, q ($\times 10^3$), in Monosubstituted Ethylenes by the CNDO/2 Method,[13] Shown as Differences from 10^3 for $C(\pi)$ and H and as Differences from 6×10^3 for C(total) (Negative Sign Indicates Increase)

Subst.	$\sum \Delta q_\pi$	$C_\alpha(\pi)$	C_α(total)	$C_\beta(\pi)$	C_β(total)	H_A	H_B	H_C
H	0	0	−31	0	−31	16	16	16
F	−53	30	216	−82	−117	2	36	42
OH	−72	54	166	−126	−135	6	22	22
CHO[a]	39	−7	−40	44	−2	30	15	19
CN	15	1	12	14	−15	28	17	25
NO_2	19	−64	−1	84	10	54	44	56
Me	−16	29	21	−45	−57	5	6	13
NMe_2	−85	46	117	−130	−127	−5	13	21
CF_3	12	−47	−85	60	15	50	25	35
OMe[b]	−72	46		−118				
CO_2Me[ab]	38	−23		61				

[a] "*trans*"-Conformer.
[b] G. P. Ford, Ph.D. Thesis, University of East Anglia (1974).

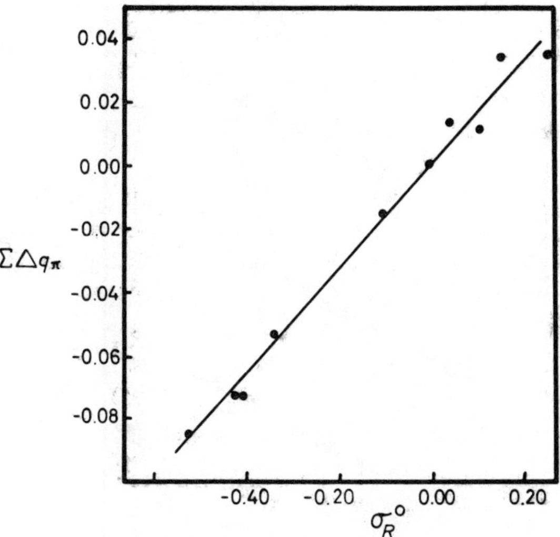

Fig. 6.1. Plot of $\sum \Delta q_\pi$ against σ_R^0 for monosubstituted ethylenes. The line shown is the least-squares plot constrained to pass through the origin.

directly σ_R^0 in terms of ground-state charge transfer. A similar result has been found[13] for monosubstituted benzenes, at the CNDO/2 level and at

$$\begin{matrix} Y & & H_C \\ & \overset{\alpha}{C}=\overset{\beta}{C} & \\ H_A & & H_B \end{matrix}$$

(I)

the *ab initio* level.[11] The overall π-charge transfer was found[13] to be similar in both series of compounds. By contrast, the change in π charge at C_β in a monosubstituted ethylene is greater[13] than at C_{para} in a monosubstituted benzene. It is later shown that "through-conjugation" in suitable disubstituted ethylenes is also more important than in *para*-disubstituted benzenes, in line with the above change in π charge and with the greater likely significance of such conjugation in the less extended ethylenic structures.

Some σ- and π-charge densities calculated at the CNDO/2 level have been published[16,17] for each of the classes of disubstituted ethylenes and compared with the analogous quantities calculated from those for monosubstituted compounds by assuming additivity of the electronic effects of substituents. The values of the total π transfer to or from the double bond or the π densities at the olefinic carbon atoms agree well with those calculated on the basis of additivity (Fig. 6.2). Thus, the substituents in 1,1- and both *cis*- and *trans*-1,2-disubstituted ethylenes appear to

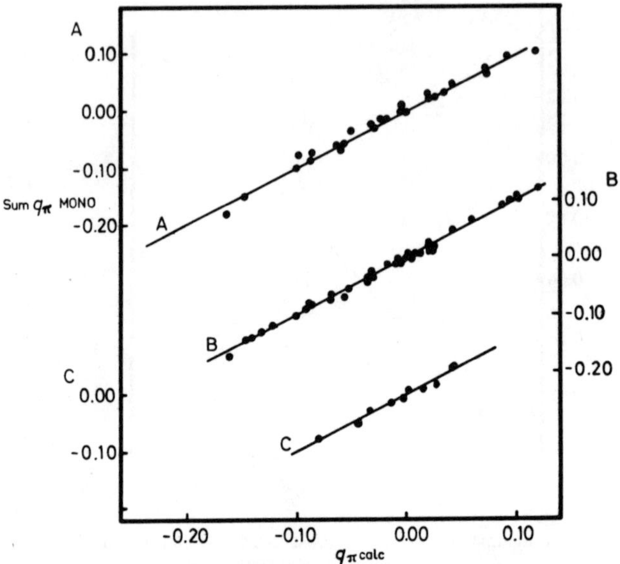

Fig. 6.2. Additivity of π-electron densities in disubstituted ethylenes. Lines are those of unit slope with zero intercept. Points refer to overall π transfer and to q_π values at individual carbon atoms (A: 1,1-compound; B: *trans* 1,2-; C: *cis* 1,2-compound).

interact with the ethylenic π system in an additive fashion. It must, however, be noted that standard geometric parameters were used throughout the work and thus no allowance was made for possible changes in geometry, particularly shortening of carbon–substituent bonds, that might be expected if through-conjugation occurs. Further, there are significant changes in the YCY angle in 1,1-disubstituted ethylenes compared with ethylene itself.[18]

Dipole transition moments have also been calculated, particularly for the $\nu_{C=C}$ stretching mode of substituted ethylenes, and these are discussed in Section 6.6.

6.3. Ground-State Properties

Unfortunately there are relatively few ground-state properties that are conveniently accessible as a guide to the interactions of substituents with unsaturated systems. In principle, geometric changes should be observed, particularly if the π system is involved, but in practice the changes occurring are usually small and hard to assign specifically. Dipole moments should also act as a measure of electronic interactions, but even if accurate measurements with a proper correction for atomic and electronic polarization terms are available, they depend on distance between and magnitude of charges, and this hinders comparisons within a series.

6.3.1. Bond Lengths

The C—Cl bond length in vinyl chloride (1.69 Å) is certainly less than that in ethyl chloride (1.77 Å). However, the hybridization of the carbon atom is sp^2 in the first case and sp^3 in the second, and thus the difference is not necessarily indicative of delocalization. Similar arguments have been invoked[19] in discussions of the delocalization in the central linkage of butadiene.

A series of disubstituted ethylenes with one fixed substituent, Z, should show changes in the C—Z bond length, but studies of sufficient accuracy are not yet available. X-ray studies[20] on (II) show that the

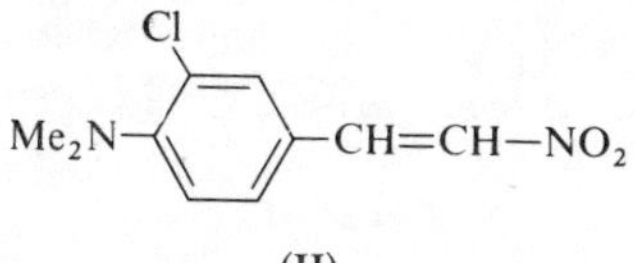

Cl

Me_2N—C_6H_3—CH=CH—NO_2

(II)

π-electron systems of the anilino, the olefinic, and the nitro group are essentially isolated. However, significant lengthening of the olefinic bond is observed[21] in structures such as (III) when two groups donating electrons

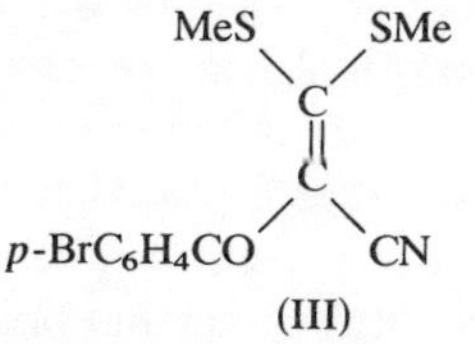

MeS SMe

C=C

p-BrC_6H_4CO CN

(III)

by resonance are attached to one carbon atom and two groups withdrawing electrons by resonance to the other.

6.3.2. Dipole Moments

In Table 6.2 we list data for some simple vinyl derivatives together with values for the corresponding substituted ethanes. The dipole moments of the vinyl compounds clearly reflect some interaction between the substituent and the double bond; thus values for resonance-donating but electronegative substituents such as halogens are less than for the corresponding ethyl derivatives, and this can, at least in part, be explained in terms of contributions of the canonical form (IV) to the overall structure. Similarly, values for resonance-withdrawing substituents such as cyano and nitro are greater in the vinyl than in the ethyl series.

Table 6.2. Dipole Moments (D) of Some Monosubstituted Ethylenes and Ethanes (in Solution, Unless Otherwise Noted)

Subst.	C_2H_5Y	$C_2H_3Y^a$	Subst.	C_2H_5Y	$C_2H_3Y^a$
H	0.00	0.00	CN	3.52^b	3.89
Me	0.00	−0.36	NO_2	3.19^b	3.44
F	1.92^c	1.43	CHO	2.57^d	3.11
Cl	1.90^b	1.44	COMe	2.8^d	3.00
Br	1.93^b	1.42	CF_3		2.45
OEt		1.27			

[a] Taken from Ref. 25.
[b] V. I. Minkin, O. A. Osipov, and Y. A. Zhdanov, *Dipole Moments in Organic Chemistry* (Plenum, New York, 1970).
[c] Gas phase, C. P. Smyth, *Dielectric Behavior and Structure* (McGraw-Hill, New York, 1955).
[d] Reference as in footnote *c*.

Although some success has been achieved in correlating dipole moments with inductive substituent parameters in certain series of substituted

$$\overset{\oplus}{Y}{=}CH{-}\overset{\ominus}{C}H_2$$

(IV)

alkanes,[22] values for monosubstituted benzenes are correlated worse,[4,23,24] partly, no doubt, because of difficulty in allowing for variation in the effective charge separation from one compound to another. Such correlations for substituted ethylenes are beset by the same difficulties and by the possibility of conformational isomerism for some substituents. Further, a set of carefully measured moments from one source (in an inert solvent or in the gas phase) is not available either for vinyl or for ethyl derivatives. A correlation[4] using equation (6.2) and data[25] for substituents that are symmetrical with respect to the substituent–carbon bond did not include any substituents donating electrons strongly by resonance and was of low precision. The quantity of interest is really the vector difference as between the vinyl and ethyl derivatives, but the factors mentioned above make attempts to use this quantity premature until more data are available.

Dipole moments for disubstituted ethylenes[4,21] do give evidence for through-conjugation, but the data are not sufficient for satisfactory analysis.

6.4. Nuclear Magnetic Resonance

Although not true ground-state properties, since excited-state wave functions are involved, nmr shifts and coupling constants have proved[26] a valuable guide to electron densities and substituent interactions, particularly in aromatic systems. In monosubstituted benzenes, the total electron density at the *para*-carbon atom seems accurately proportional to the π-electron density,[11,13] and a good linear correlation is found[11] with the corresponding ^{1}H shifts and a fair one with ^{13}C shifts. By contrast ^{13}C shifts at the *meta*-carbon atom follow the calculated total electron density, but these values are not proportional to either $q_\pi(meta)$ or to the observed *meta* ^{1}H shifts, which seem to be also a function of q_H. Thus general relationships between electron densities and nmr shifts cannot be expected; this point has also been stressed elsewhere.[27] Valid claims to such correlations apply to restricted series of compounds such as those involving the *para*-position in monosubstituted benzenes mentioned above.

6.4.1. ^{1}H *Chemical Shifts*

A considerable number of papers have reported and discussed ^{1}H shifts for substituted ethylenes and for styrenes (see Section 6.10). Charton,[4] by using equation (6.2) for simple vinyl derivatives, suggested that there was no meaningful correlation for H_A (see I), only an approximate correlation for the *cis*-vicinal proton H_C, but a moderately good correlation for the *trans*-vicinal proton H_B. Other correlations have been claimed, for example, between the shifts for the β-protons (H_B and H_C) and σ values[28,29], electronegativity,[30] dipole moments,[30,31] or calculated (INDO) electron densities.[32] The last paper[32] claims that the shifts are determined by the electron density both at the hydrogen atom and at the C_β atom to which it is attached.

In Table 6.3 we list the ^{1}H shifts for some simple vinyl derivatives together with Z values (the average shift for this arrangement of proton and substituent determined[33] from a large number of variously substituted ethylenes). The vinyl shifts are all taken from the one source,[34] as differences occur from one source to another.

We have carried out DSP analyses for all six sets of data, but none gave a correlation with goodness of fit better than 30%. The omission of the data for the unsymmetrical CO_2Me substituent gives equation (6.5) for

$$\Delta^{1}H_A = 0.78\sigma_I - 1.91\sigma_R^0 \qquad (6.5)$$

$\Delta^{1}H_A$ with a goodness of fit of 18%, and the correlation for the corresponding Z_A values is improved ($f = 22\%$). However, no significant correlation was found for the other series; this arises mainly from the greatly differing shifts found for the electronically similar F and Cl, and similar shifts for the electronically rather different OMe and NMe_2 groups. It seems unreasonable to attempt to improve the fits by omitting further data.

Table 6.3. [1]H Shifts (ppm) for Substituted Ethylenes $CH_2{=}CHX$ Relative to Ethylene[a]

	^{1}H shifts (Ref. 34)			Z Values (Ref. 33)		
X	H_A	H_B	H_C	H_A	H_B	H_C
NR_2				0.69	−1.31	−1.19
OR	1.10	−1.43	−1.29	1.18	−1.28	−1.06
SR				1.00	−0.04	−0.24
F	0.84	−1.30	−0.96	1.03	−1.19	−0.89
Cl	0.97	0.11	0.19	1.00	0.03	0.19
Br	1.16	0.70	0.55	1.04	0.55	0.40
R	0.34	−0.56	−0.48	0.44	−0.29	−0.26
H	0.00	0.00	0.00	0.00	0.00	0.00
CN	0.20	0.72	0.58	0.23	0.58	0.78
CO_2R	0.66			0.84	0.56	1.15

[a] H_A is *geminal* to the substituent, H_B is *trans*, and H_C is *cis*.

The ^{1}H shifts can also be compared directly with the calculated q_H and q_C values (Table 6.1). The only satisfactory relationship found was between ^{1}H chemical shifts for H_B and H_C and the π density at the corresponding carbon atom. This is shown in Fig. 6.3, using the Z values. The shifts are much less closely related to q_H; an analogous result was found[11] for the *para* ^{1}H chemical shifts in monosubstituted benzenes.

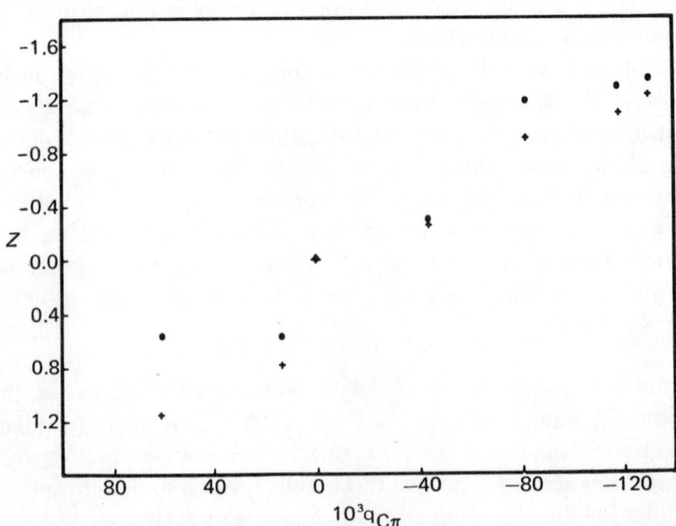

Fig. 6.3. Plot of Z values for *cis*(+) and *trans*(●) protons in vinyl derivatives versus $q_{C\pi}$ at the corresponding carbon atom.

Early workers[34] suggested a relationship between $[\delta_A - \frac{1}{2}(\delta_B + \delta_C)]$ and σ_R, but it is hard to provide an argument for the meaningfulness of this quantity apart from the possible cancellation of some effects of anisotropy.

Some degree of additivity has been observed in polysubstituted ethylenes for which substituent–substituent interactions are not expected,[33,35,36] and this has allowed the calculation of the additive Z values mentioned above. Such calculations should be limited[35] to substituents that neither distort the geometry of the ethylenic group nor interact strongly with one another. Thus the additivity principle gave good estimates of the chemical shifts in various disubstituted ethylenes containing the substituents Cl, Br, Me, or CN, but failed for methyl β-methoxyacrylate. Alternative Z values have been proposed[33,35] for electron-withdrawing substituents when they are conjugated with a strong electron donor. Niwa[28] had earlier included an interaction term in an analysis of results for *trans*-1,2-disubstituted ethylenes, but the method failed badly for simple halogeno-compounds.

On the whole it seems that the shielding mechanisms in proton nmr spectroscopy are too complex for the method to be a generally useful probe for electronic effects in these compounds.

6.4.2. ^{13}C Chemical Shifts

As mentioned in the introduction to Section 6.4, ^{13}C shifts have been widely used as measures of electronic interactions, and several workers have measured and discussed such values for substituted ethylenes and styrenes (see Section 6.10).

^{13}C Chemical shift data from one recent source[37] for a number of monosubstituted ethylenes are collected in Table 6.4. Some other papers also list such shifts.[38,39] The δC_α values parallel those for the corresponding position of attachment in monosubstituted benzenes, while those for C_β parallel those for the *ortho*-carbon atom.[37,38] No simple relationship was claimed for the corresponding proton shifts. It has been suggested that the C_α shifts follow the total charge density, which is mainly determined by σ electronic effects,[37] while the C_β shifts follow q_π as determined by an ω-HMO method including allowance for the sigma bond system, q_{tot} determined by a CNDO/2 calculation,[40] or Hammett sigma values.[35,41] Both shifts are supposed[42] to be a function of e values,† a measure of the polarization of the ethylenic bond.

We have used the data in Table 6.4(A) for DSP correlations, but no significant fit was found for the $^{13}C_\alpha$ shifts. Equation (6.6) represents the

†The e values were derived empirically from data on the copolymerization of vinyl derivatives and, it is claimed, measure the polarity of the double bond [T. Alfrey and C. C. Price, *J. Polymer Sci.*, **2**, 101 (1947)].

Table 6.4. ^{13}*C Shifts for Some Monosubstituted Ethylenes (Shifts in ppm from Ethylene, Positive is Downfield)*

Subst.	$\Delta\delta_\alpha$	$\Delta\delta_\beta$	Subst.	$\Delta\delta_\alpha$	$\Delta\delta_\beta$
		(*A*) (*From Ref.* 35)			
OMe	30.5	−38.7	COMe	14.0	6.2
OAc	18.4	−26.7	CO_2Me	5.7	6.9
Cl	2.8	−6.1	CHO	15.3	14.3
Br	−9.0	−0.9	CN	−15.1	14.7
Me	12.9	−7.4	NO_2	22.3	−0.9
CH_2Cl	11.7	−4.7			
		(*B*) *Other values for use with Table* 6.5			
I[a]	−38.4	6.7	CO_2Et[a]	6.0	6.7
Bu^n [b]	15.0	−9.5	CO_2H[b]	6.0	10.0
Et[b]	17.0	−10.4			

[a] Ref. 38.
[b] J. R. Scherer and W. J. Potts, *J. Chem. Phys.*, **31**, 1691 (1959).

correlation of $^{13}C_\beta$ shifts in monosubstituted ethylenes with $\sigma_R(BA)$ ($f =$ 23%). Substituents of low symmetry (CO_2R, COMe, OMe) are the least

$$\Delta\delta\,^{13}C_\beta = 11.9\sigma_I + 63.5\sigma_R(BA) \tag{6.6}$$

well fitted. The relative influence of inductive and resonance effects is similar to that found[11] for *para*-^{13}C shifts in monosubstituted benzenes.

A considerable amount of information is also available on the ^{13}C shifts of disubstituted ethylenes. The shifts in various methyl-substituted enones and oximes[43] and *trans*-1-substituted butadienes[44] have been correlated with electron density; for the last compounds the δ- and γ-carbon chemical shifts have been correlated with Hammett σ constants, with slopes of opposite sign in accordance with the usual picture of an alternating π charge along a conjugated chain.

The ^{13}C chemical shifts of the vinylic carbon nuclei in polysubstituted alkenes,[39,45] acrylic acids[45–48] and esters,[47] and α,β-unsaturated aldehydes and ketones[49] are related additively to the chemical shifts in the monosubstituted compounds. Sufficient data are now available to allow a general test of the additivity of these shifts for *cis*- and *trans*-1,2-disubstituted ethylenes. Data for these compounds are collected in Table 6.5; unfortunately there are data for only a limited number of 1,1-disubstituted isomers.[39] The degree of additivity observed for the *cis* and *trans* isomers is presented graphically in Fig. 6.4. Theoretically, a pair-wise rather than a direct additivity is probably more justifiable;[50] nevertheless, a significant degree of additivity in the electronic effects of the substituents is indicated,

Table 6.5. ^{13}C *Magnetic Resonance Data*[a] *for cis- and trans-1,2-Disubstituted Ethylenes*

X	Y	*cis* XHC=CHY		*trans* XHC=CHY		Calc. values[b] XHC=CHY		Ref.
OMe	CO_2Me	160.9	95.1	163.3	95.7	159.8	89.5	47
Cl	Cl	121.3	121.3	119.4	119.4	120.0	120.0	52
Cl	Me	120.3	126.7	117.7	129.4	118.6	130.2	46
Cl	CO_2Me	133.3	121.5	135.7	125.0	132.6	122.7	47
Cl	CO_2H	133.2	121.4	137.0	125.2	136.4	123.7	47
Br	Br	116.4	116.4	109.4	109.4	114.2	114.2	52
Br	Me	111.4	130.9	105.7	134.7	108.1	134.9	46
Br	CO_2Me	121.4	124.1	127.0	128.4	122.4	127.4	47
Br	CO_2H	129.1	124.7	128.3	122.0	125.9	128.4	47
I	I	96.5	96.5	79.4	79.4	92.4	92.4	52
I	Bu^n	84.2	142.2	76.6	147.7	76.2	145.8	46
I	CO_2Me	95.9	129.9	100.0	136.2	91.9	135.8	47
I	CO_2H	97.4	130.6	100.3	137.4	95.7	136.8	47
Et	Et	131.2	131.2	131.3	131.3	130.7	130.7	*c*
Me	Me	123.3	123.3	124.5	124.5	128.8	128.8	39
Me	CN	151.5	102.7	151.3	102.6	151.2	100.8	46
Me	CO_2Me	144.5	120.4	144.1	122.3	142.8	121.3	*c*
Me	CO_2H	146.2	121.0	146.0	122.8	146.6	122.3	47
CN	CN	120.8	120.8	120.2	120.2	123.2	123.2	52
CN	CO_2Me	111.7	138.3[d]	—	—	114.8	143.7	47
CO_2Me	CO_2Me	128.7	128.7	132.4	132.4	135.3	135.3	52
CO_2Et	CO_2Et	130.5	130.5	134.1	134.1	136.8	136.8	52

[a] Shifts quoted downfield from tetramethylsilane.
[b] Values calculated from the monosubstituted compound data in Table 6.4 plus 123.7, the chemical shift of the vinyl carbon atoms in ethylene itself [R. Ditchfield, D. P. Miller, and J. A. Pople, *Chem. Phys. Lett.*, **6**, 573 (1970)].
[c] R. A. Friedel and H. L. Retcofsky, *J. Amer. Chem. Soc.*, **85**, 1300 (1963).
[d] The assignment of the original reference has been reversed on the basis of the correlation presented here.

at least insofar as they determine the ^{13}C chemical shifts. The results of a statistical analysis of these data are given in Table 6.6.

The chemical shift difference between the *cis* and the *trans* isomer has been discussed by Savitsky and his co-workers.[46,51,52] In the symmetrically disubstituted compounds the electronic perturbation seemed[51] to be smaller in the *cis* series, an effect attributed[51] to steric inhibition of resonance. Later, when data for compounds carrying substituents of different types became available, exactly the opposite appeared[46] to be true. To explain this Savitsky *et al.*[46] suggested that steric repulsion between groups in the *cis* configuration leads to a weaker central bond, with a consequent stabilization of the polar structures characteristic of electronic interaction between the substituents.

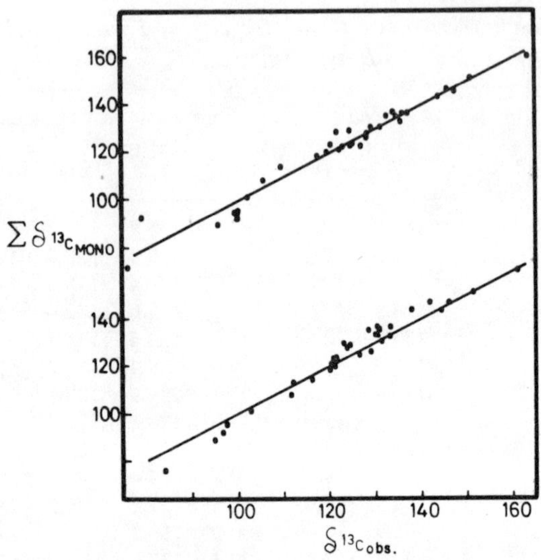

Fig. 6.4. Additivity of ^{13}C vinyl carbon chemical shifts in *cis*- (lower plot) and *trans*- (upper plot) 1,2-disubstituted ethylenes (ppm from TMS).

Whether these secondary effects can in fact be interpreted in terms of a differing electron delocalization in the two series is not entirely clear. It has, however, been established that their origin lies neither in long-range anisotropic effects[51] nor in differential solvation of the two isomers,[52] and it has recently been suggested[48,52] that a theoretical study of the mean excitation energies may be fruitful.

6.4.3. ^{19}F Chemical Shifts

In view of the extensive application of ^{19}F chemical shifts to the elucidation of substituent effects in substituted fluorobenzenes, notably by Taft and his co-workers,[53,54] it is unfortunate that studies have been limited in ethylenic compounds by synthetic and manipulative problems. A

Table 6.6. Statistical Analysis of Data in Table 6.5

Configuration	Standard deviation	Slope of regression line[a] of observed on calc. shifts	Number of points
cis	3.78	$1.12 \pm 0.07^b (\pm 0.09^c)$	44
trans	4.08	$0.93 \pm 0.07^b (\pm 0.09^c)$	42

[a] Regression line constrained to pass through the ethylene point; $\delta = 123.7$.
[b] 95% confidence level.
[c] 99% confidence level.

limited correlation of the average ^{19}F chemical shifts with ionization potentials has been reported[55] for compounds of structure $F_2C{:}CX_2$, where X was usually a halogen. Scanty data are also available in two general reviews.[56] Electron densities in *trans*-substituted fluoroethylenes have also been calculated.[13]

6.4.4. Coupling Constants

Various authors [29,34,57–59] have reported that all three J_{HH} values for monosubstituted ethylenes ($J_{H_AH_B} = J_{cis}$; $J_{H_AH_C} = J_{trans}$; $J_{H_BH_C} = J_{gem}$) are related to substituent electronegativity. The vicinal coupling constants in ethylene itself are claimed[58] to depend much more on σ-electron terms than was previously appreciated. Table 6.7 lists values of all three proton–proton coupling constants for vinyl derivatives taken from one source.[57] Also included are values of $J_{C_\alpha H_A}$ which, it has been suggested, follow electronegativity[60] or J_{CH} values in substituted methanes.[37] Recently[61] these J_{CH} values have been reported to follow better substituent σ_I values if data for resonance-withdrawing substituents are omitted. Values of $J_{C_\beta H_B}$ or $J_{C_\beta H_C}$ vary much less and are apparently affected by angular and direct interaction effects.[61]

A DSP analysis of the figures in Table 6.7 leads to equations (6.7)–(6.10) as best correlations. The J_{HH} values give moderate correlations

$$\Delta J_{H_AH_B} = -4.4\sigma_I + 9.3\sigma_R^0 \quad (f, 22\%) \tag{6.7}$$

$$\Delta J_{H_AH_C} = -5.8\sigma_I + 8.0\sigma_R^0 \quad (f, 18\%) \tag{6.8}$$

Table 6.7. Coupling Constants for Monosubstituted Ethylenes (in Hz)

Subst.	$J_{H_AH_B}$[a]	$J_{H_AH_C}$[a]	$J_{H_BH_C}$[a]	$J_{C_\alpha H_A}$[b]
OR	7.0	14.3	−1.8	182
OAc				189
F	4.7	12.7	−3.1	202[c]
Cl	7.2	14.8	−1.3	197
Br	7.3	15.1	−1.6	198
H	11.6	19.1	2.5	156.3
CN	11.7	17.9	1.2	179
CO_2H	10.4	17.2	1.3	
CO_2Me				168
COMe				164
R				150[c]
NR_2				162[c]

[a] Ref. 57.
[b] Ref. 37.
[c] From figures in Ref. 61.

$$\Delta J_{\mathrm{H_BH_C}} = -5.1\sigma_I + 7.2\sigma_R^0 \quad (f, 14.1\%) \tag{6.9}$$

$$\Delta J_{\mathrm{C_\alpha H}} = 65.6\sigma_I - 23.8\sigma_R^- \quad (f, 16.8\%) \tag{6.10}$$

depending on both inductive and resonance effects, while $J_{\mathrm{C_\alpha H}}$ changes are mainly the result of inductive effects. However, the data set is limited for the J_{HH} series. Some work has also been reported[62] on J_{CH} long-range coupling constants, and these appear to be related to J_{HH} and J_{CH} values.

6.5. Infrared Frequencies

The ir spectra of substituted ethylenes have been well studied as regards frequency measurements.[63,64] Thus, the stretching frequency of the carbon–carbon double bond at approximately 1640 cm^{-1} is altered by both the degree and geometry of substitution in the ethylene and the electrical and steric nature of the substituents. This frequency has been used mainly to diagnose the number and geometrical arrangement of substituents attached to the double bond, particularly in hydrocarbons. It is important[64,65] to consider sets that involve minimum variations in bond angles and substituent masses if the results are to reflect charge distribution effects. In such cases, the frequency has been related[64] to the electronegativity of the substituent. Variations of ν_{OH} in compounds in which a hydroxy group is hydrogen bonded to ethylene π electrons also depend on the nature of the substituents in the ethylene.[64] However, the vinylic CH out-of-plane deformation vibrations in the range 700–1000 cm^{-1} are probably the most valuable[64,66,67] as a probe for electronic effects of substituents.

CH Out-of-Plane Vibrations. Bellamy found[68] that these frequencies depended on the electronic nature of substituents and in a comprehensive study Potts and Nyquist reported[69] a strong dependence of the CH_2 wag frequency on the resonance effect of the substituent in a series of vinyl compounds; at the time adequate π-delocalization parameters were not available to test this thoroughly.

The frequency of this vibration in mono- and 1,1-di-substituted ethylenes is virtually insensitive to the mass effect of the substituent,[70] although some dependence on solvent has been demonstrated.[66] These frequencies are well calculated theoretically from CNDO/2 molecular orbital theory by using a single set of group vibrational amplitudes.[71] The observed frequencies were also found to be correlated closely with the total CNDO/2 electron density in the $=CH_2$ fragment, over a range of almost 300 cm^{-1}. When the π-electron density alone was considered, separate lines were required to correlate donor- and acceptor-substituted compounds.[71]

Russian workers[72] have attempted to study polar effects as measured by σ^*. They considered 15 compounds of the type $H_2C{:}CHCH_2X$ and obtained a rectilinear plot of the frequency ν, against σ^* for the substituent CH_2X. This was said to represent the polar effect of the substituent on the vibration; discrepancies, $\Delta\nu$, for other substituents were then considered to be a measure of the pure resonance effect and plotted against σ_R.

However, substituents of the type CH_2X show varying hyperconjugative interactions with unsaturated systems, depending on the polarity of X.[73,74] Thus a dual substituent-parameter analysis of all the data is more appropriate and by use of σ^* and σ_R^0 the data in Table 6.8 lead to equation (6.11).

$$\nu = 195\sigma_R^0 - 0.6\sigma^* + 931 \tag{6.11}$$

It is thus obvious that ν depends largely on resonance effects (including those of CH_2X groups) and the plot of ν vs. σ_R^0 is shown in Fig. 6.5 (correlation coefficient 0.944, 25 points). The polar effect (σ^*) therefore does not seem important in determining these frequencies. Our usual DSP treatment leads to equation (6.12), the fit here with σ_I and σ_R^+ ($f = 9\%$) being somewhat better than with σ_I and σ_R^0.

Table 6.8. CH Out-of-Plane Deformation Frequencies for Monosubstituted Ethylenes[a]

	ν (cm^{-1})			ν (cm^{-1})	
Subst.	twist	wag	Subst.	twist	wag
OMe	960	813	CH_2OH	987	915
OPh	944	851	CH_2OEt	991	921
OAc	950	873	CH_2Cl	983	929
F	932[b]	863	CH_2Br	981	924
Cl	938	894	H	987[c]	944[d]
Br	936	898	$CHCl_2$	975	937
I	943	905	$SiCl_3$	994	975
Bu^t	999	910	CF_3	979	965
Et	990		CO_2Me	985	964
Me	986	908	CN	960	960
Ph	989	906	COMe	989	959
$CH{:}CH_2$	1011	907	CHO	984[e]	963[e]
CH_2CN	982	926	CO_2H	982	970

[a] Measured in CS_2; from Ref. 65 unless otherwise indicated.
[b] From J. R. Scherer and W. J. Potts, *J. Chem. Phys.*, **31**, 1691 (1959).
[c] Data from *trans*-1,2-dideuterioethylene, R. L. Arnett and B. L. Crawford, *J. Chem. Phys.*, **18**, 118 (1950).
[d] Data from 1,1-dideuterioethylene, B. L. Crawford, J. E. Lancaster, and R. G. Inskeep, *J. Chem. Phys.*, **21**, 678 (1953).
[e] Average of values for s-*cis* and s-*trans* forms.

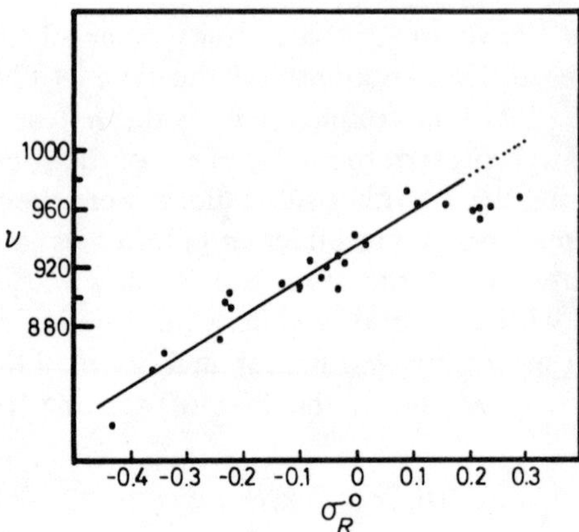

Fig. 6.5. Plot of frequency (ν in cm^{-1}) of the CH_2 wag vibration of monosubstituted ethylenes versus σ_R^0 value of the substituent.

The frequency of the CH twisting vibration in monosubstituted ethylenes is said[64] to be dominated by polar effects as measured by free energy

$$\Delta\nu_{\mathrm{CH\,wag}} = -1.5\sigma_I + 133\sigma_R^+ \tag{6.12}$$

parameters (Potts and Nyquist[69] actually used the dissociation constants of substituted acetic acids). Indeed, the data for the compounds originally studied are rather well correlated with σ^*. However, consideration of the more extensive data now available (Table 6.8) shows this correlation to be rather limited; linear regression gives a correlation coefficient of only 0.730 for 21 points.

The forms of the wag and twist vibrations in monosubstituted ethylenes are closely similar[71] to the analogous vibrations in 1,1- and *trans*-1,2-disubstituted ethylenes, respectively. It is therefore of interest to see whether the frequency shifts for the disubstituted compounds are related additively to those for the monosubstituted compounds, since deviation from additivity of the substituent electronic effects might be evident here.

The frequencies were therefore fitted to equation (6.13), the constant being chosen to minimize the sum of the squares of the deviations. In this

$$\nu(\mathrm{X, Y}) = \nu(\mathrm{X}) + \nu(\mathrm{Y}) - \mathrm{const} \tag{6.13}$$

equation ν(X, Y) is the observed frequency for the disubstituted compound and ν(X) and ν(Y) are the frequencies of the same vibration for the corresponding monosubstituted compounds. The results of this procedure are given in Table 6.9 and plotted in Fig. 6.6. The frequencies of the $=CH_2$ wag vibration are evidently rather accurately additive, suggesting a high level of additivity in the substituent electronic interaction, in this case, the

Table 6.9. CH Out-of-Plane Deformation Frequencies in 1,1- and trans 1,2-Disubstituted Ethylenes

Substituents		Twist (H–C=C–H, trans) obs.[a]	Twist calc.[b]	Wag (H_2C=) obs.[a]	Wag calc.[c]
OEt	OEt	—	—	711	699
OEt	Cl	—	—	787[d]	779
OEt	Me	—	—	795	793
OEt	CN	—	—	850[d]	845
Cl	Cl	892	881	867	859
Cl	CH_2Cl	931	928	891	894
Cl	Me	926	931	875	873
Cl	Ph	—	—	877	871
Cl	CN	920	905	916	925
Cl	CO_2H	—	—	933	935
Cl	CO_2R	—	—	925	926
Br	Br	896	879	877	867
Br	CH_2Br	935	924	896	893
Br	CF_3	—	—	929	934
F	F	872[e]	871[f]	804	797
CH_2Cl	Me	—	—	902	908
CH_2Cl	Ph	—	—	907	906
Me	Me	964	979	887	887
Me	Ph	959	982	—	—
Me	CN	953	953	930	939
Me	CO_2H	966	975	947	949
Me	CO_2R	968	975	939	940
Ph	Ph	958	985	—	—
Ph	CN	962	956	—	—
Ph	CO_2H	976	978	—	—
Ph	CO_2R	976	978	—	—
CF_3	CN	—	—	997[d]	996
CN	CN	945[g]	927	983[h]	991
CO_2H	CO_2H	983	971	—	—
CO_2R	CO_2R	976	971	—	—

[a] Data from Ref. 65 (measured in CS_2 solution).
[b] Calculated from data in Table 6.8 and equation (6.13), using a constant of 993.
[c] As in *b*, using a constant of 929.
[d] Approximate data from work of the authors (measured as a liquid film).
[e] From Ref. 68.
[f] Vapor phase measurement.
[g] From F. A. Miller, *Spectrochim. Acta*, **20**, 1233 (1964).
[h] From A. Rosenberg and J. P. Devlin, *Spectrochim. Acta*, **21**, 1613 (1965).

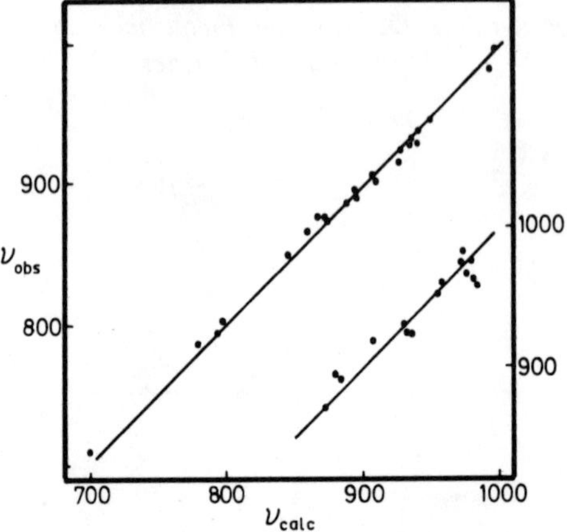

Fig. 6.6. Additivity of out-of-plane deformation frequencies (ν in cm^{-1}) for disubstituted ethylenes. Upper plot: CH_2 wag in 1,1-disubstituted ethylenes. Lower plot: HCCH twist in *trans*-1,2-disubstituted ethylenes.

resonance effect of geminal substituents. The slope of the regression line is 0.94 for 23 points, with a standard deviation of 6.2 cm^{-1}.

The relationship for the twisting vibration in *trans*-1,2-disubstituted ethylenes, for which the frequency range is smaller, is much less precise, and only very limited conclusions can therefore be reached as to the substituent electronic effect here. The slope of the regression line is 0.80 for 19 points, with a standard deviation of 13.2 cm^{-1}.

6.6. *Infrared Intensities*

Measurements of ir intensities for vibrations in several series of substituted ethylenes are restricted[16,17,74–77] to investigations of the C=C stretching mode near 1640 cm^{-1}.

6.6.1. *Monosubstituted Ethylenes*

Katritzky and Topsom and their co-workers[78,79] had shown earlier that the integrated intensity A of the ν_8 mode near 1600 cm^{-1} of monosubstituted benzenes is related to the σ_R^0 value of the substituent by equation (6.14) (for summaries of work in the benzene series see Refs. 23 and 67).

$$A = 17{,}600(\sigma_R^0)^2 + 100 \tag{6.14}$$

This was rationalized in terms of resonance theory[79] and it was shown[80] that rather accurate calculations of this intensity can be made within the framework of CNDO/2 molecular orbital theory. There is considerable formal similarity between the $\nu_{C=C}$ vibration near 1640 cm^{-1} in ethylenes and the ν_8 vibration in benzenes. Both are planar vibrations involving motion of π-bonded carbon atoms, with little motion associated with the substituent. Both vibrations are forbidden in the parent compound and derive their intensity from the influence of the substituent, yet the precise form of the vibration remains essentially independent of the nature of the substituent.[81] Indeed it was demonstrated[74] that a relation analogous to equation (6.14) held for monosubstituted ethylenes. Thus a least-squares treatment of the data for 18 compounds led to equation (6.15) with

$$A = 27{,}100(\sigma_R^0)^2 + 80 \tag{6.15}$$

a correlation coefficient of 0.998. This relationship is illustrated in Fig. 6.7 for the data in Table 6.10, where $(A-80)^{1/2}$ has been plotted against substituent σ_R^0 values inferred from those for the corresponding benzene compounds. The data conform closely to a straight line with the exception of the values for CN, CHO, and COMe. Sufficient intensity values are not available for substituents in our DSP program to establish a meaningful equation. The intensities have also been calculated[81] by using CNDO theory (see Section 6.6.3).

The good correlation found in Fig. 6.7 together with the known[11,13] proportionality of π-electron charge transfer to σ_R^0 values in monosubstituted benzenes shows that, as for the benzenes,[82] the $A^{1/2}$ values here depend almost solely on such π-electron charge effects.

Some insight into the reason for this parallel between π-electronic effects and the intensity of this band may, as in the benzene series,[79] be obtained in terms of resonance theory. The form of this vibration for a

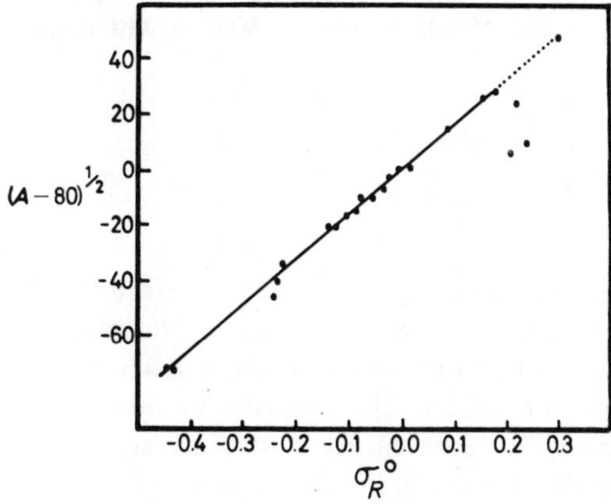

Fig. 6.7. Plot of $(A-80)^{1/2}$, where A is intensity of the $\nu_{C=C}$ band of monosubstituted ethylenes (liter mol^{-1} cm^{-2}) vs. σ_R^0 values.

Table 6.10. Intensity of the $\nu_{C=C}$ Vibration of Monosubstituted Ethylenes (A in liter mol^{-1} cm^{-2})

Subst.	$A^{1/2\,a}$	$\sigma_R^{0\,b}$	Subst.	$A^{1/2\,a}$	$\sigma_R^{0\,b}$
OEt	71.2	−0.44	CH_2Cl	6.6	−0.03[d]
OBu^n	72.0	−0.43[c]	CH_2Br	2.6	−0.02
OAc	45.1	−0.24	H	0.0	0.00
Br	39.5	−0.23	$CHCl_2$	1.0	0.02
I	33.6	−0.22	$SiCl_3$	14.9	0.09[c]
Bu^t	20.8	−0.13	CO_2Me	25.6	0.16
Bu^n	20.6	−0.12	CO_2Et	28.4	0.18
Ph	16.1	−0.10	CN	6.1	0.21[d]
CH_2CN	15.2	−0.08[d]	COMe	24.3	0.22
CH_2OH	9.8	−0.07[d]	CHO	9.8	0.24
CH_2OEt	10.2	−0.05[d]	CO_2H	47.2	0.29

[a] Corrected for an overtone contribution of 80 units.
[b] From Ref. 79 except where shown.
[c] Ref. 74.
[d] Ref. 54.

typical monosubstituted ethylene is shown in (V). This form is believed to change very little with the nature of the substituent.[81]

(V)

In the language of resonance theory the overall π-electron distribution arises from some weighted average of contributing canonical forms, the most important of which are in (VI), in which the alternative

$$\overset{\mp}{C}H_2-C(H)=\overset{\pm}{X} \leftrightarrow CH_2=CH-X$$

(VI)

signs refer to π-electron donating and accepting groups. The vibration involves principally a changing C=C bond length. Thus the polar structure in (VI) will be stabilized in that phase of vibration in which the C=C bond is lengthened, and destabilized in the opposite phase. An oscillating π-dipole moment due to the reorganization of π charge can therefore be envisaged. Assuming that bond polarity due to the substituent is essentially attenuated beyond the C—X bond (in accordance with current thinking on the inductive vs. the direct field mechanism[3,83]), substituent inductive effects will contribute to the band intensity only insofar as there is any contribution to the oscillating dipole due to the vibrational motion of the C—X bond. The form of the normal vibration does involve a small change in the length of this bond, although evidently it is so small that substituent polar effects are not manifest.

Application of Equation (6.15) to the Calculation of σ_R^0 *Values.* With the exception of the substituents noted above, the resonance effect operating in these vinylic systems closely parallels that in the corresponding substituted benzenes. In particular, σ_R^0 values deduced by use of equation (6.15) agreed well[74] with those inferred from the intensity of the ν_8 band in substituted benzenes[79] [equation (6.14)], even when discrepancies with ^{19}F values were previously noticed,[79] for example with the *d*-orbital acceptors Br, I, and SMe, and the donor–acceptor group NCO.

The σ_R^0 values deduced by use of equation (6.15) seem to be more reliable than the benzene ir values for weakly interacting substituents, since the relative importance of the overtone correction is much less.[74] Assuming a value of 80 ± 40 for this gives the following error ranges at the values of σ_R^0 quoted: 0.1 ± 0.005, 0.07 ± 0.009, 0.05 ± 0.015. The corresponding errors for the determination from monosubstituted benzenes, assuming an overtone correction of 100 ± 50 are as follows: 0.1 ± 0.015, 0.07 ± 0.025, 0.05 ± 0.05. We believe therefore that equation (6.15) is of unique value in the determination of small σ_R^0 values.

Some work has been done on γ- and δ-substituted ethylenic systems. The $A^{1/2}$ values for compounds of the type $CH_2{:}CH(CH_2)_nX$ followed the hyperconjugative ability of the substituent for $n = 1$, as expected from the discussion above, but had no significant dependence on σ^* for $n = 2$ or 3. This agrees with $A^{1/2}$ being dependent mainly on the charge-transfer ability of the substituent, which would be expected to become almost constant for $n > 1$. Work on 3- and 4-cyano-1-methoxy- and -1-chlorocyclohexene suggested[76] that the field effect of the cyano group could increase the resonance interaction between the 1-substituent and the olefinic linkage, and this finding is being further investigated[84] by ^{13}C nmr and Raman spectroscopy.

6.6.2. *Disubstituted Ethylenes*

The intensity of an ir band is proportional to the square of the quantity $\partial\mu/\partial Q$,[85] i.e., $A = \text{const} \times (\partial\mu/\partial Q)^2$, where $\partial\mu/\partial Q$ is the rate of change of dipole moment with the normal coordinate, a scalar quantity describing the progress of the molecule along a given vibrational path corresponding to one of the normal modes.

The dipole transition moment, $\partial\mu/\partial Q$, like the permanent dipole moment, is a vector quantity, and its direction can usually be qualitatively deduced from the simple valence bond picture described above, and in some cases (although not for substituted ethylenes) it is rigorously defined by symmetry.

Katritzky, Topsom, and their co-workers were able to correlate the intensities of bands for several multiply substituted systems, including *meta*-[86] and *para*-disubstituted benzenes[87,88] and disubstituted acetylenes,[89] by treating the dipole transition moment as the vector sum of those induced in the corresponding monosubstituted compounds.

These ideas were easily generalized[16] and applied[16,17,90] to more complex systems for which the angles between contributing vectors had to be found empirically. It was shown[16,17] that equation (6.16) applied to

$$A = a[\sigma_R^0(1)^2 + \sigma_R^0(2)^2 + 2\sigma_R^0(1)\sigma_R^0(2)\cos\theta] + c \qquad (6.16)$$

compounds carrying two substituents in equivalent positions and in the absence of substituent interaction, where a and c are constants, $\sigma_R^0(1)$ and $\sigma_R^0(2)$ refer to the σ_R^0 values of the two substituents 1 and 2, and θ refers to the angle between their dipole transition moments.

1,1-Disubstituted Ethylenes. The form of the $\nu_{C=C}$ vibration in a 1,1-disubstituted ethylene is shown in (VII). Comparison of the L matrix

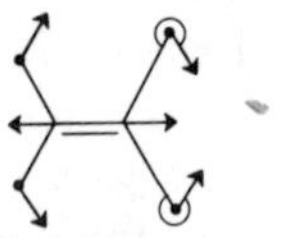

(VII)

elements for vinyl and vinylidene chloride[91] revealed[16] almost identical internal coordinate changes associated with the heavy-atom skeleton in the two systems.

According to the reasoning given above, the intensity of this vibration arises as a consequence of the changing weights of the contributing canonical forms (VIII)–(X) during the vibration, resulting in an oscillating

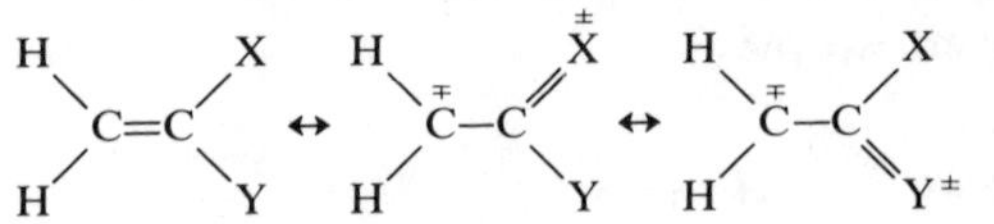

$H_2C{=}CXY \leftrightarrow H_2\overset{\mp}{C}{-}C(Y){=}\overset{\pm}{X} \leftrightarrow H_2\overset{\mp}{C}{-}C(X){=}Y^{\pm}$

(VIII) (IX) (X)

dipole moment. This dipole moment is the result of the superposition of two vectors, associated with the interaction of each of the substituents and the double bond. Unfortunately, in this case there is no way of obtaining *a priori* the angle between the vectors associated with the two substituents and it was necessary to assume that an average angle could be found applicable to the entire series. Empirically this angle was found to be close to 90°. The cosine term could therefore be neglected and the best fit gave equation (6.17). This relation is shown in Fig. 6.8. It is difficult to interpret

$$A = 24{,}100[\sigma_R^0(1)^2 + \sigma_R^0(2)^2] + 60 \qquad (6.17)$$

directly and rigorously the additivity of these intensities in terms of electron densities, but it is expected[16] insofar as they are dominated by π-electron terms which are themselves additive.

The precise value of the slope in equation (6.17) will be sensitive to small changes in the form of the vibration and to the assumption of an average angle between the contributing dipole vectors as well as to electronic effects. However, within these limits of uncertainty the combined resonance effects of geminal substituents appear to be additive. This conclusion is in accord with that reached from a consideration of the CH_2 wag frequency (Section 6.5) and CNDO/2 MO calculations (Section 6.2) on these compounds.

trans-1,2-Disubstituted Ethylenes. The form of the $\nu_{C=C}$ vibration in these compounds is shown in (XI) and again the motion of the heavy-atom skeleton is found[17,92] to be essentially identical with that of the monosubstituted compound.

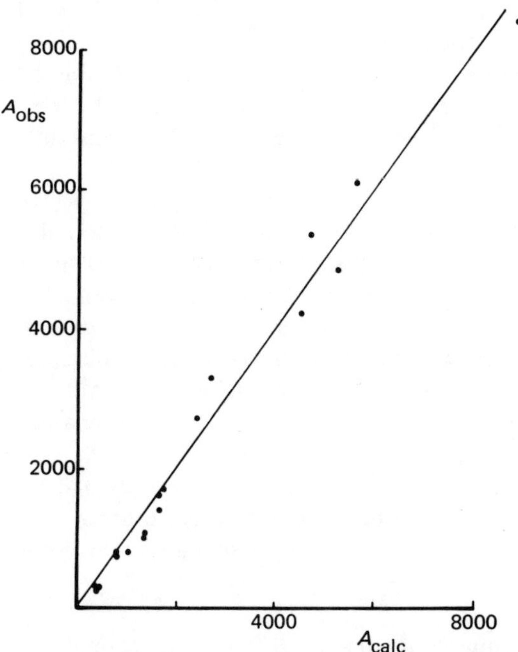

Fig. 6.8. Plot of A (liter mol^{-1} cm^{-2}) for 1,1-disubstituted ethylenes against $\sigma_R^0(1)^2 + \sigma_R^0(2)^2$; the line has the slope and intercept of equation (6.17)—see text.

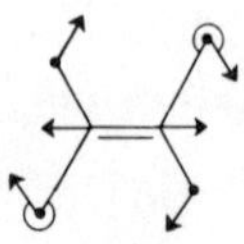

(XI)

The observed intensities can be understood in terms of the changing weights of the polar resonance structures (XII)–(XIV) during the vibration, including the through-resonance structure (XIV), which has no analog in

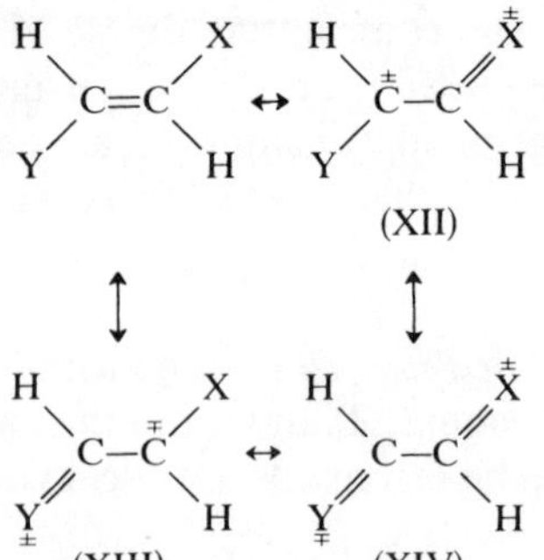

(XII)

(XIII) (XIV)

the 1,1-disubstituted ethylene series. When the substituents are identical and symmetrical, the molecule has C_{2h} symmetry and the vectors describing the interaction of each substituent with the double bond must necessarily add algebraically at 180° intervector angle. In the general case the angle θ [equation (6.16)] may differ somewhat from 180°, although large deviations are not expected.

Just as the changing weight of the polar structures characteristic of the interaction between the substituent and the double bond leads to an oscillating dipole moment, the structures characteristic of the interaction between substituents will behave analogously. In order to account for the enhancement of the $\nu_{C=C}$ intensities, equation (6.16) is modified[17] by the inclusion of a term λ [although equation (6.18) as derived involves certain simplifying assumptions,[17] including θ being taken as 180° throughout]. Here λ is an interaction constant (in the units of σ_R^0) characteristic of a given pair of substituents.

Previously[88] a series of analogous λ values had been derived appropriate to *para*-disubstituted benzenes. The ethylene intensities could be correlated[17] by equation (6.18), by assuming that the conjugation between

$$A = a[\sigma_R^0(1) - \sigma_R^0(2) + \lambda]^2 + c \tag{6.18}$$

a pair of substituents in this series was proportional to that in the corresponding disubstituted benzene. Least-squares analysis of the data for 15 compounds led to equation (6.19). Here the correction for background absorbance was not significant. Figure 6.9 shows a plot of these data, including the calculated intensities on the basis of $\lambda = 0$ (open circles). The

$$A = 26{,}500[\sigma_R^0(1) - \sigma_R^0(2) + 1.5\lambda]^2 \qquad (6.19)$$

displacement of the points from the line for a given compound illustrates graphically the contribution made by the conjugation term. In the most extreme case of the strong donor NMe_2 coupled with the strong acceptor CO_2Me, this accounts for half the observed intensity.

The coefficient of λ in equation (6.19) demonstrates that (on this criterion) the mutual conjugation present in *trans*-disubstituted ethylenes is *ca* 1.5 times as extensive as that in the *para*-disubstituted benzene series. This may reflect the greater importance of through-conjugated canonical forms here, compared to *para*-disubstituted benzenes, since the similar charge transfer is less dispersed in monosubstituted ethylenes than in monosubstituted benzenes (Section 6.2).

In fact the parallel is not wholly precise. Table 6.11 and Fig. 6.10 compare the quantity λ implied by equation (6.19) with the corresponding data[86] for p-disubstituted benzenes. The line in Fig. 6.10 is that of slope 1.5. As λ is calculated here as a "residual" quantity, it must necessarily

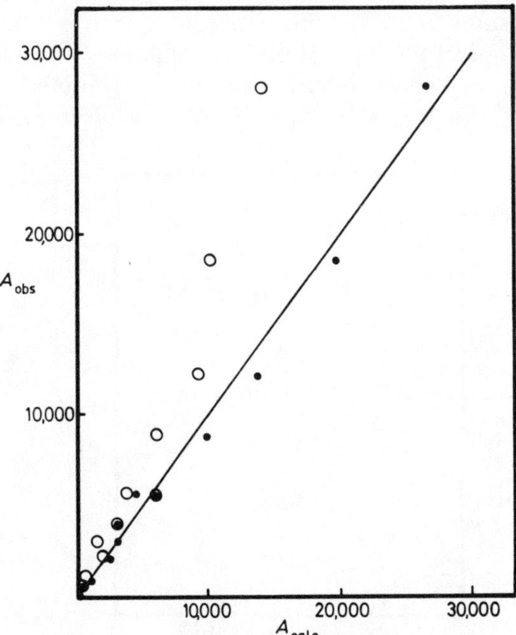

Fig. 6.9. Plot of A values for the intensity of the $\nu_{C=C}$ band in *trans*-1,2-disubstituted ethylenes against those calculated from equation (6.19) with $\lambda \neq 0$ (•), and equation (6.19) with $\lambda = 0$ (○).

Table 6.11. Comparison of λ Values Deduced in trans-1,2-Disubstituted Ethylenes and para-Disubstituted Benzenes

Substituents		λ (Et)[a]	λ (Ph)[b]
D	A		
NEt_2	CN	−0.22	−0.16
NEt_2	CO_2Me	−0.31	−0.18
OEt	Cl	−0.07	−0.09
OEt	CN	−0.11	−0.09
OMe	CO_2Me	−0.10	−0.09
Me	Cl	+0.02	+0.02
Cl	CN	−0.03	−0.01
Cl	CO_2Me	−0.09	−0.02
Me	CN	−0.04	−0.02
Me	CO_2Me	−0.04	−0.02

[a] $\lambda = -[\sigma_R^0(D) - \sigma_R^0(A)] \pm (A/26{,}500)^{1/2}$.
[b] Deduced from the data in Ref. 88 for the substituents NMe_2, Cl, Me, CN, and CO_2Et.

also contain all the uncertainties inherent both in the analysis and the experimental measurement of the intensities.

The overall slope of the correlation line [equation (6.19)] is extremely close to the slope of that correlating the monosubstituted compounds, which is indicative[17] of an almost exactly additive π-electron perturbation

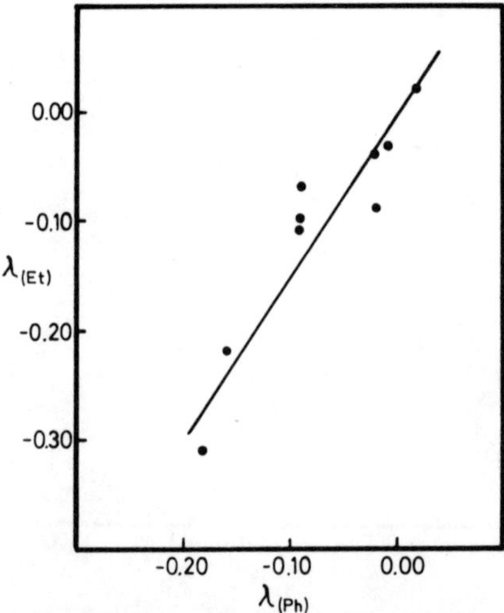

Fig. 6.10. Comparison of substituent interaction parameters (λ) derived for *para*-disubstituted benzenes and for *trans*-1,2-disubstituted ethylenes.

by the substituents. In other words, the substituents appear to exert their normal resonance effects as far as the double bond is concerned, even when there is considerable interaction between the substituents. This accords with the additivity of ^{13}C chemical shifts (6.4.2), which respond mainly to the local electronic environment, i.e., that at the carbon nuclei of the ethylenic skeleton, where π-electron densities behave additively despite the additional transfer of charge *between* the substituents. The additivity of ^{1}H chemical shifts (6.4.1), which are more susceptible to long-range effects, breaks down when this interaction is possible.

cis-1,2-Disubstituted Ethylenes. Ford *et al.*[17] have also reported intensity data for several *cis*-1,2-disubstituted ethylenes. Some correlation of these data with σ_R^0 constants was obtained according to a somewhat complex equation, but insufficient data were available to carry out any complete treatment.

6.6.3. Molecular Orbital Calculations

Although there is no way of comparing directly the calculated CNDO/2 electron densities (Section 6.2) with experiment, comparisons of the closely related dipole moments, μ, or their derivatives, $\partial\mu/\partial Q$, are possible. Experimental values of $\partial\mu/\partial Q$ for the $\nu_{C=C}$ vibration derived from the ir intensities in mono- and the three classes of di-substituted ethylenes are more numerous than values of the dipole moments themselves. This comparison is made in Fig. 6.11. The calculated values of

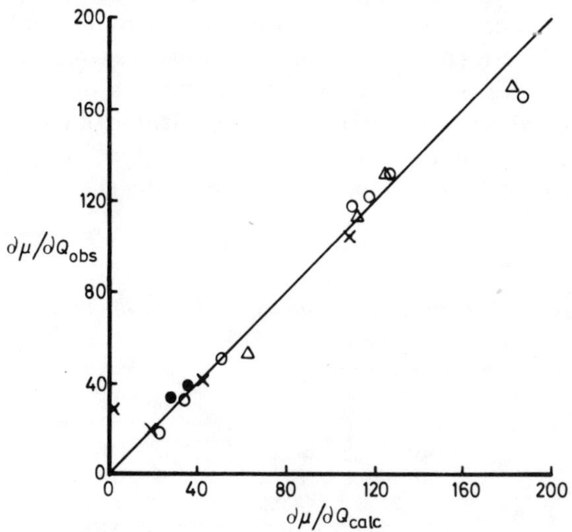

Fig. 6.11. Comparison of observed and calculated values of $\partial\mu/\partial Q$ (esu) for mono- (●), 1,1-di- (○), trans-1,2-di- (△), and *cis*-1,2-di-substituted ethylenes (×).

$\partial\mu/\partial Q$ are based on CNDO/2 wave functions by using forms of the normal vibration deduced from normal coordinate analyses of model compounds.[16,17,81] The close agreement between the calculated and experimental quantities suggests that a high level of reliance may be placed on the calculated CNDO/2 charge distributions.

The only available data excluded from Fig. 6.11 were for compounds containing the C:C·C:O group, for which calculated values of $\partial\mu/\partial Q$ were consistently too large. This discrepancy may well be due to the presence of mechanical coupling between the C=C and C=O oscillators, which was ignored in the calculations, the forms of the vibrations being based on those for simple halogenoethylenes. The line in Fig. 6.11 is the theoretical one of unit slope and not simply the "best" line through the points.

6.7. *Ultraviolet and Raman Spectroscopy. Ionization Potentials*

All three properties dealt with in this section depend on excited-state as well as on ground-state electronic energies, and thus little direct correlation is expected with substituent parameters.

No extensive data are available for the uv absorption of simple substituted ethylenes,[93] presumably partly because of experimental difficulties, and overlap by other absorptions deriving from some of the substituents. The frequencies may well be related to resonance effects,[94] but as for benzene derivatives,[95] they are probably not well correlated by resonance parameters. The intensities of $\pi \rightarrow \pi^*$ absorptions of substituted benzenes, by contrast, were found[95] to be proportional to σ_R^0, and intensity measurements on a series of monosubstituted ethylenes would be of interest. The uv intensities and the Raman intensities (as measured in the preresonance region) depend very strongly on the length of the conjugated system; a vinyl substituent, for example, has a disproportionate effect. The Raman intensities of other simple substituted ethylenes do not alter greatly.[96]

Table 6.12. Ionization Potentials of Monosubstituted Ethylenes[97] (in eV)

Subst.	I.P.	Subst.	I.P.
CN	10.91	Br	9.80
CF_3	10.90	H	10.51
Me	9.73	COMe	9.91
F	10.37	OMe	9.93
Cl	10.00		

Charton[4,97] attempted to correlate ionization potentials for twenty substituted ethylenes by using equation (6.2) but obtained a poor fit. This was improved[4] by the omission of the data for five substituents (H, F, COMe, OAc, vinyl) but there seems little justification for this. A DSP correlation of the data in Table 6.12 failed to give any significant fit. Charton[97] claimed that the resonance contribution was greater than for benzene derivatives, as might be expected in the less extended system.

6.8. *Equilibria and Reactivity*

As already mentioned, this subject has recently been discussed in detail by Charton.[4] The dissociation constants and rates of esterification of *trans*-3-substituted-acrylic acids seem to be quite generally correlated by the extended Hammett equation, resonance and polar effects operating in proportions similar to those for *para*-substituted benzoic acids.[4]

A DSP analysis of the pK_a data in Table 6.13 gives equation (6.20); other series of data are not adequate for correlation. An almost equally

$$\Delta pK_a = -2.06\sigma_I - 2.44\sigma_R^0 \quad (f, 6.8\%) \tag{6.20}$$

excellent correlation (f, 7.2%) is found with σ_I and σ_R(BA). Clearly the substituent effects here closely parallel those in substituted benzoic acids.

The situation for the corresponding *cis*-3-substituted acrylic acid series is more complex. These acids are usually stronger than the analogous *trans* compounds by about 0.3 pK_a units.[98] According to Bowden[99] this is a proximity effect analogous to that observed in *ortho*-substituted benzoic acids. On the other hand, the correlation of pK_a and reactivity data for several series of *cis*-3-substituted acrylic acids led Charton[4] to the conclusion that there were no important proximity effects in these series. The data available do not include a strongly electron-donating substituent and are inadequate for a DSP analysis.

Table 6.13. pK_a Values for trans-3-Substituted-Acrylic Acids[a] (in H_2O at 25°C)

Subst.	pK_a	Subst.	pK_a
OMe	4.85	CF_3	3.15
Cl	3.78	CO_2Et	3.40
Me	4.69	COMe	3.24
H	4.26	NO_2	2.58

[a] From Ref. 4.

Unfortunately the acidities of the disubstituted acrylic acids are probably not a useful probe for the combined *electronic* effects of the two substituents.

Additivity of substituent effects on the dissociation constants of 3,3-disubstituted acrylic acids (XV), for example, requires that equation (6.21)

$$\Delta pK_a = \Delta pK_a^{trans} + \Delta pK_a^{cis} \tag{6.21}$$

X H
C=C
Y CO_2H

(XV)

hold, where ΔpK_a^{trans} and ΔpK_a^{cis} refer to the *trans* and *cis* monosubstituted acrylic acids. However, a graphical investigation of this relationship for some data from the literature[4] (shown in Fig. 6.12) reveals a rather low level of precision. The difficulty with this approach to the study of electronic effects is that nonadditivity of the overall substituent effect may occur quite independently of the electronic behavior of the substituents if the ρ values themselves are not additive. Indeed, it has been suggested[99] that buttressing in these systems may significantly modify the ρ values with respect to those for the monosubstituted compounds.

This difficulty may be partially overcome by the use of a more flexible equation in which the ρ values for the polysubstituted compound are obtained from a correlation analysis in a manner similar to that used in a dual substituent-parameter equation, although the fundamental problem

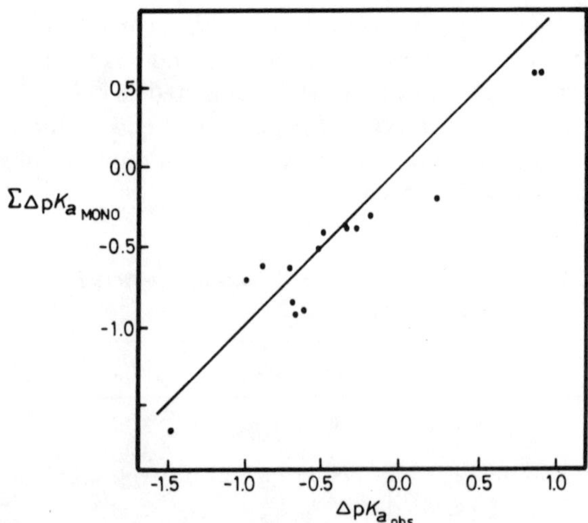

Fig. 6.12. Failure of the additivity of pK_a values in 3,3-disubstituted acrylic acids.

of substituent-dependent reaction constants in these relatively crowded systems remains.

Charton[4] has analyzed the kinetic data for various reactions of disubstituted ethylenes by use of equation (6.22); the results were in general

$$\log k = \rho_I \sum \sigma_I(\mathrm{X}) + \rho_R \sum \sigma_R(\mathrm{X}) + h \qquad (6.22)$$

not good. The rates of radical copolymerization of 2-substituted acrylates and acrylonitriles with styrene,[100] and particularly the electrophilic bromination[101] and chlorination[102] of variously substituted alkenes, have been treated with some success according to similar additive schemes. However, as far as the latter studies are concerned, this success is less real when a larger range of substituent types is considered.[103]

Interestingly, the reactivities of 1,1- and *trans*-1,2-substituted ethylenes as dienophiles were often well correlated[4,104] by equation (6.2), the rates of the Diels–Alder reactions examined increasing with the electron-withdrawing power of the substituents, thus responding predominantly to the resonance effect.

Studies on molecules such as (XVI) would be interesting, to see whether the pK_a values followed σ_R^0 as for substituted phenylacetic acids. Furthermore, while some measurements have been made on 4-substituted 1-carboxybicyclo[2.2.2]oct-2-enes (XVII),[105] no values are yet available for 4-substituted barrylcarboxylic acids (XVIII), in which the considerable

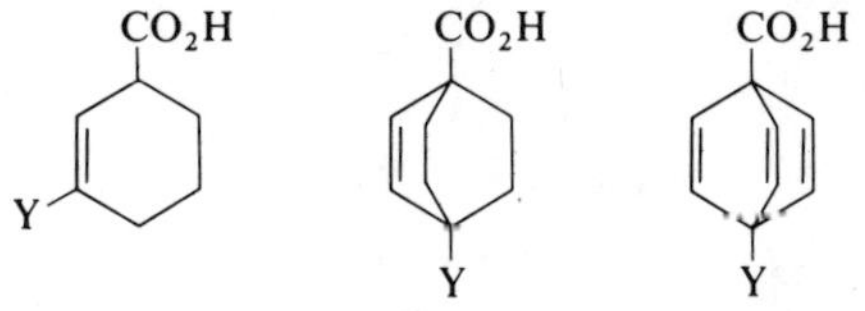

CO_2H CO_2H CO_2H

Y Y Y

(XVI) (XVII) (XVIII)

intermediate π density might significantly affect the transmission of the substituent electronic effect to the carboxy group.

6.9. Rotational Barriers

6.9.1. Rotations about Carbon–Carbon Double Bonds

The barriers hindering *cis–trans* isomerization in these simple conjugated systems form an area where linear free energy relationships should be particularly valuable, both for rationalizing the origin of the barriers and as a probe for substituent electronic effects.

The activation energy for the thermal isomerization of *trans*-1,2-dideuterioethylene[106] is 65 kcal mol^{-1}, but the activation energy is much less for compounds involving substitution by conjugating groups.[107] This effect is due in part to the weakening of the alkenic double bond associated with contributions of polar structures (XII)–(XIV) to the resonance hybrid (see Section 6.6.2). Indeed, some degree of correlation has been observed[107] between the activation energies for the isomerization of a number of compounds of this type and the heats of hydrogenation of the double bond.

When the double bond is more heavily substituted with electron-donating or -accepting substituents in a "push–pull" arrangement giving the so-called polarized ethylenes, the barrier can be low enough to be studied by variable temperature nmr,[108] i.e., 5–25 kcal mol^{-1}. These most interesting compounds have been studied extensively in recent years, and a comprehensive review through 1972 is available.[109] The low energy for rotation about what is formally a carbon–carbon double bond provides a dramatic illustration of the importance of the polar structures in the resonance hybrid, although the relief of steric compression in the planar ground state undoubtedly plays some role in lowering the barriers in these compounds. This has been specifically demonstrated for the compounds (XIX).[110] The activation free energy, $\Delta G^{\ddagger}$, for rotation about the carbon–carbon double bond fell steadily from *ca.* 27.7 kcal mol^{-1} (R = H) to 18.3 kcal mol^{-1} (R = Bu^t), with increasing bulk of the alkyl group.

$$(R)(MeO)C{=}C(CO_2Me)_2$$

R = H, Me, Et, Pr^i, or Bu^t

(XIX)

Consideration of the available data for polarized ethylenes led Kessler[109] to the following order for the efficiency with which several common substituents lower the rotational barriers:

Donors:	R_2N >	RO >	RS >	Alkyl
	−0.53	−0.43	−0.25	−0.10
Acceptors:	CN <	CO_2R <	COPh <	COMe
	0.09	0.16	0.19	0.22

The numerical values given below each substituent are the σ_R^0 parameters.[79] The barrier-lowering effects of the substituents evidently follow the order of their σ_R^0 values[111] as expected. On the other hand, Kessler has obtained[112] a good correlation ($r = 0.99$) between σ_p and the barrier hindering the C=C rotation in a series of aryl bis-(dimethylamino)vinyl cyanides (XX). The substituent set used is rather inadequate to distinguish between σ_R^0 and σ_p values (Y = H, F, Cl, Br, COMe, CO_2Me, CN, or NO_2). However, the closeness of the barrier for Y = H and for Y = F

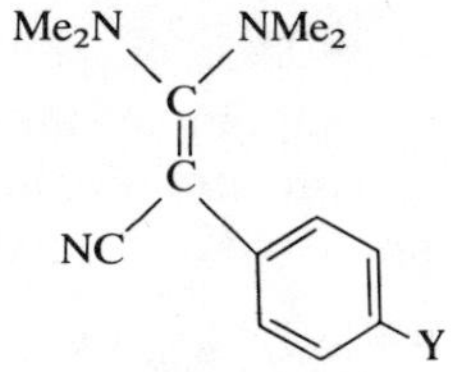

Me_2N NMe_2 C=C NC C_6H_4–Y

(XX)

clearly indicates that inductive effects operate in this particular series, whereas for simple substituted ethylenes in which the substituents are attached directly to the double bond, it is the direct resonance effects that are important. The correlation obtained by Kessler is almost certainly due to the relatively constant steric effect of the $—C_6H_4X$ group. It seems most unlikely that barriers can be understood in general by any treatment that does not specifically allow for steric interactions.

6.9.2. *Rotations about Carbon–Substituent Single Bonds*

Just as the resonance interaction of a substituent attached to a double bond tends to lower the barrier to *cis–trans* isomerization, it tends, *inter alia*, to increase the barrier to rotation about the substituent–carbon bond.[113] In more quantitative terms, the barriers hindering the rotation of several conjugating substituents attached to a benzene ring have been rationalized in terms of σ_R^0 values and strain parameters.[114] An extension of this method to substituted ethylenes has appeared.[75] The method applied to substituted benzenes will first be described.

The energy barrier to rotation of a substituent about the ring–substituent bond was equated to the difference in resonance interaction between the position of maximum energy (usually the orthogonal position of 90° twist) and that of minimum energy (usually at or near coplanarity of the substituent and the ring) less the corresponding difference in strain, including rehybridization energy. This led to equation (6.23), where the subscript

$$E = (R - R_{\text{tw}}) - S \qquad (6.23)$$

tw denotes the twisted (orthogonal) conformation and R and S are resonance and strain energies. The resonance interactions should be proportional to σ_R^0 [or σ_R^0(tw)].

By using known values for rotational barriers and for σ_R^0 for certain substituents, together with certain assumptions about steric interactions, equation (6.24) was obtained. This represents a direct relationship of σ_R^0

values to the relevant energy values as well as provides an estimate of previously unknown barriers or strain energies.

$$E = 33[|\sigma_R^0| - |\sigma_R^0(\text{tw})|] - S \tag{6.24}$$

The extension of this method to monosubstituted ethylenes where the substituent is of less than C_{2v} symmetry is complicated by the existence of s-*cis* and s-*trans* forms. Equation (6.25) was obtained[75]; the quantities

$$0.5[E_c + E_t] = b[|\sigma_R^0| - |\sigma_R^0(\text{tw})|] - 0.5(S_c + S_t) \tag{6.25}$$

referring to the two forms are identified by the subscripts c and t. However, the appropriate value of σ_R^0 is that obtained for the *benzene series*, this being equivalent to an "average" σ_R^0 value for a substituent in an environment having simultaneously characteristics of both s-*cis* and s-*trans* forms. From reasoning similar to that outlined above, the value of b in equation (6.25) was found[75] to be 25 kcal mol$^{-1}/\sigma_R^0$. Thus in energetic terms the resonance interaction of the substituent with the olefinic bond seems to be somewhat less than that in the benzene case, viz., 33 kcal mol$^{-1}/\sigma_R^0$ [equation (6.24)], in agreement with the theoretical calculations; cf. Section 6.2.

The quantitative effects of conjugation on these barriers are further illustrated by the following example.[75] The enthalpy of activation for rotation of the acetyl group (averaged over s-*cis* and s-*trans* forms) in *trans*-4-*NN*-dimethylaminobut-3-en-2-one (XXI)[115] is 13.6 kcal mol^{-1},

$Me_2N(H)C{=}C(H)C(=O)Me$

(XXI)

whereas the corresponding averaged value for the parent methyl vinyl ketone (butenone)[116,117] is 5.0 kcal mol^{-1}. The additional barrier of 8.6 kcal mol^{-1} in the enamino-ketone must therefore be due largely to the additional resonance stabilization in the ground state associated with substituent interactions. Furthermore, the barrier to rotation of the dimethylamino group should be similarly enhanced with respect to the parent enamine. Unfortunately this barrier in the parent *NN*-dimethylaminoethylene is not known, but it may be estimated[75] from equation (6.23) as 4.5 kcal mol^{-1}. The enthalpy of activation for this rotation in (XXI) is not known, although it should not be much different from $\Delta G^{\ddagger}$ ($\Delta G^{\ddagger} = 13.3$ kcal mol^{-1}).[118] On this basis an additional barrier of 8.8 kcal mol^{-1} hindering the rotation of the dimethylamino group is obtained: This compares favorably with the 8.6 kcal mol^{-1} obtained above for the acetyl group rotation.

The resonance effect is conformationally sensitive. Thus the $\nu_{C=C}$ intensity in the "*cis*" form (XXII) is found[119] to be about three times larger than in the "*trans*" form (XXIII).† Similarly the intensity for the s-*cis* conformer (XXIV) of an alkyl vinyl ketone is greater[120] than that of the s-*trans* conformer (XXV). The total $\nu_{C=C}$ intensity in a series (R = H, Me,

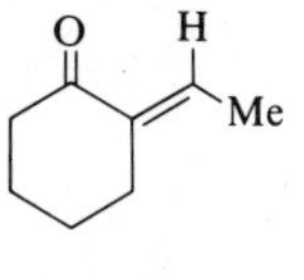

(XXII) 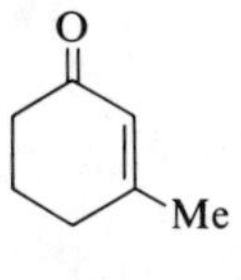 (XXIII) 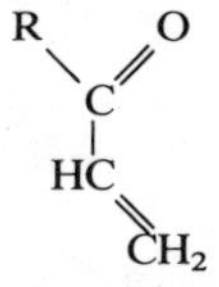 (XXIV) 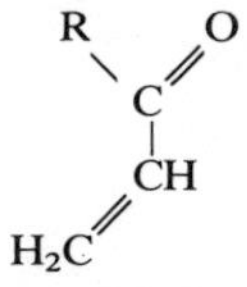 (XXV)

Et, Pr^i, or Bu^t) has been shown[120] to be explicable in terms of populations and individual integrated intensities derived for the two forms. For R = H, the s-*trans* form is predominant, but the preference changes through the series to almost pure s-*cis* form for R = Bu^t.

By contrast, results[121] for aryl vinyl ethers and sulfides suggest that conjugation between C=C and substituent OMe and SMe groups should be effectively constant in s-*cis* and s-*trans* conformers.

6.10. Transmission in Arylethylenes

There have been a considerable number of reports of measurements of physical properties, reactivities, or equilibria of systems of type (XXVI), where P is H or a probe such as a carboxy group.¶

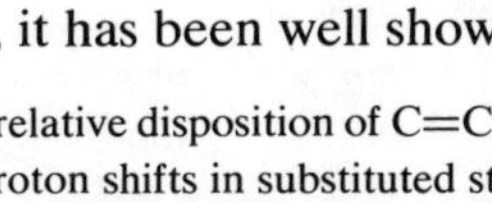

(XXVI)

Such results really refer to the overall effect of a substituted phenyl group on the ethylene unit or to the effect of Y as transmitted *via* a phenylene unit. However, it has been well shown[6,137] that, in addition, the

†"*Cis*" and "*trans*" refer to the relative disposition of C=C and C=O in (XXII) and (XXIII).

¶These studies include vinylic proton shifts in substituted styrenes,[6,122–126] in *cis*- and *trans*-cinnamic acids,[122] and in stilbenes;[127] ^{13}C nmr studies on styrenes,[6,41,123,128] coupling constant studies on substituted styrenes,[6,122,123,129] ^{19}F nmr studies on substituted *trans*-fluorostilbenes,[130] theoretical studies on styrenes,[6,124,126] ir studies on $\nu_{C=C}$ in styrenes,[74] on ν_{CF} when P is CF_3 or SO_2CF_3,[131] on $\nu_{C=O}$ in styryl ketones,[132] and in *NN*-dimethylaminocinnamides,[133]; and uv spectroscopic[134,135] and dipole moment studies[134] on stilbenes. Equilibria and reactivity studies include investigations of ionization constants of substituted cinnamic acids,[136] rates of hydrolysis of the corresponding esters[136] or of esterification of the acids,[137] rates of the Prins reaction of substituted styrenes,[138] addition of free radicals to styrenes,[139] and pK_a studies on hydroxystilbenes.[135]

field effect of the substituent Y causes direct polarization of the vinyl group. All these effects should be explicable in terms of the σ_I and $\bar{\sigma}_R$ values of the substituent. In Table 6.14 for example, we list the ^{1}H and ^{13}C shifts for the vinyl group in substituted styrenes.[6] No satisfactory fit was found for Δ^1H_A, while DSP analyses gave equations (6.26)–(6.29) as best

$$\Delta^1H_B = 0.15\sigma_I + 0.46\sigma_R^0 \quad (f, 11\%) \tag{6.26}$$

$$\Delta^1H_C = 0.27\sigma_I + 0.33\sigma_R(\text{BA}) \quad (f, 9\%) \tag{6.27}$$

$$\Delta^{13}C_\alpha = -2.71\sigma_I - 0.21\sigma_R^0 \quad (f, 19\%) \tag{6.28}$$

$$\Delta^{13}C_\beta = 5.33\sigma_I + 8.08\sigma_R^0 \quad (f, 7\%) \tag{6.29}$$

fits for the other values. A full discussion of similar analyses is given in Ref. 6; we note here that the reversal of the shift of $^{13}C_\alpha$ compared to $^{13}C_\beta$ is probably caused by the substituent-induced polarization of the π system of the vinyl group.

Studies on arylethylenes, as with those on simple ethylenes, suggest that the carbon–carbon double bond transmits electronic effects better than a *para*-phenylene system.[140] The *cis* and the *trans* arrangement seem about equal in this respect, and about the same as a carbon–carbon triple bond, although markedly better, of course, than a $-CH_2CH_2-$ linkage.[4,136]

Table 6.14. nmr Parameters for para-Substituted Styrenes[a]

Subst.	δH_A[b]	δH_B[b]	δH_C[b]	δC_α[c]	δC_β[c]
H	6.629	5.630	5.114	136.96	113.20
F	6.591	5.552	5.101	136.11	113.43
Cl	6.582	5.612	5.147	135.69	113.97
Me	6.574	5.558	5.035	136.73	112.20
CF_3	6.661	5.731	5.261	135.64	116.02
NH_2	6.494	5.404	4.891	—	—
NMe_2	6.503	5.394	4.864	136.67	108.93
OMe	6.561	5.482	4.987	136.29	110.98
NO_2	6.701	5.807	5.363	135.03	117.90
CN	6.630	5.744	5.300	135.39	117.05
COMe	6.664	5.739	5.243	136.07	115.91
CO_2Me	6.655	5.729	5.230	—	—

[a] From Ref. 6.
[b] ppm relative to TMS.
[c] ppm to low field of $^{13}CH_3Si(CH_3)_3$.

6.11. Summary

The above review attempts to present the main lines of evidence on substituent effects in olefinic systems, but is by no means complete. Recent photoelectron spectroscopic results[141] offer the possibility of estimating inductive and conjugative effects of substituents in such systems. Another important recent paper[142] is concerned with the effect of substituents on the equilibrium shown in structures (XXVII) and (XXVIII).

$$\underset{\text{(XXVII)}}{(H)(XH_2C)C{=}C(Y)(H)} \rightleftharpoons \underset{\text{(XXVIII)}}{(H)(X)C{=}C(CH_2Y)(H)}$$

Overall, however, it seems that there are many parallels between substituted ethylenes and substituted benzenes. Thus ^{1}H and ^{13}C chemical shifts at the β-carbon atom in monosubstituted ethylenes can be described in terms of the π density at that atom, which is analogous to the situation at the *para*-position in monosubstituted benzenes. Equally, substituent charge transfer, ir intensities of certain vibrational modes, rotational barriers, and σ_R^0 values are related in both series.

The physical and chemical properties measured for disubstituted ethylenes have been shown frequently to be additive, although through-conjugation occurs in suitably substituted *trans*-1,2-disubstituted ethylenes. The magnitude of this mutual substituent conjugation, as measured by λ in equations of type (6.18) and (6.19), has been shown to be proportional to, but somewhat greater than, that observed in similar *para*-disubstituted benzenes.

Resonance parameters have also been shown to give a measure of the rotational barriers about substituent–vinyl bonds; this work is in its early stages. The actual π-charge transfer and the barrier are both somewhat less in the ethylene than in the benzene series.

References

1. R. D. Topsom, *Progr. Phys. Org. Chem.*, **12**, 1 (1976) and references therein.
2. A. R. Katritzky and R. D. Topsom, *J. Chem. Educ.*, **48**, 427 (1971).
3. L. M. Stock, *J. Chem. Educ.*, **49**, 400 (1972).
4. M. Charton, *Progr. Phys. Org. Chem.*, **10**, 81 (1973).
5. M. Charton and H. Meislich, *J. Amer. Chem. Soc.*, **80**, 5940 (1958).
6. G. K. Hamer, I. R. Peat, and W. F. Reynolds, *Canad. J. Chem.*, **51**, 897 (1973).
7. S. Ehrenson, R. T. C. Brownlee, and R. W. Taft, *Progr. Phys. Org. Chem.*, **10**, 1 (1973).
8. D. A. Dawson and W. F. Reynolds, *Canad. J. Chem.*, **53**, 373 (1975).

9. R. T. C. Brownlee and R. D. Topsom, *Tetrahedron Lett.*, 5187 (1972).
10. W. J. Hehre, L. Radom, and J. A. Pople, *J. Amer. Chem. Soc.*, **94**, 1496 (1972).
11. W. J. Hehre, R. W. Taft, and R. D. Topsom, *Progr. Phys. Org. Chem.*, **12**, 159 (1976).
12. See for example: W. J. Hehre and J. A. Pople, *J. Amer. Chem. Soc.*, **92**, 2191 (1970); S. Meza and U. Wahlgren, *Theor. Chim. Acta*, **21**, 323 (1971).
13. R. T. C. Brownlee and R. W. Taft, *J. Amer. Chem. Soc.*, **92**, 7007 (1970).
14. J. A. Pople and M. S. Gordon, *J. Amer. Chem. Soc.*, **89**, 4253 (1967).
15. R. T. C. Brownlee and R. W. Taft, *J. Amer. Chem. Soc.*, **90**, 6537 (1968); G. C. Levy and G. L. Nelson, *Carbon-13 NMR for Organic Chemists*, p. 85 (Wiley, New York, 1972); G. L. Nelson, G. C. Levy, and J. D. Cargioli, *J. Amer. Chem. Soc.*, **94**, 3089 (1972); J. E. Bloor and D. L. Breen, *J. Phys. Chem.*, **72**, 716 (1968).
16. G. P. Ford, T. B. Grindley, A. R. Katritzky, and R. D. Topsom, *J.C.S. Perkin II*, 1569 (1974).
17. G. P. Ford, A. R. Katritzky, and R. D. Topsom, *J.C.S. Perkin II*, 1371 (1975).
18. N. D. Epiotis, *J. Amer. Chem. Soc.*, **95**, 3087 (1973).
19. C. K. Ingold, *Structure and Mechanism in Organic Chemistry*, 2nd edition (Bell, London, 1969).
20. T. S. Cameron, D. J. Cowley, and J. E. Thompson, *J.C.S. Perkin II*, 774 (1974).
21. E. Ericsson, T. Marnung, J. Sandström, and I. Wennerbeck, *J. Mol. Structure*, **24**, 373 (1975); S. Abrahamsson, G. Rehnberg, T. Liljefors, and J. Sandström, *Acta Chem. Scand.*, **28B**, 1109 (1974).
22. L. W. Deady, M. Kendall, R. D. Topsom, and R. A. Y. Jones, *J.C.S. Perkin II*, 416 (1973).
23. A. R. Katritzky and R. D. Topsom, *Angew. Chem. Intern. Edn.*, **9**, 87 (1970); GE **82,** 106 (1970).
24. O. Exner, *Coll. Czech. Chem. Comm.*, **25**, 642 (1960).
25. M. Charton, *J. Org. Chem.*, **30**, 552 (1965).
26. M. T. Tribble and J. G. Traynham in *Advances in Linear Free Energy Relationships*, Chap. IV, N. B. Chapman and J. Shorter, eds. (Plenum, London, 1972).
27. See, for example, G. Maciel, in *Topics in Nuclear Magnetic Resonance Spectroscopy*, Vol. 1 (Wiley, New York, 1975).
28. J. Niwa, *Bull. Chem. Soc. Japan*, **40**, 1512 (1967).
29. B. A. Trofimov, G. A. Kalabin, and O. N. Vylegjanin, *Reakts. spos. org. Soedinenii*, **8**, 943 (1971).
30. R. E. Mayo and J. H. Goldstein, *J. Mol. Spectroscopy*, **14**, 173 (1964).
31. G. S. Reddy, J. H. Goldstein, and L. Mandell, *J. Amer. Chem. Soc.*, **83**, 1300 (1961).
32. J. Niwa, *Bull. Chem. Soc. Japan*, **48**, 1637 (1975).
33. L. M. Jackman and S. Sternhell, *Applications of Nuclear Magnetic Resonance Spectroscopy in Organic Chemistry*, 2nd edition (Pergamon, New York, 1969); U. E. Matter, C. Pascual, E. Pretsch, A. Pross, W. Simon, and S. Sternhell, *Tetrahedron*, **25**, 691 (1969); C. Pascual, J. Meier, and W. Simon, *Helv. Chim. Acta*, **49**, 164 (1966).
34. C. N. Banwell and N. Sheppard, *Mol. Phys.*, **3**, 351 (1960).
35. S. W. Tobey, *J. Org. Chem.*, **34**, 1281 (1969).
36. D. F. Ewing and K. A. W. Parry, *J. Chem. Soc.* (*B*), 970 (1970).
37. G. Miyajima, K. Takahashi, and K. Nishimoto, *Org. Magnetic Resonance*, **6**, 413 (1974).
38. G. E. Maciel, *J. Phys. Chem.*, **69**, 1947 (1965).
39. D. E. Dorman, M. Jautelat, and J. D. Roberts, *J. Org. Chem.*, **36**, 2757 (1971).
40. V. R. Radeglia and E. Gey, *J. prakt. Chem.*, **314**, 43 (1972).
41. K. S. Dhami and J. B. Stothers, *Canad. J. Chem.*, **43**, 510 (1965).
42. K. Hatada, K. Nagata, and H. Yuki, *Bull. Chem. Soc. Japan*, **43**, 3267 (1970).

43. Z. W. Wolkowski, E. Vauthier, B. Gonbeau, H. Sauvaitre, and J. A. Musso, *Tetrahedron Lett.*, 565 (1972).
44. O. Kajimoto and T. Fueno, *Tetrahedron Lett.*, 3329 (1972).
45. C. Rappe, E. Lippmaa, T. Pehk, and K. Andersson, *Acta Chem. Scand.*, **23**, 1447 (1969).
46. G. B. Savitsky, P. D. Ellis, K. Namikawa, and G. E. Maciel, *J. Chem. Phys.*, **49**, 2395 (1968).
47. H. Brouwer and J. B. Stothers, *Canad. J. Chem.*, **50**, 601 (1972).
48. E. Lippmaa, T. Pehk, K. Andersson, and C. Rappe, *Org. Magnetic Resonance*, **2**, 109 (1970).
49. D. H. Marr and J. B. Stothers, *Canad. J. Chem.*, **43**, 596 (1965).
50. T. Vladimiroff and E. R. Malinowski, *J. Chem. Phys.*, **46**, 1830 (1967).
51. G. B. Savitsky and K. Namikawa, *J. Phys. Chem.*, **67**, 2754 (1963).
52. G. E. Maciel, P. D. Ellis, J. J. Natterstad, and G. B. Savitsky, *J. Magnetic Resonance*, **1**, 589 (1969).
53. R. W. Taft, *J. Amer. Chem. Soc.*, **79**, 1045 (1957); R. W. Taft, E. Price, I. R. Fox, I. C. Lewis, K. K. Andersen, and G. T. Davis, *J. Amer. Chem. Soc.*, **85**, 709 (1963).
54. R. W. Taft, E. Price, I. R. Fox, I. C. Lewis, K. K. Andersen, and G. T. Davis, *J. Amer. Chem. Soc.*, **85**, 3146 (1963).
55. T. Schaefer, F. Hruska, and H. M. Hutton, *Canad. J. Chem.*, **45**, 3143 (1967).
56. J. W. Emsley and L. Phillips, *Progr. N.M.R. Spectroscopy*, **7**, 1 (1971); K. Jones and E. F. Mooney, *Ann. Reports N.M.R. Spectroscopy*, **4**, 391 (1971).
57. T. Schaefer and H. M. Hutton, *Canad. J. Chem.*, **45**, 3153 (1967).
58. R. Ditchfield and J. N. Murrell, *Mol. Phys.*, **15**, 533 (1968).
59. V. M. S. Gil and S. J. S. Formosinho-Simões, *Mol. Phys.*, **15**, 639 (1968).
60. L. Lunazzi and F. Taddei, *Spectrochim. Acta*, **25A**, 553 (1969).
61. D. F. Ewing, *Org. Magnetic Resonance*, **5**, 567 (1973).
62. K. M. Crecely, R. W. Crecely, and J. H. Goldstein, *J. Mol. Spectroscopy*, **37**, 252 (1971).
63. L. J. Bellamy, *The Infrared Spectra of Complex Organic Molecules*, 2nd edition (Methuen, London, 1958).
64. L. J. Bellamy, *Advances in Infrared Group Frequencies* (Methuen, London, 1968).
65. S. Bank, W. D. Closson, and L. T. Hodgins, *Tetrahedron*, **24**, 381 (1968).
66. D. B. Cunliffe-Jones, *Spectrochim. Acta*, **21A**, 245 (1965).
67. A. R. Katritzky and R. D. Topsom in *Advances in Linear Free Energy Relationships*, Chap. 3, N. B. Chapman and J. Shorter, eds. (Plenum, London, 1972).
68. L. J. Bellamy, *J. Chem. Soc.*, 4221 (1955).
69. W. J. Potts and R. A. Nyquist, *Spectrochim. Acta*, **15**, 679 (1959).
70. J. K. Brown and N. Sheppard, *Trans. Faraday Soc.*, **51**, 1611 (1955).
71. N. B. Colthup and M. K. Orloff, *Spectrochim. Acta*, **27A**, 1299 (1971).
72. A. N. Egorochkin, Yu. D. Semchikov, N. S. Vyazankin, and S. Ya. Khorshev, *Izvest. Akad. Nauk. S.S.S.R., Ser. khim.*, 152 (1970); EE, 143.
73. W. Adcock, M. J. S. Dewar, and B. D. Gupta, *J. Amer. Chem. Soc.*, **95**, 7353 (1973).
74. A. R. Katritzky, R. F. Pinzelli, M. V. Sinnott, and R. D. Topsom, *J. Amer. Chem. Soc.*, **92**, 6861 (1970).
75. G. P. Ford, A. R. Katritzky, and R. D. Topsom, *J.C.S. Perkin II*, 1378 (1976).
76. T. J. Broxton, G. Butt, R. Liu, L. H. Teo, R. D. Topsom, and A. R. Katritzky, *J.C.S. Perkin II*, 463 (1974).
77. M. Podzimková, M. Procházka, and M. Paleček, *Coll. Czech. Chem. Comm.*, **34**, 2101 (1969).
78. R. T. C. Brownlee, A. R. Katritzky, and R. D. Topsom, *J. Amer. Chem. Soc.*, **88**, 1413 (1966).

79. R. T. C. Brownlee, R. E. J. Hutchinson, A. R. Katritzky, T. T. Tidwell, and R. D. Topsom, *J. Amer. Chem. Soc.*, **90**, 1757 (1968).
80. R. T. C. Brownlee, A. R. Katritzky, M. V. Sinnott, M. Szafran, R. D. Topsom, and L. Yakhontov, *J. Amer. Chem. Soc.*, **92**, 6850 (1970); R. T. C. Brownlee, D. G. Cameron, R. D. Topsom, A. R. Katritzky, and A. J. Sparrow, *J. Mol. Structure*, **16**, 365 (1973).
81. R. T. C. Brownlee, J. Munday, R. D. Topsom, and A. R. Katritzky, *J.C.S. Faraday II*, 349 (1973).
82. R. T. C. Brownlee, G. Butt, M. P. Chan, and R. D. Topsom, *J.C.S. Perkin II*, 1486 (1976).
83. M. J. S. Dewar and P. J. Grisdale, *J. Amer. Chem. Soc.*, **84**, 3539 (1962); S. Ehrenson, *Progr. Phys. Org. Chem.*, **2**, 195 (1964); R. Golden and L. M. Stock, *J. Amer. Chem. Soc.*, **94**, 3080 (1972).
84. G. Butt, E. Schmid, and R. D. Topsom, work in progress.
85. J. Overend, in *Infrared Spectroscopy and Molecular Structure*, p. 353, M. Davies, Ed. (Elsevier, Amsterdam, 1963).
86. A. R. Katritzky, M. V. Sinnott, T. T. Tidwell, and R. D. Topsom, *J. Amer. Chem. Soc.*, **91**, 628 (1969).
87. P. J. Q. English, A. R. Katritzky, T. T. Tidwell, and R. D. Topsom, *J. Amer. Chem. Soc.*, **90**, 1767 (1968).
88. R. T. C. Brownlee, D. G. Cameron, R. D. Topsom, A. R. Katritzky, and A. F. Pozharsky, *J.C.S. Perkin II*, 247 (1974).
89. T. B. Grindley, K. F. Johnson, A. R. Katritzky, H. J. Keogh, C. Thirkettle, and R. D. Topsom, *J.C.S. Perkin II*, 282 (1974).
90. G. P. Ford, T. B. Grindley, A. R. Katritzky, M. Shome, J. Morel, C. Paulmier, and R. D. Topsom, *J. Mol. Structure*, **27**, 195 (1975).
91. V. S. Kukina and L. M. Sverdlov, *Zhur. fiz. Khim.*, **40**, 2837 (1966); EE, 1523.
92. N. C. Craig and J. Overend, *J. Chem. Phys.*, **51**, 1127 (1969); N. C. Craig and J. Overend, personal communication of displacements.
93. See, for example, K. Hirayama, *Handbook of Ultraviolet and Visible Absorption Spectra of Organic Compounds* (Plenum, New York, 1967).
94. C. N. R. Rao, *Ultraviolet and Visible Spectroscopy: Chemical Applications*, 2nd edition (Butterworths, London, 1967).
95. R. T. C. Brownlee and R. D. Topsom, *Spectrochim. Acta*, **29A**, 385 (1973).
96. B. Brosa and E. Schmid, personal communication.
97. M. Charton, *Canad. J. Chem.*, **48**, 1748 (1970).
98. K. Bowden, *Canad. J. Chem.*, **43**, 3354 (1965).
99. K. Bowden, *Canad. J. Chem.*, **44**, 661 (1966).
100. B. Yamada and T. Otsu, *J. Polymer Sci., Part A-1, Polymer Chem.*, **7**, 2439 (1969).
101. J.-E. Dubois and G. Mouvier, *Compt. rend.*, **259**, 2101 (1964); J.-E. Dubois and G. Mouvier, *Bull. Soc. chim. France*, 1426 (1968); G. Mouvier and J.-E. Dubois, *Bull. Soc. chim. France*, 1441 (1968).
102. M. L. Poutsma, *J. Amer. Chem. Soc.*, **87**, 4285 (1965).
103. R. P. Bell and M. Pring, *J. Chem. Soc. (B)*, 1119 (1966).
104. M. Charton, *J. Org. Chem.*, **31**, 3745 (1966).
105. F. W. Baker, R. C. Parish, and L. M. Stock, *J. Amer. Chem. Soc.*, **89**, 5677 (1967).
106. J. E. Douglas, B. S. Rabinovitch, and F. S. Looney, *J. Chem. Phys.*, **23**, 315 (1955).
107. M. C. Lin and K. J. Laidler, *Canad. J. Chem.*, **46**, 973 (1968).
108. H. Kessler, *Angew. Chem. Intern. Edn.*, **9**, 219 (1970); GE **82**, 237 (1970).
109. H.-O. Kalinowski and H. Kessler, *Topics Stereochem.*, **7**, 295 (1972).
110. Y. Shvo, *Tetrahedron Lett.*, 5923 (1968).
111. Y. Shvo and H. Shanan-Atidi, *J. Amer. Chem. Soc.*, **91**, 6689 (1969).
112. H. Kessler, *Chem. Ber.*, **103**, 973 (1970).

113. I. Wennerbeck and J. Sandström, *Org. Magnetic Resonance*, **4**, 783 (1972).
114. T. B. Grindley, A. R. Katritzky, and R. D. Topsom, *J.C.S. Perkin II*, 289 (1974).
115. M.-L. Filleux-Blanchard, H. Durand, and G. J. Martin, *Org. Magnetic Resonance*, **2**, 539 (1970).
116. E. Wyn-Jones, K. R. Crook, and W. J. Orville-Thomas, *Adv. Mol. Relaxation Processes*, **4**, 193 (1972).
117. A. J. Bowles, W. O. George, and W. F. Maddams, *J. Chem. Soc.* (*B*), 810 (1969).
118. J. Dabrowski and L. J. Kozerski, *Org. Magnetic Resonance*, **4**, 137 (1972).
119. R. Mecke and K. Noack, *Chem. Ber.*, **93**, 210 (1960).
120. A. R. Katritzky, R. F. Pinzelli, and R. D. Topsom, *Tetrahedron*, **28**, 3449 (1972).
121. A. R. Katritzky, R. F. Pinzelli, and R. D. Topsom, *Tetrahedron*, **28**, 3441 (1972); E. Taskinen and P. Liukes, *Acta Chem. Scand.*, **B28**, 114 (1974).
122. T. A. Wittstruck and E. N. Trachtenberg, *J. Amer. Chem. Soc.*, **89**, 3803 (1967).
123. J.-E. Dubois, J. A. Miller, and J.-P. Doucet, *J. Chim. phys.*, **63**, 1283 (1966).
124. E. Gey, *Z. Chem.*, **11**, 392 (1971); J.-P. Doucet, B. Ancian, and J.-E. Dubois, *J. Chim. phys.*, **69**, 188 (1972).
125. J. W. Emsley, J. Feeney, and L. H. Sutcliffe, *High Resolution Nuclear Magnetic Resonance Spectroscopy*, p. 719 (Pergamon, Oxford, 1966).
126. G. K. Hamer and W. F. Reynolds, *J. Chem. Soc.* (*D*), 1218 (1971).
127. H. Güsten and M. Salzwedel, *Tetrahedron*, **23**, 187 (1967); H. Güsten and M. Salzwedel, *Tetrahedron*, **23**, 173 (1967).
128. Gurudata, J. B. Stothers, and J. D. Talman, *Canad. J. Chem.*, **45**, 731 (1967).
129. W. F. Reynolds, I. R. Peat, and G. K. Hamer, *Canad. J. Chem.*, **52**, 3415 (1974).
130. R. G. Pews and N. D. Ojha, *J. Amer. Chem. Soc.*, **91**, 5769 (1969); I. R. Ager, L. Phillips, T. J. Tewson, and V. Wray, *J.C.S. Perkin II*, 1979 (1972); S. K. Dayal, S. Ehrenson, and R. W. Taft, *J. Amer. Chem. Soc.*, **94**, 9113 (1972).
131. V. F. Kulik, Yu. P. Egorov, A. G. Panteleimonov, Yu. A. Fialkov, and L. M. Yagupol'skii, *Teor. i Eksp. Khim.*, **1**, 171 (1965); EE, 107.
132. W. F. Winecoff and D. W. Boykin, *J. Org. Chem.*, **37**, 674 (1972).
133. K. Spaargaren, C. Kruk, T. A. Molenaar-Langeveld, P. K. Korver, P. J. van der Haak, and Th. J. de Boer, *Spectrochim. Acta*, **28A**, 965 (1972).
134. A. E. Lutskii, L. M. Litvinenko, L. V. Shubina, L. Ya. Malkes, R. S. Cheshko, A. S. Gol'berkova, and Z. M. Kanevskaya, *Zhur. obshchei Khim.*, **35**, 2083 (1965); EE, 2073; A. E. Lutskii, L. Ya. Malkes, E. M. Obukhova, and A. I. Timchenko, *Zhur. fiz. Khim.*, **37**, 1076 (1963); EE, 565.
135. H. Veschambre, G. Dauphin, and A. Kergomard, *Bull, Soc. chim. France*, 2846 (1967).
136. A. B. Thigpen and R. Fuchs, *J. Org. Chem.*, **34**, 505 (1969) and previous papers in the series.
137. K. Bowden and D. C. Parkin, *Canad. J. Chem.*, **46**, 3909 (1968).
138. J.-P. Durand, M. Davidson, M. Hellin, and F. Coussemant, *Bull. Soc. chim. France*, 159 (1971).
139. H. Sakurai, S. Hayashi, and A. Hosomi, *Bull. Chem. Soc. Japan*, **44**, 1945 (1971).
140. M. Charton, *J. Org. Chem.*, **39**, 2797 (1974).
141. H. Bock and K. Wittel, *J.C.S. Chem. Comm.*, 602 (1972); A. Katrib and J. W. Rabelais, *J. Phys. Chem.*, **77**, 2358 (1973).
142. J. Hine and N. W. Flachskam, *J. Amer. Chem. Soc.*, **95**, 1179 (1973).

7

The Correlation Analysis of Nucleophilicity†

Claude Duboc

7.1. *Introduction*

A nucleophile is usually defined as a reagent having the tendency to form a new covalent bond by sharing an electron pair.[1] Some reducing agents are thus not included in this definition. Historically, the first nucleophilic reagents to be studied possessed a center with partial negative charge and therefore had a great affinity for a seat of partial positive charge, viz., the

†Translated from the French by the Editors.

Claude Duboc • Université de Paris VI, F-75005 Paris, France.

atomic nucleus, hence their name. The latter is called an electrophilic center.[2] Indeed, more generally, a reaction between a nucleophilic reagent and an electrophilic center in a molecule termed the substrate should be assimilated to a reaction of the acid–base type, whether in the sense of Lewis[3] or that of Pearson.[4] In the latter case, it will be seen later that the notion of polarizability plays an important part.[5] Finally, it is usual to speak of basicity in connection with thermodynamic magnitudes and of nucleophilicity in connection with kinetic magnitudes.

For a given substrate different nucleophiles have a definite order of reactivity, but this order may change from one substrate to another. For example, in organic chemistry, the order of nucleophilicity towards the substrate MeBr is as follows[6]: $S_2O_3^{2-} > I^- > OH^-$; towards *p*-nitrophenyl acetate it becomes[7] $OH^- > S_2O_3^{2-} > I^-$; in inorganic chemistry, towards the substrate Ag^+, the order is[8] $CN^- > Br^- > Cl^- > OH^-$, whereas towards $MeHg^+$ it is[9] $Br^- > Cl^- > CN^- \gg OH^-$, and towards $Fe_4S_4(SCH_2CH_2CO_2^-)_4^{6-}$ it becomes[10] $CN^- > OH^- > Br^- \approx Cl^-$.

For the same substrate the order for different nucleophiles may change if the solvent is changed. For example, passing from a dipolar aprotic solvent to a protic solvent may reverse the order.[11,12] In the first case, for example, the order for anions is $Cl^- > Br^- > I^-$, and in the second, $I^- > Br^- > Cl^-$. Finally for a given solvent ($COMe_2$) and substrate (Bu^nOBs), association of the nucleophilic anions with different cations may also reverse the order[13]: With $Bu_4^nN^+$, the order for halide anions is $Cl^- > Br^- > I^-$, but with Li^+ the order is reversed, viz., $I^- > Br^- > Cl^-$.

The prediction of the reaction between a nucleophile and a substrate in a given solvent may prove very subtle, in spite of numerous experimental studies and attempts at classification.[14,15] This is because of the large number of factors to be considered.[16,17] Among these factors, which are interdependent, thus making the resolution of the problem more difficult, six principal ones may be retained: (*a*) the basicity of the nucleophile in Brönsted's sense, (*b*) its polarizability,[18] (*c*) the nature of the solvent[11] and the role of solvation in the transition state,[19] (*d*) the strength of the bond formed in the reaction,[20] (*e*) the Coulombic interaction between the charge on the substrate and that on the nucleophile,[21] and the formation of ion pairs,[22,23] (*f*) the role of lone pairs on the atom neighboring the nucleophilic center, called the α-effect.[24–26]

The principle of hard and soft acids and bases (HSAB) also enables the course of a reaction as a function of the nature of the reagents to be predicted. Various studies and reviews have been devoted to this subject.[27–30] The HSAB principle[31,32] generalizes acid–base reactions and points up the competition between basicity and polarizability.[26] The first classifications were attempted in order to divide metallic ions into two groups[8,33,34] according to the stability of their complexes. Pearson[4] generalized this classification by considering as complexes all combinations of Lewis acids and bases, real or hypothetical, and then he developed this idea[35,36] by applying it particularly to organic chemistry.[37]

Numerous experimental studies have enabled empirical expressions to be proposed for nucleophilic power, more and more elaborate but still subject to exceptions. The theoretical attempt, particularly that of Klopman,[38] based on perturbation theory, to measure electron donor–acceptor power represents a parallel development.[39] We shall show this by considering organic chemical results as well as the most striking results from inorganic chemistry.

7.2. *Measurement of Nucleophilicity*

Pearson[28] considers the addition of nucleophile to electrophile as an example of weak generalized acid–base interaction, as in (7.1). Conventionally, as we have said, electrophile–nucleophile has the significance

$$\mathrm{A} + :\mathrm{B} \rightleftharpoons \mathrm{A}:\mathrm{B} \tag{7.1}$$

of "acid–base" for the language of kinetics.[6] Other terms signify base in different languages: base, ligand, solvent possessing an electron-rich site[40]—this last may go as far as involvement in the formation of a charge-transfer complex for π bases.

In a qualitative way, a strong base, B, and a strong acid, A, form a stable complex, A:B, whereas a weaker acid and the same base yield a less stable complex. Acid or base strength is defined by considering the substitution reaction (7.2). This shows that B′ is a stronger base than B. If one

$$\mathrm{B'} + \mathrm{A}:\mathrm{B} \rightarrow \mathrm{A}:\mathrm{B'} + \mathrm{B} \tag{7.2}$$

could classify all bases and all acids, the problem of understanding the stability of the compound formed would be resolved.

7.2.1. *Kinetic Studies*

The rate constant, k, of the reaction (7.3) is a measure of the nucleophilic power of Y. It expresses the competition between X and Y to share

$$\mathrm{Y} + \mathrm{RX} \rightarrow \mathrm{RY} + \mathrm{X} \tag{7.3}$$

an electron pair with the electrophilic center in R. Equation (7.3) is simply an example of equation (7.2).

Among heterolytic reactions, bimolecular nucleophilic substitutions (7.4), which involve the formation of a transition state (I), yield rate constants that can be linked to nucleophilic characteristics.[1] The typical

$$Y \cdots R \cdots X$$
[I]

rate law for an S_N2 reaction is (7.5), and for a given substrate RX, the greater k_2, the greater the nucleophilic power of Y.

$$Y\!:\, + R{-}X \rightarrow YR + X\!: \tag{7.4}$$

$$v = k_2[Y][RX] \tag{7.5}$$

In fact, the rate law for reaction (7.3) is rarely as simple as equation (7.5). For metallic complexes, the mechanisms are more subtle,[41] and studies of nucleophilic substitution in square-planar complexes, for example those recently carried out with palladium(II)[42] or platinum(II),[43] yielded the rate law (7.6), where k_1 represents the rate constant of the unimolecular nucleophilic substitution. Even for nucleophilic substitution at a

$$k_{obs} = k_1 + k_2[Y] \tag{7.6}$$

saturated carbon atom, Sneen[44] has suggested that there is competition between the S_N1 and the S_N2 mechanism (or S_N2' if rearrangements intervene) involving an ion pair.

Indeed if one considers, according to Miller,[45] the energies of the initial, the final, and the intermediate state, two cases arise, linked to electron configuration and its reorganization: (a) a single-stage reaction, (b) reaction in several stages, as in (7.7). The stationary state ap-

$$\begin{aligned} &Y + RX \underset{k_2}{\overset{k_1}{\rightleftharpoons}} Y{-}R{-}X \\ &Y{-}R{-}X \xrightarrow{k_3} Y{-}R + X \end{aligned} \tag{7.7}$$

proximation leads to equation (7.8).

$$v = \frac{k_1 k_3}{k_3 + k_2}[Y][RX] \tag{7.8}$$

The energy profiles are given in Fig. 7.1. The case where $k_2 \gg k_3$ is encountered with an electronically saturated substrate without empty orbitals of low energy (C of an alkyl group, O of a peroxide), and a polarizable nucleophile, but this case is often complicated by ion-pair formation.[48] The other case ($k_3 \gg k_2$) is found for a substrate whose orbitals come into play in the bond being formed and when, in the intermediate complex, the substrate is bound simultaneously to the nucleophile and to the leaving group. This case is encountered, for example, in substitution at an aromatic carbon atom,[1,49] at carbonyl carbon,[7] and at sulfenyl[50] or sulfinyl[51] sulfur. We shall take up again the results for substitutions at

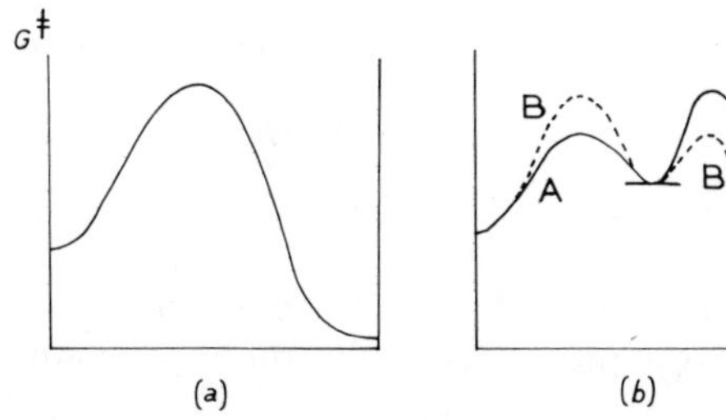

Fig. 7.1. Reaction coordinate diagram: (*a*) a one-step displacement mechanism, (*b*) an intermediate complex is formed. Case A: $k_2 > k_3$; case B: $k_2 < k_3$ (after Ref. 45). Example: 1-fluoro-2,4-dinitrobenzene reacting with PhS^- (case A) (after Ref. 46), and with MeO^- (case B) (after Ref. 47).

an unsaturated center, because in recent years they have been much studied.

The stability of the intermediate complex need not be so great that it can be isolated[52] but only sufficient for its formation constant, $K = k_1/k_2$, to be obtained by spectroscopic methods, among others.

7.2.2. Thermodynamic Studies

(a) Study of the Equilibrium (7.1)

Certain practical studies allow of the measurement of the equilibrium constant for (7.1). These include the addition of amines, alcohols, and thiols to the carbonyl group of three aldehydes studied by Jencks,[53] the addition of alcohols and thiols to the carbon–nitrogen double bond in azomethines,[54,55] and of various nucleophiles to the C=N bond in the quinazolinium cation,[56] and the Michael addition of carbanions of aryl-dinitromethanes to methyl vinyl ketone (butenone).[57]

This equilibrium constant is sometimes measured in an aromatic substitution, even when the intermediate complex cannot be isolated, as we have already said; thus Ritchie, in his study of reactions between cations and anions, measured this constant for the solvents water and methanol, for various nucleophiles reacting with aryldiazonium,[58,59] triarylmethyl,[59] and aryltropylium cations.[60,61] The addition compound between a nucleophile and an electron-deficient aromatic compound may well be isolable when the aromatic nucleus carries enough electron-attracting groups: These are the much studied 1:1 Meisenheimer-type complexes,[62] particularly those from 1,3,5-trinitrobenzene,[63] and from 1-chloro-2,4-dinitrobenzene studied by Crampton.[64]

In inorganic chemistry, the equilibrium (7.1) is related to the formation constants of metallic complexes, the subject of earlier studies than those quoted for organic chemistry. We have already mentioned these measurements[8] for Ag^+, and numerous values are assembled in various publications.[65]

(b) Study of the Equilibrium (7.2)

The equilibrium constant, K_Y^H, for the equilibrium (7.9) measures the basicity of Y towards the proton. Now this thermodynamic affinity for the

$$Y + H_2O \overset{K_Y^H}{\rightleftharpoons} YH^+ + OH^- \tag{7.9}$$

proton, which represents basicity, involves an entity exceptional in chemistry because of its small size and concentrated positive charge, and it seems probable that comparing basicity and nucleophilicity will lead to not very satisfactory conclusions, as we shall see in what follows. We may now cite the work of de la Mare and Vernon[66] on the one hand and of Hine[67] on the other, who have sought to overcome this obstacle. The former studied attack on a hydrogen atom in the elimination reaction between t-butyl chloride and phenoxide or thiophenoxide ions; the latter defined a carbon basicity linked to the equilibrium constant for (7.10). These two

$$ROH + Y \overset{K_Y^R}{\rightleftharpoons} RY^+ + OH^- \tag{7.10}$$

attempts allow the comparison of suitable magnitudes signifying in the first case basicity and nucleophilicity towards the electrophile H^+, and in the second, basicity and nucleophilicity towards a carbon center. In the HSAB concept, to which we shall return later, H^+ is a hard acid, a carbon atom in an alkyl group is a borderline case, and that of a carbonyl group is almost hard.[68] With the electrophile methyl,[67] the soft nucleophile CN^- gives a ratio K_Y^R/K_Y^H of 10^{15}, and when one passes from hard oxygen nucleophiles to soft sulfur nucleophiles (from MeO^- to MeS^-, from PhO^- to PhS^-) this ratio is increased by a factor of 10^8. Not all the examples treated by Hine are explained equally easily.

We may recall that determination of the equilibrium constants for (7.1) when they concern carbon is also an evaluation of the basicity of the nucleophile towards carbon, but unfortunately such measurements are but rarely possible.

In inorganic chemistry, Schwarzenbach[9] has chosen the methylmercuric ion, $MeHg^+$, as standard acid, and the values of the equilibrium constants for (7.11) indicate the following order of decreasing basicity of Y

$$MeHg(H_2O)^+ + YH^+ \rightleftharpoons MeHgY^+ + H_3O^+ \tag{7.11}$$

towards the acid $MeHg^+$:

$$I^- > Br^- > Cl^- > S^{--} > PEt_3 > CN^- > N_3^- > F^- > NH_3 > OH^-$$

As already stated, the order found in this competitive reaction between H^+

and $MeHg^+$ for capturing the electrons of Y differs somewhat from that determined[8] by use of Ag^+.

In a general way, it seems dubious to compare the results of a simple addition, such as (7.1), with those of a substitution, such as (7.2), because a supplementary unknown, X, is introduced, an unknown with several effects, since it intervenes through its affinity for and its bond strength with the electrophile (and the solvent). This remark partly explains why Hine's proposal[67] is of limited applicability: In fact equations (7.9) and (7.10) involve the nucleophile OH^-, which is not relevant to an absolute assessment of the nucleophilicity and basicity of a nucleophile Y towards an electrophile R.

We shall now follow the pathway of both theory and experiment to approach a knowledge, absolute we hope, of nucleophilicity.

7.3. *Prediction of Electron Donor–Acceptor Power by Means of a Two-Parameter Equation*

Pearson[28] has proposed putting the equilibrium constant of reaction (7.1) in the simplest possible form thus, (7.12), where S_A and S_B are acid- and base-strength constants that apply to the gaseous state, in the absence

$$\log K = S_A S_B \tag{7.12}$$

of the solvent, since recent results show that in the gaseous state, halides[69] or the anions[70] H^-, NH_2^-, OH^- do not show the same relative strengths as in solution.

Given the importance of water as a reaction medium, Pearson takes an arbitrary value of S_A for the proton ($S_A^{H^+} = 9.80$), which gives a series of values of S_B from $S_B^{Me^-}$ $(=5)$ to $S_B^{I^-}$ $(=-1)$: with $S_B^{OH^-} = 1.75$. This generates a series of values of S_A for aqueous media, viz., $S_A^{Hg^{2+}} = 5.9$ to $S_A^{Me^+} = 8.6$.

Unfortunately it is not possible to provide a universal order of acid or basic strength, since the strength of a group of bases, B, depends on the reference acid A, as has been known for a long time,[3] and as we have recalled in connection with Schwarzenbach's work.[8,9]

Applications of equation (7.12) are to be found in the literature, to which we now address ourselves, and which indeed show that a universal scale is hardly possible.

7.3.1. *The Brönsted Relation*

Since, by the very definition of a nucleophilic center, a connection between nucleophilicity and basicity is expected, especially for a series of

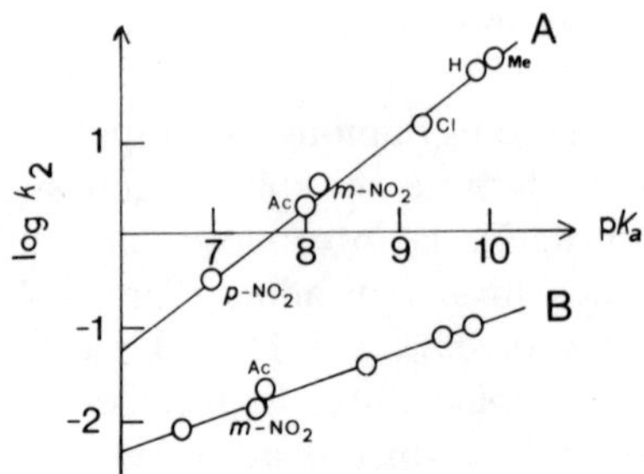

Fig. 7.2. Brönsted plots for the reactions of substituted phenoxides with (A) ethyl chloroformate in acetone (80%) at 0°C (pH = 8.0); (B) 3-bromopropan-1-ol in water at 61°C (pH = 9.5); (after Ref. 75).

reagents with the same nucleophilic center in comparable environments, an equation analogous to that proposed by Brönsted[71] to interpret acid–base catalysis is followed in certain cases. There is therefore a linear relationship between the logarithm of the rate constant and pK_a for the conjugate acid of the nucleophile (7.13). This holds good[72] for reactions in which the

$$\log k = \beta \mathrm{p}K_a + \text{const} \tag{7.13}$$

substrate is chloroacetate and the nucleophilic center is an oxygen atom, viz., $ClCH_2CO_2^- + Y^- \rightarrow YCH_2CO_2^- + Cl^-$; similarly for the series of *meta*- or *para*-substituted phenoxide or thiophenoxide ions reacting with the substrates phenethyl bromide[73] or *para*-substituted benzyl bromides,[74] or, as Fig. 7.2 shows, for substituted phenoxides in reaction with ethyl chloroformate (chlorocarbonate) or 3-bromopropan-1-ol.[75] Equation (7.13) has also been verified for thiols and glutathione[76] reacting with "benzene oxide," for a series of anions of substituted thiols and of amino acids reacting with the α,β-unsaturated compound, acrylonitrile,[77] as is indicated in Fig. 7.3, and for the addition of the anions of various 1,1-dinitroalkanes to methyl acrylate.[78]

Equation (7.13) is relatively easily used, and when it is represented by a straight line, the value of β is associated with the extent of bond formation in the transition state.[39] It can be used to support a mechanism: for example, that of the acylation of hydroxamic acids,[79] for which the points for $\log k_2$ and pK_a lie on the straight line obtained for hydroxylamines and amidoximes reacting with *p*-nitrophenyl acetate in water.

A relationship also exists, but is nonrectilinear, in other studies: for example, for the substrate *O,O*-diphenyl phosphorochloridothioate [$(PhO)_2PSCl$] reacting with anions in which the nucleophilic center is oxygen or sulfur[80] (Fig. 7.4).

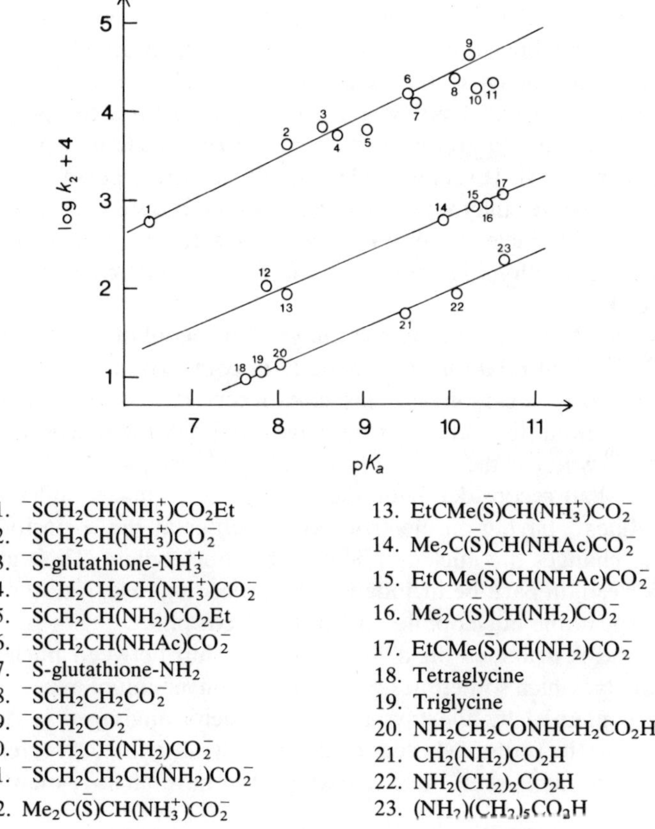

Fig. 7.3. Brönsted plots for the reactions of mercaptide groups attached to primary carbon atoms (1–11), or attached to tertiary carbon atoms (12–17), and of amino groups attached to primary carbon atoms (18–23), with acrylonitrile at 30°C (after Ref. 77).

The relationship disappears entirely if there is steric hindrance, if a Coulombic effect intervenes, or if there is a change in electronic environment, as is shown in various studies. *Ortho*-substituted compounds do not conform to and may even invert the relationship followed by *meta*- or *para*-substituted compounds.[81] Steric effects are encountered also for the reactions of alkoxides[7] with *p*-nitrophenyl acetate in water at 25°C, for those of primary, secondary, and tertiary aliphatic amines with methyl iodide in benzene,[82] for those of a series of tertiary amines with *p*-nitrophenyl acetate,[24] and for the reactions of alkyl bromides with hydroxide ions.[83] Encumbered compounds have nucleophilicity notably less than that expected from their basicity. Steric effects are similarly involved in substitutions at sulfur[84] or phosphorus.[85]

Moreover, the linear relationship (7.13) is obeyed for simple uncharged nucleophiles but not when the nucleophiles are charged for reactions of *p*-nitrophenyl acetate or *p*-nitrophenyl chloroacetate.[24,86] Also, in the attack on Sarin, O=PFMe(OPr^i), by phenoxides, the presence of a quaternary ammonio group in the benzene ring leads to a point[87] not in accord with (7.13). It seems evident that a substrate possessing a neutral phosphorus center may well react with a different rate constant according to whether the nucleophile is neutral or negatively charged. In aqueous solution, this Coulombic effect may influence the rate constant by a factor of 10–100.

Various authors have noted the great nucleophilicity of the peroxide anion[24,26,88,89] in relation to its basicity in comparison with that of OH^-. More precisely, Pearson and his co-workers have shown that in 50% aqueous acetone the ratio of rate constants is 35 for reactions of benzyl bromide,[90] whereas the ratio of basic strength[65] is 1/4.

We shall reconsider later the case of those nucleophiles or supernucleophiles[91] having an electron donor center at the α-position, which radically changes the nucleophilicity of the nucleophile. Their anomalous behavior can, in part, be included in the preceding case, viz., the influence of a charge in the nucleophile and in the electrophile.

These restrictions on a linear relationship between nucleophilicity and basicity, which sometimes go as far as an inversion of expected results, can be explained by the fact that the basicity envisaged is defined by reference to the proton whereas nucleophilicity is defined by reference to a carbon atom, as Parker[92] has remarked. We have already mentioned the

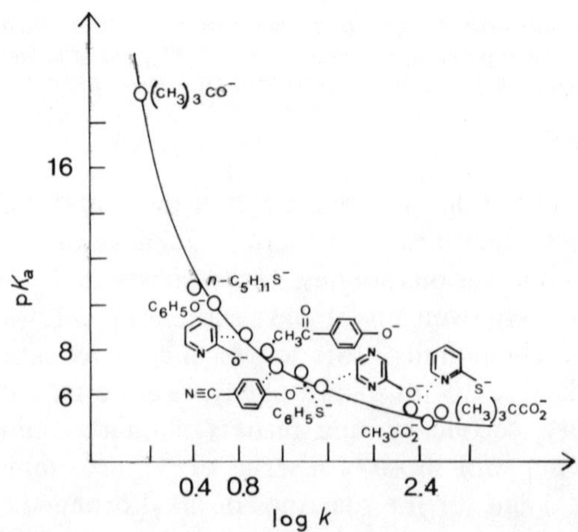

Fig. 7.4. Plot of log (second-order rate constants for anions) vs. pK_a for the reactions of *O,O*-diphenyl phosphorochloridothioate at 58°C in 90% t-butyl alcohol–dioxan (after Ref. 80).

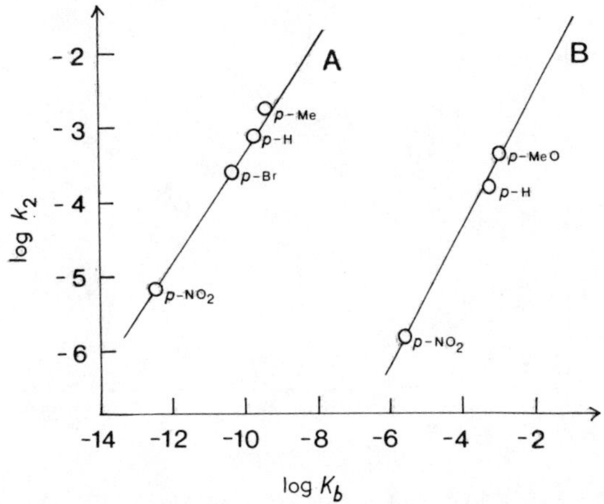

Fig. 7.5. Brönsted plots for the reactions of DDT in ethanol with (A) substituted arenethiolates at 75°C, (B) substituted phenoxides at 45°C (after Ref. 93).

two attempts to offset this drawback. In fact studies of bimolecular eliminations conform well with equation (7.13), as in the case of the dehydrochlorination of DDT in ethanol by substituted phenoxide and thiophenoxide ions[93] (Fig. 7.5), but these results require a pure E_2 mechanism.[66] We shall meet in the next section examples of the use of carbon basicity.

7.3.2. Extension of the Brönsted Relation

When the equilibrium constant for (7.1) can be measured, an equation fully analogous to (7.13) is used, viz., (7.14). Additions of cations to

$$\log K = \delta \mathrm{p}K_a + \text{const} \tag{7.14}$$

anions give a good straight line in accordance with (7.14), as for Fe^{3+} and a series of *ortho*-, *meta*-, or *para*-substituted phenoxides,[94] and for a series of metallic cations (Fe^{3+}, Cr^{3+}, In^{3+}, UO_2^{2+}) reacting with a series of anions (halides, nitrate, acetate, chloroacetate).[95] Shmidt utilized (7.14) to predict the extraction of acids by amine salts[96] or by metallic cations,[97,98] e.g., Pu^{IV}, Np^{IV}, or U^{VI}, in the form of salts with nucleophilic anions.

Equation (7.14) is obeyed by addition reactions between the *N,O*-trimethylenephthalimidium cation and various amines and alcohols.[99] It is also obeyed for the formation of 1:1 Meisenheimer-type complexes, e.g., the addition of substituted thiophenoxides to 1,3,5-trinitrobenzene.[63]

The study of these additions allows of progress in the search for a predictive linear relationship in kinetics, when the Brönsted relation (7.13)

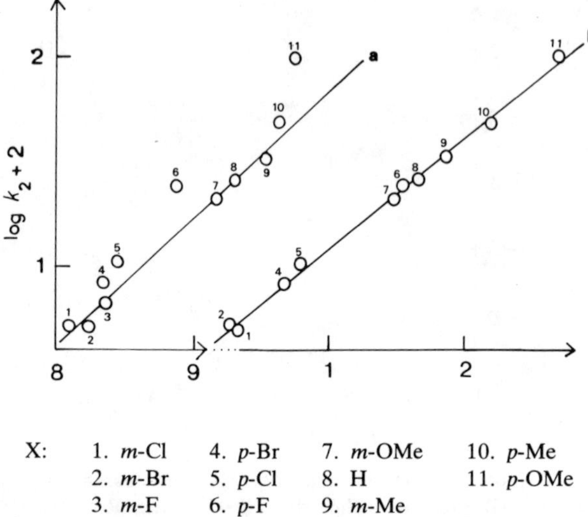

Fig. 7.6. Plot of the rate constants for the reactions of *p*-nitrophenyl acetate with X-substituted arenethiols in ethanol at 22°C vs. (*a*) pK_a of the corresponding arenethiols, (*b*) $\log K_{\mathrm{TNB}}$, where K_{TNB} is the equilibrium constant for formation of the adduct between arenethiolate and 1,3,5-trinitrobenzene (after Ref. 101).

is not obeyed. Thus, whereas for the addition of various amines[100] or substituted thiophenoxides[64] to 1-chloro-2,4-dinitrobenzene, equation (7.13) is not obeyed, Crampton, on the contrary, found the linear relation (7.15) to be obeyed.

$$\log k = \eta \log K + \text{const} \tag{7.15}$$

Equation (7.15) is also obeyed for the reactions of substituted arenethiols with *p*-nitrophenyl acetate,[101] and is obtained from (7.13) in which pK_a for the conjugate acid of the nucleophile is replaced by log K for the solvent used for the reaction (ethanol). (K is the equilibrium constant for the addition of the arenethiol to 1,3,5-trinitrobenzene.) These results are displayed in Fig. 7.6. We note that whenever substituted nucleophiles are studied, the investigator approaches them also from the viewpoint of Hammett relations. We shall not develop the study of nucleophilicity from this viewpoint and we refer the reader to the papers concerned and to Ref. 102.

Although of much interest, these recent studies on measuring carbon basicity are not yet very numerous, and we now proceed to the history of the utilization of the two-parameter equation (7.12).

7.3.3. *The Swain–Scott Equation*

Swain and Scott[6] sought to define an order of nucleophilicity empirically. To this end they took a standard substrate, methyl bromide, and as the zero of the nucleophilicity scale, water. The suggested equation is (7.16), where s is a constant characteristic of the substrate ($s=1$ for

$$\log k/k_0 = sn \tag{7.16}$$

MeBr), k_0 the rate constant for the reaction of water, and n the nucleophilicity of the compound reacting with MeBr. Thus when MeBr is the substrate, equation (7.16) leads directly to the value of n for each nucleophile in aqueous solution at 25°C.

Then, given the values of n for a number of nucleophiles, the slope of the rectilinear plot obtained for the measured values of $\log k/k_0$ for a given substrate yields the constant s for the substrate in question, e.g., $s=0.715$

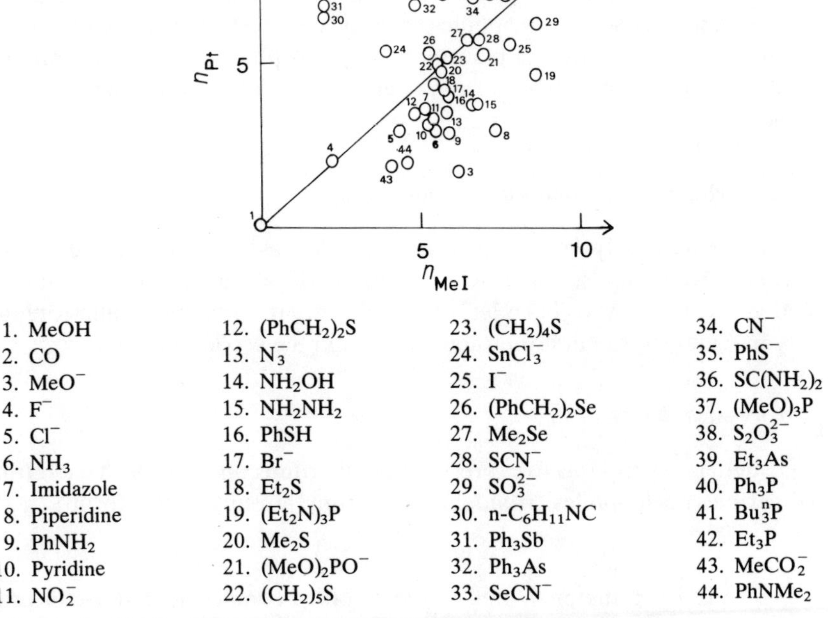

Fig. 7.7. Swain–Scott nucleophilic parameters for the reactions *trans*-$Pt(py)_2Cl_2 + Y \rightarrow$ *trans*-$Pt(py)_2ClY^+ + Cl^-$, and $MeI + Y \rightarrow MeY^+ + I^-$ (after Ref. 36).

for 3-methoxysulfonyl-1-methylpyridinium perchlorate.[103] Knowing the value of s other nucleophilic constants n can be determined, e.g., in the preceding examples those of various phosphonate anions. It is possible to construct tables that directly permit *a priori* calculation of k. This equation has been extended to inorganic substitution reactions,[104] in which case the standard substrate is *trans*-dichlorodipyridineplatinum(II) [*trans*-$Pt(py)_2Cl_2$]. Finally, in organic chemistry the standard substrate has been changed by taking MeI[36,105] instead of MeBr, and in both inorganic and organic chemistry methanol instead of water has been taken as reference nucleophile.[36,106]

Equation (7.16) is merely a kinetic translation of the equation (7.12) proposed by Pearson,[28] and it presents the same drawbacks. For a given substrate the values of n permit the prediction of yields for the relevant reactions,[107] but these values only persist from one substrate to another when these are closely similar, as for various alkyl chlorides in the case of *O*- and *C*-alkylation of enolates in dimethyl sulfoxide,[108] or as when one passes from Pt^{II} to Pd^{II} tetracoordinated in given ways.[109] The deviations increase when the substrates are different, and in particular there is no linear relationship[36] between n_{MeI} and n_{Pt}, as Fig. 7.7 shows. This is comparable to the fact that, in the application of equation (7.12), the order of nucleophilic reactivity varies when the standard acid, Ag^+, is replaced[8,9] by $MeHg^+$. This has led to the drawing up of tables giving for a given solvent and for each electrophilic center of the substrate a value of n for a given nucleophile: n_P (displacement on phosphorus), n_O (oxygen), n_S (sulfur), n_{Car} (aromatic carbon), n_{H^+}, and $n_{C=O}$ (Refs. 6 and 24). This last scale now requires our attention.

7.3.4. *Jencks's Equation and Ritchie's Equation*

Independently of one another, Jencks, as a result of studies of carbonyl compounds, and Ritchie, in his study of the reactions of anions and aryl cations, have been led to introduce new comparable nucleophilicity parameters, the first thermodynamic and the second kinetic.

Jencks's Equation

Jencks[53] proposes to express the equilibrium constant for the addition of various nucleophiles to aldehydic carbonyl groups in the form (7.17),

$$\log K_0 = \Delta\gamma + A \tag{7.17}$$

where Δ is a parameter related to the carbonyl compound and γ concerns the nucleophile. More precisely, γ is a measure of the affinity of the nucleophile for carbon in comparison with that for hydrogen. K_{HX} and

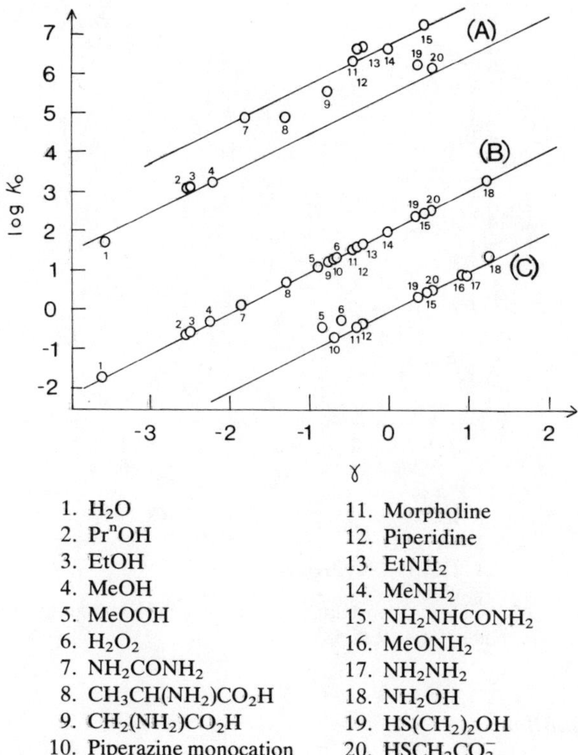

Fig. 7.8. Jencks plots for the addition of nucleophiles to (A) formaldehyde, (B) pyridine-4-carboxaldehyde, (C) *p*-chlorobenzaldehyde (after Ref. 53).

K_{MeNH_2} are the respective equilibrium constants for addition of the

$$\gamma = \log K_{HX}/K_{MeNH_2} \tag{7.18}$$

nucleophilic compound studied and of monomethylamine to the reference carbonyl group in pyridine-4-carboxaldehyde.

The experimental determination of the parameter γ for amines, alcohols, and thiols leads to a verification of equation (7.17) for 3 aldehydes, as Fig. 7.8 shows. The parameter Δ has the value unity whatever the aldehyde and the nucleophilic center (N, O, S) within the limits of experimental error. These values of γ have been reproduced in the study of the addition of nucleophiles to the C=N bond of the quinazolinium cation.[56] The equation (7.17), where K_0 becomes the equilibrium constant for the addition of these nucleophiles to this C=N double bond, is obeyed perfectly (Fig. 7.9) with $\Delta = 1$, showing that for the nucleophiles involved, the circumstances of the addition to C=O and C=N are very similar.

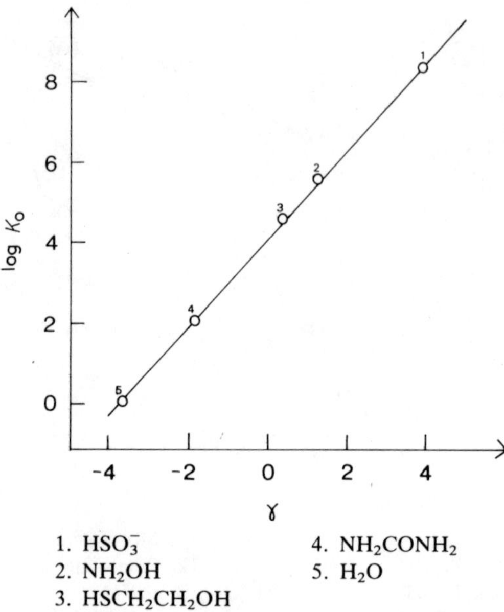

1. HSO_3^-
2. NH_2OH
3. $HSCH_2CH_2OH$
4. NH_2CONH_2
5. H_2O

Fig. 7.9. Plot of log of equilibrium constants for addition of nucleophiles to the quinazolinium cation vs. Jencks parameters (after Ref. 56).

Ritchie's Equation

As a consequence of work bearing essentially on a kinetic study of the reactions between nucleophiles and aryldiazonium, triarylmethyl, and aryltropylium cations in various solvents, Ritchie and Virtanen[61] introduced the parameter N_+ of a nucleophile in a given solvent,[110] as in (7.19),

$$N_+ = \log k_N^{PNMG}/k_{H_2O}^{PNMG} \tag{7.19}$$

where k_N^{PNMG} is the rate constant for the reaction of the nucleophile, N, in the solvent considered with the cation of p-nitromalachite green [bis-(p-dimethylaminophenyl)-p-nitrophenylmethyl cation], and $k_{H_2O}^{PNMG}$ is that for the reaction of water with the same substrate.

The group of results obtained shows that this equation is independent of the particular cation (as indicated by Fig. 7.10) and can even be generalized to uncharged nucleophiles such as amines (always reacting with the three classes of cations already specified).[111] Figure 7.11 shows that equation (7.20) is confirmed over a wide range (10^{12}) of values of the rate

$$N_+ = \log k_N^R/k_{H_2O}^R \tag{7.20}$$

constant for widely varied nucleophiles; only water, hydrogen, and methoxylamine engender unsatisfactory points, whereas α-effect nucleophiles such as semicarbazide and phenylhydrazine conform to the line.

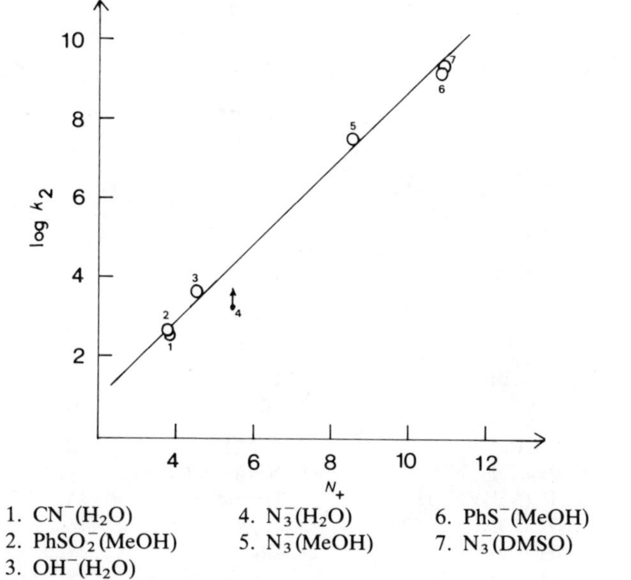

Fig. 7.10. Ritchie plots for the reactions of benzenediazonium ion with nucleophiles in different solvents (after Ref. 61).

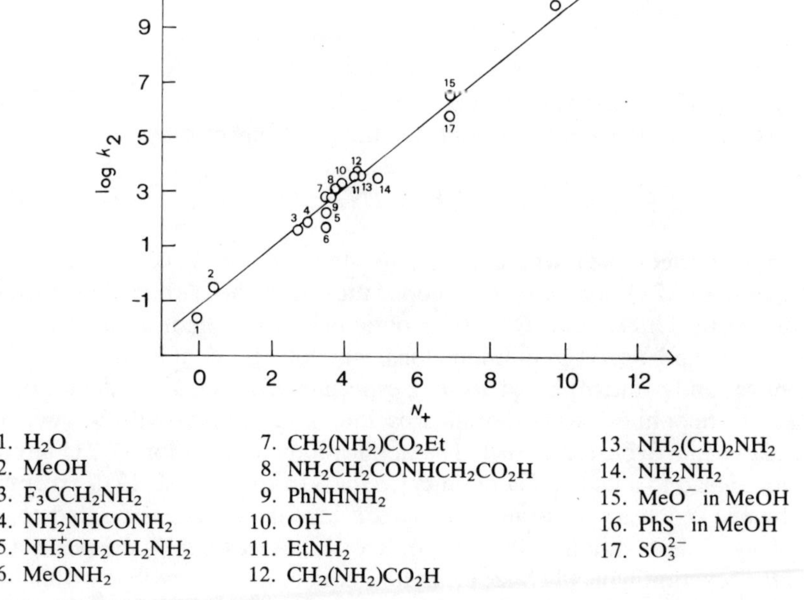

Fig. 7.11. Ritchie plots for the reactions of *p*-dimethylaminophenyltropylium ion with nucleophiles (after Ref. 111).

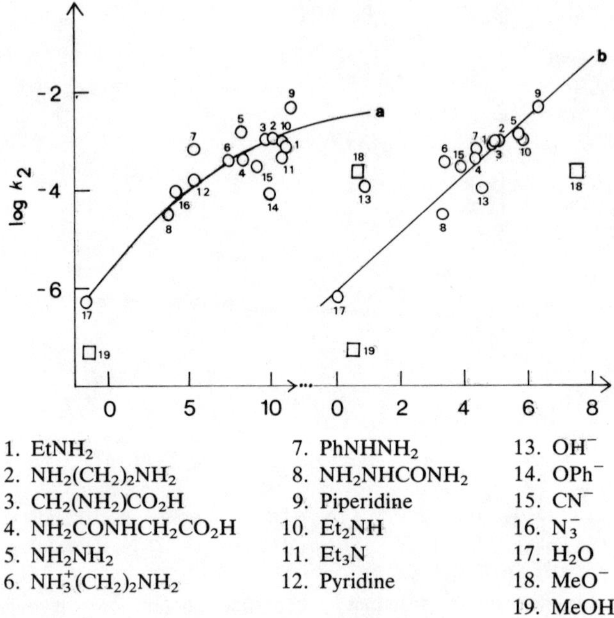

Fig. 7.12. Plot of the rate constants for the reactions of ethylene oxide with nucleophiles as a function of (*a*) the basicity of pK_a of the nucleophile, (*b*) the Ritchie parameter N_+. Solvents: water (1–17); methanol (18–19). (After Ref. 112.)

Equation (7.21) derived from (7.20), in the way that equations (7.17) and (7.18) are interrelated, is obeyed for the reactions of ethylene oxide with

$$\log k_N = \phi N_+ + B \tag{7.21}$$

various nucleophiles, whereas the Brönsted equation is not[112] (Fig. 7.12). Equation (7.21) finds a field of application with the carboxylic substrates studied by Jencks, and Ritchie[111] obtained an excellent straight line for acetoxy- and acetyl-pyridinium ions, but less good, it is true, for 2,4-dinitro- and *p*-nitro-phenyl acetate, especially with regard to the following four nucleophiles: water, semicarbazide, 2,2,2-trifluoroethylamine, and azide (weak Brönsted bases). The applications of equation (7.21) already quoted lead to a value of ϕ of unity [as for Δ in equation (7.17)], except for ethylene oxide as substrate. If it is considered that N_+ is related to the energy of desolvation of the nucleophile,[110] ϕ measures[112] the percentage of the reaction typically S_N2; or, better, equation (7.21) with $\phi = 1$ characterizes a reaction in which the attack of the nucleophile is rate-determining, as Ritchie[113] has shown, in close collaboration with Jencks, by calculations for esters. This work indicates that the application of equation (7.20) (until then to 26 electrophiles and 52 nucleophiles) is only a beginning and warrants a critical and attentive follow-up.

Before the full development of these recent studies, other attempts appeared in the literature to express donor–acceptor power, particularly by increasing the number of parameters characteristic of the acceptor and of the donor.

7.4. *Prediction of Electron Donor–Acceptor Power by Means of a Four-Parameter Equation*

Instead of characterizing each acid or each base by a single parameter S_A or S_B, Pearson also characterizes them by a second parameter expressing their "softness", σ_A or σ_B. Thus instead of equation (7.12) Pearson[28] proposes equation (7.22).

$$\log K = S_A S_B + \sigma_A \sigma_B \tag{7.22}$$

This numerical relation involving a double scale is analogous to the oxi-base scale proposed by Edwards[5] and by Davis,[114] which takes account of polarizability as well as basicity, and it is reminiscent of the proposal of Winstein *et al.*[115] to express the effect of the solvent.

7.4.1. *A Relation for Solvolysis*

The Swain–Scott equation[6] (7.16) is derived from an equation proposed to take account of the effect of varying the solvent on the rate constant of solvolysis, a four-constant relation[116] (7.23), where k is the rate

$$\log (k/k_0) = lN + mY \tag{7.23}$$

constant of solvolysis for a given solvent, k_0 is that for 80% aqueous ethanol, l and m are constants characteristic of the substrate, N is the nucleophilicity of the solvent, and Y its ionizing power. In general, two substrates are taken as standards: either t-butyl chloride and methyl bromide for solvents derived from carboxylic acids,[117] or methyl tosylate and 2-adamantyl tosylate for various acids and alcohols studied by Schleyer.[118] Tables of N and Y parameters serve to identify solvents of low nucleophilicity and of strong ionizing power, such as 2,2,2-trifluoroethanol[119] or hexafluoropropan-2-ol.[20] In conjunction with tables of the parameters m and l, they also allow the effects of substituents in vinylic substrates to be understood.[121]

7.4.2. The Edwards Equation

In view of the limited applicability of equation (7.16), instead of considering an empirical relationship based on experiment, an attempt has been made to relate nucleophilicity to thermodynamic magnitudes characteristic of the species considered,[122] and once again four parameters have been suggested.

In the first formulation, Edwards[18] introduced the electrode potential of the nucleophile, i.e., a quantity expressing greater or less tendency of the nucleophile to be oxidized, and the relative basicity of the nucleophile with respect to the proton. The equation is (7.24), where k_0 is the rate constant

$$\log (k/k_0) = \alpha E_n + \beta H \tag{7.24}$$

of reaction (7.3), in which the nucleophile is H_2O, taken as the standard nucleophile, α and β are the substrate constants, and H and E_n characterize the nucleophile and are defined to take zero value for the nucleophile H_2O. $H = pK_a + 1.74$ and is the relative basicity of the nucleophile with respect to the proton (K_a in terms of concentrations); $E_n = E^0 + 2.60$, and E^0 is the standard potential (American convention) of the oxidative dimerization $2X^- \rightleftharpoons X_2 + 2e^-$, and α and β are parameters adjusted to express best the experimental results.

Davis has further related α to the reduction potential of the substrate and β to its acidity constant. This symmetrization of equation (7.24) applies well when the electrophilic center is a saturated carbon atom, and finds important verification in the formation of metallic complexes.[123,124] Equation (7.24) presents the drawback of leading in certain cases to negative values of β, which would mean that nucleophilicity and basicity have opposite influences, which appears impossible. Also, in a second stage, Edwards[5] has suggested that E_n depends on the basicity, and may be regarded as consisting of two terms, one related to the polarizability, the other to basicity with respect to the proton, viz., (7.25), where P is the

$$E_n = aP + bH \tag{7.25}$$

polarizability of the nucleophile relative to that of water, and by definition $P = \log R/R_{H_2O}$, R being the molar refraction at infinite wavelength. Equation (7.24) takes the form (7.26), where A and B are new substrate

$$\log (k/k_0) = AP + BH \tag{7.26}$$

constants, related to the old ones by $A = 3.60\alpha$ and $B = 0.0624\alpha + 3.60\beta$. This time the coefficients, B, are never negative, which supports the idea of a basicity contribution in the term E_n.

Elements giving abnormally elevated nucleophilicity in relation to their basicity, e.g., I^- and S^{2-} compared with Cl^- and O^{2-}, respectively,

Table 7.1. Edwards Constants

Lewis Acids[124]	α	β	Lewis Acids[124]	α	β
Hg^{2+}	5.786	−0.031	Ba^{2+}	1.786	0.411
Cu^{+}	4.060	0.143	Pb^{2+}	1.771	0.110
I_2[a]	3.04	0.000	Mn^{2+}	1.438	0.166
Ag^{+}	2.812	0.171	Mg^{2+}	1.401	0.243
Au^{3+}	2.442	0.353	Zn^{2+}	1.367	0.252
Cu^{2+}	2.259	0.233	Ca^{2+}	1.073	0.327
Cd^{2+}	2.132	0.171	H^{+}	0.000	1.000
Fe^{3+}	1.939	0.523	Al^{3+}	−0.749	1.339

[a] Ref. 18.

possess an increased polarizability, while the feebly nucleophilic F^- and OH^- have a low polarizability. Edwards and Pearson[26] and Bunnett[16] have discussed, with the aid of numerous examples, how reaction velocity may be explained in terms of two properties: basicity and polarizability, which operate with different weights for different substrates.

7.4.3. *Pearson's Equation: The Principle of Hard and Soft Acids and Bases*

(*a*) *Classification of Acids and Bases*

The parallelism between Pearson's equation (7.22) and Edwards's equation (7.24) is close. Table 7.1 indicates that β is in the same proportion greater as the acid is more polarizing; it thus varies as does S_A. The quantities H and S_B characterize the nucleophilic reagent and βH may be identified with $S_A S_B$. If αE_n is identified with $\sigma_A \sigma_B$, this involves σ_B being large for bases that are easily oxidizable, such as I^-, and small (or even negative) for difficultly oxidizable bases such as F^-. Likewise σ_A varies with α, that is to say that it is in the same proportion greater as the acid is more polarizable (large ion with low charge) and possesses empty *p* or *d* atomic orbitals (e.g., Ag^+).

If we recall the results of Schwarzenbach,[9] who interpreted the competition between the two acids H^+ and $MeHg^+$ in combining with bases, we can classify these bases into two categories:

(i) Those whose donor atom is N, O, or F and which prefer to bind to the proton. These donor atoms have the following characteristics: strongly electronegative, of low polarizability, oxidized with difficulty, good hydrogen-bond acceptors. Their occupied frontier orbitals are of low energy. The term "hard base" is given to such donors to convey the fact that they hold their electrons very strongly.

Table 7.2. Lewis Bases[28]

Hard	Soft	Borderline
H_2O, OH^-, F^-	R_2S, RSH, RS^-	$PhNH_2$, C_5H_5N, N_3^-,
$CH_3CO_2^-$, PO_4^{3-}, SO_4^{2-}	I^-, SCN^-, $S_2O_3^{2-}$	Br^-, NO_2^-,
Cl^-, CO_3^{2-}, ClO_4^-, NO_3^-	R_3P, R_3As, $(RO)_3P$	SO_3^{2-}, N_2
ROH, RO^-, R_2O	CN^-, RNC, CO	
NH_3, RNH_2, N_2H_4, NH_2^-	C_2H_4, C_6H_6	
	H^-, R^-	

(ii) Those whose donor atom is P, S, I, Br, Cl, or C and which prefer to bind to the mercury cation. These donors possess characteristics that are the inverse of the former: slightly electronegative, of high polarizability, and easily oxidized. Their occupied frontier orbitals are of higher energy. These bases are described as "soft," expressing the looseness of binding of their valence electrons.

Although Pearson has described elsewhere[32] the principle of hard and soft acids and bases, we give in Table 7.2 the classification of bases and in Table 7.3 that of acids. The acids that do not present pairs of electrons to participate in the valence state, whose vacant frontier orbitals are of high energy, such as metallic ions of class A,[33] are hard acids; soft acids, of which the metallic ions of class B[33] are a part, as well as organic cations and certain organic molecules, have a low positive charge and a large size, and possess available pairs of electrons in the valence state. In diverse independent ways acids and bases may be placed in three categories: hard, soft, and borderline, and Tables 7.1 and 7.3 show well the remarkable parallelism between the results of Edwards and those of Pearson. High values[5] of α and low values of β correspond to soft acids,[37] while lower values of α and higher values of β relate to hard acids.

Thus soft acids form stable adducts with soft bases, i.e., those that are strongly polarizable and good reducing agents, but that are not necessarily strong bases with respect to the proton. Hard acids, on the other hand, of which H^+ is one, give stable addition compounds with hard bases, which are strong bases with respect to the proton. In this case the polarizability or the ease of reduction of the base plays a negligible role.

This qualitative treatment might perhaps be made quantitative with the aid of equation (7.22), but this four-parameter equation has proved insufficient, like the Edwards equation (7.24), to account for the complex phenomenon of the combination of an electron donor and an electron acceptor, since it depends on an extremely sensitive reference system. Instead of identifying the intrinsic strength parameter S with pK_{H^+} and the softness parameter σ with pK_{MeHg^+} quantitatively, it is preferable to leave

Table 7.3. Lewis Acids[28]

Hard	Soft	Borderline
H^+, Li^+, Na^+, K^+	Cu^+, Ag^+, Au^+, Tl^+, Hg^+	Fe^{2+}, Co^{2+}, Ni^{2+},
Be^{2+}, Mg^{2+}, Ca^{2+}, Sr^{2+}, Mn^{2+}	Pd^{2+}, Cd^{2+}, Pt^{2+}, Hg^{2+}	Cu^{2+}, Zn^{2+}, Pb^{2+},
Al^{3+}, Sc^{3+}, Ga^{3+}, In^{3+}, La^{3+}	$MeHg^+$, Pt^{4+}, Te^{4+}, $Co(CN)_5^{2-}$	Sn^{2+}, Bi^{3+}, Rh^{3+},
Gd^{3+}, Lu^{3+}, N^{3+}, Cl^{3+}		Sb^{3+}
Cr^{3+}, Co^{3+}, Fe^{3+}, As^{3+}, $MeSn^{3+}$	Tl^{3+}, $TlMe_3$, BH_3, $GaMe_3$	Ir^{3+}, SO_2, NO^+,
Si^{4+}, Ti^{4+}, Zr^{4+}, Th^{4+}, U^{4+}	$GaCl_3$, GaI_3, $InCl_3$	Ru^{2+}, Os^{2+}, R_3C^+,
Pu^{4+}, Ce^{3+}, Hf^{4+}, WO^{4+}, Sn^{4+}	RS^+, RSe^+, RTe^+	$C_6H_5^+$, GaH_3,
UO_2^{2+}, Me_2Sn^{2+}, VO^{2+}, MoO^{3+}	I^+, Br^+, HO^+, RO^+	BMe_3
BF_3, $B(OR)_3$, $BeMe_2$	I_2, Br_2, ICN, etc.	
$AlMe_3$, $AlCl_3$, AlH_3	Trinitrobenzene, chloranil, quinones, tetracyanoethylene, etc.	
RPO_2^+, $ROPO_2^+$		
RSO_2^+, $ROSO_2^+$, SO_3		
Cl^{7+}, I^{7+}, Cr^{6+}, I^{5+}	O, Cl, Br, I, N, RO, RO_2	
RCO^+, CO_2, NC^+	Metallic atoms	
Molecules capable of forming hydrogen bonds	Carbenes, CH_2	

the parameters undetermined and to use qualitatively the properties of acids and bases.[125] However that may be, the Edwards equation is seen to be verified in numerous examples, and the introduction of the HSAB concept that arose next permits their discussion in a new light.

(*b*) *Competition between Basicity and Polarizability*

In reaction (7.3) the basicity of Y plays an important role in governing the rate of reaction, if, in the transition state (I) (p. 316), R resembles a proton, that is to say, an increasing charge at the electrophilic center of the substrate causes the rate to increase. In fact, the basicity of the nucleophile is the dominant factor governing the rates of substitution in which the electrophilic center of the substrate is hydrogen, carbon of a carbonyl group,[126] trigonal boron, tetrahedral boron, sulfur in a sulfonyl group,[127] or tetrahedral phosphorus. We have just listed some rather hard acids.

Of course, increase in polarizability of Y entails a decrease of activation energy. Strongly polarizable nucleophiles possess empty orbitals of relatively low energy, so displacement reactions are in the same proportion more rapid as the substrates present fully occupied outer orbitals, which may easily overlap those of Y. Thus the polarizability of the nucleophile dominates when the electrophilic atom is tetrahedral carbon,[126] trivalent nitrogen,[128] peroxidic oxygen, or Pt(II); these are rather soft acids.

As far as displacements at divalent sulfur or aromatic carbon are concerned, basicity and polarizability of the nucleophile govern reaction velocity in an identical way, although certain work is rather in favor of polarizability in the case of nucleophilic substitution at aromatic carbon.[129]

Ambident Ions. One may try to apply the idea of competition between basicity and polarizability, that is to say the hard–soft concept, to ambident ions, which are nucleophilic reagents possessing two centers, one of which may be preferentially attacked by an electrophile. There are numerous examples in the literature and we take only some from among the most recent. The sulfinate anion reacts predominantly with a soft alkylating agent (MeI) through its soft S atom and with a hard reagent through the harder oxygen atom.[130] Recently a literature study[131] discusses the case of bridged carbonyl compounds, where CO is bound through oxygen to a hard acid, such as Al^{3+}, and through C to iron or cobalt cations, which are less hard acids. Sulfur is a softer base than nitrogen; so a process involving S appears more probable than one via N in the action of the soft nucleophiles CN^- and PBu^n_3 on *N*-*p*-tosylsulfilimine.[132] Likewise RSCN forms a more stable adduct with phenol, a hard acid, than with iodine, a soft acid, whereas RNCS gives inverse results but with a very small difference.[133] Moreover, in $Mo(NCS)_5$, Mo is bound to N.[134] Examples taken from metal complexes of class B[33] provide a contradiction. Thus it is the N that is bound to Ir(I), a soft acid, and not the S, in the complex *trans*-$Ir(PPh_3)(CO)NCS$,[135] and Pearson[136] has been led by examples of this type to impose a restriction to the symbiotic effect advanced by Jørgensen.[137] According to Jørgensen, if a given center is surrounded by several ligands, the addition product will be the more stable when the bases are of the same nature. Thus in inorganic chemistry the stability of complexes, in spite of the high state of oxidation of the central atom, such as AsS_4^{3-} or $Mo(SCN)_6^-$,[138] may be explained by the intervention of a large number of soft groups for stabilization,[139] except when these soft ligands are in the *trans* position, for then they produce an antisymbiotic effect.[136]

Solvation. Parker[11] divided solvents into two classes: protic solvents, such as H_2O, MeOH, or $HCONH_2$, solvate anions by an ion–dipole interaction and by a hydrogen bond that is the more stable when the anion is small and highly charged, that is to say is a hard base; and dipolar aprotic solvents, such as DMF, DMA, DMSO,[140] Me_2CO, MeCN, $PhNO_2$, sulfolan, and HMPT,[141] in which the anions are solvated by interactions of the soft-acid–soft-base type, the more strongly when the anion is larger, that is to say highly polarizable.

The specific solvation of anions influences the rate of the reaction and its products.[142] Thus, in protic solvents, solvation decreases strongly in the following series:

$$OH^-, F^- \gg Cl^- > Br^- > N_3^- > I^- > SCN^- > \text{picrate}$$

whereas in dipolar aprotic solvents it decreases slightly in the opposite direction.

The transition state, produced by the reaction of an anionic nucleophile and a neutral substrate, behaves as a large anion; it will be more highly solvated in dipolar aprotic solvents.[143] Thus the S_N2 reactions of nucleophilic anions are much more rapid in dipolar aprotic solvents than in protic solvents, as is indicated on the one hand by the comparison of results for DMF and, respectively, methanol[144] and ethanol,[145] or for benzene and methanol,[146] and, on the other hand, by general reviews.[147,148] The rate will be the more augmented on passing from a protic solvent to a dipolar aprotic solvent when the anion is smaller and hence has a denser charge. These various results are confirmed by studies on 2-fluoropyridine or 2-bromopyridine with MeS^- and PhS^- on the one hand in methanol, and on the other hand in HMPT.[149] As regards the formation of ion pairs between the anionic nucleophile and the cation that is inevitably in the reaction medium, an increase in the polarity of this medium can increase the reactivity of the ionic nucleophile by promoting the disaggregation of groups of ion pairs or the formation of reactive solvated ion-pairs or even by facilitating ionic dissociation.[23] This increase in polarity can occur on changing dipolar aprotic solvents or by adding a protic solvent, although this latter method may be difficult to quantify. In fact in the majority of cases the reactivity begins by increasing when one adds a protic solvent but diminishes thereafter.[150] This diminution of reactivity in protic solvents is more or less general,[23] except for nucleophiles, which are the anion of the solvent: OEt^- in ethanol,[151] OH^- in water,[113] where probably there is a mechanism of internal catalysis by proton transfer,[23,151] acid–base catalysis playing an important part for these protic solvents.[152,153]

For a protic solvent (hard) the nucleophilic order with respect to tetrahedral carbon (soft) is:

$$CN^- > I^- > SCN^- > \dot{N}_3^- > Br^- > Cl^- > F^-$$

the order of decreasing softness. In a dipolar aprotic solvent the order is reversed, in particular there is found the classical order[13,154]:

$$F^- > Cl^- > Br^- > I^-$$

Delpuech[155] has produced evidence for the linear dependence within one class of solvents of the rate constant for a given nucleophile on the basicity of the solvent and the reciprocal of its dielectric constant, if hydrogen bonding with the nucleophile[156] is completely excluded.

Hudson has shown the value of studying the energetics of solvation; he proposes to characterize the order of nucleophilicity of various nucleophiles towards a single substrate in terms of several quantities.[157] These are (a) the desolvation energy of the nucleophile, $\alpha \Delta H_N$, where α is the fraction desolvated in the transition state; (b) the energy required to remove an electron from the nucleophile, βE_N, where E_N is the electron affinity of the nucleophile and β the electron fraction removed; (c) the energy gained in the formation of a new bond between the electrophilic center C, and the nucleophilic center N, i.e. νD_{N-C}, where ν represents the fractional bond formation in the transition state.

After several hypotheses and transformations Hudson's equation reduces to (7.27), where ΔE is the activation energy and a^n is the simplest function indicating the variation of D_{N-C} with the distance N—C.

$$\Delta E = \alpha(\Delta H_N + E_N - a^n D_{N-C}) + \text{const} \qquad (7.27)$$

In contrast to Edwards's equation, that of Hudson not only takes account of the nucleophile (ΔH_N term) and the electrophile (E_N term) but also brings into consideration the transition state by the quantity[158] D_{N-C}.

In fact, other factors connected with solvation intervene: (*a*) the loss in energy by desolvation of the substrate; (*b*) the gain in energy by solvation either of the activated complex or of the final product.

Arnett and his colleagues[159,160] have shown how the effects of solvent on the enthalpy of activation for a simple reaction may be separated into the effect of solvent on the initial state and the effect on the transition state.

Other studies deal with thermodynamic activation functions. The entropy of activation is always negative, of the order of -10 to -30 cal mol^{-1} K^{-1}; for example, -15 for CN^-, -27 for diazabicyclo[2.2.2]octane, both in reaction with malachite green.[161] Likewise the volume of activation is negative. These negative values are explained by the fact that two independent reactants give birth to a single species, the activated complex. Given the values of the enthalpy of activation in the literature,[162,163] it is impossible to predict values for other examples; but the distinction between protic solvents and dipolar aprotic solvents coincides with an energy greater than 20 kcal $mole^{-1}$ for the first, and less than 17 kcal $mole^{-1}$ for the second.[155] The diminution of this value in a dipolar aprotic solvent can be interpreted as being due to the greater solvation of the activated complex.[143]

(*c*) *Applications of Pearson's Equation*

Before the rule of greater stability of an adduct and the HSAB concept were inferred from a multitude of experimental results by Pearson, other relations of the same kind as (7.22) had been studied, although applied to enthalpy and not to free energy. Thus Drago and Wayland[164] utilize an empirical relation (7.28), where ΔH is the enthalpy of formation

$$-\Delta H = E_A E_B + C_A C_B \qquad (7.28)$$

Table 7.4. Enthalpies ΔH (kcal mole^{-1}) of Acid–Base Addition in Poorly Solvating Solvents[166]

			I_2 $C_A = 1$, $E_A = 1$		Phenol $C_A = 0.44$, $E_A = 4.33$		$AlMe_3$ $C_A = 1.43$, $E_A = 16.9$	
Base	C_B	E_B	calc.	exp.	calc.	exp.	calc.	exp.
NH_3	3.46	1.36	4.8	4.8	7.4	7.8	28.0	27.6
Pyridine	6.40	1.17	7.6	7.8	7.9	8.0	28.9	26.7
Me_3N	11.54	0.81	12.3	12.1	8.6	8.8	30.2	30.0
$(CH_2)_4S$	7.90	0.34	8.2	8.3	5.0	4.9	17.1	17.0
Et_2O	3.25	0.96	4.2	4.2	5.6	6.0	20.9	20.2

of the addition compound in (7.1) in a poorly solvating solvent such as carbon tetrachloride, hexane, or cyclohexane.[165] In equation (7.28) the parameters E_A and E_B vary with the polarity of the acids and bases, and the product E_AE_B expresses in some measure the ionic binding; the parameters C_A and C_B express the susceptibilities of acid and of base to form covalent bonds, so the product C_AC_B measures the covalent binding between A and B. The reactions to which (7.28) applies are addition reactions between neutral bases and weak acids, such as the reaction of ammonia with iodine or with phenol. It is necessary to choose a reference acid for which one makes $E_A = C_A = 1$; iodine has been chosen. Results for addition compounds given by ammonia and some bases with iodine, phenol, or $AlMe_3$ are assembled in Table 7.4.

This identification of the product of softness parameters, $\sigma_A\sigma_B$, with the covalent contribution to the enthalpy of formation of the addition compound has been confirmed in certain cases. Pearson and Mawby[167] have calculated enthalpies of formation of metallic halides in the gas phase, equation (7.29), and for the acid they defined a softness parameter, σ_p, as in (7.30). Ahrland[168] has extended equation (7.28) to cases of ionic species,

$$A^{n+}(g) + nB^-(g) \rightleftharpoons AB_n(g) \tag{7.29}$$

$$\sigma_p = [\Delta H_0(F^-) - \Delta H_0(I^-)]/\Delta H_0(F^-) \tag{7.30}$$

defining[169] σ_A as the $\Delta H_0/n$ value of the reaction $A(g) \rightarrow A^{n+}(aq) + ne^-$ and σ_B that of the reaction $B(g) + ne^- \rightarrow B^{n-}(aq)$.

Lohmann[170] has presented another softness parameter calculated from standard free energies of solvation, μ, in two solvents, water and acetonitrile, as in (7.31). Klopman,[171] in a theoretical study, to which we

$$\sigma = 100(\mu_{H_2O} - \mu_{MeCN})/\mu_{H_2O} \tag{7.31}$$

will return in the following section, has been led to define softness

Table 7.5. Various Softness Parameters[167–171]

A	σ_A	σ_K	σ_p	σ	B	σ_B	σ_L	σ
Ca^{2+}	0.9	−2.33	0.181	5.27	Cl^-	−7.38	−9.94	63.9
Na^+	0.93	0.00	0.211	3.92	Br^-	−6.84	−9.22	62.4
Zn^{2+}	3.1		0.115	4.45	I^-	−6.13	−8.31	57.7
Cu^{2+}	3.1	0.55	0.104	−2.82				
Cd^{2+}	3.5	2.04	0.081	2.72				
Pb^{2+}	4.1		0.131	4.25				
Ag^+	4.2	2.82	0.074	−5.25				

parameters, σ_K for the acceptor, σ_L for the donor, from the energies of frontier orbitals (the lowest empty orbital of the acceptor, and the highest occupied orbital of the donor) and of desolvation. Various results are shown in Table 7.5.

The order of softness signified by σ_A and σ_K is preserved almost the same from one definition to another. However, σ_p defined from the nucleophile F^-, a hard entity, and σ, defined from the protic solvent H_2O, a hard entity also, contain in some manner deviations from softness.

We shall now turn to the theoretical development that has permitted Klopman recently to support the HSAB concept and the idea of softness, an extremely interesting idea for the generalization of acid–base reactions and their prediction, and in consequence for the problem of nucleophilicity.

7.5. An Approach to a Theoretical Prediction of Donor–Acceptor Power

The HSAB principle summarizes a large number of chemical phenomena. Without doubt there exist fundamental theoretical reasons for this, and *a priori* it may be foreseen that they will not be simple, since the stability of an addition compound between donor and acceptor depends on the many factors that govern the strength of the chemical bond.[29]

7.5.1. Available Theories

(a) First attempts

Ionic–Covalent Bonding. The oldest explanation is based on the ionic–covalent theory. The interaction of hard entities makes ionic forces intervene, these being favored by a greater hardness of the reactants; the interaction of soft entities produces covalent bonding, the more so as the two atoms bound together have similar size and electronegativity. These generalizations must be regarded with caution, for very hard acids such as I(VII) and Mn(VII) certainly produce covalent bonds in IO_4^- and MnO_4^-.

π Bonding. Chatt's theory of π bonding[172] appears very appropriate for metal cations of group B, since it envisages the intervention of π bonding through the vacant orbitals of the base, and it may be extended to other bases: P, As, S, I, unsaturated organic molecules (CO, isonitriles). Likewise, metallic cations of group A, with empty orbitals of sufficiently low energy, associate with bases that possess one pair of electrons (O, F), with the formation of a π bond in the opposite sense.

London Forces, van der Waals Forces, and Orbital Hybridization. Contributions from London or van der Waals forces explain the stabilization of the adduct by the intervention of polarizabilities, as Pitzer[173] has pointed out. This point of view accounts for the affinity between soft entities: Energies of the F—F, O—O, N—N single bonds are markedly less than those of the corresponding elements in the next row of the Periodic Table. For the latter, Mulliken[174] explains the greater stability by the hybridization of *p* and *d* orbitals, which augments the overlap of bonding π orbitals and diminishes that of π* antibonding orbitals. This study of the different overlaps in π and π* orbitals has been taken further with more sophistication by Hudson *et al.*, and by a perturbation method they[175] explain the influence of alkyl substituents in alcohols and amines on the acid and base strengths.

(*b*) *Molecular Orbital Theory*

Instead of studying the transition state with the aid of the perturbation method (as we shall see in the following section) various authors study it by the molecular orbital method.

Wheland[176] was the first to propose a transition state in which some delocalization of π electrons remains. The difference in energy between the two systems involved is equal to the energy necessary for localizing two electrons at the reaction center. The Wheland model may be illustrated by scheme (II) in the case of a nucleophilic aromatic substitution.

$$\text{C}_{10}\text{H}_8 + X^- \longrightarrow [\text{C}_{10}\text{H}_8X]^-$$

(II)

This model has subsequently been improved by Mulliken[177] by extending the delocalization by introducing hyperconjugation between the H,X system and the tetrahedral carbon. This amounts to putting a pseudoatom in the place of the H,X grouping, and to attributing to it a Coulomb integral, α, and a resonance integral, β, with the neighboring carbon atom.

Finally, Daudel and his colleagues[178,179] have extended this model by making the integrals β vary according to the reaction studied, the energy levels of the transition state being determined by the Hückel method. This model has been applied to the attack of a nucleophile on the quinolinium ion.[180] For this, one studies the variation of the energy difference between the transition state produced by attack at the 4- and at the 2-position [see (III)] as a function of the Coulomb integral, α_X, of the orbital of the

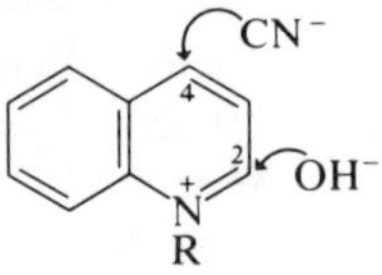

(III)

pseudoatom. It is thus found that if one writes α_X in the form $\alpha_X = \alpha^0 + h\beta^0$ (where α^0 is the normal value of the Coulomb integral of carbon, and β^0 that of the resonance integral for the C=C double bond), for $h < -0.1$ the 4-position is substituted in preference to the 2-position, and for $h > -0.1$, the 2-position reacts more easily than the 4-position. Such a result is in accord with observations, which indicate attack at position 4 by the soft base CN^- (corresponding to a small absolute value of α) and attack at position 2 by the hard base OH^- (corresponding to a large absolute value of α). The Hückel method has similarly been used in the theoretical determination of the sites of protonation of thiazine dyes,[181] and it is found that the proton, a hard acid, prefers the harder ring nitrogen to the exocyclic nitrogen.

7.5.2. Klopman's Perturbation-Based Theory

Before presenting these theoretical studies, we point out that, from their results, it is possible to work back to the empirical formulas inferred from experimental studies carried out to characterize nucleophilic reactivity, as Hudson has recently described in a brilliant way.[39]

(a) Development of This Theory

In calculations of reactivity it is the practice to try to establish correlations between reactivity and a particular index of molecular orbital theory. One does not allow for relative variations in reactivity of different positions in an electron donor according to the nature of the electrophile.

In regarding the formation of the transition state as a mutual perturbation of the molecular orbitals of the two reactants, Klopman[171,182,183] has shown that the relative reactivity of different reaction centers varies with the importance of the perturbation.

The theoretical development is entirely similar to a classical perturbation theory in which one allows for electron interaction. One considers two reactants R and S, which approach each other in such a way that an atom r of R and an atom s of S interact. Let us denote by ψ_m the various molecular orbitals of the molecule R, and by ψ_n those of S; it is supposed that these orbitals may be represented by LCAO expressions of the type (7.32) and (7.33). In (7.32) and (7.33) ϕ_r and ϕ_s are the atomic orbitals of

$$\psi_m = \sum_{\rho \neq r} C_\rho^m \phi_\rho + C_r^m \phi_r \tag{7.32}$$

$$\psi_n = \sum_{\sigma \neq s} C_\sigma^n \phi_\sigma + C_s^n \phi_s \tag{7.33}$$

atoms r and s which enter into a bond during the reaction studied, ϕ_ρ and ϕ_σ being, respectively, the atomic orbitals of atoms ρ of R, and σ of S which will remain unchanged after the reaction. The two *ensembles* of orbitals ψ_m and ψ_n combine by interaction of the atoms r and s to give new perturbed molecular orbitals, which are linear combinations of unperturbed orbitals. To determine these orbitals, one uses a new Hamiltonian H, which can be written in the form (7.34), where H_0^R and H_0^S are

$$H = H_0^R + H_0^S + H_1 \tag{7.34}$$

the Hamiltonian operators acting in the isolated systems, and H_1 is the additional term arising from the union of R and S.

In a first series of studies[184–186] Klopman did not allow for solvation. If E_m^* is the energy of an electron in the orbital ψ_m, taking account of the interaction between the total charge carried by the atom s and the charge of r in the orbital ψ_m, and E_n^* the energy of an electron in the orbital ψ_n, taking account also of the interaction between the total charge carried by the atom r and the charge of s in the orbital ψ_n, the perturbation energy may be written as in (7.35), where q_r and q_s are the net initial charges of

$$\Delta E = -q_r q_s \Gamma_{rs} + \sum_{\substack{m \\ \text{occ}}} \sum_{\substack{n \\ \text{unocc}}} \frac{2(C_r^m)^2 (C_s^n)^2 \beta_{rs}^2}{E_m^* - E_n^*} \tag{7.35}$$

atoms r and s, Γ_{rs} is an integral whose value depends on the distance between the atoms r and s, and β_{rs} is given by (7.36).

$$\beta_{rs} = \int \phi_r H_1 \phi_s \, d\tau \tag{7.36}$$

Klopman and Hudson[183] have shown that there are two limiting cases following from the relative magnitude of the quantities (i) $2(C_r^m C_s^n \beta_{rs})^2$ and (ii) $(E_m^* - E_n^*)$. If (ii) is very large for all the orbitals m and n, it is found numerically that the perturbation is principally due to the two atoms r and s carrying the largest total charge, and that the variation of the energy ΔE is given approximately by (7.37). The reaction is then called a charge-controlled reaction.[171] If (i) is very much larger than (ii) and if, moreover,

$$\Delta E = -q_r q_s \Gamma_{rs} \tag{7.37}$$

the reaction centers are neutral ($q_r = q_s = 0$), one obtains equation (7.38).

$$\Delta E = 2C_r^m C_s^n \beta_{rs} \tag{7.38}$$

In this second case, the reaction takes place between the two atoms carrying the largest density of charge in the highest occupied orbital of R and in the lowest unoccupied orbital of S. These orbitals are usually called "frontier orbitals,"[187] respectively, HOMO and LUMO. The reaction is then called a frontier-controlled reaction.[38,171]

Given equation (7.35) one may seek to predict in which examples the two foregoing limiting cases will occur.

The reaction is controlled by the effect of charge, i.e., (ii) ≫ (i) if (1) the reacting species are very polar ($q_s = +1$, $q_r = -1$); (2) the Coulombic interaction term, Γ_{rs}, between these ionic species is very large—this occurs for small interatomic distances and small orbitals; (3) $E_m^* > E_n^*$—this is the case when the donor has a large ionization potential and the electron acceptor has a small electron affinity; (4) β_{rs} is small—this occurs particularly when the orbitals are of different size or the distance r–s is large.

An example of this case is the formation of the ion $[F\cdots H—F]^-$, where there is effectively strong polarity of reactants, short distances between F^- and H, small polarizability of F, large ionization potential of the donor, and small β_{rs} because of the different nature of the overlapping orbitals of F and H.

The reaction is controlled by the frontier effect, i.e., (ii) ≪ (i) if (1) the reactants are radicals or neutral species or very feebly polar; (2) the Coulomb interaction term, Γ_{rs}, is small—this occurs for orbitals of large radius and for a large polarizability; (3) the electron-donor atom has a very low ionization potential and the electrophilic atom has a large electronegativity; (4) β_{rs} is large.

The frontier effect occurs for charge-transfer complexes between two neutral species of zero dipole moment, for example (IV) and (V). In fact,

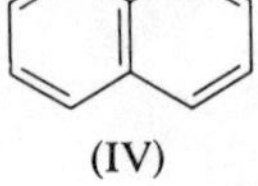

NO_2

O_2N NO_2

(IV) 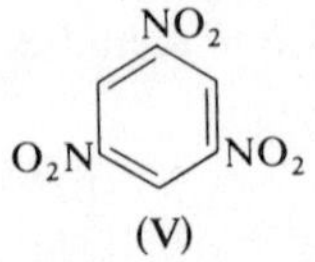 (V)

Mulliken's theory[188] had already shown that the bonding energy of these charge-transfer complexes depends on the ionization potential of the donor and the electron affinity of the acceptor.

The influences of orbital radius, polarizability, and electronegativity recall the classification of acids and bases according to their hard or soft character.

(b) Comparison of the Results of Klopman and Hudson with Those of Pearson

When the reacting species form ionic transition states, or when $E_m^* \gg E_n^*$, the interaction is charge-controlled and is equivalent to a hard-acid–hard-base interaction. Further, an important perturbation, produced by reactants of large polarizability which gives a transition state of covalent character (frontier-controlled), is typical of a soft-base–soft-acid interaction.

It is seen that a base (that is to say, the nucleophile) may be characterized by the value E_m^*, the energy of the highest occupied frontier orbital (HOMO); it is described as hard if E_m^* is low, and soft if E_m^* has a higher value. Likewise an acid (that is to say, the electrophile) may be characterized by the value E_n^* of the lowest unoccupied frontier orbital (LUMO), and it is described as hard if E_n^* has a high value, and soft if E_n^* is low. Table 7.6 gives the various cases of reaction according to the soft or hard character of the donor and of the acceptor (or according to the values of E_m^* and of E_n^*). In fact the majority of reactions correspond to two cases shown in Table 7.6, and are frontier- or charge-controlled.

Table 7.6. Reactions between Soft and Hard Donors and Acceptors[171]

Donor (base) E_m^*	Acceptor (acid) E_n^*	Conclusion			
		$E_m^* - E_n^*$	Γ	β	Reactivity
High (soft) large orbital	High (hard) small orbital	Medium	Small	Very small	Weak
	Low (soft) large orbital	Small	Very small	Large	Considerable, frontier-controlled
Low (hard) small orbital	High (hard)	Large	Large	Small	Considerable, charge-controlled
	Low (soft)	Medium	Small	Very small	Weak

Table 7.7. *Klopman's Scale of Hardness and Softness*[171]

Reactant	Orbital energy (eV)	Solvation energy (in water) (eV)	E_n‡ (eV)	
Al^{3+}	26.04	32.05	6.01	hard
La^{3+}	17.24	21.75	4.51	
Ti^{4+}	39.46	43.81	4.35	
Be^{2+}	15.98	19.73	3.75	
Mg^{2+}	13.18	15.60	2.42	
Ca^{2+}	10.43	12.76	2.33	
Fe^{3+}	26.97	29.19	2.22	
Sr^{2+}	9.69	11.90	2.21	
Cr^{3+}	27.33	29.39	2.06	
Ba^{2+}	8.80	10.69	1.89	
Ga^{3+}	28.15	29.60	1.45	
Cr^{2+}	13.08	13.99	0.91	
Fe^{2+}	14.11	14.80	0.69	borderline
Li^{+}	4.25	4.74	0.49	
H^{+}	10.38	10.8	0.42	
Ni^{2+}	15.00	15.29	0.29	
Na^{+}	3.97	3.97	0	
Cu^{2+}	15.44	14.99	−0.55	soft
Tl^{+}	5.08	3.20	−1.88	
Cd^{2+}	14.93	12.89	−2.04	
Cu^{+}	6.29	3.99	−2.30	
Ag^{+}	6.23	3.41	−2.82	
Tl^{3+}	27.45	24.08	−3.37	
Au^{+}	7.59	3.24	−4.35	
Hg^{2+}	16.67	12.03	−4.64	
			E_m‡	
F^{-}	6.96	5.22	−12.18	hard
H_2O	15.8	(−5.07)	−(10.73)	
OH^{-}	5.38	5.07	−10.45	
Cl^{-}	6.02	3.92	−9.94	
Br^{-}	5.58	3.64	−9.22	
CN^{-}	6.05	2.73	−8.78	soft
SH^{-}	4.73	3.86	−8.59	
I^{-}	5.02	3.29	−8.31	
H^{-}	3.96	3.41	−7.37	

Klopman has continued his studies by introducing the influence of solvation,[189] and equation (7.35) then takes the form (7.39),[171] where the

$$\Delta E = -q_r q_s \frac{\Gamma_{rs}}{\varepsilon} + \Delta_{solv} + \sum_{\substack{m \\ occ}} \sum_{\substack{n \\ unocc}} \frac{2(C_r^m)^2 (C_s^n)^2 \beta_{rs}^2}{E_m^* - E_n^*} \tag{7.39}$$

quantities q_r, q_s, Γ_{rs}, C_r^m, C_s^n, β_{rs}, E_m^*, and E_n^* have previously been defined, ε is the dielectric constant of the solvent and Δ_{solv} the energy of desolvation which follows the union of R and S.

By supposing that the quantity Γ_{rs} varies little from one molecule to another, Klopman defines a quantity $E^\ddagger$, obtained from E_m^* or from E_n^* by taking $\Gamma_{rs} = 0$, for all acids and bases. The hard or soft character of the reactants thus includes an arbitrary term, Γ_{rs}. Starting from ionization potentials and electron affinities, Klopman has calculated the energy of intrinsic softness for different reactants (Table 7.7).

Tables 7.6 and 7.7 show on the one hand that reactants group themselves in Klopman's theory as do the acids or bases in Pearson's concept; on the other hand the tables show that the ideas of charge-controlled or frontier-controlled reactions are an illustration of the HSAB rule of highest stability.

In this numerical interpretation [equation (7.39)] the variation of the nucleophilicity when the electrophile is charged appears evident, since to determine the nucleophilicity order, it is necessary to know not only E_m^*, E_n^*, but also Γ and β, variables for the electrophile–nucleophile pair. The influence of the solvent appears clearly through the intervention of the term Δ_{solv}.

Klopman has calculated the quantity ΔE for a series of nucleophiles under the attack of three electrophiles in water (very soft acid, soft acid, and hard acid), and the orders of nucleophilicity thus obtained may be compared with the experimental values of rate constants (Table 7.8). The agreement is excellent.

The good agreement between Klopman's theoretical approach and the experimental properties of various ions suggests that the simple explanation based on a hard–hard bond of electrostatic type and a soft–soft bond of covalent type is to a first approximation satisfactory. Moreover, the prediction of nucleophilic reactivity in the second case ($E_m^* \approx E_n^*$) improves with knowledge of numerical values of the energies of the frontier orbitals of the donor and of the acceptor, as shown by recent work.

(c) *Contribution of HOMO and LUMO Energy Levels*

After the establishment of the empirical rule of greatest stability of an addition compound given by Pearson, and its promising theoretical support

Table 7.8. Experimental Order of Nucleophilicity[a] and as Calculated by Klopman[171]

E_n^*	Method	Order of nucleophilicity						
−7	Calculated	HS^- >	I^- >	NC^- >	Br^- >	Cl^- >	HO^- >	F^-
	10^4k, liter $mole^{-1}$ sec^{-1}, for reaction with peroxidic oxygen	Very rapid	6900	10	0.23	0.0011	0	
−5	Calculated	HS^- >	NC^- >	I^- >	HO^- >	Br^- >	Cl^- >	F^-
	10^6k, liter $mole^{-1}$ sec^{-1}, for attack on saturated carbon	(25)	10	12	1.2	0.5	0.11	
	E (Edwards)		2.79	2.06	1.65	1.51	1.24	1.0
+1	Calculated	HO^- >	NC^- >	HS^- >	F^- >	Cl^- >	Br^- >	I^-
	k, liter $mole^{-1}$ min^{-1}, for attack on carbonyl carbon	890	10.8		0.001		No reaction	
	pK_a	15.7	9.1	7.1	3.2		(−4.3)	(−7.3)

[a] See Ref. 171 for sources of data.

advanced by Klopman about 1968, and following the work of Woodward and Hoffmann, many theoretical studies appeared. In view of their large number, we consider only some which follow from studies already cited.

Pearson has expressed new selection rules for a reaction according to the orbital symmetry properties of the reactants, both for inorganic chemistry[190,191] and organic chemistry,[192] in unimolecular reactions[193] or reactions of various molecularities with, among others, examples from nucleophilic displacements.[194]

This examination of the orbital symmetry of the donor and of the acceptor has been taken up in detail and in a most complete way by Klopman[38] in the framework of the generalized perturbation method.

An interesting area of application of this latter method is the study of the alpha effect. We have already had occasion to present those nucleophiles where two donors occur side by side,[195] entailing orbital splitting, which raises the energy of the highest occupied orbital (HOMO). This has as a consequence the enhancement of the frontier effect[91] and gives an explanation of the powerful reactivity of these supernucleophiles with soft acids, as due to a predominantly covalent interaction, and the markedly weaker response of the proton, a hard acid, as due to intervention through its charge effect. Diagrams of the Brönsted type in general indicate this [equation (7.13)].[196] In a more precise fashion, Hudson[197] has been able to set up the orbital origin of the alpha effect by following the influence of groups carried by carbon in the oximes. An electron-attracting group lowers the energy of the π orbitals of the conjugated system C—N—O, while a donor group elevates it, and the orbitals arising from the orbital splitting of the two donor atoms N and O may be placed with regard to the π orbitals according to the scheme in Fig. 7.13 (after Ref. 197). This scheme shows in a very satisfying manner the existence of the alpha effect for molecules possessing electron-attracting substituents and its absence for those that include donor groups. Thus a knowledge of the HOMO level offers the hope of reaching an explanation of donor power, that is, nucleophilicity.

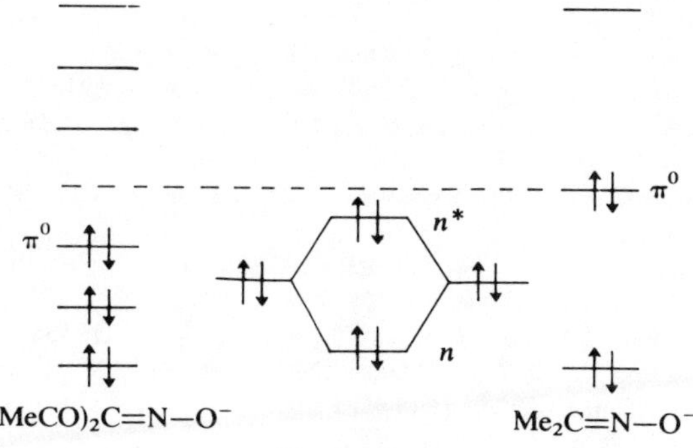

Fig. 7.13.

7.6. Conclusion

The problem of predicting chemical reactivity (and in particular of the understanding of nucleophilic power) challenges many workers. One need only consult reviews (notably Ref. 198) to recognize the impressive quantity of publications concerning nucleophilic substitution and addition, contributing to developing the state of knowledge of and to changing the outlook of the chemist; it is thus, for example, with nucleophilic aromatic substitution.[199] The way travelled owes much to theory which daily permits the further explanation of experimental phenomena—thus, for example, the influence of alkyl substituents on acidic or basic strengths of alcohols or of amines.[175]

If we represent the bimolecular nucleophilic displacement reaction as $Y + RX + \text{solvent} \rightarrow Y \cdots R \cdots X\text{–solvent} \rightarrow YR + X + \text{solvent}$, this expresses the competition between four components to associate in pairs by an exchange, according to two generalized acid–base reactions: nucleophile–electrophile association on the one hand, solvent–solute association on the other. For a given solvent, the prediction of the rate of reaction depends on the relative hardness or softness of Y and of R with respect to X, of which Klopman's tables give some idea. This viewpoint[200] is in favor of the influence of the framework of the substrate R, and of that[201] of the leaving group X on the reactivity of the nucleophile Y, rather than the independence of the various participants.[113]

One is, however, always shocked by the impossibility of defining an intrinsic scale of acid–base strength or of hardness–softness, and the various relations proposed to express nucleophilicity (Sections 7.3 and 7.4) are valid only for a limited number of closely related reactions. Hope for the future resides in the theoretical understanding of HOMO and LUMO energy levels.

References

1. N. L. Allinger, M. P. Cava, D. C. DeJongh, C. R. Johnson, N. A. LeBel, and C. L. Stevens, *Organic Chemistry* (Worth Publishers, New York, 1971).
2. M. Julia, *Mécanismes Electroniques en Chimie Organique* (Gauthier–Villars, Paris, 1972).
3. G. N. Lewis, *Valence and the Structure of Atoms and Molecules* (Chemical Catalog Co., New York, 1923).
4. R. G. Pearson, *J. Amer. Chem. Soc.*, **85**, 3533 (1963).
5. J. O. Edwards, *J. Amer. Chem. Soc.*, **78**, 1819 (1956).
6. C. G. Swain and C. B. Scott, *J. Amer. Chem. Soc.*, **75**, 141 (1953).
7. W. P. Jencks and M. Gilchrist, *J. Amer. Chem. Soc.*, **84**, 2910 (1962).
8. G. Schwarzenbach, *Experientia Suppl.*, **5**, 162 (1956); *Adv. Inorg. Chem. and Radiochem.*, **3**, 257 (1961).

9. G. Schwarzenbach and M. Schellenberg, *Helv. Chim. Acta*, **48**, 28 (1965); G. Schwarzenbach, *Chem. Eng. News*, **43** (22), 92 (1965).
10. R. C. Job and T. C. Bruice, *Proc. Nat. Acad. Sci. U.S.A.*, **72**, 2478 (1975).
11. A. J. Parker, *Quart. Rev.*, **16**, 163 (1962).
12. J. Miller and A. J. Parker, *J. Amer. Chem. Soc.*, **83**, 117 (1961).
13. S. Winstein, L. G. Savedoff, S. G. Smith, I. D. R. Stevens, and J. S. Gall, *Tetrahedron Lett.*, (9) 24 (1960).
14. J. O. Edwards, *J. Chem. Educ.*, **45**, 386 (1968).
15. K. M. Ibne-Rasa, *J. Chem. Educ.*, **44**, 89 (1967).
16. J. G. Bunnett, *Ann. Rev. Phys. Chem.*, **14**, 271 (1963).
17. R. F. Hudson, *Structure and Mechanism in Organo-Phosphorus Chemistry*, p. 90 (Academic Press, New York, 1965).
18. J. O. Edwards, *J. Amer. Chem. Soc.*, **76**, 1540 (1954).
19. E. S. Gould, *Mechanism and Structure in Organic Chemistry*, p. 277 (Holt, New York, 1959).
20. W. P. Jencks, *Progr. Phys. Org. Chem.*, **2**, 104 (1964).
21. J. Epstein, H. O. Michel, D. H. Rosenblatt, R. E. Plapinger, R. A. Stephani, and E. Cook, *J. Amer. Chem. Soc.*, **86**, 4959 (1964).
22. C. D. Ritchie, *Accounts Chem. Res.*, **5**, 348 (1972).
23. F. Guibe and G. Bram, *Bull. Soc. chim. France*, 933 (1975).
24. W. P. Jencks and J. Carriuolo, *J. Amer. Chem. Soc.*, **82**, 1778 (1960).
25. A. Streitwieser, *Solvolytic Displacement Reactions* (McGraw-Hill, New York, 1962).
26. J. O. Edwards and R. G. Pearson, *J. Amer. Chem. Soc.*, **84**, 16 (1962).
27. R. Gompper, *Angew. Chem. Internat. Edn.*, **3**, 560 (1964); GE, **76**, 412 (1964).
28. R. G. Pearson, *J. Chem. Educ.*, **45**, 581 (1968).
29. R. G. Pearson, *J. Chem. Educ.*, **45**, 643 (1968).
30. J. Seyden-Penne, *Bull. Soc. chim. France*, 3871 (1968).
31. J. B. Dence, H. B. Gray, and G. S. Hammond, *Chemical Dynamics* (Benjamin, New York, 1968).
32. R. G. Pearson, in *Advances in Linear Free Energy Relationships*, p. 281, N. B. Chapman and J. Shorter, eds. (Plenum, London, 1972).
33. S. Ahrland, J. Chatt, and N. R. Davies, *Quart. Rev.*, **12**, 265 (1958).
34. K. B. Harvey and G. B. Porter, *Introduction to Physical Inorganic Chemistry* (Addison-Wesley, Reading, Massachussetts, 1963).
35. R. G. Pearson, *Science*, **151**, 172 (1966); *Chem. in Britain*, **3**, 103 (1967).
36. R. G. Pearson, H. Sobel, and J. Songstad, *J. Amer. Chem. Soc.*, **90**, 319 (1968).
37. R. G. Pearson and J. Songstad, *J. Amer. Chem. Soc.*, **89**, 1827 (1967).
38. G. Klopman, in *Chemical Reactivity and Reaction Paths*, p. 55, G. Klopman, ed. (Wiley, New York, 1974).
39. R. F. Hudson, Ref. 38, p. 167.
40. R. S. Drago and K. F. Purcell, *Progr. Inorg. Chem.*, **6**, 271 (1964).
41. F. Basolo and R. G. Pearson, *Mechanisms of Inorganic Reactions* (Wiley, New York, 2nd edn., 1967).
42. D. A. Palmer and H. Kelm, *Inorg. Chim. Acta*, **14**, L27 (1975).
43. Yu. N. Kukushkin and V. B. Ukraintsev, *Zhur. neorg. Khim.*, **18**, 1063 (1973); EE, 560.
44. R. A. Sneen and H. M. Robbins, *J. Amer. Chem. Soc.*, **94**, 7868 (1972).
45. J. Miller, *J. Amer. Chem. Soc.*, **85**, 1628 (1963).
46. K. C. Ho, J. Miller, and K. W. Wong, *J. Chem. Soc.* (*B*), 310 (1966).
47. D. L. Hill, K. C. Ho, and J. Miller, *J. Chem. Soc.* (*B*), 299 (1966).
48. F. G. Bordwell and T. G. Mecca, *J. Amer. Chem. Soc.*, **97**, 127 (1975).
49. S. D. Ross, in *Comprehensive Chemical Kinetics*, Vol. 13, p. 407, C. H. Bamford and C. F. H. Tipper, eds. (Elsevier, Amsterdam, 1972).

50. J. L. Kice and G. B. Large, *J. Amer. Chem. Soc.*, **90**, 4069 (1968).
51. J. L. Kice and G. Guaraldi, *J. Amer. Chem. Soc.*, **90**, 4076 (1968).
52. G. H. Langford and H. B. Gray, *Ligand Substitution Processes* (Benjamin, New York, 1965).
53. E. G. Sander and W. P. Jencks, *J. Amer. Chem. Soc.*, **90**, 6154 (1968).
54. T. R. Oakes and G. W. Stacy, *J. Amer. Chem. Soc.*, **94**, 1594 (1972).
55. Y. Ogata and A. Kawasaki, *J.C.S. Perkin II*, 134 (1975).
56. M. J. Cho and I. H. Pitman, *J. Amer. Chem. Soc.*, **96**, 1843 (1974).
57. I. V. Tselinskii and G. I. Kolesetskaya, *Reakts. spos. org. Soedinenii*, **8**, 79 (1971).
58. C. D. Ritchie and D. J. Wright, *J. Amer. Chem. Soc.*, **93**, 6574 (1971).
59. C. D. Ritchie and P. O. I. Virtanen, *J. Amer. Chem. Soc.*, **94**, 1589 (1972).
60. C. D. Ritchie and H. Fleischhauer, *J. Amer. Chem. Soc.*, **94**, 3481 (1972).
61. C. D. Ritchie and P. O. I. Virtanen, *J. Amer. Chem. Soc.*, **94**, 4963 (1972).
62. M. J. Strauss, *Chem. Rev.*, **70**, 667 (1970).
63. M. R. Crampton, *J. Chem. Soc.*, (*B*), 2112 (1971).
64. M. R. Crampton and M. J. Willison, *J.C.S. Perkin II*, 238 (1974).
65. G. Charlot, *Les Réactions Chimiques en Solution, l'Analyse Qualitative Minérale* (Masson, Paris, 6th edn., 1969).
66. P. B. D. de la Mare and C. A. Vernon, *J. Chem. Soc.*, 41 (1956).
67. J. Hine and R. D. Weimar, *J. Amer. Chem. Soc.*, **87**, 3387 (1965).
68. C. Duboc, *Bull. Soc. chim. France*, 1768 (1970).
69. R. C. Dougherty and J. D. Roberts, *Org. Mass. Spectrometry*, **8**, 81 (1974).
70. L. B. Young, E. Lee-Ruff and D. K. Bohme, *J.C.S. Chem. Comm.*, 35 (1973).
71. R. P. Bell, *The Proton in Chemistry*, p. 155 (Cornell University Press, Ithaca, New York, 1959).
72. G. F. Smith, *J. Chem. Soc.*, 521 (1943).
73. R. F. Hudson and G. Klopman, *J. Chem. Soc.*, 5 (1964).
74. R. F. Hudson and G. Klopman, *J. Chem. Soc.*, 1062 (1962).
75. R. F. Hudson and G. Loveday, *J. Chem. Soc.*, 1068 (1962).
76. D. M. E. Reuben and T. C. Bruice, *J.C.S. Chem. Comm.*, 113 (1974).
77. M. Friedman, J. F. Cavins, and J. S. Wall, *J. Amer. Chem. Soc.*, **87**, 3672 (1965).
78. V. K. Krylov, I. V. Tselinskii, and L. I. Bagal, *Reakts. spos. org. Soedinenii*, **6**, 959 (1969); EE, 413.
79. J. D. Aubort and R. F. Hudson, *J. Chem. Soc.* (*D*), 938 (1970).
80. B. Miller, *J. Amer. Chem. Soc.*, **84**, 403 (1962).
81. D. L. Dalrymple, J. D. Reinheimer, D. Barnes, and R. Baker, *J. Org. Chem.*, **29**, 2647 (1964).
82. K. Okamoto, S. Fukui, I. Nitta, and H. Shingu, *Bull. Chem. Soc. Japan*, **40**, 2350 (1967).
83. J. Hine, *Physical Organic Chemistry*, p. 161 (McGraw-Hill, 2nd edn., New York, 1962).
84. W. A. Pryor, *Mechanisms of Sulfur Reactions*, p. 62 (McGraw-Hill, New York, 1962).
85. J. R. Cox and O. B. Ramsay, *Chem. Rev.*, **64**, 317 (1964).
86. K. Koehler, R. Skora, and E. H. Cordes, *J. Amer. Chem. Soc.*, **88**, 3577 (1966).
87. J. Epstein, R. E. Plapinger, H. O. Michel, J. R. Cable, R. A. Stephani, R. J. Hester, C. Billington, and G. R. List, *J. Amer. Chem. Soc.*, **86**, 3075 (1964).
88. J. O. Edwards, ed., *Peroxide Reaction Mechanisms* (Wiley, New York, 1962).
89. K. M. Ibne-Rasa and J. O. Edwards, *J. Amer. Chem. Soc.*, **84**, 763 (1962).
90. R. G. Pearson and D. N. Edgington, *J. Amer. Chem. Soc.*, **84**, 4607 (1962).
91. G. Klopman, K. Tsuda, J. B. Louis, and R. E. Davis, *Tetrahedron*, **26**, 4549 (1970).
92. A. J. Parker, *Proc. Chem. Soc.*, 371 (1961).
93. B. D. England and D. J. McLennan, *J. Chem. Soc.* (*B*), 696 (1966).
94. K. E. Jabalpurwala and R. M. Milburn, *J. Amer. Chem. Soc.*, **88**, 3224 (1966).

95. V. S. Shmidt and V. S. Sokolov, *Zhur. neorg. Khim.*, **15**, 1208 (1970); EE, 620.
96. V. S. Shmidt and K. A. Rybakov, *Radiokhimiya*, **16**, 440 (1974).
97. V. S. Shmidt and V. N. Shesterikov, *Radiokhimiya*, **13**, 815 (1971).
98. V. S. Shmidt and K. A. Rybakov, *Radiokhimiya*, **16**, 580 (1974).
99. N. Gravitz and W. P. Jencks, *J. Amer. Chem. Soc.*, **96**, 507 (1974).
100. G. Biggi and F. Pietra, *J. Chem. Soc.* (*B*), 44 (1971).
101. G. Guanti, C. Dell'erba, F. Pero, and G. Leandri, *J.C.S. Perkin II*, 212 (1975).
102. O. Exner, in *Advances in Linear Free Energy Relationships*, N. B. Chapman and J. Shorter, eds., p. 1 (Plenum, London, 1972).
103. A. B. Ash, P. Blumbergs, C. L. Stevens, H. O. Michel, B. E. Hackley, and J. Epstein, *J. Org. Chem.*, **34**, 4070 (1969).
104. F. Basolo, in *Mechanisms of Inorganic Reactions, Advances in Chemistry Series*, No. 49, p. 95 (American Chemical Society, Washington D.C., 1965).
105. J. Koivurinta, A. Kyllonen, L. Leononen, K. Valaste, and J. Koskikallio, *Finn. Chem. Lett.*, 239 (1974).
106. U. Belluco, L. Cattalini, F. Basolo, R. G. Pearson, and A. Turco, *J. Amer. Chem. Soc.*, **87**, 241 (1965).
107. W. L. Petty and P. L. Nichols, *J. Amer. Chem. Soc.*, **76**, 4385 (1954).
108. H. D. Zook and J. A. Miller, *J. Org. Chem.*, **36**, 1112 (1971).
109. M. Cusumano, G. Faraone, V. Ricevuto, R. Romeo, and M. Trozzi, *J.C.S. Dalton*, 490 (1974).
110. C. D. Ritchie and P. O. I. Virtanen, *J. Amer. Chem. Soc.*, **94**, 4966 (1972).
111. C. D. Ritchie and P. O. I. Virtanen, *J. Amer. Chem. Soc.*, **95**, 1882 (1973).
112. P. O. I. Virtanen and R. Korhonen, *Acta Chem. Scand.*, **27**, 2650 (1973).
113. C. D. Ritchie, *J. Amer. Chem. Soc.*, **97**, 1170 (1975).
114. R. E. Davis, *Tetrahedron Lett.*, 5021 (1966).
115. S. Winstein, E. Grunwald, and H. W. Jones, *J. Amer. Chem. Soc.*, **73**, 2700 (1951).
116. S. Winstein, A. H. Fainberg, and E. Grunwald, *J. Amer. Chem. Soc.*, **79**, 4146 (1957).
117. P. E. Peterson and F. J. Waller, *J. Amer. Chem. Soc.*, **94**, 991 (1972).
118. T. W. Bentley, F. L. Schadt, and P. v. R. Schleyer, *J. Amer. Chem. Soc.*, **94**, 992 (1972).
119. D. J. Raber, M. D. Dukes, and J. Gregory, *Tetrahedron Lett.*, 667 (1974).
120. F. L. Schadt, P. v. R. Schleyer, and T. W. Bentley, *Tetrahedron Lett.*, 2335 (1974).
121. R. H. Summerville, C. A. Senkler, P. v. R. Schleyer, T. E. Dueber, and P. J. Stang, *J. Amer. Chem. Soc.*, **96**, 1100 (1974).
122. E. M. Kosower, *An Introduction to Physical Organic Chemistry* (Wiley, New York, 1968).
123. R. E. Dessy, R. L. Pohl, and R. B. King, *J. Amer. Chem. Soc.*, **88**, 5121 (1966).
124. A. Yingst and D. H. McDaniel, *Inorg. Chem.*, **6**, 1067 (1967).
125. B. Saville, *Angew. Chem. Internat. Edn.*, **6**, 928 (1967); GE, **79**, 966 (1967).
126. R. V. Vizgert and I. M. Ozdrovskava. *Reakts. spos. org. Soedinenii*, **3**(2), 16 (1966); EE, 111.
127. J. L. Kice, G. J. Kasperek, and D. Patterson, *J. Amer. Chem. Soc.*, **91**, 5516 (1969).
128. J. H. Krueger, B. A. Sudbury, and P. F. Blanchet, *J. Amer. Chem. Soc.*, **96**, 5733 (1974).
129. L. Di Nunno and P. E. Todesco, *Tetrahedron Lett.*, 2899 (1967).
130. D. R. Hogg and A. Robertson, *Tetrahedron Lett.*, 3783 (1974).
131. A. E. Crease and P. Legzdins, *J. Chem. Educ.*, **52**, 499 (1975).
132. T. Aida, M. Nakajima, T. Inoue, N. Furukawa, and S. Oae, *Bull. Chem. Soc. Japan*, **48**, 723 (1975).
133. B. B. Wayland and R. H. Gold, *Inorg. Chem.*, **5**, 154 (1966).
134. P. C. H. Mitchell, *Quart. Rev.*, **20**, 103 (1966).
135. N. J. DeStefano and J. L. Burmeister, *Inorg. Chem.*, **10**, 998 (1971).
136. R. G. Pearson, *Inorg. Chem.*, **12**, 712 (1973).

137. C. K. Jørgensen, *Inorg. Chem.*, **3**, 1201 (1964).
138. J. L. Burmeister and F. Basolo, *Inorg. Chem.*, **3**, 1587 (1964).
139. C. K. Jørgensen, *Structure and Bonding*, **1**, 234 (1966).
140. C. Agami, *Bull. Soc. chim. France*, 1021 (1965).
141. A. Kirrmann and J.-J. Delpuech, *Compt. rend. Acad. Sci.*, **260**, 6600 (1965).
142. B. Tchoubar, *Bull. Soc. chim. France*, 2069 (1964).
143. P. Haberfield, A. Nudelman, A. Bloom, R. Romm, H. Ginsberg, and P. Steinherz, *Chem. Comm.*, 194 (1968).
144. A. J. Parker, *Adv. Phys. Org. Chem.*, **5**, 173 (1967).
145. A. J. Fry, *J. Org. Chem.*, **31**, 1863 (1966).
146. K. Okamoto, S. Fukui, I. Nitta, and H. Shingu, *Bull. Chem. Soc. Japan*, **40**, 2354 (1967).
147. F. Madaule-Aubry, *Bull. Soc. chim. France*, 1456 (1966).
148. A. J. Parker, *Chem. Rev.*, **69**, 1 (1969).
149. R. A. Abramovitch and A. J. Newman, *J. Org. Chem.*, **39**, 3692 (1974).
150. H. Ginsburg, G. Le Ny, O. Parguez, and B. Tchoubar, *Bull. Soc. chim. France*, 301 (1969).
151. A. Brändström, *Arkiv. Kemi*, **11**, 527 (1957).
152. N. Gravitz and W. P. Jencks, *J. Amer. Chem. Soc.*, **96**, 489 (1974).
153. A. C. Satterthwait and W. P. Jencks, *J. Amer. Chem. Soc.*, **96**, 7031 (1974).
154. W. M. Weaver and J. D. Hutchison, *J. Amer. Chem. Soc.*, **86**, 261 (1964).
155. J. -J. Delpuech, *Tetrahedron Lett.*, 2111 (1965).
156. C. Béguin and J.-J. Delpuech, *Bull. Soc. chim. France*, 378 (1969).
157. R. F. Hudson, *Chimia*, **16**, 173 (1962).
158. R. F. Hudson and M. Green, *J. Chem. Soc.*, 1055 (1962).
159. E. M. Arnett and D. R. McKelvey, *Rec. Chem. Progr.*, **26**, 185 (1965).
160. E. M. Arnett, W. G. Bentrude, J. J. Burke, and P. McC. Duggleby, *J. Amer. Chem. Soc.*, **87**, 1541 (1965).
161. C. D. Ritchie, D. J. Wright, D.-S. Huang, and A. A. Kamego, *J. Amer. Chem. Soc.*, **97**, 1163 (1975).
162. E. J. Behrman and J. O. Edwards, *Progr. Phys. Org. Chem.*, **4**, 93 (1967).
163. A. Fava, A. Iliceto, and S. Bresadola, *J. Amer. Chem. Soc.*, **87**, 4791 (1965).
164. R. S. Drago and B. B. Wayland, *J. Amer. Chem. Soc.*, **87**, 3571 (1965).
165. G. C. Vogel and R. S. Drago, *J. Amer. Chem. Soc.*, **92**, 5347 (1970).
166. R. S. Drago, *J. Chem. Educ.*, **51**, 300 (1974).
167. R. G. Pearson and R. J. Mawby, in *Halogen Chemistry*, Vol. 3, p. 55, V. Gutmann, ed. (Academic Press, New York, 1967).
168. S. Ahrland, *Chem. Phys. Letters*, **2**, 303 (1968).
169. S. Ahrland, *Structure and Bonding*, **5**, 118 (1968).
170. F. Lohmann, *Chem. Phys. Letters*, **2**, 659 (1968).
171. G. Klopman, *J. Amer. Chem. Soc.*, **90**, 223 (1968).
172. J. Chatt, *J. Inorg. Nuclear Chem.*, **8**, 515 (1958).
173. K. S. Pitzer, *J. Chem. Phys.*, **23**, 1735 (1955).
174. R. S. Mulliken, *J. Amer. Chem. Soc.*, **77**, 884 (1955).
175. R. F. Hudson, O. Eisenstein, and Nguyen Trong Anh, *Tetrahedron* **31**, 751 (1975).
176. G. W. Wheland, *J. Amer. Chem. Soc.*, **64**, 900 (1942).
177. N. Muller, L. W. Pickett, and R. S. Mulliken, *J. Amer. Chem. Soc.*, **76**, 4770 (1954).
178. J. Bertran, O. Chalvet, R. Daudel, T. F. W. McKillop, and G. H. Schmid *Tetrahedron*, **26**, 339 (1970).
179. O. Chalvet, R. Daudel, C. Ponce, and J. Rigaudy, *Internat. J. Quantum Chem.*, **2**, 521 (1968).
180. O. Chalvet, R. Daudel, and T. F. W. McKillop, *Tetrahedron*, **26**, 349 (1970).

181. O. Chalvet, J. Hoarau, J. Joussot-Dubien, and J.-C. Rayez, *J. Chim. phys.*, **69**, 630 (1972).
182. G. Klopman, *J. Amer. Chem. Soc.*, **86**, 4550 (1964).
183. G. Klopman and R. F. Hudson, *Theor. Chim. Acta*, **8**, 165 (1967).
184. G. Klopman, *J. Amer. Chem. Soc.*, **87**, 3300 (1965).
185. M. J. S. Dewar and G. Klopman, *J. Amer. Chem. Soc.*, **89**, 3089 (1967).
186. R. F. Hudson and G. Klopman, *Tetrahedron Lett.*, 1103 (1967).
187. K. Fukui and H. Fujimoto, *Bull. Chem. Soc. Japan*, **39**, 2116 (1966).
188. R. S. Mulliken, *J. Chim. phys.*, **61**, 20 (1964).
189. G. Klopman, *Chem. Phys. Letters*, **1**, 200 (1967).
190. R. G. Pearson, *J. Amer. Chem. Soc.*, **91**, 1252 (1969).
191. R. G. Pearson, *J. Chem. Phys.*, **52**, 2167 (1970).
192. R. G. Pearson, *Theor. Chim. Acta*, **16**, 107 (1970).
193. R. G. Pearson, *J. Amer. Chem. Soc.*, **92**, 8287 (1972).
194. R. G. Pearson, *Accounts Chem. Res.*, **4**, 152 (1971).
195. N. J. Fina and J. O. Edwards, *Internat. J. Chem. Kinetics*, **5**, 1 (1973).
196. M. Dessolin, M. Laloi-Diard, and M. Vilkas, *Tetrahedron Lett.*, 2405 (1974).
197. J. D. Aubort, R. F. Hudson, and R. C. Woodcock, *Tetrahedron Lett.*, 2229 (1973).
198. *Organic Reaction Mechanisms*, B. Capon and C. W. Rees, eds. (Wiley, New York, 1970 *et seq.*).
199. J. F. Bunnett, *J. Chem. Educ.*, **51**, 312 (1974).
200. A. Pross, *Tetrahedron Lett.*, 637 (1975).
201. L. Cattalini, M. Cusumano, V. Ricevuto, and M. Trozzi, *J.C.S. Dalton*, 771 (1975).

8

Correlation of nmr Chemical Shifts with Hammett σ Values and Analogous Parameters

D. F. Ewing

8.1. Introduction

This chapter is devoted to an examination of the correlation of nmr chemical shifts with Hammett's structure–reactivity parameters and analogous parameters developed by others. Correlation analysis in nmr spectroscopy as a whole is, of course, a much wider field, involving as it does, exploration of the interrelationships between nmr data and the relationships between nmr data and molecular parameters (such as electronegativity, ionization potential, electron density, and bond order), solvent parameters, and classical magnetic- and electric-field effects. Although a few of these other aspects of correlation analysis in nmr studies will be mentioned, the major emphasis is given to correlations with accepted LFER parameters.

D. F. Ewing • Chemistry Department, The University, Hull, HU6 7RX, England.

Correlation of coupling constants (J) is not covered in this review. Although a wide range of linear relationships have been developed between J and other nmr parameters (e.g., between J and δ) and between J and substituent electronegativity, correlation with Hammett's σ has not been developed to any appreciable extent. Two factors are responsible for this major difference from chemical shifts: the dependence of J on the s-electron density at the nucleus, and on the nature of the orbital interactions along the coupling pathway. Both of these factors reduce the likelihood of correlation with a parameter (σ) which is essentially a measure of local total electron density. That is not to say that in particular systems a parallelism between J and σ (particularly σ_I) cannot be expected. In view of the fact that correlations of J and σ are intrinsically more complex and less meaningful in their present state of development than those for δ, it was decided to exclude this area from this chapter and to devote the available space to a detailed appraisal of chemical shift correlations.

Part of the attraction of Hammett and related parameters for the nmr spectroscopist lies in their empirical character. Until very recently an empirical approach was the only way to achieve even a semiquantitative rationalization of nmr data, and the success of the Hammett equation in other directions suggested its application to the phenomenon of nuclear shielding and, to a less extent, to nuclear spin coupling. In spite of the enormous potential of nmr data for the complete characterization of the electronic organization in molecules, the development of an effective theoretical basis has lagged a long way behind data acquisition, and, particularly in the area of substituent effects in aromatic compounds, Hammett correlations have proved extremely valuable. Not only has an interim, empirical rationalization been widely achieved, but our understanding of the theoretical foundations has itself been advanced.

However, and perhaps inevitably, much of the published work concerned with the correlation of nmr data with LFER parameters suffers from two deficiencies. There is often a lack of a proper appreciation of the underlying limitations of this approach to the rationalization of nmr data; but even more significant is a lamentable disregard for proper statistical characterization of any supposed correlation. In many cases the primary statistical information (standard deviation of the estimate and/or correlation coefficient) is absent and it is exceptional to find a more sophisticated treatment. Throughout this review the two parameters, n, the number of data sets, and r and R, the simple or multiple correlation coefficient, are given where possible for the correlations discussed. In many cases r has been recalculated from the published data.

To discover whether there can be any real expectation of a meaningful *linear* correlation between chemical shifts and structure–reactivity parameters we should examine the basis of each, in particular their relationships to electron distribution in the ground states of molecules, and the way that this distribution can respond to the introduction of a substituent. It is not possible here to give a detailed discussion of the theoretical background to the Hammett relation, and readers are referred elsewhere.[1–4] It is sufficient for present purposes to note that although Hammett σ constants are derived either from equilibrium or kinetic studies in the main, and hence characterize either final or transition states differentially with respect to initial (or ground) states, there is some theoretical justification for supposing that these experimental constants *are* a measure of the ground-state electronic perturbations introduced by substituents.

Support for this supposition comes from classical electrostatic arguments and from quantum mechanical analysis, and we can assume that σ-parameters are related in a quasi-theoretical way to electron densities. Furthermore the polar and resonance parts into which σ values may be separated do often correspond to σ- and π-electron densities in aromatic compounds, although the correspondence is probably not exact.

8.2. Chemical Shifts

Factors contributing to nuclear shielding, σ, are usually discussed directly, but the following discussion will be conducted in terms of the chemical shift, δ, to avoid the obvious clash with the accepted symbolism of structure–reactivity relationships. Since a rigorous treatment must yield equations in terms of nuclear shielding, it is implicit in the present discussion that the shielding of a reference nucleus has been taken into account.

The chemical shift for any nucleus, N, can be conveniently divided into the four parts shown in equation (8.1), representing contributions to the

$$\delta^N = \delta_d^{NN} + \delta_p^{NN} + \delta_{lr}^{NN'} + \delta_m^N \qquad (8.1)$$

nuclear shielding from electrons on the atom of interest (δ^{NN}), from electrons on other atoms in the molecule ($\delta_{lr}^{NN'}$) and from interactions with the medium (δ_m^N). (The subscripts d and p signify dia- and paramagnetic, respectively, in this context, and lr long-range.)

Diamagnetic circulation of local electrons, i.e., on the atom containing nucleus N, induced by the applied field, gives rise to a negative, upfield contribution, δ_d^{NN}. This term has significant magnitude for all nuclei and, certainly for common nuclei, is numerically the largest. Its value may be represented by the well-known Lamb equation (8.2), where $\langle r_i^{-1} \rangle$ is the mean inverse distance of electron i from the nucleus, e is the charge on the electron, m is its mass, and c is the velocity of light. Equation (8.2) indicates a dependence of δ_d^{NN} on electron density around the nucleus, and if other factors are not important δ should be correlated linearly with

$$\delta_d^{NN} = -\frac{e^2}{3mc^2} \sum_i \langle r_i^{-1} \rangle \tag{8.2}$$

substituent electronegativity, or with the inductive constant, σ_I. Such correlations have been widely sought for simple aliphatic compounds, particularly for proton chemical shifts. Although many apparently "good" correlations have been found,[5] the correct interpretation is often more complex, since the other factors in equation (8.1) are seldom negligible.

The paramagnetic contribution, δ_p^{NN}, can be conveniently expressed in the simplified equation (8.3) due to Karplus and Pople,[6] where $\langle r^{-3} \rangle_{2p}$ is

$$\delta_p^{NN} = \frac{e^2 \hbar^2}{2m^2c^2 \, \Delta E} \langle r^{-3} \rangle_{2p} \sum_{N'} Q_{NN'} \tag{8.3}$$

related to the dimensions of the atomic p orbitals, $\sum Q_{NN'}$ contains elements of the charge density–bond order matrix in the MO description of the ground state of the molecule and reflects the unsymmetrical distribution of orbital charge. The parameter ΔE is a mean excitation energy, and the other symbols have the above significance, and $\hbar = h/2\pi$. This paramagnetic term is positive and represents a perturbation of the diamagnetic screening arising from coupling of the electronic orbital angular momentum with the nuclear spin momentum. Exact evaluation of δ_p^{NN} is difficult since it requires knowledge of excited-state wave functions, and equation (8.3) is obtained by application of the closure approximation to the full expression given in terms of individual excitation energies.[6] Although now regarded as inadequate for the calculation of the paramagnetic contribution to chemical shifts, equation (8.3) does illustrate the relationship of δ_p^{NN} to electron density. Clearly this relationship as a whole *cannot be linear* if all the variable factors are important. The mean excitation energy, ΔE, an empirical parameter, is usually taken to be constant, at least for closely related compounds such as a series of benzene derivatives. The magnitude of the term $\langle r^{-3} \rangle_{2p}$ has been shown[6,7] to depend on electron density; an increase in electron density results in a decrease in the effective nuclear charge with a concomitant expansion of the $2p$ orbitals (e.g., for ^{13}C and ^{19}F) giving rise to a decrease in δ_p^{NN} linear with respect to electron density (an increase in overall shielding). Variations in $Q_{NN'}$ are probably less important, particularly for compounds that differ only in the nature of the substituent. It is generally accepted that the dependence of δ_p^{NN} on electron density is contained principally in the $\langle r^{-3} \rangle_{2p}$ term, and this is supported by the existence of many tolerably good linear correlations of ^{13}C and ^{19}F chemical shifts with electron density.

However, the possibility of variations in the terms ΔE or $Q_{NN'}$ cannot be discounted, and sometimes these terms may be important.

For protons, δ_p^{NN} is small since the distortion of the $1s$ orbital is not large and the nearest appropriate excited orbital ($2p$) has too high an energy. For all other nuclei, low-lying p or d orbitals are available, and the paramagnetic term is large. For ^{13}C and ^{19}F, δ_d^{NN} is numerically larger than δ_p^{NN} (and of opposite sign) but the latter term is usually more sensitive to electronic effects and dominates changes in chemical shift. Recent work[8] suggests that, at least for ^{13}C, variations in δ_d^{NN} may be more significant than has usually been assumed.

The third term in equation (8.1) represents long-range magnetic perturbation at nucleus N, associated with electronic circulation induced around other nuclei N' by the applied field. Such contributions can be effectively handled in terms of localized groups of electrons, usually individual bonds, but some compounds contain electrons that must be regarded as delocalized, and a separation is often made on this basis, cf. equation (8.4). The local term can be related to the anisotropy in the

$$\delta_{lr}^{NN'} = \delta_{\text{loc}}^{NN'} + \delta_{\text{deloc}}^{NN'} \tag{8.4}$$

diamagnetic susceptibility of the individual bonds by the well-known McConnell relation.† In principle, summation over all bonds is required, but in practice many bonds have negligible effect. Furthermore, when substituent effects are being considered only the new bonds will be important. The magnetic perturbations at a nucleus N depend on the distance from it of the anisotropic bonds or groups, and in aromatic compounds only nuclei *ortho* to a substituent will be significantly affected.

The anisotropy in the diamagnetic susceptibility of electrons in delocalized orbitals is not the sum of local classical bond contributions, and an additional factor must be considered for such systems. For aromatic rings the so-called ring-current effect, arising from the delocalized π orbitals, has been widely accepted as an indicator of aromaticity, but a perfect parallel between these properties is unlikely.[9] The extent to which substituents can modify the π orbitals of a benzene or heterocyclic ring and hence the ^{1}H and ^{13}C chemical shifts through $\delta_{\text{deloc}}^{NN'}$ is clearly important for Hammett correlations, since structure–reactivity parameters cannot themselves reflect purely magnetic effects.

The terms in equation (8.4) do not depend upon the nature of the nucleus N, and $\delta_{lr}^{NN'}$ will have the same magnitude for any nucleus with a given orientation with respect to nuclei N'. This means that the effects of magnetic anisotropy in bonds or delocalized orbitals are *relatively* more important for proton chemical shifts, which exhibit only a small range

†$\delta_{\text{loc}}^{NN'} = \sum \Delta\chi(3\cos^2\theta - 1)/3r^3$, where $\Delta\chi$ is the anisotropy in the magnetic susceptibility of a bond, r is the distance of N from the bond center, and θ is the angle between r and the bond.

(arising from changes in δ_d^{NN}). Early work on ^{13}C and ^{19}F chemical shifts tended to neglect contributions from $\delta_{lr}^{NN'}$ for these nuclei, but with improving precision in experimental results such a procedure is becoming increasingly dangerous.

The final term in equation (8.1) takes account of the existence of the medium, i.e., it is concerned with intermolecular interactions of the solute–solvent and solute–solute type. Such interactions can be dissected into several contributions,[10,11] but we need be concerned only with the overall perturbation of the electron distribution in the solute arising directly from dipolar or dispersion effects, or indirectly from changes in geometry. Both δ_d^{NN} and δ_p^{NN} can be modified by medium effects, and these changes can usually be related to the physical properties of the medium (e.g., dielectric constant, hydrogen bonding ability, and polarizability).

It is very unlikely that there is any precise parallel between medium effects on δ and medium effects on reaction rates and equilibria. Since different types of solvent are normally used for the determination of δ values and of σ values (chloroform or carbon tetrachloride for the former, and methanol or aqueous solvents for the latter) it is evident that solvent effects are very important for correlations between these parameters. Perversely, they are perhaps the most neglected consideration. While it is impossible to remove the absolute effect of the different solvents used for the determination of δ and σ, this is probably not as important as using a common solvent thoughout a series of chemical shift measurements on the same system, and indeed for corresponding series in other systems, when comparisons are to be made via Hammett correlations.

Moreover, solute–solute interactions may often be significant, particularly for proton nmr, and much of the early data suffers from lack of attention to this point. Many correlations could be usefully re-examined by using infinite dilution data. However, all nuclei are sensitive to solute and solvent effects and strenuous efforts should be made by all workers to reduce the term δ_m^N to a constant for all Hammett correlations of chemical shift data, by aiming for the condition of infinite dilution in a common solvent (probably $CHCl_3$ or $CDCl_3$).

The totality of the dependence of δ on electron density is evidently complex if all the relevant terms in equation (8.1) are considered. The minimum requirement for a linear correlation of δ with electron density is that the terms $\delta_{lr}^{NN'}$ and δ_m^N are negligible or constant, and for ^{13}C and ^{19}F the further restriction must be applied that only the orbital dimension term in δ_p^{NN} is important. With respect to correlation with electron density, the separation into σ and π contributions raises some problems. Diamagnetic and paramagnetic shielding contributions are sensitive to both σ- and π-electronic charge, but in principle this sensitivity is different for the two types of orbital and separation into σ and π contributions to δ is not simple. This problem is reduced in some measure for correlations of δ with Hammett parameters, since the latter represent a combination of σ and π effects. However, this raises the question of whether δ and σ respond in the same way to the notionally separate electronic contributions. Further complications may be introduced by using parameters of the type developed by Taft and others to divide substituent effects into polar and resonance effects for aromatic compounds. Dual-parameter analyses of chemical shifts or electron densities add another dimension to the exploration of the interrelationships of these quantities, and much dispute centers around this area.

Any discussion of correlations between δ and σ must consider the quality of such correlations. *Some* kind of broadly linear correlation is certain to be found in most cases on the basis of the general dependence of both parameters on electron density. To support any serious claim that there is a *precise* parallel between δ and σ, the linear correlation must be extremely good, i.e., significant to at least the 0.1% level ($r = 0.991$ for $n = 5$, $r = 0.925$ for $n = 8$), and should summarize the data so as to give $r > 0.95$ generally. A final point concerns the presentation of correlation equations. Chemical-shift data may be referenced to some standard compound such as tetramethylsilane or referenced internally to the compound in the series with H as substituent. No consistent approach is used in the literature and we shall refer to all data as simply $\delta(N_x)$ (N signifies the nucleus under investigation and x indicates its position relative to the variable substituent) irrespective of the reference used. This will result in a regression line with an intercept that depends on the reference used. Furthermore, downfield shifts are taken as positive so that ρ for Hammett correlations is always positive.

8.3. Aromatic Compounds

Any aromatic compound suitable for a study of substituent chemical shifts can be represented by the generalized formula (I). The three basic components are a substituent X, variable in type and orientation with respect to N, a probe nucleus attached to S, a transmitting molecular framework. Unquestionably the most important framework is a benzene ring, i.e., $S = \phi$, and the simplest system, in which the probe is ^{13}C in the benzene ring, is represented by (II). Many possible modifications to (I) can

$$\underset{\text{(I)}}{\text{X—S—N}} \qquad \underset{\text{(II)}}{\text{X—}\phi\text{(C)}}$$

be envisaged and these may be summarized by (III) and (IV), where Z is a bridging group such as a single atom (e.g., O or S), a small group (e.g., methylene or vinyl), or another benzene ring. In all these systems if the probe is ^{13}C or ^{31}P, other groups must be attached to the nucleus, and this may modify the substituent interaction.

$$\underset{(III)}{X{-}Z{-}\phi{-}N} \qquad \underset{(IV)}{X{-}\phi{-}Z{-}N}$$

This classification of systems, which will broadly form the basis of our treatment, is undoubtedly somewhat artificial, but it is preferable to any of the alternatives; a classification in terms of a particular nucleus neglects the underlying unity of the chemical shift phenomenon and reduces the opportunity for evaluating the similarity of correlations with different nuclei at the same position in a system. For similar reasons a classification in terms of particular structure–reactivity parameters is also undesirable.

8.3.1. Systems of the Type $X{-}\phi(C)$

Extensive correlations of ^{13}C chemical shifts in monosubstituted benzenes with Hammett σ constants were first reported by Spiesecke and Schneider,[12] following the pioneering studies of Lauterbur.[13] A close parallel was found between substituent effects at the *para* carbon atom, $\delta(C_p)$, and at its attached proton, $\delta(H_p)$. Neither set of substituent effects was correlated particularly well with σ_p or with the Taft resonance parameters, σ_R, but it was concluded that inductive and anisotropy contributions were absent at the *para*-position. Changes in π-electron density were attributed solely to resonance effects of substituents. Some similarity in the trends in $\delta(C_p)$ in monosubstituted benzenes and $\delta(F_p)$ in *para*-substituted fluorobenzenes was also noted, but this correlation was poorer than that with $\delta(H_p)$.By using an extended series of substituents, Maciel and Natterstad[14] confirmed the existence of a broadly linear relation between $\delta(C_p)$ and σ_R or σ_R^0. Slightly improved correlations were evident with the quantity $[\delta(C_p)-\delta(C_m)]$, which was rationalized in terms of a separation of the resonance and inductive contributions to $\delta(C_p)$. However, none of the correlations was particularly good. A dual-parameter treatment (σ_I and σ_R) appears to give slightly better results.[15]

This early work suffered from a lack of precise data since the available instrumentation for ^{13}C work made use of rapid-passage conditions and required high concentrations (i.e., neat liquids). With the advent of Fourier Transform techniques, studies at low concentration became feasible and data are now available for monosubstituted benzenes in CCl_4 (10%),[16] cyclohexane (33%),[17] and acetone (10%).[18] The most extensive data set is that of Nelson *et al.*,[16] and a good correlation was found between $\delta(C_p)$ and the σ_p^+ values of Swain and Lupton[19] ($n = 13$, $r = 0.98$). With σ_p the correlation was distinctly poorer ($r = 0.91$). By using the regression line for the $\delta(C_p)$ correlation, σ_p^+ values were calculated for new substituents such as N:C:O, COCl, and C : CH. Similarly σ_p and σ_p^+ have been evaluated[20] for several substituents of the types PX_2 and POX_2 by using the regression line for earlier data.[12] Precise determination of $\delta(C_p)$ may be the only practicable method of evaluation of σ_p^+ for reactive or unstable substituents.

Data for monosubstituted benzenes in acetone solution are summarized[17] extremely well by equation (8.5) ($n = 14$, $r = 0.992$). The plot of

$$\delta(C_p) = 9.809\sigma_p^+ - 0.814 \tag{8.5}$$

these data (Fig. 1) is typical of good correlations and shows that many of the deviations from the regression line are less than the experimental error in $\delta(C_p)$. Surprisingly $\delta(C_p)$ values for cyclohexane solutions are correlated[18] slightly less well with σ_p^+ ($n = 11$, $r = 0.970$; data from Ref. 18).

Attempts to find analogous relationships for $\delta(C_m)$ for monosubstituted benzenes have met with varied success. No correlation exists with substituent electronegativity,[12] σ_I,[14] σ_m,[12] or σ_m^+,[17] but Schulman *et al.*[17] have shown that if substituents with nonbonding electron pairs (i.e., those with a $+M$ effect) are treated separately then $\delta(C_m)$ values are well correlated with σ_m^+ ($n = 6$, $r = 0.978$), but the remaining substituents (CO_2Me, CO_2H, CN, NO_2, Me) are correlated less well ($n = 5$, $r = 0.902$). The significance of this behavior is not clear but it does suggest that resonance effects may be more important for the *meta* position than is usually assumed. Further confirmation of this is found in the improved correlations obtained[17] with σ_m^+ and σ_p^+, by comparison with σ_m and σ_p, thus reflecting the greater resonance contribution in the former pair of constants (33% and 66%)[19] relative to the latter (22% and 53%).[19]

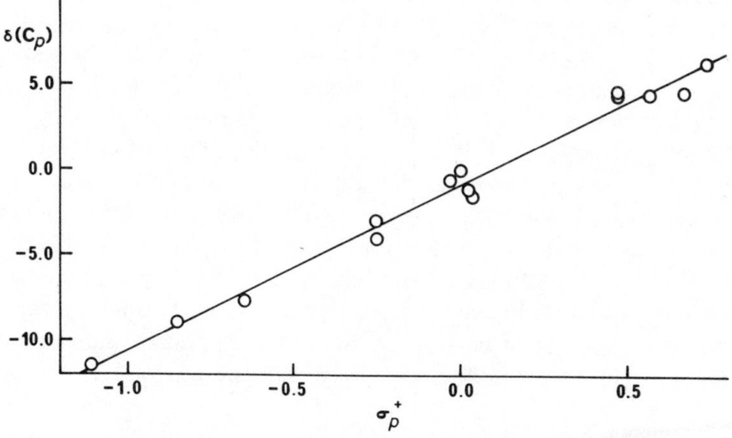

Fig. 8.1. Correlation of $\delta(C_p)$ in monosubstituted benzenes (in acetone)[17] with σ_p^+.

Values of $\delta(C_p)$ for 1-substituted naphthalenes and 9-substituted anthracenes parallel the values of $\delta(F_p)$ in the corresponding fluorocompounds, but, just as for benzene derivatives, there are significant discrepancies and these two parameters are not intimately related.[21] The naphthalene system is better behaved ($n = 9$, $r = 0.960$) than the anthracene system ($n = 9$, $r = 0.907$) in correlations with σ_p^+. It is not surprising that annelation of further benzene rings should perturb the substituent interactions, and a more detailed examination would probably indicate that the resonance interaction was most affected.

Substituent chemical shifts at the directly bound carbon and at the *ortho* carbons are more complex in nature and unlikely to be correlated with Hammett parameters. However, quite good linear plots have been obtained[18] between $\delta(C_{ipso})$ or $\delta(C_o)$ and the semiempirical parameter Q,† first defined by Schaefer and his co-workers.[22] By using a combination of field and resonance parameters and Q, the substituent shifts for all ring carbons in a series of *ortho*-substituted fluorobenzenes can be fitted to straight lines.[18] These correlations are of little consequence for $\delta(C_p)$ and $\delta(C_m)$, since the introduction of a third parameter is unlikely to be statistically significant, but, as equation (8.6) shows, for the carbon bearing the substituent, Q probably has real significance ($n = 11$, $R = 0.909$). This may also be true for the carbon *ortho* to the variable substituent, cf. equation (8.7) ($n = 11$, $R = 0.968$), but detailed statistical confirmation is required.

$$\delta(C_{ipso}) = 0.38\mathfrak{F} + 37.40\mathfrak{R} - 19.73Q + 44.40 \qquad (8.6)$$

$$\delta(C_o) = -5.518\mathfrak{F} + 3.043\mathfrak{R} + 6.491Q - 10.870 \qquad (8.7)$$

The parameter Q has been loosely related[18,22,23] to a paramagnetic contribution to $\delta(H_o)$ in benzene and other systems, and to $\delta(F_o)$ in substituted fluorobenzenes, but this may be too naïve an interpretation since the dependence of δ on Q and $\mathfrak{F}$ changes sign on going from C_{ipso} to C_o, and for $\mathfrak{F}$ changes again on going from C_o to H_o or F_o. This suggests that Q may in large measure represent an alternating through-bond inductive effect (see Section 8.4), only partly accommodated by $\mathfrak{F}$.

Little systematic work has appeared on the effect of solvent on ^{13}C substituent effects in benzene. Nelson *et al.*[16] compared $\delta(C_p)$ values determined in CCl_4 and trifluoroacetic acid (TFA) for six substituents and used the regression line for the former to evaluate σ_p^+ for these substituents in TFA (Table 8.1). Although this might appear to be a useful method for determining σ_p^+ and hence relative reactivity in strongly interacting solvents, it requires ρ to be solvent independent; and taking the solvent shift in $\delta(C_p)$ to be contained entirely within the σ_p^+ values is probably

†$Q = P/Ir^3$, where P is the polarizability and r the length of the C—X bond, and I is the first ionization potential of X; cf. equation (8.16).

Table 8.1. Solvent Dependence of σ_p^+ as Determined from ^{13}C Substituent Shifts

	σ_p^+ Values					
Solvent	NH_2	OMe	Me	COMe	CHO	NO_2
CCl_4	−0.99	−0.73	−0.20	0.57	0.73	0.75
TFA	0.35	−0.49	−0.26	0.90	1.10	0.92

incorrect. An analysis of the solvent-shift data of Nelson *et al.*[16] by Dayal and Taft[24] revealed that a dual-parameter treatment (σ_I and σ_R) accommodated the observed variations entirely in ρ_I, the ρ_R value remaining constant. This treatment permits no variation in the σ values with solvent and in that respect may be inadequate. The influence of a second invariant substituent on correlations of $\delta(C)$ has not been studied systematically, but several interesting reports are available. In view of the importance of ^{19}F correlations, *para*-substituted fluorobenzenes (V) and polyfluorobenzenes (VI) are obvious candidates for ^{13}C studies. In series (V) $\delta(C_p)$ is correlated

p-XC_6H_4F (V) C_6F_5X (VI)

distinctly less well with the corresponding $\delta(F_p)$ values than is $\delta(C_p)$ for monosubstituted benzenes; this was taken[14] to indicate that the local π-electron density of C_p in (V) is complicated by the presence of a directly bound fluorine atom. However, other workers[25] reported a fairly good correlation with σ_R [equation (8.8), $n=17$, $r=0.970$], in spite of using

$$\delta(C_p)=14.7\sigma_R-0.8 \tag{8.8}$$

data for neat liquids. The $\delta(C_p)$ values for the series of compounds (VI) parallel those for series (V) and for monosubstituted benzenes; such correlations are helpful in the assignment of the spectra.[26] A similar correlation was found with $\delta(F_p)$ for (VI), but no statistical data were given[26] and there is probably no close parallel between $\delta(C_p)$ and $\delta(F_p)$ for either (V) or (VI).[14] A dominant resonance effect at C_p in (VI) was apparent from a dual-parameter treatment [equation (8.9)], but the validity of this correlation was not reported.

$$\delta(C_p)=10.41\sigma_R+4.01\sigma_I+3.07 \tag{8.9}$$

Substituted biphenyls provide an interesting system (VII) and values of $\delta(C_p)$ for dilute solution in acetone show[17] an excellent correlation with σ_p^+ ($n=12$, $r=0.993$), but just as for monosubstituted benzenes, $\delta(C_m)$ values are correlated extremely badly ($n=12$, $r=0.406$ for σ_m^+), unless substituents with nonbonding electron pairs are treated separately.

Although the number of data sets is small (5 or 6), correlation with σ_m^+ appears to be more successful than that with σ_m. For a series of 4,4′-dihalogenobiphenyls, $\delta(C_p)$ and $\delta(C_m)$ vary linearly with substituent electronegativity.[27] *Para*-substituted phenylferrocenes (VIII) have been

p-XC_6H_4Ph

(VII)

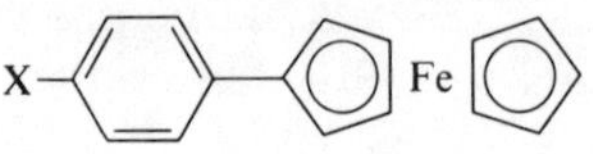

(VIII)

compared[28] with the corresponding biphenyls. Excellent correlations (n = 8, $r > 0.99$) were found between the δ(C) values for the *ortho*, *para*, and *ipso* carbons, but those for the cyclopentadienyl ring carbons compared less successfully with those for the carbons in the second phenyl ring of (VII). The correlation of $\delta(C_p)$ in (VIII) with σ_p or σ_p^0 was surprisingly poor ($n = 10$, $r = 0.95$), but better results were found with a dual-parameter analysis ($\mathfrak{F}$ and $\mathfrak{R}$, $n = 8$, $R = 0.995$). Some *meta*-substituted phenylferrocenes were also investigated.[28] For series (IX) $\delta(C_p)$ was again correlated better with σ_p^+ ($n = 7$, $r = 0.985$) than with σ_p ($r = 0.895$).[29] A plot of $\delta(C_p)$ against σ_R for some *para*-substituted bromobenzenes[30] gave an excellent fit ($n = 7$, $r = 0.994$, calculated from the data of Ref. 30), but this may be fortuitous since $\delta(C_m)$ was reported to show negligible variation, suggesting some imprecision in the data. For the triarylcarbinols (X)

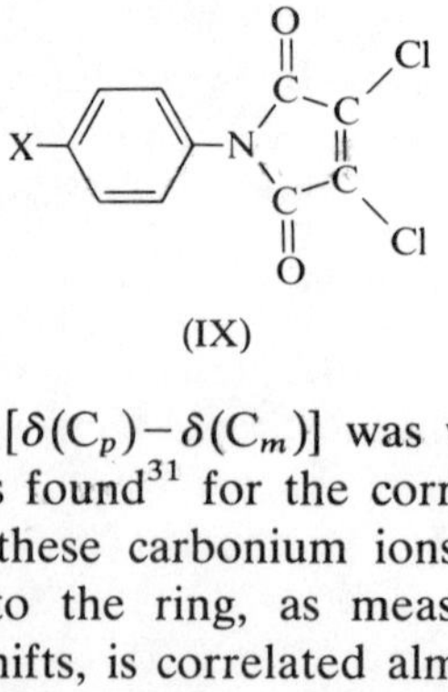

(IX)

$(p$-$XC_6H_4)_3COH$

(X)

the quantity $[\delta(C_p) - \delta(C_m)]$ was well correlated with σ_R, and an even better fit was found[31] for the corresponding carbonium ions ($n = 7$, r = 0.980). For these carbonium ions the extent of positive charge delocalization into the ring, as measured by the deviation from normal substituent shifts, is correlated almost perfectly with σ_R for $+M$ substituents ($n = 5$, $r = 0.999$). In spite of a claim[32] to the contrary, *para*-substituent effects in benzoylium cations exhibit[33] normal trends in relation to σ_p and σ_p^+ ($n = 5$, $r = 0.995$; data from Ref. 33).

In a very detailed study of proton and carbon shifts in the side chain of *para*-substituted styrenes (XI) and the analogous systems (XII, R = Me or Bu^t), Reynolds and his co-workers[34] have included some results of a

p-$XC_6H_4CH{:}CH_2$ p-$XC_6H_4CR{:}CH_2$

(XI) (XII)

dual-parameter correlation of $\delta(C_p)$. Equation (8.10) shows how the data for the styrenes are related to the field and resonance parameters of Swain

$$\delta(C_p) = 2.98\mathfrak{F} + 15.78\mathfrak{R} \tag{8.10}$$

and Lupton ($n = 22$, $R = 0.985$). For the α-substituted styrenes (XII) the coefficient of $\mathfrak{F}$ is little changed but the coefficient of $\mathfrak{R}$ increases to 18.03 (α-Me) or 18.79 (α-Bu^t), as would be expected for increasing limitation to interaction with the benzene ring of the resonance effect of the substituent, as the phenyl ring and the side chain become orthogonal. It is notable that a correlation of $\delta(C_m)$ for (XI) with $\mathfrak{F}$ and $\mathfrak{R}$ gave a single regression line ($n = 22$, $R = 0.898$) in contrast to σ_m^+ correlations for benzene and for biphenyl.[17] Good correlations were also found for $\delta(C_p)$ in a series of substituted phenylacetylenes.[35] In the biphenyls (VII) the substituted ring can be regarded as a bridge between the substituent and the second ring. Values of $\delta(C_p)$ for the second ring are correlated[17] extremely well with σ_p^+ ($n = 12$, $r = 0.991$) or σ_p ($r = 0.997$) but $\delta(C_m)$ is correlated less well ($n = 12$, $r = 0.867$). Significant correlations with σ_p^+ were also found for the *ortho* carbons ($n = 12$, $r = 0.95$) and surprisingly, for the ipso carbon, with σ_m ($n = 12$, $r = 0.947$). However, a negative slope for this last correlation makes its meaning somewhat obscure.

It is evident from the above discussion that the resonance effect predominates at the *para*-carbon atom. To compare the relative effectiveness of the different reactivity parameters in rationalizing $\delta(C_p)$ values a wide range of published data have been correlated[36] with the σ_p and σ_p^+ values tabulated by Swain and Lupton,[19] and with the σ_R values tabulated by Exner.[1] Correlation coefficients for these analyses (Table 8.2) clearly confirm that σ_p^+ offers the best measure of the combined polar and resonance effects of substituents. Given the substantial experimental uncertainties in some of the data the range of r values is not excessive. The intercept was significantly different from zero (in the range -0.8 to -1.5 ppm) in most cases, indicating that the point for H was consistently one of the worst. Notably σ_R was a poor measure of the total electronic effect at C_p.

The rationale for the use of the σ_p^+ parameter is the postulate that it takes account of the direct conjugation of the substituent with an electron-acceptor reaction center in the side chain (XIII). Such conjugation is important for reactions that require the development of a positive change in the transition state, a good example being the electron-deficient transition state (XIV) characteristic of electrophilic aromatic substitution. In

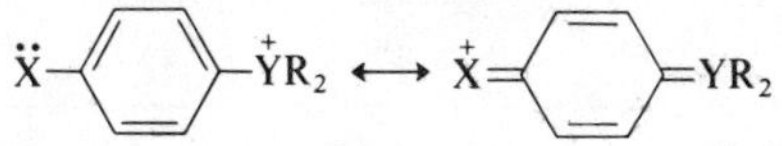

(XIII) 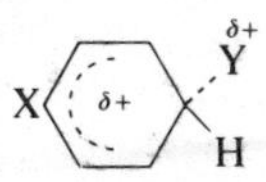 (XIV)

Table 8.2. Correlation Coefficients (r) for the Correlation of $\delta(C_p)$ in p-X—C_6H_4—Y with σ_p, σ_p^+, and σ_R

Substituent, Y	Solvent	n[a]	$r(\sigma_p)$	$r(\sigma_p^+)$	$r(\sigma_R)$
H	Cyclohexane	10	0.954	0.983	0.940
H	CCl_4	17	0.929	0.962	0.942
H	$CDCl_3$	13	0.861	0.951	0.913
H	Me_2CO	14	0.959	0.992	0.957
H	Neat	19	0.913	0.936	0.928
Me	CCl_4	10	0.962	0.974	0.932
Bu^t	CCl_4	9	0.926	0.967	0.928
$SiMe_3$	CCl_4	8	0.860	0.965	0.918
$GeMe_3$	CCl_4	6	0.912	0.980	0.921
$SnMe_3$	Neat	8	0.906	0.970	0.985
CH_2SnMe_3	Neat	4	0.899	0.985	0.752
$C(OH)(C_6H_4X\text{-}p)_2$	THF	7	0.916	0.968	0.972
$CH{=}CH_2$	CCl_4	18	0.921	0.970	0.950
$CMe{=}CH_2$	CCl_4	9	0.906	0.972	0.937
$CBu^t{=}CH_2$	CCl_4	9	0.929	0.972	0.942
$CH{=}C(CN)_2$	$(CD_3)_2CO$	10	0.958	0.966	0.926
Ferrocenyl	$CDCl_3$	11	0.942	0.966	0.948
Ph	Me_2CO	14	0.960	0.992	0.954
COMe	Neat	11	0.918	0.973	0.885
CO_2H	DMSO	7	0.948	0.990	0.907
CO_2Me	Neat	5	0.970	0.996	0.968
$CONMe_2$	$CDCl_3$	4	0.942	0.995	0.968
$CH{=}NPh$	$CDCl_3$	6	0.956	0.990	0.959
$CMe{=}NOH$	$CDCl_3$	6	0.928	0.990	0.915
$C(OMe){=}NH$	Neat	5	0.886	0.979	0.950
CN	$CDCl_3$	9	0.959	0.992	0.952
NH_2	$CDCl_3$	9	0.946	0.975	0.967
NMeCN	Neat	4	0.752	0.975	0.887
$N{=}CHPh$	$CDCl_3$	5	0.916	0.904	0.941
NC	CCl_4/CD_3CN	4	0.708	0.959	0.915
NO_2	$CDCl_3$	9	0.923	0.976	0.969
OMe	$CDCl_3$	9	0.942	0.976	0.969
$OCH{=}CH_2$	Neat	4	0.700	0.996	0.918
OCN	Neat	5	0.868	0.968	0.961
SH	$CDCl_3$	6	0.813	0.965	0.902
SMe	$CDCl_3$	7	0.990	0.988	0.918
$SCH{=}CH_2$	Neat	4	0.777	0.983	0.868
SOMe	$CDCl_3$	7	0.908	0.989	0.920
SO_2Me	$CDCl_3$	7	0.889	0.982	0.925
F	Neat	16	0.878	0.922	0.884
Br	Neat	7	0.950	0.994	0.960
$C^+(C_6H_4X\text{-}p)_2$	THF	7	0.939	0.986	0.931
CO^+	SbF_5	5	0.808	0.994	0.758
$N_2^+(BF_4^-)$	SO_2	8	0.909	0.963	0.900

[a] n is the number of compounds in the set.

terms of this definition σ^+ *differs from* σ to the extent that it contains a contribution that is largely irrelevant to ground-state properties, and hence σ_p^+ should be correlated less well with $\delta(C_p)$. Furthermore, while quinonoid conjugation in the ground state is expected to influence the electron density at C_p [and hence $\delta(C_p)$] inasmuch as normal resonance interactions are reduced, this should be significant only for substituents (Y in XIII) of the electron-accepting ($-M$) type. The results in Table 8.2 show that not only is $r(\sigma_p^+)$ good in most cases but there is no discernible trend in the difference between the σ_p and σ_p^+ correlations, depending upon whether Y is of the $-M$ or $+M$ type. It seems likely[36] that no great significance can be attached to the fact that σ_p^+ provides a better correlation than σ_p for all the series of $\delta(C_p)$ values; apparently σ_p^+ fortuitously approaches more closely to that combination of polar and resonance effects that is reflected in $\delta(C_p)$. Consequently $\delta(C_p)$ does not measure a substituent effect of the same composition (in terms of field and resonance effects) as does σ_p.

Dual-parameter correlations of $\delta(C_p)$ are not available for a wide range of substituents, Y. Taft and his co-workers[37] have shown that $\delta(C_p)$ for monosubstituted benzenes (Y = H) is correlated best with a combination of σ_I and σ_R^0. Alternative resonance parameters, $\sigma_{R(\text{BA})}$, $\sigma_{R(\text{A})}^-$, σ_R^+ (defined in Ref. 37) or the $\mathfrak{F}$ and $\mathfrak{R}$ parameters of Swain and Lupton,[19] performed less well. An attempted triple-parameter correlation (Dewar's FMMF method[38]) of $\delta(C_p)$ and $\delta(C_m)$ for the first ring of biphenyl, together with all four values of $\delta(C)$ for the second, revealed[17] no unified relationship if the correct geometrical terms are used. A more widely based regression analysis did produce a good fit ($n \approx 50$, $R = 0.93$) but of course lost the exactitude of Dewar's model. Nevertheless this analysis indicated that the mesomeric, mesomeric–field, and field effects were contributing in the ratio 41 : 4 : 1, supporting the general trend evident in Hammett correlations.

Some insight into the nature of the field effect has come[39] from an examination of the various $\delta(C)$ values in the phenyl group of compounds (XV, $n = 1$ or 2) with a charged ($R = NH_3^+$) or polar (R = Cl or Br)

$$\mathrm{Ph\overset{\alpha}{C}H_2\overset{\beta}{C}H_2\overset{\gamma}{C}H_3}$$

or

$$\mathrm{Ph(CH_2)_nR}$$

(XV)

side chain. Comparison with the corresponding $\delta(C)$ values for phenylalkanes (R = Me) permits isolation of the field effect (from the intrinsic α, β, and γ effects of the side chain). The distribution of $\delta(C)$ values is consistent with a through-space π-polarization mechanism, i.e., polarization of the benzene π electrons in the sense shown in (XVI) and (XVII). Similar results were obtained for phenylalanine (XVIII) as it was titrated

$\overset{\delta+\quad\delta-}{C_6H_5}CH_2\overset{+}{N}H_3$ (XVI)

$\overset{\delta+\quad\delta-}{C_6H_5}CH_2\overset{\delta+}{C}H_2-\overset{\delta-}{Cl}$ (XVII)

$PhCH_2CH(\overset{+}{N}H_3)CO_2^-$ (XVIII)

from the cationic form to the anionic form. This work appears to impugn the idea that π-polarization effects are small, but other interpretations of the data may be possible.

Since the common ground between σ values and chemical shifts is the electronic response of the ground state (or an early transition state) to substituent perturbations, correlations of either parameter (σ or δ) with calculated charge densities, q, are of considerable importance. For ring carbons in simple substituted benzenes, several workers have demonstrated that δ(C) is linearly related to q, equation (8.11), but the most

$$\delta(\mathrm{C}) = a(q^{\pi} + q^{\sigma}) + b \tag{8.11}$$

notable aspect of these correlations is the variation of the slope a of the regression line. (The intercept is usually in the range ± 3 ppm.) Some representative values of a are collected in Table 8.3. Although a parallel is observed[25,39,40] between δ(C) and q^{π}, inclusion of q^{σ} usually improves the correlation.[25,39] Hamer *et al.*[34] found that q^{π} for the carbon *para* to the substituent in the styrenes (XI) could be correlated with a combination of $\mathfrak{F}$ and $\mathfrak{R}$ values and that the $\mathfrak{R}$ parameter was clearly dominant. This is also evident, for C_p in substituted benzenes, in the good dual-parameter correlations (σ_I and σ_R^0) which have been noted with q^{π} and q^{σ} at both the semiempirical (CNDO/2)[41,42] and the *ab initio* (STO-3G)[43] level.

Table 8.3. Correlation of δ(C) with Charge Density q [Equation (8.11)] for Ring Carbon Atoms in Substituted Benzenes

System	Carbon atoms included	MO method	Slope	Ref.
PhX	all	HMO/Del Re	127	44
PhX	all	CNDO	180	16
PhX	all	CNDO	166	45
o-, *m*-, or *p*-$O_2NC_6H_4X$	all	CNDO	148	46
p-FC_6H_4X	*para*	HMO	160	25
PhX	*para*	CNDO	290	16
p-NCC_6H_4X	*para*	HMO	334	40
PhX	*para*	*Ab initio*	450	43
PhX	*meta*	*Ab initio*	150	43

At present, it appears that calculations of q are too dependent on the MO method used and its parameterization to reveal any meaningful quantitative correlation with $\delta(C)$. *Ab initio* methods fail to produce any good overall correlation of q and δ for the *meta* and the *para* positions although these are independently well correlated. Local π charges are found to depend on both σ_R^0 and σ_I.

8.3.2. Systems of the Type $X—\phi—N$

The principal nuclei N in this category of system are ^1H and ^{19}F for obvious reasons. For $N = {}^{13}C$ or ^{14}N other groups must be attached to N and a large range of compounds can be included, but the study of such compounds has scarcely begun.

The two nuclei ^1H and ^{19}F present an interesting contrast in the development of linear free energy relationships in nmr spectroscopy. Early work on $\delta(H)$ in substituted benzenes was inevitably characterized by imprecise data and a proliferation of qualitative correlations with various parameters, but rarely with Hammett σ constants. In spite of the apparently quite good correlation of $\delta(H_p)$ with σ_p displayed in the notable paper of Spiesecke and Schneider,[12] hardly any attempt was made to examine rigorously the relationship between $\delta(H)$ and Hammett σ values. In some measure this was due to the discovery by Taft that $\delta(F)$ in simple substituted fluorobenzenes was apparently a well-behaved parameter with respect to structure–reactivity correlations and was especially valuable in the study of the separation of polar and resonance contributions. With hindsight, this concentration on $\delta(F)$ at the expense of $\delta(H)$ was unfortunate, since ^{19}F has turned out to be a more difficult nucleus to study than ^1H in many respects.

The attraction of ^{19}F over ^1H as a probe nucleus is due to the relatively large range of $\delta(F)$ values (*ca.* 30 ppm in fluorobenzenes) compared to the range in $\delta(H)$ values (*ca.* 2 ppm) and to the concomitant reduction in the importance of anisotropy and ring-current effects. Nonetheless Hayamizu and Yamamato[47] have shown that $\delta(H_p)$ values for monosubstituted benzenes at infinite dilution in CCl_4 are correlated extremely well with $\delta(F_p)$. An excellent correlation was also found with σ_p^+ ($n = 11$, $r = 0.997$)[48] and a somewhat poorer one with σ_p ($n = 13$, $r = 0.951$).[49] Dual-parameter analyses[47] of the same data gave equations (8.12) and (8.13). Equation

$$\delta(H_p) = 0.45\sigma_I + 1.14\sigma_R - 0.08 \quad (8.12)$$

$$\delta(H_m) = 0.28\sigma_I + 0.41\sigma_R - 0.03 \quad (8.13)$$

(8.12) is very different from that given by Taft *et al.*[51] for crude ^1H data but similar to that found by this group in a more recent examination.[43] This correlation appears to give an excellent account of the ^{1}H data (rms error = 0.04 ppm), comparable in fact with that demonstrated for ^{19}F data.[43,52] Similar results have been reported[50] for the Yukawa–Tsuno equation for values determined in cyclohexane and in *NN*-dimethylacetamide. Although $\delta(H_m)$ was found originally to show no correlation with σ_m, after correction for substituent anisotropy and ring-current effects, the correlation had $r = 0.971$ ($n = 8$).[53] However, the significance of this improvement is not clear since equation (8.13) (and an almost identical expression given by Taft and his co-workers[43]) represents a good correlation without such corrections. Poor data must account in some measure for the lack of fit in the original work, but the dual-parameter treatment is probably intrinsically superior. Application of the Swain–Lupton parameters[19] to a lamentably small number of data sets ($n = 5$) revealed[54] that $\delta(H_p)$ was only slightly more sensitive to resonance effects (36%) than was $\delta(H_m)$ (32%). This unlikely result can be compared more directly with the above correlations by using the ratios of the coefficients of the resonance and polar parameters in the correlation equation (i.e., $\gamma = \rho_R/\rho_I$, or r/f): with $\mathfrak{F}$ and $\mathfrak{R}$, $\gamma_p = 0.92$ and $\gamma_m = 0.76$[54]; with σ_I and σ_R, $\gamma_p = 2.53$ and $\gamma_m = 1.46$[47]; and with σ_I and σ_R^0, $\gamma_p = 4.63$ and $\gamma_m = 1.83$.[43]

Extension of the Hammett relationship to $\delta(H_o)$ in substituted benzenes is likely to be less successful because of the possible contribution of steric effects and local magnetic effects. The first notable correlation[47] of infinite-dilution values of $\delta(H_o)$ gave excellent results in an extended Hammett equation (8.14) (rms error = 0.07 for $n = 10$) provided that the substituents such as NMe_2, CMe_3, CN, or I were excluded. A more extensive analysis of reported data gave equation (8.15).[55] All groups were

$$\delta(H_o) = 0.85\sigma_I + 1.96\sigma_R + 0.06 \quad (8.14)$$

$$\delta(H_o) = 0.64\sigma_I + 1.27\sigma_R + 0.186 \quad (8.15)$$

included in the correlation ($n = 24$, $R = 0.936$), which was not improved by inclusion of a steric parameter. In general, steric effects were negligible for other series investigated (polysubstituted benzenes, substituted naphthalenes) and deviations from the regression line are due to local magnetic effects at the *ortho* hydrogen atoms.

An elegant demonstration of this has been reported by Hermann and Rae.[56] In the series of substituted *o*-dinitrobenzenes (XIX), $\Delta\delta$ [$=\delta(H_3) - \delta(H_5)$] should remain constant if the conformation of the 2-NO_2 group does not change. However, $\Delta\delta$ varies broadly in parallel with σ_p^+ for X, the 4-substituent, a reflection of electronic "buttressing" of the 1-NO_2 group† with concomitant rotation of the 2-NO_2 group. The downfield shift of H_3 in

†Conjugative interaction of X and NO_2 is maximal when the ring and NO_2 are coplanar.

2-NO_2-4-X-anilines (XX), associated with acylation of the amino group, appears to give an excellent correlation with σ_p, and this is taken[57] to indicate an increase in hydrogen bonding between the AcNH and NO_2 groups, the result of mesomeric interaction of the nitrogen nonbonding

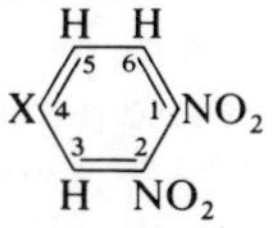

(XIX) 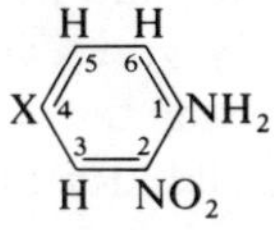 (XX)

electron pair with the ring, as mediated by substituent X. However, one badly fitting point (X = H) was ignored. A rather better correlation ($n = 10$, $r = 0.94$) can be found with σ_p^+, implying through-conjugation and stabilization of the planar deshielding conformation of NHAc.

The parameter, Q, has been widely correlated with $\delta(H_o)$[22,23,58–60] and was in fact introduced for this purpose. Originally, Q for a substituent X was defined[22] in terms of the ionization potential (I) of X and the length (r) and polarizability (P) of the bond C—X [equation (8.16)], but values of Q

$$Q = P/Ir^3 \qquad (8.16)$$

for polyatomic substituents have been evaluated on a purely empirical basis. Q is best regarded as an empirical parameter probably corresponding only in some measure to local magnetic effects.

Smith and his co-workers[23,58] have explored the utility of Q for the correlation of $\delta(H_o)$ for monosubstituted benzenes, *o*-disubstituted benzenes (XXI, R = OH, OMe, NH_2, NO_2, CH_3, or CHO), *p*-disubstituted benzenes, and 1-substituted naphthalenes. For substituents without axial symmetry Q appears to be conformation dependent. Inclusion of Q as third parameter in a correlation of $\delta(H_3)$ with $\mathfrak{F}$ and $\mathfrak{R}$ for 2-X-chlorobenzenes (XXII) produced an exceptionally good fit of the observed values [equation (8.17), $n = 12$, $R = 0.997$].[23]

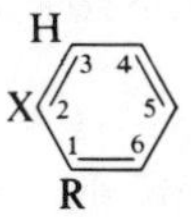

(XXI) 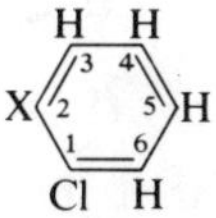 (XXII)

Analogous correlations for the other ring protons revealed a negligible contribution from Q, except perhaps for $\delta(H_4)$. Although no details are

$$\delta(H_3) = 0.101\mathfrak{F} + 0.110\mathfrak{R} + 0.253Q + 6.70 \qquad (8.17)$$

available of similar correlations for other series of compounds (XXI), Q is probably always a significant third parameter for $\delta(H_3)$. There is need, however, for a proper statistical evaluation of these triple-parameter correlations, particularly since the inclusion of Q may in part enable the electronic effects at the *ortho*-position to be isolated (in $\mathfrak{F}$ and $\mathfrak{R}$) from local magnetic effects.

Evidently $\delta(H_p)$ and $\delta(H_m)$ values for substituted benzenes are correlated satisfactorily with several types of reactivity parameter, and this is of little value in assessing the relative merits of these parameters. In contrast, the corresponding data for fluoro compounds, $\delta(F_p)$ and $\delta(F_m)$, have been taken to offer convincing support for (*a*) normal correlations with σ_p, (*b*) dual substituent-parameter correlations with σ_I and σ_R, and (*c*) the complete denial of any meaningful correlation with σ_p or with a combination of σ_I and σ_R. Irreconcilable though these conclusions may seem, these differences in interpretation are to some extent more apparent than real. Since many data sets have been accumulated for the *p*-fluorophenyl system and related polycyclic systems, detailed evaluation is not possible and only the major conclusions and criticisms will be discussed.

About 1960 Taft and his group,[51,52,61] in their exploration of the division of the Hammett parameters into separate parts for polar and resonance effects, found that values of $\delta(F_m)$ and $\delta(F_p)$ for substituted fluorobenzenes fitted equations (8.18) and (8.19), respectively.

$$\delta(F_m)=7.1\sigma_I-0.05 \tag{8.18}$$

$$\delta(F_p)=7.1\sigma_I+29.5\sigma_R^0+0.60 \tag{8.19}$$

These results indicated that resonance effects were negligible at the *meta*-position (σ_I was taken to measure only the classical field effect) and that polar effects were relatively unimportant at the *para*-position. Much controversy has surrounded these conclusions and they have been widely criticized, sometimes unfairly.[62] Recently Taft[43,63] has introduced some resonance dependence for $\delta(F_m)$, and equations (8.20) ($f=0.08$)† and (8.21) ($f=0.21$ were found (for CCl_4 solutions).

$$\delta(F_m)=5.26\sigma_I+0.81\sigma_R^0 \tag{8.20}$$

$$\delta(F_p)=7.02\sigma_I+30.55\sigma_R^0 \tag{8.21}$$

It is noteworthy that these expressions use parameters, σ_I and σ_R^0, which are now defined essentially by kinetic and equilibrium data, and hence criticism of (8.20) and (8.21) does not necessarily undermine the validity of the dual-parameter approach itself.

A major criticism of equation (8.18) has been the lack of any resonance dependence, and Phillips and his co-workers[64] claim that $\delta(F_m)$

†For the statistical evaluation of the quality of correlations Taft and his group prefer the quantity f (the ratio of the standard deviation to the rms of the data) to the more usual correlation coefficient, r.

can be largely accounted for by a "π contribution", but it is doubtful whether their method leads to an effective separation of polar and resonance effects. Both Phillips[64–66] and Taft[37,67] have examined a wide range of compounds of type (XXIII), in which the bridging group, Z, may

$$X-C_6H_4-Z-C_6H_4-F$$

(XXIII)

be saturated ($-CH_2-$, $-CHOH-$, $-CHPh-$), unsaturated ($-CH{=}CH-$, $-CH{=}N-$, $-N{=}N-$), a heteroatom ($-O-$, $-S-$), or a charged moiety ($-\overset{+}{C}H-$, $-CH{=}\overset{+}{N}H-$, $-\overset{-}{C}Ph-$). Solvent effects on δ(F) in these systems have also been studied.[24,79]

Good correlations were found with σ_I and σ_R^0 for most systems in all solvents, but when Z was a neutral group results were not notably worse for correlations with $\mathfrak{F}$ and $\mathfrak{R}$. With charged groups in the bridging position, data were correlated best with the appropriate inductive and modified resonance parameters (σ_I and σ_R^+, or σ_I and σ_R^-). Although the ^{19}F data for this series of compounds (XXIII) extend and support Taft's dual-parameter approach, the necessity for the separation of σ_p into σ_I and σ_R has been challenged with the aid of the same data. Phillips[64–66] contends that, because the values of $\delta(F_p)$ in 4-fluoro-4′-X-*trans*-stilbenes (XXIII, $Z = -CH{=}CH-$) are correlated with σ_p [equation (8.22), $n = 12$, $r = 0.992$] and $\delta(F_p)$ values for all the other series precisely parallel those for the stilbenes, Taft's σ_R parameters (and those of Swain and Lupton) are of doubtful significance. However, for the fluorobenzenes $\delta(F_p)$ is not correlated with σ_p and equation (8.23) is necessary.

$$\delta(F_p) = 2.68\sigma_p \tag{8.22}$$

$$\delta(F_p) = 15.75\sigma_I - 21.06\sigma_p \tag{8.23}$$

Furthermore Phillips' assumption[64] (subsequently modified[66]) that changes in $\delta(F_p)$ for the stilbenes are wholly attributable to resonance interactions, is unreasonable. Some contribution from field effects must be involved, and, on the whole, Taft's dual-parameter approach is the more satisfactory. In fact there is little fundamental difference between these two interpretations of ^{19}F data and Taft's attempt to separate polar and resonance perturbations has obvious attractions.

A more significant criticism of the correlation of $\delta(F_p)$ with free-energy parameters has come from the work of Dewar and his group. These workers have evolved[68,70] an empirical method of calculating the σ value for a substituent X in terms of parameters characterizing the dipole field effect, $F(\mathrm{X})$, the resonance (mesomeric) effect, $M(\mathrm{X})$, and the secondary field effects of the charges resulting from resonance interaction, $M_F'(\mathrm{X})$ parameter [equation (8.24)]. This so-called FMMF approach, when

$$\sigma(\mathrm{X}) = GF(\mathrm{X}) + qM(\mathrm{X}) + \sum q'G'\{M_F'(\mathrm{X})\} \tag{8.24}$$

modified to apply to ^{19}F chemical shifts, takes account of the geometrical dependence of the field effect [the terms G and G' in equation (8.24)] and of the specific sensitivity of $\delta(\mathrm{F})$ to charges, q, on the adjacent carbon and q' on the other carbon atoms in the ring system [through values of $M(\mathrm{X})$ and $M'(\mathrm{X})$ different from those required for correlation of equilibrium data]. Taft's dual-parameter approach can make no specific allowance for the geometrical dependence of the field effect and cannot account for $\delta(\mathrm{F})$ in naphthalene derivatives as adequately as does Dewar's method. Correlation of $\delta(\mathrm{F})$ values for benzene,[70] biphenyl,[70] terphenyl,[70] naphthalene,[60,70,72] and other bicyclic compounds[71] with $\sigma(\mathrm{X})$ from equation (8.24) requires a different ρ value for each series, but ρ for p-substituted fluorobenzenes [equation (8.25), $n = 13$, $r = 0.955$] is virtually identical with that for 4-substituted-1-fluoronaphthalenes.

$$\delta(\mathrm{F}) = -23.2\sigma(\mathrm{X}) \tag{8.25}$$

However, the success of the FMMF method does not mean that Taft's analyses can be dismissed as fortuitous and irrelevant[69] nor does it imply that the whole basis of δ *vs.* σ correlations is invalid.[70] For most of the p-fluorophenyl systems of type (XXIII) the field term is a function only of distance (since the angular relationships are very similar) and the charge distribution associated with a given substituent will vary over only a narrow range. To a first approximation the field and "resonance-field" terms in equation (8.24) can be combined and equation (8.25) can be written in the general form (8.26). Equation (8.26) may be expressed in the form (8.27),

$$\delta(\mathrm{F}_p) = \rho GF(\mathrm{X}) + qM(\mathrm{X}) \tag{8.26}$$

$$\delta(\mathrm{F}_p) = \rho_F F(\mathrm{X}) + \rho_M M'(\mathrm{X}) \tag{8.27}$$

formally analogous to the Taft treatment. The parameter ρ_F accommodates variations in G from system to system, while corresponding variations in the sensitivity of $\delta(\mathrm{F}_p)$ to resonance effects are reflected in ρ_M. The substituent parameter $M'(\mathrm{X})$ contains the terms $M(\mathrm{X})$ and q, which have different values, in Dewar's treatment, for reactivity data and chemical shifts. Taft's treatment contains this intrinsic difference in the resonance response in the ρ_R factor rather than in the substituent resonance constant, but this is no less reasonable than other approaches.

Certainly for correlations of ^{19}F data in most compounds and perhaps generally for reactivity data, Taft's dual-parameter method can be viewed as a limiting case of the more general (and, perhaps after modification, the more valuable) method of Dewar, but the former method has the virtue of simplicity in understanding and application. It may be noted that the FMMF method does not account for δ(C) in biphenyls.[17]

Although the individual approaches of Taft and of Dewar are not mutually exclusive, their respective proponents continue to claim otherwise.[73–75] An unresolved problem concerns the mechanism of the transmission of the polar effect, i.e., the relative importance of through-space and through-bond contributions. Taft[75] has advanced further evidence in support of his belief that the latter contribution is the more important. In a comparison of the ketones (XXIII, Z = CO) with the analogous series (XXIV), *both* the inductive and the resonance effect are reduced by about

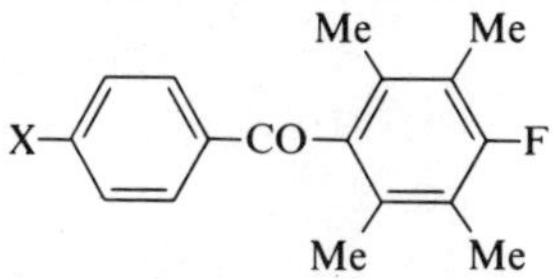

(XXIV)

50%, and in the corresponding BCl_3 adducts by 400% or more. Lack of coplanarity of the rings is thought to reduce polar transmission via the π-system. Adcock and Gupta[74] have criticized this choice of model system, maintaining that δ(F) in systems (XXV) and (XXVI) reflects a dominant

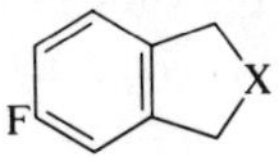

(XXV) 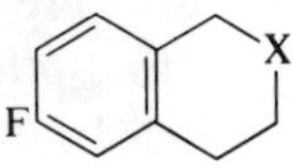 (XXVI)

polar (field) effect.[71,73,76] It is difficult to define precisely the role of localized and delocalized polarization arising from a field effect, and, together with the effect of geometrical distortion,[73] this area is in urgent need of unbiased investigation with a wider range of more suitable model systems.

Chemical shifts in a series of pentafluorobenzenes have been correlated[77] by Taft's method, and the results reveal that the resonance sensitivity for a m-fluoro-group is much higher than in the monofluorophenyl system [equation (8.28), $n = 24$, $R = 0.99$]. In contrast, $\delta(F_p)$ is similar in both cases, cf. equations (8.29) and (8.21). An intersting extension of the Taft method [equation (8.30)] has been used to account for $\delta(F_p)$ values for the silicon derivatives (XXVII) ($n = 21$, $R = 0.984$).[78]

$$\delta(F_m) = 5.21\sigma_I + 6.26\sigma_R^0 - 0.7 \quad (8.28)$$

$$\delta(F_p) = 8.91\sigma_I + 31.52\sigma_R^0 - 1.8 \quad (8.29)$$

$$\delta(F_p) = 5.26\sum\sigma_I + 2.69\sum\sigma_R^0 + 2.53 \quad (8.30)$$

For *meta*-substituted fluorobenzenes coordinated to chromium tricarbonyl, $\delta(F_m)$ is not correlated with σ_I, and this has been taken[80] to

$ZYXSiC_6H_4F$-*p*

(XXVII)

X–C_6H_4–$\overset{\alpha}{C}H{:}\overset{\beta}{C}HMe$

(XXVIII)

indicate electron withdrawal from σ orbitals, but the perturbation of all orbitals of the benzene ring in such compounds probably makes any appeal to Taft parameters invalid.

Very few studies of the correlation of $\delta(C_\alpha)$ values have appeared and these are largely restricted to the styrenes (XI) and (XII)[34] and similar compounds with an unsaturated side chain [(XXVIII)–(XXX)].[35,81,82] The

p-$XC_6H_4\overset{\alpha}{C}H{:}\overset{\beta}{C}H.\overset{\gamma}{C}H{:}\overset{\delta}{C}H_2$

(XXIX)

p-$XC_6H_4C{:}CR$

(XXX)

variation in $\delta(C_\alpha)$ is always small and in the styrenes (XI) and (XII) only a field effect appears to operate [$\delta(C_\alpha)=-1.30\mathfrak{F}$ $(r>0.95)$]. In contrast $\delta(C_\beta)$ for the styrenes shows a substantial dependence on resonance [equation (8.31), $n=22$, $R=0.99$)] and a normal positive response to the

$$\delta(C_\beta)=2.82\mathfrak{F}+7.14\mathfrak{R} \tag{8.31}$$

field effect. Similar results were found for the substituted phenylacetylenes (XXX, R = H).[35] For the other systems Hammett correlations were obtained with small data sets, but $\delta(C_\alpha)$ had a negative ρ value in all cases [$\rho=-5.8$ for (XXIX)], whereas ρ for the C_β correlation was always positive [$\rho=5.07$ for (XXIX), $\rho=5.38$ for (XXX, R = Me), $\rho=6.64$ for (XXVIII)]. For the phenylbutadienes ρ continues to alternate in sign along the chain, i.e., negative for C_γ and positive for C_δ.

π-Polarization[34,39] may be responsible for this negative field dependence at C_α and C_γ, which was also observed for the terminal vinyl carbon in (XXXI) and its sulfur analog.[83] For the compounds (XXXII, Y = C, Si,

p-$XC_6H_4OCH{:}CH_2$

(XXXI)

p-$XC_6H_4YMe_3$

(XXXII)

or Ge) the opposite trend was observed.[84] For these compounds both ρ_I and ρ_R are positive for C_α and negative for C_β. Carbonium ions (XXXIII, R = H, Me, Et, or Pr^i)[85] showed reasonably good correlations of $\delta(C_\alpha)$ with σ_p^+, with a nearly constant ρ value of 0.027. Chemical shifts for ^{14}N or ^{15}N in substituted anilines are reported to be correlated with the analogous ^{19}F shifts,[87] ^{13}C shifts (in toluenes),[87] or ^1H shifts (of the attached protons),[88] with Hammett constants,[86] and with the resonance parameter, $\mathfrak{R}$,[87] but there has been no thorough study. The existence of a pair of polarizable nonbonding electrons on nitrogen may introduce complications into the correlation of $\delta(N_p)$, through contributions from quinonoidal resonance and orbital polarization.

$$p\text{-}XC_6H_4\overset{+}{C}RMe \qquad p\text{-}XC_6H_4Y$$

(XXXIII) (XXXIV)

Scattered reports exist of chemical shift correlations for other nuclei attached to the benzene ring (e.g., silicon,[89] phosphorus,[90] and boron[91,92]) but no general conclusions can be reached.

For the two intensively studied nuclei ^1H and ^{19}F good but different correlations are possible with σ_I and σ_R, but the position for ^{13}C is unresolved. There are many unanswered questions concerning the important type of compound (XXXIV) in which the nmr probe (Y) can interact directly with the π system, and it seems likely that the next major advance in this area will come from a ^{13}C nmr study of a wide range of compounds. It is perhaps an unfortunate accident that $\delta(F)$ is and has been a more accessible nmr parameter than $\delta(C)$, since the latter is certain to be the more revealing.

Correlations of $\delta(H_p)$,[34,35,42–44,93] $\delta(C_\alpha)$,[34,35,43,93,94] and $\delta(F_p)$[25,43] with calculated electron densities are sometimes tolerably good, and the same data can often be correlated satisfactorily with structure–reactivity parameters. Comparison between these studies is not possible without evaluating the MO methods, so this work will not be further discussed here.

8.3.3. *Systems of the Type X—ϕ—Z—N*

By far the most widely studied groups of compounds are those in which the probe nucleus, *N*, is a proton attached to the side chain Z of an aromatic system (XXXV), and the general feeling of chemists about relating δ to free-energy parameters is based firmly on this class of compound. Many correlations have been reported, mainly with Hammett parameters, ranging from excellent to very bad in quality, and extensive compilations of these correlations have been published.[5,54,96] Rather than duplicate such compilations, it is more useful to try to establish just what these correlations can be expected to reveal about the relationship between δ and free-energy parameters for this class of compound. One essential prerequisite to the examination of the compounds (XXXV) is to exclude

$$X\text{—}\phi\text{—}Z\text{—}H \qquad X\text{—}\phi\text{—}H$$

(XXXV) (XXXVI)

those that properly belong to type (XXXVI), for which good correlations can be expected with various parameters (Section 8.3.2).

Generally it can be assumed that in any series of compounds of type (XXXV) both field and resonance effects will operate, but it is extremely unlikely that these two effects will make the same relative contribution to series such as (XXXVII) and (XXXVIII). In (XXXVIII) the side chain will

$X{-}\phi{-}CH_2R$ (XXXVII) $\qquad$ $X{-}\phi{-}CH{=}CHR$ (XXXVIII)

itself have a significant conjugative interaction with the group X—ϕ, which will be quite different from any such interaction in (XXXVII). Usually the sensitivity to substituent resonance effects of a probe ^{1}H nucleus in (XXXV) will be a function of the nature of Z, and hence structure–reactivity parameters that do not permit the ratio of field and resonance effects to vary *cannot* be expected to be correlated equally well with δ(H) for all types of compounds (XXXV).

A further crucial difference between the ^{1}H probes in (XXXVII) and (XXXVIII) (and generally between XXXV and XXXVI) is their geometrical relationship to the substituent. There is ample evidence from many sources[68,70,95] that the field effect of a substituent at a given site must be a function of its distance from [$f(r)$] and orientation towards [$f(\theta)$] that site, and correlation of δ should take account of this fact. This requirement clearly rules out Hammett parameters, which not only cannot permit independent variation of the field contribution, but also embody values of $f(r)$ and $f(\theta)$ appropriate only to benzoic acids.

In principle, independent field and resonance parameters are essential to cope with the independent response of δ(H) of side-chain protons to these two effects, and hence the conclusion of Yokoyama *et al.*[96] that correlation analysis of side-chain proton chemical shifts with Hammett parameters is of little value is undoubtedly true in a general sense. Indeed dual-parameter analysis must be more effective for the correlation of data for any nucleus in any system and only the similarity of the forms of $f(\theta)$ and $f(r)$ for *para*-substituted benzoic acids, for substituted benzenes, and for *para*-substituted fluorobenzenes permits σ_p (or σ_p^+) to be even reasonably successful in correlations of $\delta(H_p)$, $\delta(F_p)$, and $\delta(C_p)$ (see previous sections).

It is not surprising that in the correlations for a wide range of side-chain protons examined by Yokoyama *et al.*[96] the ρ values failed to make any chemical sense. A simple relationship[96] of the value of ρ to the number bonds between the ring and the probe is not to be expected.

A further shortcoming of most Hammett correlations (including those recalculated by Tribble and Traynham[5]) is the inclusion of both *meta*- and *para*-substituents in the regression analysis. There can be no exact parallel in the change in field effect for the two types of substituent on going from (XXXVI) to (XXXV), and hence the basic deficiencies of a single-parameter analysis are magnified by coanalysis. The need to treat *meta*- and *para*-substituents separately must also apply to dual-parameter analyses, particularly the Yukawa–Tsuno equation.

Excellent correlations of the values for β-protons in styrene[34] with the $\mathfrak{F}$ and $\mathfrak{R}$ parameters of Swain and Lupton[19] demonstrate the improvement over single-parameter analysis [equations (8.32) and (8.33), $n=13$, $R=0.996$ and 0.997, respectively]. Equally good linear relationships with $\mathfrak{F}$

$$\delta(\mathrm{H}_{cis}) = 0.100\mathfrak{F} + 0.414\mathfrak{R} \tag{8.32}$$

$$\delta(\mathrm{H}_{trans}) = 0.166\mathfrak{F} + 0.420\mathfrak{R} \tag{8.33}$$

and $\mathfrak{R}$ or with σ_I and σ_R^0, were found for α-methyl- and α-*t*-butylstyrenes,[34] and for phenylacetylenes.[35] It is notable that in equations (8.32) and (8.33) the resonance coefficient is almost the same, the observed variations in δ being entirely accommodated by a changed field dependence, whereas for the phenylacetylenes [equation (8.34)] both coefficients are different.

$$\delta(\mathrm{H}) = 0.193\mathfrak{F} + 0.332\mathfrak{R} \tag{8.34}$$

In general it can be expected that dual-parameter correlations (preferably with σ_I and σ_R^0) will be of high quality (provided that the raw data are also of high quality,[34,35,54,97,98] i.e., infinite-dilution chemical shifts in a noninteracting solvent, with corrections for the purely magnetic effects of strongly anisotropic groups[34,35,98] such as phenyl or cyano), and there is need for a complete reassessment of the many published Hammett correlations.[5] Until such reexamination has been carried out, interpretation of the variation in the ρ_I and ρ_R values is not warranted, but a full rationalization will be possible only along the lines of Dewar's FMMF approach.[68,70] Explicit consideration of the dependence of the field effect [$f(r)$ and $f(\theta)$], and an evaluation of the orbital interaction between the side chain and the aromatic ring as required for conjugative or hyperconjugative transfer of charge, are likely to be essential to any explanation of trends in ρ_I and ρ_R. Data for some of the few available correlations (with $\mathfrak{F}$ and $\mathfrak{R}$) are given in Table 8.4.

Benzene solvent shifts for substituted cyclopropanes, oxiranes, and aziridines are also correlated[100] better with $\mathfrak{F}$ and $\mathfrak{R}$ than with σ_p, but the number of data sets is too small for this result to be acceptable. The existence of a good correlation for solvent shifts[98,101,102] is of no particular significance in itself, since normal solvent effects are not specific to substituents, and it merely implies linear behavior for δ in both solvents. A nonzero slope is, of course, a clear indication that the solvent pair differ in

Table 8.4. Correlation of δ Values for Side-Chain Protons with $\mathcal{F}$ and $\mathcal{R}$

Side chain	Solvent	Number of points	f	r	R	Ref.
p-COMe	CCl_4	10	0.05	0.17	0.978	99
p-OH	DMSO	14	0.95	1.88	0.994	54
m-$\overset{+}{N}Me_3$	MeCN	7	0.04	0.11	0.999	54
p-C≡CH	C_6D_{12}	12	0.19	0.33	0.988	35

their contributions to the field effect (through-space transmission, charge stabilization, dipole enhancement) and the resonance effect (stabilization of resonance-induced charges) and perhaps in their effect on side-chain conformation. With few exceptions,[98] the treatment of solvent effects is poor, but their study can provide valuable information.

Lynch[102] has shown that improved correlations can be obtained with σ_p^+ instead of σ_p for protons in the side-chain groups, $\overset{+}{N}Me_3$, $\overset{+}{N}Me_2H$, CH_2F, CH_2Cl, CH_3, and $COCH_3$ and has reasoned that the application of dual-parameter analysis offers no significant advance. Whilst *some* general improvement can be expected by analogy with the relative effectiveness of σ_p and σ_p^+ for the correlation of $\delta(C_p)$ (Table 8.2), correlations with σ_p^+ must be less successful than dual-parameter analyses, for the same reasons as apply to σ_p. An increase in the resonance contribution can be reflected in a better correlation with σ_p^+, but otherwise the improvement is fortuitous. Examples[100,103] of the reverse trend (σ_p better than σ_p^+) include the substituted phenylsilanes (XXXIX, R = H or Me). Some aspects of side-chain conformation have also been investigated.[104,105]

p-$XC_6H_4SiR_3$ (XXXIX) m-XC_6H_4R (XL)

In *meta*-substituted systems (XL, R = Me, OMe, CH═CHPh, COMe, or C≡CH) poor correlations were found[53] with σ_m, and only a slight improvement was obtained with the Yukawa–Tsuno equation. Application of corrections for substituent anisotropy and ring-current effects gave excellent results. Similar problems exist for *ortho*-substituted toluenes, anilines, and anisoles, for which correlation with σ_I and σ_R did not give rise to any uniform pattern in the correlation coefficients.[55] These ranged from excellent to poor even with allowance for a steric effect. Earlier work on *ortho*-substituents is summarized by Tribble and Traynham.[5]

Side-chain protons such as those of OH, NH_2, and SH show only moderate correlations in normal noninteracting solvents, but excellent results are found for data for strong hydrogen-bonding solvents such as dimethyl sulfoxide. This is due to the quenching of self-association and has been discussed in detail by Tribble and Traynham.[5] Best correlation results for phenols are found with σ^-, but the Yukawa–Tsuno equation gave a correlation coefficient of 0.999 for 30 *meta*- and *para*-substituted phenols.[107] These types of δ(H) value are easily measured, and further reports have appeared concerning phenols (Yukawa–Tsuno analysis,[106–108] Hammett correlations,[110,113] correlations with pK_a[111–113]), anilines and related systems (Hammett correlations,[114–116] Yukawa–Tsuno equation[109,117]) and thiophenols (correlation with σ_R).[118]

The reason for the high quality of correlations of δ(H) for phenols (compared with other side-chain protons) is a fortunate coincidence of the contributions to field and resonance effects embodied in σ^-, and those reflected in the chemical shifts in polar solvents. An N*H* proton in aniline is, of course, electronically similar to the O*H* proton in phenol and has much the same orientation with respect to substituents in the benzene ring. There is no reason why good infinite-dilution data for phenols in noninteracting solvents should not show an excellent fit to a dual-parameter equation (σ_I and σ_R, or $\mathfrak{F}$ and $\mathfrak{R}$), although this does not appear to have been demonstrated. In DMSO the resonance effect is twice as large as the field effect on the basis of a Swain–Lupton analysis,[54] but correlation with σ_I and σ_R^- gave $\gamma = 1.09$ [equation (8.35), $n = 15$, $f = 0.13$], a disturbingly

$$\delta(\mathrm{OH})_p = 1.453\sigma_I + 1.579\sigma_R^- \tag{8.35}$$

large difference between the two approaches.† Taft parameters were consistently better than those of Swain and Lupton for the correlation of a wide range of equilibrium and rate data[37] and are probably the more effective. The improvement with σ^- rather than σ_R^0 for δ(OH) was not large, and enhanced resonance parameters may not be necessary for correlation (δ(OH), δ(NH), and δ(SH) values.

Investigation of other nuclei as probes of side-chain interaction with ring substituents has been largely neglected. Excellent dual-parameter analysis of δ(C) for styrenes and phenylacetylenes have been discussed above (p. 380). Taft[37] has noted similar results for δ(F) values in benzyl fluorides and the series (XLI) when using σ_I and $\sigma_{R(\mathrm{BA})}$, or $\mathfrak{F}$ and $\mathfrak{R}$. Hammett correlations of δ(F) for substituted 2-fluoro-2-phenylethanols (σ_p, $n = 6$, $r = 0.906$),[119], 2-fluoro-1-phenylethanols (σ_p, $n = 6$, $r = 0.996$),[119] β,β-difluorostyrenes [$\delta(\mathrm{F}_{cis})$, σ^-, $n = 7$, $r = 0.976$; $\delta(F_{trans})$, $r = 0.990$],[120] and series (XLII), (σ_R, $n = 7$, $r = 0.905$)[90] have also been noted.

p-$XC_6H_4SF_5$	p-$XC_6H_4POF_2$
(XLI)	(XLII)

†In some measure, this difference can be accounted for by the wrong scaling of their $\mathfrak{F}$ values by Swain and Lupton.

Popular skepticism of δ *vs.* σ correlations as summarized by Yokoyama *et al.*[96] is clearly well founded, and virtually all of the published discussion of such correlations with respect to the sensitivity of different groups, the effectiveness of π as opposed to σ transmission, the role of electric fields, anisotropy effects, inductive effects, and resonance effects, the parallel with excess charge density, etc., rest on a basis lacking in conceptual rigor and experimental precision. There is need for a reappraisal of the whole area of side-chain proton correlations, a daunting task since most of the raw data are inadequate.

Pessimism about the usefulness of dual-parameter correlations[54] for side-chain protons is perhaps premature. After extensive re-analysis (preferably in terms of the Taft equation with careful testing of the need for different σ_R parameters) has been carried through, some semblance of order may appear from the present chaos.

8.3.4. Heterocyclic Systems

Little attempt has been made systematically to extend the application of LFER to chemical shifts in heterocyclic aromatic systems. Pyridine is the compound most similar to benzene and a parallel has been noted for the two systems as regards $\delta(H_p)$,[121] $\delta(H_m)$,[122] and $\delta(C_p)$[123] (in 2- and 3-substituted pyridines) and $\delta(F_p)$ in series (XLIII).[124] Some of these data [$\delta(H_p)$ and $\delta(C_p)$] have been plotted against σ_p but no correlation coefficients were given.[121,122] Both $\delta(H_p)$ and $\delta(H_o)$ for the pyrazines (XLIV) are correlated with σ_p^+ ($n = 12$, $r = 0.980$ and $r = 0.985$, respectively),[125] and for a series of N-methylpyridinium iodides[126] δ(H) and δ(C)

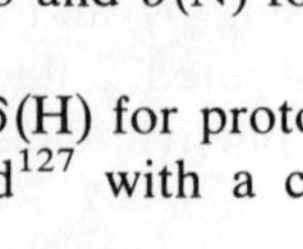

X–, F, N
(XLIII)

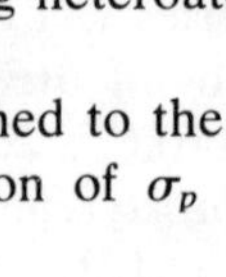

N, N, X
(XLIV)

for the methyl group and δ(N) for the ring heteroatom give linear plots with σ_p or σ_p^+.

Correlation of δ(H) for protons attached to the heterocyclic ring in (XLV) was achieved[127] with a combination of σ_p and σ_m for 5- and

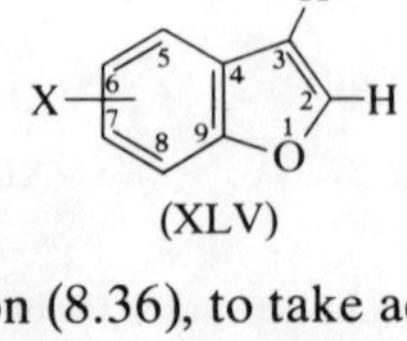

H, X, 5, 6, 7, 8, 9, 4, 3, 2, 1, O, H
(XLV)

6-substituents, e.g., equation (8.36), to take account of transmission of the electronic effects through C_3 and through oxygen. Not unexpectedly, the best correlations were found for $\delta(H_2)$ and $\delta(C_2)$ but the number of data sets was too small for these results to be very reliable. A similar criticism

$$\delta(H_2) = 0.175\sigma_1 + 0.036\sigma_2 + 7.477 \tag{8.36}$$

applies to the Hammett correlation of $\delta(H_{Me})$ in the four heterocyclic systems (XLVI, Z = O, S, or Se; and XLVII),[128,129] and it is difficult to

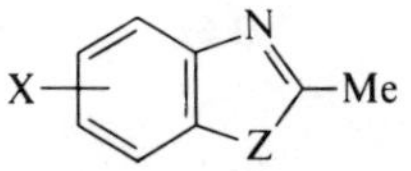

(XLVI) 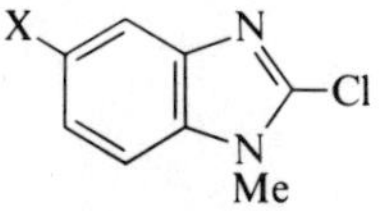 (XLVII)

judge the significance of the observed relationships. However, the probe nuclei in such systems, and in (XLV), can probably be regarded as part of a side chain and a similarity to analogous benzenes can be expected.

In a series of papers Gronowitz and his co-workers[130–133] have explored the application of benzene substituent parameters ($\mathfrak{F}$ and $\mathfrak{R}$) to the five-membered heterocycles (XLVIII) (see Table 8.5). For 2-substituted compounds (XLVIII, X = O, S, Se, or Te) correlations of $\delta(H_3)$

X = O, S, Se, or Te

(XLVIII)

and $\delta(H_5)$ are uniformly excellent ($n = 7$ to 9, $R = 0.99$). Although some individual series were good, the correlations of $\delta(H_4)$ in the 2-substituted series and $\delta(H_4)$ and $\delta(H_5)$ in the 3-substituted series were poorer on the whole ($R = 0.90$ to 0.96). In the latter series, except for furan, $\delta(H_2)$ also gave good results ($n = 7$, $R = 0.99$). These results conform to classical resonance ideas, which allow 2-substituents to produce electronic perturbations at C_3 and C_5, but only at C_2 for 3-substituents, and are supported by good correlations of $\delta(C)$ only at these three positions (Table 8.5) in all series (XLVIII, X = O, S, or Se).

The relative magnitude of the field and resonance terms in the equation of the regression line for the different series has been taken[133] to indicate some change in the contributions from the various resonance forms. However, these results must be interpreted with caution since field contributions seem anomalously low compared with those for benzene. In 2-substituted thiophens, for instance, $\gamma(=r/f)$ is 84 for $\delta(C_5)$ compared with 7.6 for $\delta(C_p)$ in benzene, a change much larger than is likely for the ratio of resonance to polar effects. A further curious feature is the negative field contribution for the *ortho*-carbon in both the 2- and the 3-substituted

Table 8.5. Correlation of δ(H) and δ(C) in Furan, Thiophen, and Selenophen by the Equation[a] $\delta(N) = f\mathfrak{F} + r\mathfrak{R} + c$

Substituent[b] interaction	Ring heteroatom	f	r	n	R
	O	0.63	2.89	7	0.99
$\delta(H_3)^2$	S	0.49	2.21	8	0.99
	Se	0.34	2.67	8	0.99
	O	−11.2	58.7	7	0.93
$\delta(C_3)^2$	S	−5.8	46.7	8	0.93
	Se	−5.7	46.6	8	0.93
	O	0.23	1.24	7	0.99
$\sigma(H_5)^2$	S	0.27	1.73	9	0.99
	Se	0.18	1.99	9	0.99
	O	0.3	20.4	7	0.97
$\delta(C_5)^2$	S	0.4	33.5	8	0.99
	Se	2.4	34.0	8	0.99
	O	0.78	1.81	7	0.91
$\delta(H_2)^3$	S	0.59	2.67	7	0.99
	Se	0.57	3.23	7	0.99
	O	−2.6	36.9	7	0.97
$\delta(C_2)^3$	S	−5.2	55.5	8	0.97
	Se	−3.7	61.7	7	0.98

[a] The value of c for proton correlations is usually less than 0.1 ppm and for carbon correlations less than 3 ppm, but those for $\delta(C_3)^2$ range up to 8 ppm.
[b] The notation $\delta(H_3)^2$ signifies changes at H_3 produced by a 2-substituent.

series and for $\delta(C_o)$ in benzene. This may reflect factors (such as anisotropy effects and an alternating through-bond inductive effect) which have been included by other workers in Q. Correlation of $\delta(C_o)$ with Q directly was poorer than with $\mathfrak{F}$ and $\mathfrak{R}$, but no triple-parameter analyses were carried out (cf. p. 366).

A rather closer similarity between thiophen and benzene can be observed[130] in the values of $\delta(F_5)$ in series (XLIX). Equation (8.37)

$$\delta(F_5) = 7.5\mathfrak{F} + 25.0\mathfrak{R} + 3.8 \tag{8.37}$$

($n = 12$, $R = 0.96$) indicates a value of γ of 3.3 close to the value of 4.5 found for $\delta(F_p)$ in benzene. For the corresponding 4-fluoro series, $\delta(F_4)$ is badly correlated ($n = 11$, $R = 0.89$).

Long-range substituent effects on H_5 and H_6 in polycyclic sulfur heterocycles, e.g., (L),[134] show some parallel with σ_p^0 [$\rho = 0.15$, $n = 7$, $r = 0.98$ for $\delta(H_5)$ in (L)], but the number of data sets is too small for this result to be very significant.

Although little work on heterocyclic compounds has appeared, the application of dual-parameter analysis to chemical shifts has considerable

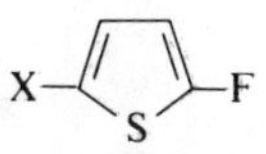

(XLIX) 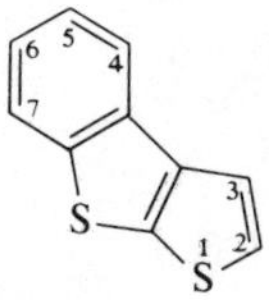 (L)

potential, particularly in the investigation of the mode of resonance interaction in polycyclic systems. However, proper statistical evaluation of the significance of the separate terms in such correlations will be important.

8.4. Aliphatic Compounds

Correlation analysis of chemical-shift data for aliphatic compounds† with genuine free energy parameters such as σ_I or σ^* has not been conducted in any systematic way and very little work of value has appeared. The concept most commonly invoked is electronegativity, and both δ(H) and δ(C) are widely reported as showing a broad correlation with substituent electronegativity (E_x).[62,135–138] Many of these correlations, however, are merely internal comparisons of nmr data, since E_x is frequently derived from the chemical shifts of methyl and ethyl derivatives. Electronegativity is a somewhat ill-defined concept, particularly for composite groups, and it seems much more meaningful to seek correlation with the inductive constants, σ_I, which have the virtue of being clearly defined on a common basis[1,139] which does not involve nmr data. A full account of the background to the inductive parameters σ_I and σ^* has been given by Shorter.[139]

The absence of a resonance effect in saturated aliphatic compounds means that substituent effects are much smaller than in aromatic compounds and hence extraneous contributions become relatively more important. Quite apart from contributions from substituent anisotropy, steric effects, hyperconjugative interactions, and solvent effects, the intrinsic problem of conformation is also certain to complicate the interpretation of δ(H) values. Even in rigid systems such as (LI) ($Z = CH_2$, CCl_2, or O) or (LII), or various similar systems[140] the values of δ(H) are not amenable to any simple general analysis.

Substituent effects at the geminal proton (H_α) follow E_x for a restricted range of X, but depend on the system. This is thought to reflect the

†For the present purpose aliphatic compounds are those in which the *substituent* is attached to saturated carbon.

ability of the rest of the molecule to compensate the substituent inductive effect.

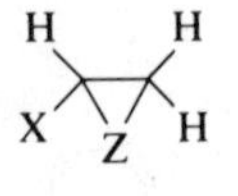

(LI) 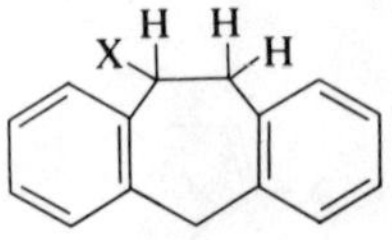 (LII)

Proton chemical shifts for cyclopropanes,[141] benzyl derivatives,[142] and 2-halogenobutanes[143] have been examined recently for correlation with E_x, but this work throws no fresh light on the intrinsic problems of this area. From the examination of an enormous number of alkyl halides Gouverneur *et al.*[144] derived equation (8.38) ($n = 419$, $r = 0.993$), where

$$\delta(\mathrm{H})_{\mathrm{corr}} = 1.15 \sum \sigma^* - 1.26 \tag{8.38}$$

such corrections were made for "anisotropy effects" as totally to obscure the inherent angular dependence of substituent field effects.

Slightly more interesting are the correlations observed between σ_I and $\delta(F_p)$ for (LIII)[145] and between $\sum \sigma^0$ and $\delta(F)_p$ for (LIV).[146] A poorer

p-$FC_6H_4CPh_2X$ (LIII)

$(p$-$XC_6H_4)_3SnC_6H_4F$-p (LIV)

correlation ($r = 0.935$) for the silicon analog of (LIII) suggests some resonance participation by d orbitals, but there appears to be no similar interaction in the tin compounds.[145]

Central to the problem of relating chemical shifts for aliphatic systems to electronic effects (supposing that perturbing factors can be allowed for) is the nature of the inductive effect itself. For correlations for aromatic compounds only a dipole-field effect (including π polarization) is important (four or more bonds between probe nucleus and substituent). But over the shorter pathways of aliphatic compounds (two to four bonds usually) the classical through-bond effects must be considered, especially since large resonance perturbations are absent. This raises the vexing question of whether through-bond inductive perturbations are simply attenuated or alternate in sign at successive atoms. Evidence from molecular orbital calculations[147–149] clearly suggests that the latter mode of interaction may be real.

Recent interest in $\delta(C)$ values for aliphatic and alicyclic compounds has led to a wealth of data,[150–156] but little clear evidence about the nature of inductive effects. ^{13}C chemical shifts are not entirely free from complicating factors such as anisotropy effects[153,155,156] and steric effects, and generally the values of $\delta(C_\alpha)$ and $\delta(C_\beta)$ will only broadly parallel σ_I or E_x.

For substituted cyclohexanes[150] and 1-substituted adamantanes,[151] $\delta(C_\delta)$ values are correlated with σ^* with a negative ρ value in each case. Effects at C_γ are usually obscured by 1,3-steric interactions, but in the 1-substituted adamantanes, in which such effects are absent, deshielding of C_γ is observed but the correlation with σ^* is poor. Nevertheless, the ρ value is positive and these results for $\delta(C_\gamma)$ and $\delta(C_\delta)$ are in agreement with an alternating inductive effect. Similar conclusions have been reached from a study of the changes induced in δ(C) by protonation for various substituted piperidines. A useful study[142] of δ(C) for benzyl derivatives (LV) showed that $\delta(C_\alpha)$ is correlated quite poorly with E_x. However,

$$\delta(C_p) = 5.34\sigma_I + 125.8 \tag{8.39}$$

further investigation of these δ(C) values finds no correlation of $\delta(C_1)$ with σ_I, but a reasonable linear fit for $\delta(C_4)$ [equation (8.39), $n = 15$, $r = 0.94$].

$X\overset{\alpha}{C}H_2-C_6H_5$ (ring positions 1–6)

(LV)

A positive ρ value would be expected from either type of inductive effect, and $\delta(C_3)$ values might be more revealing. Unfortunately the $\delta(C_3)$ resonances were not assigned.

Correlation analysis offers an interesting view of substituent effects in inorganic and organometallic compounds, but little application of it in this area is evident. In the series (LVI) δ(C) values for the carbonyl group were

$X-Fe(CO)_2(\pi\text{-}Cp)$

(LVI)

correlated with σ_I, giving a negative value of ρ. This has been ascribed[157] to *d*-orbital contraction in iron, with concomitant enhanced (*sic*) paramagnetic screening on carbon, but it is obvious that an alternating inductive effect would produce a similar trend from increased diamagnetic shielding. The validity of such an interpretation remains to be established.

Phillips and Wray[158,159] have described an interpretation of substituent effects on δ(H), δ(C), and δ(F) values for simple polysubstituted aliphatic compounds in terms of additive modifications to atomic electronegativities. Application of this empirical approach is relatively restricted with respect to substituent type, but does reproduce observed shifts quite well (although this is in some measure a reflection of a substantial degree of internal definition of the interaction constants). This approach does not, however, offer much new insight into the nature of the inductive effect itself.

8.5. Conclusions

The present necessarily selective examination of the relationships between chemical shifts and substituent constants has shown that, in spite of extensive investigation, only a little work of value has appeared, and there is much scope for further study.

Not only can δ *vs.* σ correlations be regarded as valuable and worthwhile (if the basic experimental rigor has been applied), but also it seems likely that, since the chemical shift is one of the best physical parameters for probing substituent interactions, these correlations can be of fundamental importance in bridging the gap between theoretical substituent perturbations and empirical substituent constants. Nonempirical MO calculations can now successfully account[160] for the effect of substituents in electrophilic aromatic substitution, and chemical shifts can clearly extend the scope of such analysis to systems for which the MO calculations are prohibitively complex and expensive. It is to be hoped that this survey can stimulate fresh interest in an area that has been regarded as somewhat played out and even as of doubtful value.

References

1. O. Exner, in *Advances in Linear Free Energy Relationships*, Chap. 1, N. B. Chapman and J. Shorter, eds. (Plenum, London, 1972).
2. S. Ehrenson, *Progr. Phys. Org. Chem.*, **2**, 195 (1964).
3. M. Charton, *Chem. Tech.*, 502 (1974); 245 (1975).
4. C. D. Ritchie and W. F. Sager, *Progr. Phys. Org. Chem.*, **2**, 323 (1964).
5. M. T. Tribble and J. G. Traynham, in *Advances in Linear Free Energy Relationships*, Chap. 4, N. B. Chapman and J. Shorter, eds. (Plenum, London, 1972).
6. M. Karplus and J. A. Pople, *J. Chem. Phys.*, **38**, 2803 (1963).
7. R. J. Pugmire and D. M. Grant, *J. Amer. Chem. Soc.*, **90**, 697 (1968).
8. J. Mason, *J. Chem. Soc.* (*A*), 1038 (1971).
9. R. B. Mallion, in *Nuclear Magnetic Resonance*, Vol. 4, Chap. 1, R. K. Harris, ed. (Chemical Society, London, 1975).
10. A. D. Buckingham, T. Schaefer, and W. G. Schneider, *J. Chem. Phys.*, **32**, 1227 (1960).
11. P. H. Weiner and E. R. Malinowski, *J. Phys. Chem.*, **75**, 3971 (1971).
12. H. Spiesecke and W. G. Schneider, *J. Chem. Phys.*, **35**, 731 (1961).
13. P. C. Lauterbur, *Ann. New York Acad. Sci.*, **70**, 841 (1958); *J. Amer. Chem. Soc.*, **83**, 1846 (1961); *Tetrahedron Lett.*, 274 (1961).
14. G. E. Maciel and J. J. Natterstad, *J. Chem. Phys.*, **42**, 2427 (1965).
15. G. P. Syrova, V. F. Bystrov, V. V. Orda, and L. M. Yagupol'skii, *Zh. obshch. Khim.*, **39**, 1395 (1969); EE, 1364.
16. G. L. Nelson, G. C. Levy, and J. D. Cargioli, *J. Amer. Chem. Soc.*, **94**, 3089 (1972).
17. E. M. Schulman, K. A. Christensen, D. M. Grant, and C. Walling, *J. Org. Chem.*, **39**, 2686 (1974).
18. W. B. Smith and D. L. Deavenport, *J. Magnetic Resonance*, **7**, 364 (1972).
19. C. G. Swain and E. C. Lupton, *J. Amer. Chem. Soc.*, **90**, 4328 (1968).

20. H. L. Retcofsky and C. E. Griffin, *Tetrahedron Lett.*, 1975 (1966).
21. W. Adcock, M. Aurangzeb, W. Kitching, N. Smith, and D. Doddrell, *Austral. J. Chem.*, **27**, 1817 (1974).
22. F. Hruska, H. M. Hutton, and T. Schaefer, *Canad. J. Chem.*, **43**, 2392 (1965).
23. W. B. Smith, A. M. Ihrig, and J. L. Roark, *J. Phys. Chem.*, **74**, 812 (1970), and references therein.
24. S. K. Dayal and R. W. Taft, *J. Amer. Chem. Soc.*, **95**, 5595 (1973).
25. G. Miyajama, H. Akiyama, and K. Nishimoto, *Org. Magnetic Resonance*, **4**, 811 (1972).
26. J. M. Briggs and E. W. Randall, *J.C.S. Perkin II*, 1789 (1973).
27. M. Imanari, M. Kohno, M. Ohuchi, and K. Ishizu, *Bull. Chem. Soc. Japan*, **47**, 708 (1974).
28. S. Gronowitz, I. Johnson, A. Maholanyiova, S. Toma, and E. Solcániová, *Org. Magnetic Resonance*, **7**, 372 (1975).
29. H. M. Relles and R. W. Schleunz, *J. Org. Chem.*, **37**, 1742 (1972).
30. J. F. Hinton and B. Layton, *Org. Magnetic Resonance*, **4**, 353 (1972).
31. G. J. Ray, R. J. Kurland, and A. K. Colter, *Tetrahedron*, **27**, 735 (1971).
32. J. W. Larsen and P. A. Bouis, *J. Amer. Chem. Soc.*, **97**, 4418 (1975).
33. G. A. Olah and P. W. Westerman, *J. Amer. Chem. Soc.*, **95**, 3706 (1973).
34. G. K. Hamer, I. R. Peat, and W. F. Reynolds, *Canad. J. Chem.*, **51**, 897, 915 (1973).
35. D. A. Dawson and W. F. Reynolds, *Canad. J. Chem.*, **53**, 373 (1975).
36. D. F. Ewing, unpublished observations.
37. S. Ehrenson, R. T. C. Brownlee, and R. W. Taft, *Progr. Phys. Org. Chem.*, **10**, 1 (1973).
38. M. J. S. Dewar, R. Golden, and J. M. Harris, *J. Amer. Chem. Soc.*, **93**, 4187 (1971).
39. W. F. Reynolds, I. R. Peat, M. H. Freedman, and J. R. Lyerla, *Canad. J. Chem.*, **51**, 1857 (1973).
40. F. W. Wehrli, J. W. de Haan, A. I. M. Keulemans, O. Exner, and W. Simon, *Helv. Chim. Acta*, **52**, 103 (1969).
41. R. T. C. Brownlee and R. W. Taft, *J. Amer. Chem. Soc.*, **90**, 6537 (1968).
42. R. T. C. Brownlee and R. W. Taft, *J. Amer. Chem. Soc.*, **92**, 7007 (1970).
43. W. J. Hehre, R. W. Taft, and R. D. Topsom, *Progr. Phys. Org. Chem.*, **12**, 159 (1976).
44. P. Lazzeretti and F. Taddei, *Org. Magnetic Resonance*, **3**, 283 (1971).
45. J. E. Bloor and D. L. Breen, *J. Phys. Chem.*, **72**, 716 (1968).
46. R. G. Jones and P. Partington, *J.C.S. Faraday II*, 2087 (1972).
47. K. Hayamizu and O. Yamamato, *J. Mol. Spectroscopy*, **28**, 89 (1968); **29**, 183 (1969).
48. B. M. Lynch, *Org. Magnetic Resonance*, **6**, 190 (1974).
49. Reference 5, Table 4.1.
50. Y. Yukawa, Y. Tsuno, and N. Shimizu, *Bull. Chem. Soc. Japan*, **44**, 2843 (1971).
51. R. W. Taft, S. Ehrenson, I. C. Lewis, and R. E. Glick *J. Amer. Chem. Soc.*, **81**, 5352 (1959).
52. R. W. Taft, *J. Amer. Chem. Soc.*, **79**, 1045 (1957).
53. H. Yamada, Y. Tsuno, and Y. Yukawa, *Bull. Chem. Soc. Japan*, **43**, 1459 (1970).
54. G. R. Wiley and S. I. Miller, *J. Org. Chem.*, **37**, 767 (1972).
55. M. Charton, *J. Org. Chem.*, **36**, 266 (1971).
56. R. I. Herrmann and I. D. Rae, *Austral. J. Chem.*, **25**, 811 (1972).
57. J. M. Appleton, B. D. Andrews, I. D. Rae, and B. E. Reichert, *Austral. J. Chem.*, **23**, 1667 (1970).
58. W. B. Smith, D. L. Deavenport, and A. M. Ihrig, *J. Amer. Chem. Soc.*, **94**, 1959 (1972).
59. A. R. Tarpley and J. H. Goldstein, *J. Phys. Chem.*, **75**, 421 (1971).
60. G. Socrates and M. W. Adlard, *J. Chem. Soc. (B)*, 733 (1971).
61. R. W. Taft, E. Price, I. R. Fox, I. C. Lewis, K. K. Andersen, and G. T. Davis, *J. Amer. Chem. Soc.*, **85**, 709, 3146 (1963).

62. J. W. Emsley and L. Phillips, *Progr. NMR Spectroscopy*, **7**, 1 (1971).
63. R. T. C. Brownlee, S. K. Dayal, J. L. Lyle, and R. W. Taft, *J. Amer. Chem. Soc.*, **94**, 7208 (1972).
64. I. R. Ager, L. Phillips, T. J. Tewson, and V. Wray, *J.C.S. Perkin II*, 1979 (1972).
65. I. R. Ager, L. Phillips, and S. J. Roberts, *J.C.S. Perkin II*, 1988 (1972).
66. P. J. Mitchell and L. Phillips, *J.C.S. Perkin II*, 109 (1974).
67. S. K. Dayal, S. Ehrenson, and R. W. Taft, *J. Amer. Chem. Soc.*, **94**, 9113 (1972).
68. M. J. S. Dewar and P. J. Grisdale, *J. Amer. Chem. Soc.*, **84**, 3548 (1962).
69. W. Adcock and M. J. S. Dewar, *J. Amer. Chem. Soc.*, **89**, 379 (1967).
70. M. J. S. Dewar, R. Golden, and J. M. Harris, *J. Amer. Chem. Soc.*, **93**, 4187 (1971).
71. W. Adcock, M. J. S. Dewar, and B. D. Gupta, *J. Amer. Chem. Soc.*, **95**, 7353 (1973).
72. W. Adcock, S. Q. A. Rizvi, W. Kitching, and A. J. Smith, *J. Amer. Chem. Soc.*, **94**, 369 (1972).
73. W. Adcock, M. J. S. Dewar, R. Golden, and M. A. Zeb, *J. Amer. Chem. Soc.*, **97**, 2198 (1975).
74. W. Adcock and B. D. Gupta, *J. Amer. Chem. Soc.*, **97**, 6871 (1975).
75. J. Fukunaga and R. W. Taft, *J. Amer. Chem. Soc.*, **97**, 1612 (1975).
76. W. Adcock, P. D. Bettess, and S. Q. A. Rizvi, *Austral. J. Chem.*, **23**, 1921 (1970).
77. L. N. Pushkina, A. P. Stepanov, V. S. Zhukov, and A. D. Naumov, *Org. Magnetic Resonance*, **4**, 607 (1972).
78. J. Lipowitz, *J. Amer. Chem. Soc.*, **94**, 1582 (1972).
79. I. R. Ager and L. Phillips, *J.C.S. Perkin II*, 1975 (1972).
80. J. L. Fletcher and M. J. McGlinchey, *Canad. J. Chem.*, **53**, 1525 (1975).
81. K. Izawa, T. Okuyama, and T. Fueno, *Bull. Soc. Chem. Japan*, **46**, 2881 (1973).
82. T. Okuyama, K. Izawa, and T. Fueno, *Bull. Soc. Chem. Japan*, **47**, 410 (1974).
83. O. Kajimoto, M. Kobayashi, and T. Fueno, *Bull. Soc. Chem. Japan*, **46**, 1422 (1973).
84. C. D. Schaeffer, J. J. Zuckerman, and C. H. Yoder, *J. Organometal. Chem.*, **80**, 29 (1974).
85. G. A. Olah, R. D. Porter, C. L. Jeuell, and A. M. White, *J. Amer. Chem. Soc.*, **92**, 2044 (1972).
86. P. Hampson, A. Mathias, and R. Westhead, *J. Chem. Soc.* (*B*), 397 (1971).
87. R. L. Lichter and J. D. Roberts, *Org. Magnetic Resonance*, **6**, 636 (1974).
88. T. Axenrod, P. S. Pregosin, M. J. Wieder, E. D. Becker, R. B. Bradley, and G. W. A. Milne, *J. Amer. Chem. Soc.*, **93**, 6536 (1971).
89. C. R. Ernst, L. Spialter, G. R. Buell, and D. L. Wilhite, *J. Amer. Chem. Soc.*, **96**, 5375 (1974).
90. L. L. Szafraniec, *Org. Magnetic Resonance*, **6**, 565 (1974).
91. A. R. Siedle and G. M. Bodner, *Inorg. Chem.*, **11**, 3108 (1972).
92. J. T. Vandeberg, C. E. Moore, and F. P. Cassaretto, *Org. Magnetic Resonance*, **5**, 57 (1973).
93. G. R. Howe, *J. Chem. Soc.* (*B*), 984 (1971).
94. G. A. Olah and D. A. Forsyth, *J. Amer. Chem. Soc.*, **97**, 3137 (1975).
95. L. M. Stock, *J. Chem. Educ.*, **49**, 400 (1972).
96. T. Yokoyama, G. R. Wiley, and S. I. Miller, *J. Org. Chem.*, **34**, 1859 (1969).
97. R. R. Fraser and R. N. Renaud, *Canad. J. Chem.*, **49**, 800 (1971).
98. G. A. Caplin, *Org. Magnetic Resonance*, **5**, 169 (1973).
99. G. A. Caplin, *Org. Magnetic Resonance*, **6**, 99 (1974).
100. A. B. Turner, R. E. Lutz, N. S. McFarlane, and D. W. Boykin, *J. Org. Chem.*, **36**, 1107 (1971).
101. N. E. Alexandrou and A. G. Varvoglis, *Org. Magnetic Resonance*, **3**, 293 (1971).
102. B. M. Lynch, *Org. Magnetic Resonance*, **6**, 190 (1974).

103. Y. Nagai, M. A. Ohtsuki, T. Nakano, and H. Watanabe, *J. Organometal. Chem.*, **35**, 81 (1972).
104. R. R. Fraser, Gurudata, R. N. Renaud, C. Reyes-Zamora, and R. B. Swingle, *Canad. J. Chem.*, **47**, 2767 (1969).
105. M. D. Bentley and M. J. S. Dewar, *J. Org. Chem.*, **35**, 2707 (1970).
106. Y. Tsuno, M. Fujio, Y. Takai, and Y. Yukawa, *Bull. Chem. Soc. Japan*, **45**, 1519 (1972).
107. M. Fujio, M. Mishima, Y. Tsuno, Y. Yukawa, and Y. Takai, *Bull. Chem. Soc. Japan*, **48**, 2127 (1975).
108. Y. Tsuno, M. Fujio, and Y. Yukawa, *Bull. Chem. Soc. Japan*, **48**, 3324 (1975).
109. M. Fujio, Y. Tsuno, Y. Yukawa, and Y. Takai, *Bull. Chem. Soc. Japan*, **48**, 3330 (1975).
110. A. Fischer, M. C. A. Opie, J. Vaughan, and G. J. Wright, *J.C.S. Perkin II*, 319 (1972).
111. Y. Murakami and J. Sunamoto, *J.C.S. Perkin II*, 1231, 1235 (1973).
112. J.-C. Halle and R. Schaal, *Bull. Soc. chim. France*, 3785 (1972).
113. U. Folli, D. Iarossi, and F. Taddei, *J.C.S. Perkin II*, 848 (1973).
114. M. Kasai, M. Hirota, Y. Hamada, and H. Matsuoka, *Tetrahedron*, **29**, 267 (1973).
115. C. J. Giffney and C. J. O'Connor, *J. Magnetic Resonance*, **18**, 230 (1975).
116. A. Wu, E. R. Biehl, and P. C. Reeves, *J. Organometal. Chem.*, **33**, 53 (1971).
117. Y. Asabe and Y. Tsuzuki, *Bull. Chem. Soc. Japan*, **44**, 3482 (1971).
118. S. Kawamura, T. Horii, and J. Tsurugi, *J. Org. Chem.*, **36**, 3677 (1971).
119. G. Aranda and H. de Luze, *Org. Magnetic Resonance*, **4**, 847 (1972).
120. I. D. Rae and L. K. Smith, *Austral. J. Chem.*, **25**, 1465 (1972).
121. W. B. Smith and J. B. Roark, *J. Phys. Chem.*, **73**, 1049 (1969).
122. H.-H. Perkampus and U. Krüger, *Chem. Ber.*, **100**, 1165 (1967).
123. H. L. Retcovsky and R. A. Friedel, *J. Phys. Chem.*, **72**, 2619 (1968).
124. C. S. Giam and J. L. Lyle, *J. Amer. Chem. Soc.*, **95**, 3235 (1973).
125. G. S. Marx and P. E. Spoerri, *J. Org. Chem.*, **37**, 111 (1972).
126. F. W. Wehrli, W. Giger, and W. Simon, *Helv. Chim. Acta*, **54**, 229 (1971).
127. T. Okuyama and T. Fueno, *Bull. Chem. Soc. Japan*, **47**, 1263 (1974).
128. G. Di Modica, E. Barni, and A. Gasco, *J. Heterocyclic Chem.*, **2**, 457 (1965).
129. P. Dembech, G. Seconi, P. Vivarelli, L. Schenetti, and F. Taddei, *J. Chem. Soc.* (*B*), 1670 (1971).
130. S. Rodmar, S. Gronowitz, and U. Rosén, *Acta Chem. Scand.*, **25**, 3841 (1971).
131. S. Gronowitz, I. Johnson, and S. Rodmar, *Acta Chem. Scand.*, **26**, 1726 (1972).
132. F. Fringuelli, S. Gronowitz, A.-B. Hörnfeldt, I. Johnson, and A. Taticchi, *Acta Chem. Scand. B*, **28**, 175 (1974).
133. S. Gronowitz, I. Johnson, and A.-B. Hörnfeldt, *Chem. Scripta*, **7**, 76, 111, 211 (1975).
134. D. F. Ewing, D. N. Gregory, and R. M. Scrowston, *Org. Magnetic Resonance*, **6**, 293 (1974).
135. H. Spiesecke and W. G. Schneider, *J. Chem. Phys.*, **35**, 722 (1961).
136. J. R. Cavanaugh and B. P. Dailey, *J. Chem. Phys.*, **34**, 1099 (1961).
137. P. Bucci, *J. Amer. Chem. Soc.*, **90**, 252 (1968).
138. M. J. Lacey, C. G. Macdonald, A. Pross, J. S. Shannon, and S. Sternhell, *Austral. J. Chem.*, **23**, 1421 (1970).
139. J. Shorter, in *Advances in Linear Free Energy Relationships*, Chap. 2, N. B. Chapman and J. Shorter, eds. (Plenum, London, 1972).
140. C. K. Fay, J. B. Grutzner, L. F. Johnson, S. Sternhell, and P. W. Westerman, *J. Org. Chem.*, **38**, 3122 (1973).
141. K. B. Wiberg, D. E. Barth, and P. H. Schertler, *J. Org. Chem.*, **38**, 378 (1973).
142. L. Zetta and G. Gatti, *Org. Magnetic Resonance*, **4**, 585 (1972).
143. G. Schrumpf, *Chem. Ber.*, **106**, 246 (1973).
144. P. Gouverneur, O. B. Nagy, J. Ph. Soumillion, T. Burton, and A. Bruylants, *Org. Magnetic Resonance*, **4**, 391 (1972).

145. S. Yolles and J. H. R. Woodland, *J. Organometal. Chem.*, **54**, 95 (1973).
146. D. N. Kravtsov, B. A. Kvasov, T. S. Khazanova, and E. I. Fedin, *J. Organometal. Chem.*, **61**, 219 (1973).
147. J. A. Pople and M. Gordon, *J. Amer. Chem. Soc.*, **89**, 4253 (1967).
148. N. C. Baird, M. J. S. Dewar, and R. Sustmann, *J. Chem. Phys.*, **50**, 1275 (1969).
149. G. R. Howe, *J. Chem. Soc.* (*B*), 984 (1971).
150. T. Pehk and E. Lippmaa, *Org. Magnetic Resonance*, **3**, 679 (1971).
151. T. Pehk, E. Lippmaa, V. V. Sevostjanova, M. M. Krayuschkin, and A. I. Tarasova, *Org. Magnetic Resonance*, **3**, 783 (1971).
152. I. Morishima, K. Yoshikawa, K. Okada, T. Yonezawa, and K. Goto, *J. Amer. Chem. Soc.*, **95**, 165 (1973).
153. J. G. Batchelor, R. J. Cushley, and J. H. Prestegard, *J. Org. Chem.*, **39**, 1698 (1974).
154. G. Miyajima and K. Nishimoto, *Org. Magnetic Resonance*, **6**, 313 (1974).
155. C. Charrier, D. E. Dorman, and J. D. Roberts, *J. Org. Chem.*, **38**, 2644 (1973).
156. D. E. Dorman, M. Jautelat, and J. D. Roberts, *J. Org. Chem.*, **38**, 1026 (1973).
157. O. A. Gansow, D. A. Schexnayder, and B. Y. Kimura, *J. Amer. Chem. Soc.*, **94**, 3406 (1972).
158. L. Phillips and V. Wray, *J. Chem. Soc.* (*B*), 2068, 2074 (1971).
159. L. Phillips and V. Wray, *J.C.S. Perkin II*, 214, 220, 223 (1972).
160. J. M. McKelvey, S. Alexandratos, A. Streitwieser, J.-L. M. Abboud, and W. J. Hehre, *J. Amer. Chem. Soc.*, **98**, 244 (1976).

9

Recent Advances in Biochemical QSAR

Corwin Hansch

Corwin Hansch • Department of Chemistry, Pomona College, Claremont, California 91711, U.S.A.

9.1. Introduction

The object of this report is to survey the development of quantitative structure–activity relationships (QSAR) in biochemistry and medicinal chemistry since the reviews by Cammarata and Rogers[1] and Kirsch.[2] Other recent reviews[3–12] have appeared including those in French[4,11] and German.[5]

The QSAR literature is now growing so rapidly that it is not possible to consider all the articles that have appeared in the last five years. Indeed, it is almost impossible to keep track of them since they appear in such an unusual variety of journals.

The QSAR paradigm is developing out of the interaction of the fields of physical organic chemistry, statistics, computer science, quantum chemistry, biomedicinal and pesticide chemistry, and molecular biology. A view of the area of the battlefield can be had by perusal of the kinds of journals the references in this report cover. We have rapidly become enmeshed in an extremely complex endeavor, in which each individual feels increasingly inadequate to keep abreast of the new advances in all of the related areas, let alone the proper orchestration of the various disciplines. This review necessarily neglects many aspects of the above complex set of fields. The primary effort will be to consider the building of structure–activity models by using regression analysis with physicochemical constants and indicator variables. Other approaches can only be mentioned briefly.

Since the pioneering work of the Pullmans, there has been an increasing interest in the use of quantum chemical calculations in relating chemical structure to biological activity. Kier's book[13] serves as a good introduction to this large field. Other reviews cover the more recent literature.[14–16]

Another approach to complex structure–activity relationships that is receiving more attention than heretofore is that of computerized pattern recognition.[17–21] Jurs has written an introductory book from the point of view of the chemist.[19] There has been much experience with pattern recognition, mostly outside the fields of chemistry and biomedicinal chemistry, so highly sophisticated techniques have been evolved. The most difficult problem for the biomedicinal chemist is to formulate meaningful descriptors; Unger[22] has discussed some of these difficulties.

Once one has developed some kind of abstract relationship between chemical structure and activity, it must be examined from a variety of points of view to be sure, so far as possible, that the model conforms with known biochemistry and physical organic chemistry.[23] This is a particularly difficult problem with pattern recognition.

Another technique for SAR beginning to receive attention by chemists is that of cluster analysis.[24] This method appears to be especially well suited to the selection of structural modifications in the design of sets of congeners.

Martin and her colleagues have recently demonstrated[25] the advantages of discriminate analysis in SAR work. Weiner and Weiner[26] have applied factor analysis, and Schiffman[27] has explored the use of multidimensional scaling in olfaction studies. Bustard[28] and Santora and Auyang[29] have discussed the use of a Fibonacci search in uncovering an optimum structural feature for a congeneric series of drugs. This technique has the advantage that it is not computer-dependent.

One of the most serious problems in the development of QSAR is that of designing a set of congeners that will avoid collinearity among the parameters, maximize the variance in the system, and provide a well-spaced exploration of data space. Craig[30] first emphasized the importance of plotting π vs. σ to obtain a set of substituents with minimal collinearity and, at the same time, good coverage of substituent space. The collinearity problem has been discussed by others[23,24,31,32]; a good example comes from a study by Hellenbrecht *et al.*[33]

Topliss[34,35] has discussed the use of a decision tree to assist in the selection of derivatives in drug design. While the "Topliss Tree" can be of value in the early stages of drug design and is not computer-dependent, one cannot rely on it to avoid the collinearity problem, which begins to develop as the number and complexity of derivatives under study grows. Very recently, an interesting algorithm has been devised for the systematic exploration of substituent space.[36]

The most difficult problem in SAR work in biomedicinal chemistry, which has received little systematic discussion in the literature, is finding completely new structures for drugs or enzyme inhibitors. Two interesting discussions of the problem have, however, appeared recently.[37,38]

While many so-called retrospective studies have been successful in the formulation of QSAR, there has been relatively little follow-up on the predictive value of the equations. A review of successful predictions shows that QSAR can be relied on to make accurate forecasts and be of great help in avoiding redundancy in derivatization of a parent molecule.[39]

The greatest pressure for research in QSAR techniques comes from medicinal chemistry. Annual world-wide research budgets are probably approaching $2 billion. There is every reason to expect efforts to find new and better drugs to increase. Despite the greater effort in drug research in recent years, higher safety standards have actually slowed the appearance of new drugs. Sarett[40] has analyzed the problem and pointed out that the cost of developing a new drug in the United States has risen from $1.2 million in 1962 to $11.5 million in 1972. He projects a cost of $40 million by 1977. Spinks[41] has also discussed the difficulties in finding new drugs and estimates that, today, one new drug arises out of each 200,000 new compounds. He suggests that one can expect to find an anticancer drug out of each 400,000,000 randomly tested compounds.

The challenge of finding new drugs that are quite safe is forcing medicinal chemists to employ all available tools. QSAR will no doubt profit greatly in the coming years from this effort.

9.2. Parameterization of Physicochemical Properties of Organic Compounds

9.2.1. Hydrophobic Constants

Ever since the work of Meyer and Overton around 1900, it has been recognized that the lipophilic character of organic compounds plays an important role in their ability to perturb biochemical systems. However, it was only recently that systematic efforts have been made to obtain numerically defined constants to assess hydrophobic character.[42]

Early workers attempted to correlate biochemical activity with oil/water partition coefficients (P). Today, $\log P$ is generally used and interest in $\log P$ values is growing rapidly. In the first compilation of $\log P$ values,[42] a list of 5800 was made; today, over 10,000 values are known.[43] The hydrophobic substituent constant π_X is defined as $\pi_X = \log P_{R-X} - \log P_{R-H}$ where P_{R-X} refers to the partition coefficient of a derivative and P_{R-H} that of a parent compound.

Since the hydrophobic interaction of ligands with receptors depends on the geometry of the reactants, it is important to consider various parts of the ligand independently[12]; π constants are of help with such problems.

With the recognition of its importance in hydrophobic interactions of ligands and macromolecules or membranes, $\log P$ has become a subject for structure–activity studies.[42,44–70] When polar effects and hydrogen bonding can be held constant, $\log P$ appears to depend on the cavity size required by the partitioning species. Harris *et al.*[50] showed that the surface area of hydrocarbons is most significant in setting water solubility (i.e., the partitioning of the solute between itself and water). Tanford[64] has applied the idea of Harris *et al.*,[50] and Hermann,[60,61] in a more sophisticated approach, showed that $\log S$ (S = molar solubility in water) of hydrocarbons is linearly related to the surface area of a cavity, and that aromatic rings are more soluble, but the relevant line displays nearly the same slope as that for aliphatic hydrocarbons. Amidon *et al.*[62,63] simplified Hermann's method of surface area calculation and extended it to include alcohols.

Leo *et al.*[65] have shown that $\log P$ is highly dependent on the molecular volume of the solute but that with the data now available, the collinearity between volume and area is too great to allow one to make a sound choice between the two parameters; they also observed a distinct difference in log P values of solutes having essentially a hydrogen surface (aliphatic hydrocabons) and those with a surface of lone-pair electrons (e.g., noble gases, perhalogenated hydrocarbons, and aromatic hydrocarbons). They point out that different types of interactions between the surface of the solutes and water appear to account for two kinds of hydrophobic interaction.

Moriguchi[58] has attempted to factor log P into hydrophilic and hydrophobic components and relate these to physicochemical parameters.

Nys and Rekker[69,70] have recently formulated an alternative approach to the use of π by splitting molecules into hydrophobic fragments. In their method, log P can be defined as in (9.1). In equation (9.1), f is in effect the

$$\log P = \sum_{1}^{n} a_n f_n \tag{9.1}$$

log P value of a molecular fragment such as CH_3 or NO_2; it differs from π in that π is referenced to H. Until recently it has been assumed that, in addition to being defined as zero, π_H was so small that it could be assumed to be zero. Nys and Rekker[69,70] pointed out the error in this assumption and, by using regression analysis, estimated f_H as being 0.2. Log P_{H_2} has since been found[59] to be 0.45, so that $f_H = \frac{1}{2}(0.45) \approx 0.23$. The value of f_{CH_3} can be calculated from log $P_{CH_3CH_3}$, i.e., 1.82; $\frac{1}{2}(1.82) = 0.91 = f_{CH_3}$. By using statistical techniques, Nys and Rekker [69,70] calculated f_{CH_3} as 0.705. This difference is significant and illustrates that one must proceed with caution in calculating log P values from either π or f. Since it can be time-consuming as well as expensive to measure log P values for a large set of congeners, one is tempted to calculate them. A better procedure is to measure log P for at least a few members of the series before attempting calculations. Fragment constants (f) are more convenient to work with than π in calculating log P values for large molecules.

Chromatographic Hydrophobic Constants

A shortcut to hydrophobic constants is to measure R_m values by using some form of chromatography. In certain instances, good correlation between R_m and log P from the octanol/water system has been found.[56,71–78] While the increased use of R_m values in correlation analysis establishes their validity,[71–92] correlation equations are generally obtained by using quite diverse types of solvents or mixed solvents, so that equations from one laboratory cannot be easily compared with those for other systems from other laboratories. Another limitation of R_m is that a single solvent system is not able to cover large differences in lipophilicity; so far, the practical range is about 3 log $P_{octanol}$ units,[78] while log $P_{octanol}$ values have been measured over a range of 9 log units. Biagi and his colleagues have turned their attention to the problem of obtaining standard R_m values.[80]

Considerable interest has developed in obtaining $\log P$ values *via* high-pressure liquid chromatography.[93–97] It appears likely that, at least within a series of similar structures, retention times can be related to $\log P_{\text{octanol}}$ values. Even more than with other types of chromatography, high-pressure methods allow one to work with rather impure compounds when only very small amounts of sample are available.

In retrospect, it is astonishing that so obviously important a parameter for biochemical systems as the partition coefficient has received so little serious study. The first systematic review of partition coefficients was not published until 1971.[42] Since that time, interest in the subject has expanded greatly; however, it is still an undeveloped subject of very great importance demanding attention by physical organic chemists. Additional careful studies of the type published by Kaufman and her colleagues[98] and Huyskens and Tack[99] are needed.

9.2.2. Hydrophilic Parameters

Hine and Mookerjee[100] have formulated a constant ($\log \gamma$) for the intrinsic hydrophilic character of organic compounds, where $\gamma = C_w/C_g$. The partition coefficient γ is defined as the ratio of the concentration in dilute aqueous solution at 25°C (C_w) to the concentration in the gas phase (C_g), where both concentrations are in mol l^{-1}. Values of $\log \gamma$ have been calculated for 292 compounds and values for bond contributions have been developed. It remains to be seen how $\log \gamma$ can be employed in biochemical QSAR.

9.2.3. Electronic Parameters

The Hammett–Taft σ constants have received much more thorough study than hydrophobic constants over the past 30 years; nevertheless, much remains to be done before such constants can be mechanically applied in the formulation of biomedicinal chemical QSAR. The recent paper by Norrington *et al.*[101] on the machine loading of σ and π constants is a step in the right direction, but it will be some time before any but the simpler problems can be handled in such fashion.

Two major problems confront those using σ constants: (*a*) should σ be dissected into inductive and resonance components (σ_I and σ_R)? (*b*) how should "through-resonance" (i.e. direct resonance between substituent and reaction center) be dealt with?

Taft and his colleagues[102–105] have pioneered the effort to dissect σ. Their efforts seem to show that while a single set of σ_I constants should give satisfactory results in almost any aliphatic or aromatic system, *at least* three types of σ_R values are needed to deal with cases (i) where through-resonance is absent, (ii) where a negative charge is delocalized (σ_R^-), and (iii) where a positive charge is involved (σ_R^+).

Swain and Lupton[106] have also proposed a means for dissecting σ into $\mathfrak{F}$ and $\mathfrak{R}$, field/inductive and resonance parameters analogous to Taft's σ_I and σ_R. Swain and Lupton analyzed the great variety of σ constants that have been proposed and showed that they were highly interrelated. They reached the overoptimistic conclusion that a single inductive and a single resonance parameter should suffice to treat all types of electronic effects. It has since become evident to those who have examined the problem more closely that $\mathfrak{R}$ will not in general do a good job of correlating data where through-resonance is involved; cf. Chapter 4.

A much extended list of $\mathfrak{F}$ and $\mathfrak{R}$ values has been collected; $\mathfrak{F}$ and $\mathfrak{R}$ were rescaled to correct for improper scaling used by Swain and Lupton.[107] When through-resonance is not involved, $\mathfrak{F}$ and $\mathfrak{R}$ can be employed to assess the relative importance of polar and resonance factors in QSAR. A disadvantage in the dissection procedure is that the greater number of constants needed requires a larger number of data points to ensure meaningful correlation equations. For example, in a normal Hammett relationship involving substituents in the 3- and 4-position of a benzene ring, equation (9.2) can be expected to apply. At present σ, σ^-, and σ^+ can be used for σ_i. Equation (9.3) applies for the dissection of σ by using the

$$\log k = \rho\sigma_i + \text{const} \tag{9.2}$$

$$\log k = a\mathfrak{F}\text{-}3 + b\mathfrak{R}\text{-}3 + c\mathfrak{F}\text{-}4 + d\mathfrak{R}\text{-}4 + e \tag{9.3}$$

Swain–Lupton constants, and similarly for Taft constants. In equation (9.3), the parameters a–e must be evaluated by the least-squares method. The descriptors 3 and 4 attached to $\mathfrak{F}$ and $\mathfrak{R}$ are for substituents in the *meta*- and *para*-position, respectively. One would normally want at least 20 data points for a proper evaluation of a–e; however, the coefficients a and c often turn out to be the same, especially in biochemical QSAR where there is usually considerable "noise" in the data. Sometimes the $\mathfrak{R}$–3 term is not significant and this yields a further simplification.

The following example illustrates such an application. Inhibition constants (K_i) (from Kakeya *et al.*[108]) for the inhibition of carbonic anhydrase by $XC_6H_4SO_2NH_2$ yielded equation (9.4). Reformulation of the data in terms of equation (9.3) yields (9.5).[12] It is clear from equation (9.5)

$$\log 1/K_i = 0.24\pi + 0.83\sigma + 0.48 \tag{9.4}$$

n	R	s
16	0.959	0.188

that the coefficients of the two $\mathfrak{F}$ terms are essentially the same and hence can be merged into a single term, $\mathfrak{F}$-3, 4. The 95% confidence interval on

$$\log 1/K_i = 0.17(\pm 0.21)\pi + 1.05(\pm 0.56)\mathfrak{F}\text{-}3 - 0.36(\pm 1.2)\mathfrak{R}\text{-}3 + 1.22(\pm 0.50)\mathfrak{F}\text{-}4 + 0.53(\pm 0.40)\mathfrak{R}\text{-}4 + \text{const} \quad (9.5)$$

n	R	s
16	0.974	0.173

the $\mathfrak{R}$-3 coefficient shows that this term can be dropped. Refitting the data, we obtain (9.6). Not only is equation (9.6) a superior correlation, but it also

$$\log 1/K_i = 0.18\pi + 1.07(\pm 0.35)\mathfrak{F}\text{-}3,4 + 0.57(\pm 0.36)\mathfrak{R}\text{-}4 + \text{const} \quad (9.6)$$

n	R	s
16	0.971	0.165

affords a better view of the electronic role of the substituents. Resonance effects are of little importance and polar effects are of major importance.

The Taft values[104] of σ_R appear to be better defined than $\mathfrak{R}$, while in this respect $\mathfrak{F}$ and σ_I are about the same. By using pK_a values of benzoic acid in 50% ethanol from Exner's extensive study,[109,110] equations (9.7) and (9.8) give a comparison of Swain–Lupton and Taft constants. The Taft

$$\text{p}K_a = -1.64\sigma_{I\text{-}3} - 0.34\sigma_{R\text{-}3} - 1.76\sigma_{I\text{-}4} - 1.15\sigma_{R\text{-}4} + 5.76 \quad (9.7)$$

n	R	s
27	0.992	0.083

$$\text{p}K_a = -1.60\mathfrak{F}\text{-}3 - 0.34\mathfrak{R}\text{-}3 - 1.57\mathfrak{F}\text{-}4 - 1.22\mathfrak{R}\text{-}4 + 5.75 \quad ((9.8)$$

n	R	s
27	0.992	0.082

equation (9.7) is improved considerably if the 4-OPh data point is dropped ($n = 26$, $R = 0.997$, $s = 0.056$). In equation (9.7), σ_R is from the benzoic acid system [σ_R(BA)].[104]

Unger and Swain have also dissected σ to yield S and P constants.[23] While the derivation of S and P is based on quite different principles from those for $\mathfrak{F}$ and $\mathfrak{R}$ or σ_I and σ_R, the actual values are similar to those of $\mathfrak{F}$ and $\mathfrak{R}$. Sjöström and Wold[111] have initiated a statistical analysis of σ constants, cf. Chapter 1.

One is often unable to decide what type of electronic effect of substituents to expect in the formulation of biomedicinal QSAR; hence, when aromatic systems are involved, one should explore the use of σ^- and σ^+ to allow for through-resonance effects. These can be quite significant even in biochemical systems.[112]

The radical parameter E_R has received some attention in correlation analysis, but Kieboom[113] has criticized the use of this electronic constant.

9.2.4. *Steric Parameters*

The pioneering work of Taft[114] showed that a parameter (E_s) could be formulated to correlate steric effects of substituents on the rate or equilibrium constants of homogeneous organic reactions. A review of E_s, which lists many new values determined since Taft's initial study, has been published recently.[115] E_s is a specific constant derived from the effect of substituents on the rates of acid hydrolysis of substituted acetates. In effect, this measures mainly the steric inhibition by the substituent of the formation of the transition state for the attack of water on the protonated ester (i.e., a differential effect as between the activated complex and the reactants). There are, however, many instances where E_s has been shown to correlate intermolecular steric effects of the kind that occur between ligands and macromolecules.[115]

Verloop and his associates have shown that calculated steric constants based on effective van der Waals radii give good results in biochemical QSAR.[116]

Simon[117–119] has devised an approach to the steric fit of ligands to macromolecules. The minimal steric difference (MSD) is estimated within certain limits, by superimposing the structural diagram of the ideal substrate on that of each of its congeners. Hydrogen atoms are neglected and second-row atoms that are not superimposed are counted, particularly carbon; Ref. 118 should be consulted for details. A serious stumbling block in this method is deciding what molecule should be taken as the reference compound.

Pratesi *et al.*[120] have formulated an *ad hoc* steric factor for β-adrenergic inhibitors.

9.2.5. *Molar Refractivity (Polarizability)*

A parameter first suggested by Pauling and Pressman[121] which shows promise of generality in biochemical QSAR is molar refractivity (MR).[12,39,122–131] Pauling and Pressman viewed MR as a means of assessing the polarizability of substituents which they considered to be of critical importance in the attachment of ligands to macromolecules. MR values are obtained from the Lorentz–Lorenz equation (9.9), where n is the refractive index for the D-line of sodium, M the molecular weight, and d the density. It is an additive-constitutive property of organic compounds and values are known for the common substituents.[107] The results obtained using MR in correlation analysis can be ambiguous. While this parameter is a measure

$$\mathrm{MR} = \frac{n^2 - 1}{n^2 + 2} \frac{M}{d} \tag{9.9}$$

of polarizability, it is highly collinear with molar volume; in addition, MR and π for a given set of substituents may also be highly interrelated. This difficulty can easily be avoided by proper experimental design.[31] If the coefficient of an MR term in a correlation equation is negative in sign, a steric effect in the ligand–macromolecule interaction is suggested. A positive coefficient would suggest a role for dispersion forces and/or production of an essential conformational change.

An interesting example of the use of MR in correlation analysis comes from the studies of Williams and his colleagues on papain, (cf. Ref. 122). K_m values for two sets of ligands [(I) and (II)] binding to papain were

$XC_6H_4OCOCH_2NHSO_2Me$ (I)

$XC_6H_4OCOCH_2NHCOPh$ (II)

determined. The QSAR is given in equation (9.10). I in this expression is an indicator variable (cf. Section 9.2.8) needed to merge the two data sets.

$$\log 1/K_m = 0.57\mathrm{MR} + 0.56\sigma - 1.92I + 3.74 \tag{9.10}$$

n	R	s
20	0.990	0.148

The parameters π and MR are almost orthogonal and MR yields a good correlation while π gives a poor one.[122] The structure of papain is known from X-ray crystallographic studies. It appears that X in (I) and (II) is binding on a bank of polar amino acids and not in a hydrophobic pocket.

Franks[132] has summarized evidence for two types of "hydrophobic bonding." In addition to the usual type where desolvation of an aqueous "flickering cluster" of water molecules provides the binding force, there is considerable evidence to show that two solvated groups may be held together in solution by mutual stabilization of their water "clathrates." Franks's view is one way of explaining the efficacy of MR in correlating certain types of enzyme–ligand interactions and his idea has been used to rationalize equation (9.10).[122]

Hubbard and Kirsch[133] attempted to correlate the rates of acylation of α-chymotrypsin by various aryl benzoates by using σ constants. The correlations are quite poor but can be greatly improved by using MR. This parameter may be just as important as π in enzymic studies.

It is surprising that nonspecific parameters such as π and MR can correlate diverse sets of ligand–macromolecular interactions. The fact that they do must mean that enzymes as well as other macromolecules are more flexible and pliable in solution or even when embedded in their cellular domains than traditional "lock-and-key" thinking suggests. This discovery promises enormously to simplify the problem of structure–activity relationships in biomedicinal chemistry.

9.2.6. Miscellaneous Parameters

Since the initial studies of McGowan,[134] the parachor has interested those doing correlation analysis on biochemical systems. A recent application comes from the study of Ahmad *et al.*[135]

Mullins[136] first explored the use of Hildebrand's solubility parameter in rationalizing the narcotic action of organic compounds. The applicability of these constants is still under investigation.[46]

A novel substituent constant for charge-transfer complexation has been formulated by Hetnarski and O'Brien[137–139] as in (9.11). The charge-transfer constant (C_T) is defined from the dissociation constant of a

$$C_T = \log K_X - \log K_H \tag{9.11}$$

complex between tetracyanoethylene and a parent molecule (K_H), and a derivative (K_X). The definition is analogous to that used for σ and π. Hetnarski and O'Brien have shown that C_T is of value in formulating QSAR for cholinesterase inhibitors.[137–139] Some care must be exercised in formulating C_T values since steric effects between tetracyanoethylene and the various derivatives needed to get K_X values may not parallel those of the ligand–macromolecular complex one wishes to model. Considerable work in physical organic chemistry indicates that equilibrium constants for charge-transfer complexation can be correlated with Hammett σ constants.[140–144]

Interest continues[145,146] in the use of dipole moments for QSAR work.

Also the adsorption of organic compounds by carbon is correlated with log *P*.[78] Carbon-binding constants have been employed in developing QSAR.[147–149]

Many studies have been made of the protein-binding constants of organic compounds. Such binding constants have been considered as a reference system to evaluate hydrophobic interactions.[150] Nishimura *et al.*[151] have used binding constants obtained with nylon powder to derive QSAR for triazines inhibiting the Hill reaction in chloroplasts.

9.2.7. Spectroscopic Constants

There has been a long-standing interest in the use of spectroscopic data (ir, uv, nmr) to correlate biochemical interactions. An ambitious effort has been made recently[152] by Seth-Paul and van Duyse to formulate substituent constants from the ir carbonyl stretching frequencies of simple compounds of the type, R^1COR^2. Tollenaere[153,154] has shown that these constants can yield good QSAR. It remains to be seen how much new information is contained in the ir-based constants beyond that in σ. Other recent reports include the use of ir[155,156] and uv[157] data in correlation analysis. Chemical shift constants from nmr spectra have also yielded good QSAR.[72,158]

9.2.8. *Indicator Variables*†

Organic and medicinal chemists have a long tradition of considering substituent effects to be, in some respects, additive. Work with the Hammett equation has done much to advance this view. In the area of biomedicinal QSAR, substituent effects are far more complex and one might expect simple additivity to be encountered rarely; fortunately, the facts are quite different.

Bruice *et al.*[159] first showed that one could derive novel additive substituent constants for thyroxine analogs. Free and Wilson[160] generalized this thinking so that one now speaks of the Free–Wilson method of analysis.

The big advantage of the Free–Wilson approach is that one needs no physicochemical constants to correlate a set of congeners; only the biological data are required. The details of the procedure have been discussed by Craig[161] and Purcell *et al.*[9] Interest in the Free–Wilson method continues to grow.[55,162–179]

The basis of the Free–Wilson analysis rests on the additive contribution to activity (+ or −) of each substituent, independently of other substituents in the molecule. This can be expressed as in (9.12). A in this

$$A_n = \sum_p \sum_s a_{n,ps} + \mu \qquad (n = 1, 2, \ldots, N) \tag{9.12}$$

expression is the biological activity of a member of a congeneric series. The activity contribution of the substituent (s) is position-dependent (p); that is, one must assume, unless there is evidence to the contrary that, for example, a *meta*-NO_2 group would make a contribution to activity different from that of an *ortho*-NO_2 group. Also, μ represents the activity of the constant portion (parent structure) of the series. Following equation (9.12) strictly, one faces a dilemma: Either one makes a rather large number of derivatives to explore substituent space widely, in which case there are few data points to support each derived constant (a), or one limits the exploration to obtain more data points per variable. In practice, it is

†See p. 411.

often observed that similar substituents may make (within the precision of the biological assay) identical contributions. Under such conditions, one can take advantage of the situation by formulating much simpler expressions. A recent example[179] illustrates the point. Equation (9.13) was

$$\log 1/C = 0.36I\text{-}1 - 1.01I\text{-}8 - 0.78I\text{-}9 + 0.42I\text{-}13 - 0.22I\text{-}15 + 0.51I\text{-}20 + 0.67I\text{-}4 \cdot I\text{-}8 + 7.17 \quad (9.13)$$

n	R	s
105	0.903	0.229

formulated from data for 105 derivatives of type (III). C in equation (9.13) is the molar concentration causing 50% *reversible* inhibition of dihydrofolate reductase. Only 105 out of 111 data points were used to formulate equation (9.13). The types of structural changes are shown in Table 9.1.

(III)

There are 36 different structural changes listed in Table 9.1 so that 36 constants would be needed in a strict Free–Wilson treatment to correlate the data. Since only one or two data points are available for 14 features, one could place little or no confidence in the a values for these groups. Equation (9.13) with seven variables accounts for 82% ($R^2 = 0.82$) of the variance in the data. A 26-variable equation accounted for 87% of the variance. This is a large price to pay for the small improvement obtained with 19 more variables.

The variables of equation (9.13) are defined below.

I-1	$\omega = CH_2$		I-15	4′-SO_2F	
I-8	4-NHCONH	Bridge	I-20	L-1210/0	Enzyme
I-9	4-NHCO	Bridge	I-4	Y = 3-Me	
I-13	4-$NHSO_2$	Bridge			

Note that of the five different Y groups in Table 9.1, only one (I-4) is parameterized. The others make no significant contribution to activity. The same kind of result is observed with Z. A number of cross-product terms were considered, only one of which ($I\text{-}4 \cdot I\text{-}8$) proved to be significant.

Of course, the major shortcoming of the Free–Wilson method is that the novel constants obtained have no direct physicochemical meaning;

Table 9.1. Structural Changes Involved in the Formulation of Equation (9.13)

ω	Number of occurrences	Y	Number of occurrences	Bridge	Number of occurrences	Z	Number of occurrences	SO_2F	Number of occurrences
O	92	2-Cl	20	4-NHCONH	27	2′-Cl	13	5′-	17
CH_2	19	3-Cl	8	4-NHCO	13	3′-Cl	2	4′-	45
		2-Me	1	4-CH_2NHCONH	15	2′-Me	1	3′-	47
		3-Me	16	4-CH_2NHCO	20	3′-Me	5		
		6-Me	1	3-CH_2NHCONH	5	4′-Me	7		
		2-OMe	6	3-CH_2NHCO	3	2′-OMe	3		
		4-SO_2F	2	4-CH_2CH_2NHCO	2	3′-OMe	1		
				4-CH_2CH_2NHCONH	2	4′-OMe	2		
				4-$NHSO_2$	6	4′-OEt	1		
				4-CH_2NHSO_2	7	4′-Pr^i	1		
				4-$CH_2CH_2NHSO_2$	2				
				3-CH_2NHSO_2	4				
				3-CH_2CH_2	1				
				4-CH_2CH_2	2				

hence, it is difficult to decide what new types of substituents should be studied. Where possible, it is much wiser to employ physicochemical constants in deriving QSAR; however, the necessary constants are often missing. Even worse, no obvious means exists of deriving them. Highly specific interactions between ligand and macromolecular receptor may not be parameterizable in a general way. The combination of novel constants and substituent constants from suitable reference systems greatly extends one's reach in the formulation of QSAR. The term *indicator variable* seems most appropriate for such constants.[180] The interest in such "mixed QSAR" is increasing.[39,122–129,164,181–190]

The formulation of mixed QSAR can be illustrated with data from the extensive studies of Baker on dihydrofolate reductase inhibitors.[126] About 260 derivatives of type (IV) were tested as inhibitors of dihydrofolate

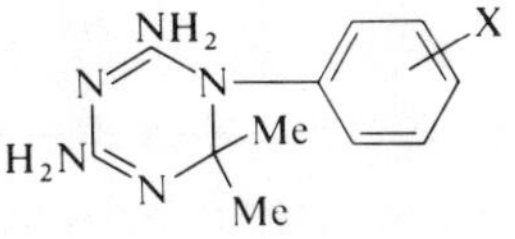

(IV)

reductase. Samples of the enzyme from two sources, L1210 leukemia and Walker 256 tumor, were used. Equation (9.14) was derived in a preliminary analysis.[191] This equation was based on results from Walker 256

$$\log 1/C = 0.89\pi\text{-}3 - 0.13(\pi\text{-}3)^2 + 0.15\text{MR-}4 + 6.62 \qquad (9.14)$$

n	R	s
83	0.905	0.328

enzyme: C is the molar concentration of compound giving 50% reversible inhibition of dihydrofolate reductase, π-3 refers to the hydrophobic character of substituents in the 3-position of the N-phenyl moiety, and MR-4 refers to the molar refractivity of groups in the 4-position.

When only the two continuous variables, π and MR, were employed, very many derivatives were beyond the reach of QSAR. However, equation (9.14) shows that there is a large amount of internal self-consistency in $\log 1/C$ and that enzyme space around the 3-position is hydrophobic while that around the 4-position is not.

Starting with equation (9.14), other sets of derivatives were added and it was possible, from a study of the residuals, to develop indicator variables to account for special substituent features. Equation (9.15), correlating 244 data points, was developed[126] by such a process.

The "substituents" X in these data constitute an enormous range of structures considerably beyond the usual meaning of the term. Some of the small substituents used are H, F, Cl, Br, CN, OMe, CF_3, NO_2, SO_2F,

$$\log 1/C = 0.68\pi\text{-}3 - 0.12(\pi\text{-}3)^2 + 0.23\text{MR-}4 - 0.024(\text{MR-}4)^2 + 0.24I\text{-}1 - 2.53I\text{-}2 - 2.00I\text{-}3 + 0.88I\text{-}4 + 0.69I\text{-}5 + 0.70I\text{-}6 + 6.49 \quad (9.15)$$

n	R	s
244	0.923	0.377

ideal π-3 = 2.9; ideal MR-4 = 4.7

CH_2CN. Equation (9.15) also embraces *many* gross functions such as those below:

4-$OCH_2C_6H_4$-4′-$SO_2OC_6H_4$-3″-$CONMe_2$

3-$CH_2CH_2CH_2CH_2C_6H_4$-3′-SO_2F

4-$CH_2CH_2CONHC_6H_4$-2′-Me-4′-SO_2F

3-$CH_2NHCONHC_6H_4$-3′-SO_2F

4-$OCH_2CH_2OCH_2CH_2OC_6H_4$-4′-$SO_2F$

3-$CH_2CH(CH_2NHCOCH_2Br)Ph$

When one considers the enormous number of ways these large substituents could interact with an enzyme, it is astonishing that any kind of correlation can be achieved at all. Equation (9.15) accounts for 85% of the variance in log $1/C$ and C spans an over 200,000-fold concentration range.

However one might view equation (9.15), it is of tremendous help in organizing a large mass of data so that one can begin to study it in an objective fashion. Besides the two continuous variables π-3 and MR-4, six indicator variables have been used. I-1 is assigned the value of 1 for Walker enzyme and 0 for L1210 enzyme. The positive coefficient with this term shows that Walker enzyme is more sensitive to the inhibitors, which may be more a matter of experimental conditions than intrinsic activity. I-2 takes the value of 1 for *ortho*-substituents which decrease activity 300-fold, other factors remaining constant. I-3 is assigned the value of 1 for rigid groups (i.e., —Ph, —CONH, —CH=CHCONH) attached directly to the N-phenyl ring, and brings out the deleterious effect of such groups in structures which are much less active than those with the more flexible substituents; the CH_2Ph group, for example, is well predicted without I-3. I-4 is given the value of 1 for those congeners containing the highly active leaving group $SO_2OC_6H_4Y$; these derivatives are about eight times more active than use of π and MR alone would predict. Other substituents sensitive to nucleophilic substitution (SO_2F, $NHCOCH_2Br$, and $COCH_2Cl$) required no special parameterization. I-5 takes the value of 1 for flexible bridges [—CH_2—, —CH_2CH_2—, —$(CH_2)_4$—, —$(CH_2)_6$—, —$(CH_2)_4O$—] between the N-phenyl moiety and a second phenyl ring. I-6 is given the value of 1 for groups of the type: $—CH_2NHCONHC_6H_4X'$ or $—CH_2CH_2CONRC_6H_4X'$.

Ortho-substitution is almost always a difficult problem with which to cope. From a practical point of view, it was soon obvious in the present instance that the negative effect of 2-substituents did not warrant further study. Only a few congeners with a 2-substituent plus a second or third group were prepared, so that the constancy of additivity of *I*-2 in the presence of other groups cannot be properly assessed.

There are nine examples where *I*-3 is needed and, as one might expect from the rather loose definition of *I*-3, log $1/C$ is not well predicted for these congeners. Six of the congeners have deviations between 1 and 2 standard deviations and two have deviations of essentially 3 standard deviations; nevertheless, in working on a SAR of over 200 complex structures, it is valuable to include such features in the analysis even though they are not accommodated as nicely as one would like. *I*-3 is a reflection of the topography of dihydrofolate reductase which must be reckoned with in designing other ligands. There are 20 instances where *I*-4 applies and these congeners are well fitted, the average deviation being 0.24, which is considerably less than the standard deviation.

The examples where *I*-4 applies are almost all of the type 3-Cl and 4-$OCH_2C_6H_4SO_2OC_6H_4X'$. Despite considerable variation in X′, no parameterization of X′ other than by MR-4 is necessary to achieve the good fit for these congeners. *I*-4 is remarkably independent of changes in X′.

There are 47 examples where *I*-5 applies. Two of these data points are so poorly fitted that they were not used in deriving equation (9.15). The other 45 yield predicted log $1/C$ values with an average deviation of 0.25. The moderately large coefficient with *I*-5 brings out the importance of this variable. *I*-5 applies to substituents in either the 3- or 4-position, and there is a large amount of concurrent structural variation in the congeners containing these flexible bridges. It is impressive that the contribution of *I*-5 remains notably independent of a wide variety of other structural variations.

I-6 is utilized with 28 congeners and the average deviation from predicted log $1/C$ is 0.28. Again, large variations have been made in other parts of these congeners in addition to the amide linkages accounted for by *I*-6.

In a Free–Wilson analysis, there are often relatively few data points supporting each of the novel constants and, since these analyses have not been much followed up, one is left with the nagging thought that novel constants may not hold up in more complex circumstances. Mixed QSAR give one another view of novel constants which is indeed reassuring; in fact, good predictions have been realized with them.[39] For example, equation (9.16) was formulated for a set of 69 pyridinium derivatives (V)

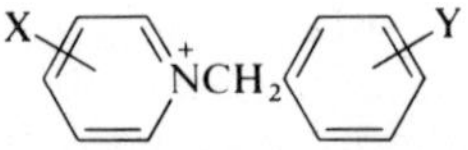

(V)

$$\log 1/C = 0.18\pi_X + 0.46\pi_Y + 1.01\sigma_X^+ + 0.72D + 2.50 \qquad (9.16)$$

n	R	s
69	0.939	0.198

inhibiting complement.[127] D in equation (9.16) is an indicator variable assigned a value of 1 for congeners where $Y = 2\text{-}SO_2F$. After the publication of equation (9.16), 63 new derivatives became available and equation (9.16) did a good job of predicting their activities.[39] Fitting all 132 data points to the variables of equation (9.16) gave (9.17). The coefficients

$$\log 1/C = 0.16\pi_X + 0.38\pi_Y + 0.91\sigma_X^+ + 0.71D + 2.58 \qquad (9.17)$$

n	R	s
132	0.945	0.213

in equation (9.17) are close to those of equation (9.16), although the standard deviation is a little higher. Adding a new indicator variable for a feature not contained in the first set gave a correlation with identical standard deviation.

There are 76 instances where $D = 1$ is used. Structural modification in other parts of these congeners is widely varied, again illustrating the independent additivity of the contribution of $2\text{-}SO_2F$ to activity.

Considerable space has been devoted to equation (9.15). The justification for this is that it is the most extensive QSAR published so far and encourages one to believe that very extensive QSAR (>1000 congeners) can be formulated for complex compounds. Such expressions will be of the utmost value in efficiently developing an on-going search for a new drug. QSAR of enzymes such as equation (9.15) will be helpful in working out *via* X-ray crystallography the interactions of ligands (e.g. drugs) with macromolecules.

Equation (9.15) should not be considered the final answer for this set of congeners. The complexity of the interactions involved is so great that even though a substantial effort went into its development, improvements can be expected. In fact, this collection of data should be a good testing ground for new QSAR techniques.

9.3. Regression Analysis

It will now be clear that biomedicinal chemists regard the formulation of QSAR as very important. Various ideas about the basic theory and strategy in the development of QSAR are now subjects for discussion[192–201]; nevertheless, we must admit that our ignorance about the factors involved is so abysmal, even in the simple case of the properties of purified enzymes, that progress must depend upon the frank use of the grossest type of empiricism. Useful new parameters such as π and MR can be refined by theoretical studies.

Undoubtedly the most powerful tool at our disposal is computerized regression analysis coupled with proper statistics. Even if one knew precisely their positions with respect to one another, there is no deterministic or rigorous method of calculating all of the atom-by-atom interactions in a set of large molecules, such as those on which equation (9.15) is based, with the atoms of a macromolecule (e.g. dihydrofolate reductase). Difficulties become serious even for the naive kind of model-building implied in equation (9.15). Consider, for example, the simple case (VI). Assume that

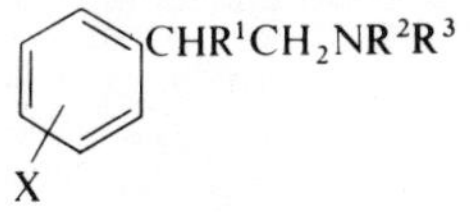

$CHR^1CH_2NR^2R^3$

X

(VI)

one has made and tested 100 derivatives in which considerable variation exists in R^1, R^2, and R^3. Consider also that X has been varied widely in the 2-, the 3-, and the 4-position. A typical set of variables that one might study is π, $\mathfrak{F}$, $\mathfrak{R}$, MR, E_s. To consider all variables at the six positions (ring and side chain) would mean 30 terms; all possible linear combinations without considering higher power or cross-product terms would imply $2^{30}-1$ equations. This calculation is out of the practical range of even the largest computers, using the usual least-squares technique. There are alternatives to this brute force approach; one could divide the set into, say, three sets of all *ortho*-, all *meta*-, and all *para*-substituted congeners. Even in these cases, one can become involved in trying to find the best subset of 15–20 variables when considering some higher-power terms and a few indicator variables, plus the applicability of those mentioned above. The situation gets worse rapidly when cross-product terms are considered. For pairwise cross-product terms (e.g. $\pi \cdot \sigma$), one is faced with $n(n-1)/2$ terms where n is the number of primary variables; when n is 8, we are faced with 28 cross-product terms plus 8 primary terms plus higher-power terms. Until recently, such problems were attacked by means of stepwise regression analysis; this procedure, which involves computerized selection of the best single variable, then considers the remaining variables one at a time until the best two-variable equation is selected. The process is continued until addition of a variable is not justified by the F statistic. It is now widely recognized that such an approach can easily overlook significant variables, which may only show up when two or more variables are together added to the equation under study.

A promising alternative to stepwise regression is regression by "leaps and bounds," developed by Furnival and Wilson.[202] These authors have developed an algorithm for computing the residual sums of squares for all possible regressions. This does not give the regression coefficients but enormously speeds up calculations so that one can readily examine all sums of squares for, say, 20 variables (e.g. *ca.* 10^6 equations) and in this way select the most important variable for detailed study. Furnival and Wilson show that by combining this algorithm with a second one which quickly limits the consideration of variables to the significant ones, it is possible to select the best subset out of as many as 40 variables. Regression by leaps and bounds promises to be of great value in complex QSAR work.

Another technique which is of particular value when high collinearity results in unstable regression coefficients is ridge regression.[203]

9.4. Models for QSAR

The ultimate problem in drug research is to account for the principal interactions of about 10^{20} molecules of drug injected into a mouse or human with an unknown but incredibly large number of macromolecules that make up the living organism. To make matters worse, one gets only a few highly integrated signals out of the black box (mouse); that is, there may be scores of different interactions with different macromolecular systems in the mouse (not the least of which is a complex pathogen) but the final visible response is often only the difference between a dead and a live mouse! From the chemical point of view, the cause of this relatively small but rather significant change in the mouse is in general too complicated to delineate at the molecular level. Despite the incredible odds, the undaunted medicinal chemists continue to develop better and better drugs.

Equation (9.18) suggests a beginning on the formidable problem of rationalizing biological response to a set of drugs in terms of physical

$$\text{rate of biological response} = ACk \qquad (9.18)$$

organic chemistry, where C is moles/kg of drug injected, A is a probability factor that the drug will reach the site of action in time to register a visible response in the allotted time of the test, and k is a rate or equilibrium constant for a rate-controlling reaction which ultimately leads to the observable response.[204,205] In much biological testing, the rate is set constant (so many minutes or hours before the reading of the result is taken) and C is varied to yield a constant response.

Two avenues of study are open to the development of the hypothesis of equation (9.18). One can devise a means for studying the form of A and techniques for accounting for differences in k. It was suggested[204] initially that, as a first approximation, A can be assumed to be a Gaussian function of $\log P$ in the non-steady-state random walk of drug molecules to the sites of action.[205] Equation (9.18) is usually converted into the form of (9.19),

$$\log 1/C = a(\log P)^2 + b \log P + c \log k + d \tag{9.19}$$

where the parameters a, b, c, and d are found by the method of least squares. When k appears to be more or less constant, many good correlations have been found to be parabolic between $\log 1/C$ and $\log P$.[197] McFarland[193] has offered a probabilistic rationalization of the parabolic dependence of $\log 1/C$ on $\log P$. Flynn and Yalkowsky have undertaken a program better to understand the passive diffusion of organic compounds through membranes and to show how this could rationalize the parabolic dependence of $\log 1/C$ on $\log P$.[196,206–209] Lin has attempted[199] a statistical mechanical treatment of QSAR which also justifies the parabolic dependence of $\log 1/C$ on $\log P$ of equation (9.19).

Equation (9.19) is based on a non-steady-state assumption. Higuchi and Davis[192] have assumed a quasi-equilibrium of free drug and that on the receptor sites. Assuming receptors having different degrees of hydrophobicity, they show that a great variety of nonlinear relationships between $\log 1/C$ and $\log P$ can be accounted for. Hyde[210] has also assumed an equilibrium model and shows that nonlinear dependence of activity on hydrophobicity can be expected.

The third term of equation (9.19), $\log k$ can be studied separately in isolated systems. It seems reasonable to correlate variation in this parameter in the usual Taft–Hammett fashion by the linear combination of free-energy-based physicochemical constants; as in (9.20). Many studies

$$\log k = a\pi + b\sigma + cE_s + d \tag{9.20}$$

with enzymes or relatively simple isolated biological systems yield good correlations with such an approach. As discussed in Section 1.2.5, MR sometimes replaces π.

9.5. *Optimum Lipophilicity* ($\log P_0$)

A practical aspect of equation (9.19) is that one can calculate the optimum degree of hydrophobic character ($\log P_0$) for a set of congeners from the partial derivative $[\partial \log (1/C)/\partial \log P]$. As our file of $\log P_0$ values grows, this should provide guidance in drug development. It was first observed[211] that $\log P_0$ is a constant of about 2 for neutral molecules (16 sets), which penetrate the central nervous system (CNS). This finding has been confirmed by the study of a set of phenylacetamide hypnotics.[212]

An ingenious study by Kutter *et al.*[213] also brings out the importance of $\log P_0$ for drugs entering the CNS. This study concerns a set of morphine-like analgesics in which the stereochemistry was so varied that it was not possible to formulate a QSAR directly. Two types of activity were determined for the analgesics: For one, the drugs were injected directly [intraventricularly (iventr)] into the rabbit brain; for the second, the drugs were given intravenously (iv). The postulates of (9.21)–(9.23) were made.

$$\log 1/C_{\text{iventr}} = b \log k + \text{const} \tag{9.21}$$

$$\log 1/C_{\text{iv}} = a \log A + b \log k + \text{const} \tag{9.22}$$

$$\log A = a'(\log P)^2 + b' \log P + \text{const} \tag{9.23}$$

As a first approximation, it was assumed that biological response from an essentially direct partitioning of drugs onto receptor sites in the brain [iventr injection, equation (9.21)] is determined by the intrinsic activity of each drug (k). When the drug is injected iv, a random walk process must occur with the drug crossing the so-called blood–brain barrier, which is then followed by the highly specific reaction (k) with the receptors of the CNS. Substitution from equation (9.23) into equation (9.22) and then subtracting equation (9.21) from (9.22) yields (9.24). Fitting the data to equation (9.22) produces (9.25). P' in equation (9.25) refers to

$$\log 1/C_{\text{iventr}} - \log 1/C_{\text{iv}} = \alpha(\log P)^2 + \beta \log P + \text{const} \tag{9.24}$$

$$\log (C_{\text{iventr}}/C_{\text{iv}}) = -0.090(\log P')^2 + 0.036 \log P' - 0.67 \tag{9.25}$$

n	R	s
11	0.970	0.297

heptane/water partition coefficients. Attempting to fit $\log 1/C_{\text{iv}}$ to the right-hand side of equation (9.25) gave a poor correlation, while that from equation (9.25) is quite good. Other factors being eliminated, equation (9.25) shows that an optimum $\log P'$ exists for crossing the blood–brain barrier, even for a set of drugs of widely different stereochemistry.

While many $\log P_0$ values have been determined, there are not enough data in hand to establish many useful generalizations. An effort has been made to compare $\log P_0$ for antifungal and antibacterial agents.[214]

9.6. Characterization of Hydrophobic Space

While log P_0 gives some inkling of the character of hydrophobic space at the reaction site, the study of simpler systems is necessary to establish the broad outlines of what one can expect to encounter. In an extensive study (137 examples)[215] of QSAR from relatively simple systems fitting the expression (9.26), it was observed that the values of the slope a tended to

$$\log 1/C = a \log P + b \tag{9.26}$$

be grouped around 1 and 0.7; that is, 57 examples had a slope of mean value and standard deviation of 1.0 ± 0.13 and 71 examples were found with a mean of 0.66 ± 0.12. The equations with slopes near 1 were associated with interaction of organic compounds with nerve membranes or erythrocyte membranes, both of which are known to have high lipid content. The lower slopes were found to occur in correlations involving proteins, enzymes, and bacteria.

These simple equations can be used to bridge the gap between physical organic and biochemical systems. The following equations[215] illustrate the point.

ROH causing disaggregation of 0.2-mm silanized glass beads:

$$\log 1/C = 0.98 \log P - 0.80 \tag{9.27}$$

n	r	s
4	0.995	0.077

ROH causing change in resistance of black lipid membrane:

$$\log 1/C = 1.16 \log P - 0.51 \tag{9.28}$$

n	r	s
7	0.985	0.262

ROH causing −10-mV change in rest potential of lobster axon:

$$\log 1/C = 0.87 \log P - 0.24 \tag{9.29}$$

n	r	s
5	0.993	0.100

The slopes of equations (9.27)–(9.29) are approximately 1, indicating the hydrophobic perturbation by alcohols to be essentially the same in each of the grossly different systems. C in equation (9.27) is the molar concentration of alcohol which barely disperses tiny glass beads held together by a lipophilic coating (a crude model membrane). Equation (9.28) correlates the concentration of alcohol necessary to change the resistance of a black lipid membrane (BLM) from 10^8 to $10^6\ \Omega/\text{cm}^2$. The BLM membranes are formed by dissolving lipid of sheep erythrocytes in a hydrocarbon and painting this solution over a small orifice. The difference in intercepts between equations (9.27) and (9.28) indicates that it takes twice the concentration of alcohol to disperse the beads that it does to lower the resistance of the BLM. No doubt one could develop two equations with essentially identical slopes and intercepts by varying experimental conditions. The intercept of equation (9.29) indicates that about double the concentration of alcohol is required to change the rest potential of the lobster nerve as to produce a standard perturbation in a BLM. Again, conditions can be adjusted to bring together the parameters of these equations. Hence, by means of correlation equations one can compare model systems with living systems, and learn more about the structure of membranes.

9.7. Applications of QSAR

9.7.1. Introduction

New QSAR are beginning to appear very frequently for increasingly diverse kinds of biological systems. Frank and Oehme list 375 QSAR in a recent review.[5] There seems to be no area, from the simplest proteins to the most complex organisms (humans themselves), where QSAR do not apply. Also, there appears to be no limitation on the type of organic compound amenable to QSAR treatment.

9.7.2. Simple Proteins

A study of the binding of simple organic compounds by plasma proteins is important from the point of view of drug distribution. These studies also provide a better understanding of the hydrophobic interactions between micro- and macromolecules in a system where penetration of the micromolecule to the binding site is not a serious problem. Many such studies have been made.[56,216–230]

Generally, a linear relationship is found between log (binding constant) and π, $\log P$, or R_m. Dunn,[229,230] however, has found parabolic dependence on $\log P$. It has been shown recently[231] that the stabilization of serum albumin against denaturation by 2-aryl-1,3-indandiones affords a QSAR.

The binding of glucosides by concanavalin A has also been rationalized by correlation equations.[232,233] The first analyses indicated a hydrophobic interaction correlated by π. A more recent study[12] finds that MR is the significant variable rather than π.

Tichý has shown that the binding of aromatic compounds to methemoglobin can be correlated with a variety of partition parameters.[234] Much of the earlier work on protein binding has been reviewed by Franke and Schmidt.[235] The dissolution of blood clots (fibrinolysis) by lipophilic anions is correlated very well with log *P*.[236]

9.7.3. *Membranes*

As it has become more apparent that the cell membranes have highly important functional aspects other than that of acting as container membranes, membrane research has taken on new importance. There are two aspects of the QSAR for membranes under study. A long-standing problem is that of simple passive diffusion of molecules through these potential barriers. A number of theoretical and practical studies have been made with simple membranes.[215,237–244] Gastric and intestinal absorption,[77,237,245,246] buccal absorption,[238,240,241] red-cell membrane rupture phenomena,[242,247] depolarization of plant membranes,[244] membrane-controlled macrophage spreading,[248] penetration of cockroach integument,[249] and stratum uptake[250] all depend heavily on log *P* or π.

Seeman[251–253] has studied extensively the interaction of a variety of simple organic compounds and drugs with membranes from erythrocytes. Good relationships were found between log *P* and measures of membrane stabilization. Seeman also found a strong correlation between octanol/water and membrane/water log *P* values.

9.7.4. *Nerve Potentials*

Since the Meyer–Overton work at the turn of the century, it has been appreciated that nerve function is inhibited by lipophilic compounds. In recent years, the focus of study has shifted from whole animals to single cells.[215,254–258] There is a strong correlation between log *P* and the ability of organic compounds to increase membrane potential or to alter permselectivity. Good correlations have been reported of log *P* with anesthetic potency of the gaseous anesthetics.[190,259]

9.7.5. *Enzymes*

The study of the interactions of ligands with purified enzymes offers possibly the most interesting area for QSAR studies. The systems are so simple that one can normally expect good correlations if a reasonable range of activity can be found, but they are challenging enough to test out every weapon in the QSAR armamentarium. With the increased knowledge of the details of enzyme structure being obtained from

X-ray crystallographic studies, unusual opportunities are developing to check conclusions drawn from QSAR. Many enzymic QSAR have appeared.†

Enzymes studied include: cholinesterase,[84,87,138,139,161,261–272] α-chymotrypsin,[123,124,130,273–280] β-D-xylosidase,[281,282] Na^+K^+ATPase,[283–285] carboxypeptidase,[286–288] aminopeptidase,[289] taka-*N*-acetyl-β-D-glucosaminidase,[290] sucrase and isomaltase,[291] trypsin,[124] plasmin,[292] complement,[39,187,292] papain,[122] alcohol dehydrogenase,[293,294] glutamate dehydrogenase,[124,295] succinate dehydrogenase,[296] malate dehydrogenase,[124] lactate dehydrogenase,[124,297] glyceraldehyde phosphate dehydrogenase,[124] dihydrofolate reductase,[124–126,131] monoamine oxidase,[298–301] dopamine β-hydroxylase,[164] *N*-methyltransferase,[164,302,303] thymidine phosphorylase,[124,304] uridine phosphorylase,[124] thymidilate synthetase,[124] cytosine nucleoside deaminase,[124] xanthine oxidase,[125] guanine deaminase,[125] hexokinase,[297] NADH–ubiquinone reductase,[305] histidine decarboxylase,[306] adenosine cyclic-3′,5′-monophosphate phosphodiesterase,[307] ATP_i-L-methionine *S*-adenosyltransferase,[308] carbonic anhydrase,[158,299,309] urease,[310] adenosine deminase,[311] choline acetyltransferase,[312] glutamate decarboxylase,[313] lipoxygenase,[299] hydroxyindole methyltransferase,[299] D-amino acid oxidase,[299] mixed function oxidase,[314] neuroaminidase,[146] and uridine diphosphate glucose dehydrogenase.[315] For this last enzyme the authors found [315] some correlations between K_m and van der Waals radius. We have found a better relationship with molar refractivity ($\log 1/K_m = -1.00\text{MR} - 0.04$; $n = 8$, $r = 0.910$, $s = 0.192$). The negative coefficient with the MR term indicates a steric effect in the binding of the α-D-glucopyranosyl pyrophosphates of 5-X-uridines with the enzyme.

Many of the above-mentioned QSAR are not very sophisticated. The field is so new that many laboratories have not had time to catch up with a rapidly moving subject. Often, too few data points have been used and not enough attention paid to the problem of collinearity, and few investigators have considered the possible importance of MR. Nevertheless, it is quite certain that QSAR techniques are going to play an important role in biochemistry.

9.7.6. *Organelles*

The great importance of oxidative phosphorylation and electron transport, coupled with the ease of isolation of mitochondria, has made these organelles popular for study.[86,154,225,316–323] Chloroplasts have also

†See references 8, 12, 39, 78, 84, 85, 87, 122–131, 138, 139, 158, 162, 164, 179, 187, 191, 197, 215, 260.

been studied.[151,324,325] The inhibition of alcohols of norepinephrine uptake by synaptosomes is correlated well with log *P*.[326]

9.7.7. Viruses and Micro-organisms

A few attempts have been made to formulate QSAR for the inhibition of viruses.[175,327,328]

A major area of drug research is that of the antimicrobial compounds. Studies in this area are often of the *in vitro* type where metabolism and elimination are not such serious problems as in *in vivo* work.† A number of satisfactory QSAR have been formulated for antifungal agents.[8,214,336–338]

Although there has been an enormous effort over the past 35 years to develop better antimalarials, relatively few QSAR have been formulated for such compounds.[169,339–343] The antimalarial problem is, short of cancer, one of the most complicated problems facing the medicinal chemist interested in QSAR. The malarial protozoon with its complex life cycle in the mouse is a difficult target for the drug designer. Even so, a set of over 100 phenanthrene carbinols (VII) has yielded[169] a relatively simple QSAR. *C* in equation

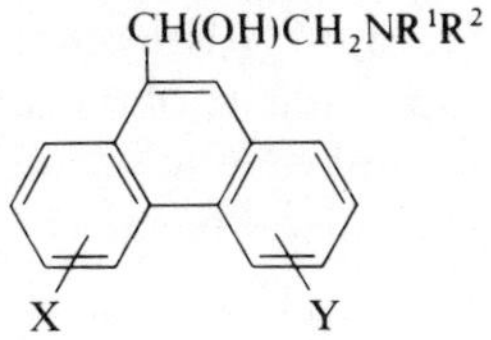

$CH(OH)CH_2NR^1R^2$

X Y

(VII)

(9.30) is the ED_{50} in mol kg^{-1}, π_{X+Y} and σ_{X+Y} refer to the sum of substituents on the two rings, while π_{sum} is $(\pi_{X+Y}+\pi_{R_1R_2})$. There is an overall optimum log *P* that appears to be independent of the hydrophobic interactions of the ring substituents.

$$\log 1/C = 0.31\pi_{X+Y} + 0.79\sigma_{X+Y} + 0.13\pi_{sum} - 0.015(\pi_{sum})^2 + 2.35 \tag{9.30}$$

n	*R*	*s*
102	0.908	0.263

An interesting QSAR for trichomonocidal nitrothiazoles has been formulated by using $E_{1/2}$ from polarographic measurements for the electronic term.[344] An attempt has been made to formulate a QSAR for schistosomicidal tetrahydroquinolines[345] and the antiprotozoal activity of imidazoles, both of which gave a partial correlation with log *P*.[346]

†See References 8, 75, 81, 88, 90, 153, 155, 165, 171, 181, 183, 184, 329–335.

9.7.8. Central Nervous System and Local Anesthetic Agents

Drugs acting as CNS depressants or anticonvulsants are often amenable to QSAR treatment.[213,347–350] An equation relating hallucinogenic potency in humans to the structure of amphetamines and phenylethylamines has recently been derived.[351] The correlation from a simple [log P + (log P)^2] equation was not strong, but log P_0 was found to be 3.1. The antiserotonin activity of LSD analogs provided a QSAR from which log P_0 was found[352] to be 2.9, in close agreement with that of the hallucinogens. Locomotor activity of amphetamines in mice is E_s-dependent.[353] The neuroleptic activity of dibenzoxazepines is correlated[92] with π or R_m, as are the analgesic activity of morphine-like compounds in guinea pig ileum and the analgesic activity of acetanilides.[354] Local anesthetic activity often depends strongly on log P.[355,356] Barbiturates inhibit H^+ release in frog gastric secretion and the process is dependent on the hydrophobicity of the drugs.[357]

9.7.9. Microsomal Action

Understanding the metabolism of organic compounds caused by microsomal oxidation is important in the design of better drugs. Brodie and his colleagues first pointed out[358] that microsomes attack compounds at rates in proportion to the lipophilicity of the latter. Since their observations, more precise evidence has linked microsomal action with log P.[185,359–366] It is likely that microsomal metabolism is an important determinant of log P_0.

9.7.10. Acute Toxicity

The acute toxicity of prospective therapeutic agents in mice is determined in drug studies. The few published QSAR of LD_{10} or LD_{50} in rodents highlight the importance of log P.[80,167,367] It seems to this author that in drug research one should always attempt to formulate a QSAR for toxicity (e.g., LD_{10}) alongside the QSAR for potency. The *two* QSAR should guide further research.

9.7.11. Steroids and Adrenergic Agents

Relatively few papers have appeared in which biological activity and chemical structure of steroids have been related quantitatively.[71,91,368,369] These limited results at least show that this group of complex drugs is open to QSAR. Adrenergic agents have also yielded quantitative relationships.[23,370,371]

9.7.12. Miscellaneous

Correlation equations have been formulated for anti-arrhythmic activity,[156,355] endocytosis,[284] hypoglycaemic activity,[71] radioprotective activity,[176] histoplasmosis,[174] CNS stimulation,[145] locomotor activity of mice,[178] anti-inflammatory activity,[166,372] bronchodilator activity,[182] anti-spasmodic activity,[170] vasopressin-stimulated water loss from toad bladder,[373] inhibition of nucleoside transport,[374] antihypertensive activity,[375] spasmolysis,[376] flavor-enhancing activity,[377] inhibition of tryptophan-into-serotonin conversion by synaptosomes,[378] skin sensitization,[379] inhibition of fibroblast cells,[380] porphyrin-inducing activity in chick-embryo cells,[381] anaphylactic reaction,[382] antihistamine activity,[383] inhibition of cell proliferation,[384] effect of organoselenium compounds against dietary liver necrosis,[385] effectiveness of X-ray contrast media,[386] hapten inhibition of hemagglutination by benzylpenicilloyl-specific antibodies,[387] inhibition of prostaglandin biosynthesis,[388] and increased contraction of locust retractor muscle caused by ROH.[389] Lien *et al.*[390] have formulated a QSAR for the penetration of a wide variety of antibiotics and sulfa drugs into the prostate fluid and into milk in whole-animal systems.

An unusual publication is that of H. L. Holmes.[391] This two-volume (1500 pages) report summarizes QSAR work of about 15 years on several hundred heteroenoid and morphine congeners. Several biological tests have been employed and a variety of physicochemical constants have been determined for most of the several hundred compounds studied. Many partition coefficients in the octanol/water as well as in the cyclohexane/water system have been reported. Although a large number of correlation equations are reported, they do not give a consistent overview of the SAR. This very large data set deserves further study.

9.7.13. Pesticides

Many of the more fundamental QSAR with enzymes (e.g., cholinesterase) are directed toward the development of better herbicides and insecticides. A number of whole-organism studies have also been made. Correlation equations have been found for insecticidal activity of alkyl phenyl phosphates,[392] for biodegradable DDT analogs acting on mice and insects,[393] toxicity of phenols to *Daphnia magna*,[394] herbicidal activity of salts of amino-4-benzylpiperidine-4-carboxylic acids,[395] and the herbicidal activity of diphenylmethanes.[396]

In an interesting study of the toxicity of cinnamate esters against rice and barnyard grass, two equations were formulated and then combined to obtain a single equation for estimating selective toxicity.[397]

9.7.14. Cancer

Correlation analysis is beginning to prove of value in the study of cancer chemotherapy.[83,188,398–410] The most interesting application so far is that to nitrosoureas of the structure $ClCH_2CH_2N(NO)CONHR$, where R is an alkyl or substituted alkyl group, or is often cyclohexyl. The QSAR of a set of these congeners acting against L1210 leukemia in mice yielded a simple parabolic relationship with log P.[398] A similar set gave a parabolic relationship between LD_{10} and log P. Log P_0 for the toxicity curve (LD_{10}) was 0.4, while log P_0 for antitumor activity was −0.6. These curves suggested that more hydrophilic nitrosoureas would be slightly more potent and, in addition, less toxic. Preliminary results support this prediction[406,409] and bring out the importance of developing QSAR for both efficacy and toxicity in the search for more effective drugs. A similar relationship of the activity of nitrosoureas has been more firmly established for Lewis lung carcinoma.[403] Wheeler *et al.*[404] have developed a more elaborate QSAR for the nitrosoureas and have shown that in addition to lipophilic character, the carbamoylating and alkylating action are also important determinants of biological activity.

The major effort in cancer chemotherapy over the past 20 years has been in the rapid screening of compounds for activity and in the development of suitable test systems. The problem of suitable animal models for predicting drug activity in humans is far from solved, but enough experience has been gained from the screening of a random selection of a quarter-million compounds to know that it is unlikely that we shall find an antibiotic against cancer by such means. Moreover antitumor activity is not as rare a property of organic compounds as was once thought. Higher-quality testing of fewer, more carefully designed compounds, and the application of the QSAR paradigm is bound to increase our understanding of antitumor-drug design.

9.7.15. Environmental Effects

During the last century, chemists have been busy isolating (and concentrating), as well as creating, millions of new compounds with relatively little concern for the effect of such substances on themselves or their environment. As better statistical and analytical techniques have evolved in the biomedical sciences, it has become all too clear that the toxicity of many simple compounds to plants, animals, and humans can be severe over a long period of low-level contact. Suddenly we are faced with the enormous problem of testing the known compounds, as well as the several hundred thousand new compounds discovered annually, for a wide variety of toxic effects. The size of the problem is staggering when it is considered that the testing of one compound for carcinogenic activity now costs \$75,000. Testing only the 600 new compounds put into commercial production each year is a large project. What to do about all the known organic compounds boggles the imagination.

Our vast search for new drugs, pesticides, plastics, etc. has opened a Pandora's box, releasing a multitude of structures, long familiar and seemingly benign, which are now changing into ominous spirits lurking within ourselves as well as our surroundings.

A much larger effort must be made to understand the relationship between chemical structure and biological activity. The QSAR paradigm can be of help in limiting toxicity studies to well-selected congeners; to do toxicity studies on hundreds of thousands of compounds is out of the question. It has proved feasible to formulate meaningful QSAR[204] even in the area of carcinogenic activity with the "noisy" test data at hand. Better data will yield better QSAR, which can then suggest which compounds are most likely to be carcinogenic.

It has been apparent for some time that the lipophilicity of compounds has a strong influence on how they distribute themselves in the environment. Neely and his colleagues have recently shown[411] that bioaccumulation in trout can be quantitatively related to $\log P$. The partition coefficient is a parameter that should be determined for any new compound destined for commercial production.

9.8. *Conclusions*

The above report has been optimistic in tone, some might say evangelistic! We must not lose sight of the fact that there is a *long way* to go before we reach the stage of plugging in constants to a standard equation to get valuable solutions. We have only a few years' background in a modest number of laboratories by scientists studying QSAR as a sideline. As stressed in the Introduction, this is a most complex endeavor, best undertaken by a group having a variety of specialities. At present we are formulating QSAR based on, at most, a few hundred compounds; there is no reason why this cannot be extended to thousands to constitute a most valuable method of molecular bookkeeping, leaving completely aside the forecasting ability of QSAR.

By persistent chipping away, some of the more formidable problems are yielding. For example, a topic that always comes up for discussion is, "Can QSAR deal with the divergent activity of stereoisomers?" A first step in answering this question has been the derivation of a single equation which correlates the binding of both (+)- and (−)-acylamino-acid amides to α-chymotrypsin.[123] Enough background in physical organic chemistry has accumulated to enable us to do meaningful work on almost any data set having a tenfold or greater range in activity.

The formulation of individual equations is only the first step in the global QSAR problem. Keeping account of these equations and especially their interrelationships is vital to the development of biomedicinal chemistry. Some few first steps have been taken in this direction.[31,412]

In summary, while much of the work cited above cannot be called elegant, taken altogether it should be enough to assure anyone considering entering the field that important and satisfying results can be obtained, even with our present underdeveloped techniques.

References

1. A. Cammarata and K. S. Rogers, in *Advances in Linear Free Energy Relationships*, Chap. 9, N. B. Chapman and J. Shorter, eds. (Plenum Press, London, 1972).
2. J. F. Kirsch, in *Advances in Linear Free Energy Relationships*, Chap. 8, N. B. Chapman and J. Shorter, eds. (Plenum Press, London, 1972).
3. M. S. Tute, *Adv. Drug. Res.*, **6**, 1 (1971).
4. B. Duperray, *Chim. Therap.*, **6**, 305 (1971).
5. R. Franke and P. Oehme, *Pharmazie*, **28**, 489 (1973).
6. W. J. Dunn III, *Ann. Rep. Med. Chem.*, **8**, 313 (1973).
7. C. Hansch, in *Structure–Activity Relationships*, Vol. 1, Chap. 5, C. F. Cavallito, ed. (Pergamon Press, London, 1973).
8. *Adv. Chem. Series*, #114 (American Chemical Society, 1972).
9. W. P. Purcell, G. E. Bass, and J. M. Clayton, *Strategy of Drug Design* (Wiley–Interscience, New York, 1973).
10. A. Verloop in *Drug Design*, Vol. III, Chap. 2, E. J. Ariëns, ed. (Academic Press, New York, 1972).
11. C. Hansch, in *Relations Structure–Activité*, p. 9 (Société de Chimie Thérapeutique, Paris, 1974).
12. C. Hansch, *Adv. Pharmacol. Chemotherap.*, **13**, 45 (1975).
13. L. B. Kier, *Molecular Orbital Theory in Drug Research* (Academic Press, New York, 1971).
14. J. P. Green, C. L. Johnson, and S. Kang, *Ann. Rev. Pharmacol.*, **14**, 319 (1974).
15. J. J. Kaufman and W. S. Koski, in *Drug Design*, Vol. V, Chap. 6, E. J. Ariëns, ed. (Academic Press, New York, 1976).
16. R. E. Christoffersen, *Cancer Chemother. Rep.*, Part 2, **4** (4), 247 (1974).
17. B. R. Kowalski and C. F. Bender, *J. Amer. Chem. Soc.*, **96**, 916 (1974).
18. A. J. Stuper and P. C. Jurs, *J. Amer. Chem. Soc.*, **97**, 182 (1975).
19. P. C. Jurs and T. L. Isenhour, *Chemical Applications of Pattern Recognition* (Wiley–Interscience, New York, 1975).
20. K. C. Chu, R. J. Feldmann, M. B. Shapiro, G. F. Hazard, Jr., and R. I. Geran, *J. Med. Chem.*, **18**, 539 (1975).
21. K. C. Chu, *Analyt. Chem.*, **46**, 1181 (1974).
22. S. H. Unger, *Cancer Chemotherap. Rep.*, Part 2, **4** (4), 45 (1974).
23. S. H. Unger and C. Hansch, *J. Med. Chem.*, **16**, 745 (1973), especially footnote on p. 746.
24. C. Hansch, S. H. Unger, and A. B. Forsythe, *J. Med. Chem.*, **16**, 1217 (1973).

25. Y. C. Martin, J. B. Holland, C. H. Jarboe, and N. Plotnikoff, *J. Med. Chem.*, **17**, 409 (1974).
26. M. L. Weiner and P. H. Weiner, *J. Med. Chem.*, **16**, 655 (1973).
27. S. S. Schiffman, *Science*, **185**, 112 (1974).
28. T. M. Bustard, *J. Med. Chem.*, **17**, 777 (1974).
29. N. J. Santora and K. Auyang, *J. Med. Chem.*, **18**, 959 (1975).
30. P. N. Craig, *J. Med. Chem.*, **14**, 680, 1251 (1971).
31. C. Hansch, *J. Med. Chem.*, **19**, 1 (1976).
32. A. Cammarata, R. C. Allen, J. K. Seydel, and E. Wempe, *J. Pharm. Sci.*, **59**, 1496 (1970).
33. D. Hellenbrecht, B. Lemmer, G. Wiethold, and H. Grobecker, *N. S. Arch. Pharmakol.*, **277**, 211 (1973).
34. J. G. Topliss, *J. Med. Chem.*, **15**, 1006 (1972).
35. J. G. Topliss and Y. C. Martin, in *Drug Design*, Vol. V, Chap. 1, E. J. Ariëns, ed. (Academic Press, New York, 1975).
36. R. Wootton, R. Cranfield, G. C. Sheppey, and P. J. Goodford, *J. Med. Chem.*, **18**, 607 (1975).
37. T. Fujita, *Adv. Chem. Series*, No. 108, p. 81 (American Chemical Society, 1971).
38. G. Redl, R. D. Cramer, and C. E. Berkoff, *Chem. Soc. Rev.*, **3**, 273 (1974).
39. C. Hansch, M. Yoshimoto, and M. Doll, *J. Med. Chem.*, **19**, 1089 (1976).
40. L. H. Sarett, *Research Management*, **18** (1974).
41. A. Spinks, *Chem. Ind.* (London), 885 (1973).
42. A. Leo, C. Hansch, and D. Elkins, *Chem. Rev.*, **71**, 525 (1971).
43. A. Leo, Pomona College Medicinal Chemistry Project.
44. C. Hansch, A. Leo, and D. Nikaitani, *J. Org. Chem.*, **37**, 3090 (1972).
45. A. Cammarata and K. S. Rogers, *J. Med. Chem.*, **14**, 269 (1971).
46. S. S. Davis, *Experientia*, **26**, 671 (1970).
47. F. M. Plakogiannis, E. J. Lien, C. Harris, and J. A. Biles, *J. Pharm. Sci.*, **59**, 197 (1970).
48. S. S. Davis, T. Higuchi, and J. H. Rytting, *J. Pharm. Pharmacol.*, **24** (Suppl.), 30P (1972).
49. R. Vochten and G. Pétré, *Bull. Soc. chim. belges*, **81**, 583 (1972).
50. M. J. Harris, T. Higuchi, and J. H. Rytting, *J. Phys. Chem.*, **77**, 2694 (1973).
51. S. S. Davis, *J. Pharm. Pharmacol.*, **25**, 769 (1973).
52. S. S. Davis, *J. Pharm. Pharmacol.*, **25**, 293 (1973).
53. S. S. Davis, *J. Pharm. Pharmacol.*, **25**, 1 (1973).
54. S. S. Davis, *J. Pharm. Pharmacol.*, **25**, 982 (1973).
55. S. Inoue, A. Ogino, M. Kise, M. Kitano, S. Tsuchiya, and T. Fujita, *Chem. Pharm. Bull.* (*Tokyo*), **22**, 2064 (1974).
56. E. Tomlinson and J. C. Dearden, *J. Chromatog.*, **106**, 481 (1975).
57. G. L. Flynn, *J. Pharm. Sci.*, **60**, 345 (1971).
58. I. Moriguchi, *Chem. Pharm. Bull.* (*Tokyo*), **23**, 247 (1975).
59. A. Leo, P. Y. C. Jow, C. Silipo, and C. Hansch, *J. Med. Chem.*, **18**, 865 (1975).
60. R. B. Hermann, *J. Phys. Chem.*, **76**, 2754 (1972).
61. R. B. Hermann, *J. Phys. Chem.*, **79**, 163 (1975).
62. G. L. Amidon, S. H. Yalkowsky, and S. Leung, *J. Pharm. Sci.*, **63**, 1858 (1974).
63. G. L. Amidon, S. H. Yalkowsky, S. T. Anik, and S. C. Valvani, *J. Phys. Chem.*, **79**, 2239 (1975).
64. J. A. Reynolds, D. B. Gilbert, and C. Tanford, *Proc. Nat. Acad. Sci., U.S.A.*, **71**, 2925 (1974).
65. A. Leo, C. Hansch, and P. Y. C. Jow, *J. Med. Chem.*, **19**, 611 (1976).
66. R. N. Smith, C. Hansch, and M. M. Ames, *J. Pharm. Sci.*, **64**, 599 (1975).
67. P. Seiler, *European J. Med. Chem.—Chim. Therap.*, **9**, 473 (1974).

68. N. Kurihara, M. Uchida, T. Fujita, and M. Nakajima, *Pestic. Biochem. Physiol.*, **2**, 383 (1973).
69. G. G. Nys and R. F. Rekker, *European J. Med. Chem.—Chim. Therap.*, **8**, 521 (1973).
70. G. G. Nys and R. F. Rekker, *European J. Med. Chem.—Chim. Therap.*, **9**, 361 (1974).
71. G. L. Biagi, A. M. Barbaro, O. Gandolfi, M. C. Guerra, and G. Cantelli-Forti, *J. Med. Chem.*, **18**, 873 (1975), and other references cited therein.
72. J. K. Seydel, H. Ahrens, and W. Losert, *J. Med. Chem.*, **18**, 234 (1975).
73. A. E. Bird and A. C. Marshall, *J. Chromatog.*, **63**, 313 (1971).
74. M. Kuchař, B. Brůnová, V. Rejholec, and V. Rábek, *J. Chromatog.*, **92**, 381 (1974).
75. G. L. Biagi, A. M. Barbaro, M. C. Guerra, G. Cantelli–Forti, and M. E. Fracasso, *J. Med. Chem.*, **17**, 28 (1974).
76. M. C. Bonjean, J. Alary, and M. C. Luu Duc, *Chim. Therap.*, **8**, 93 (1973).
77. J. Pla-Delfina, J. Moreno, J. Duran, and A. del Pozo, *J. Pharmacokinet. Biopharm.*, **3**, 115 (1975).
78. W. J. Dunn III and C. Hansch, *Chem.-Biol. Interactions*, **9**, 75 (1974).
79. D. Brown and D. Woodcock, *J. Chromatog.*, **105**, 33 (1975).
80. G. L. Biagi, O. Gandolfi, M. C. Guerra, A. M. Barbaro, and G. Cantelli-Forti, *J. Med. Chem.*, **18**, 868 (1975).
81. T. Miyagishima, *Chem. Pharm. Bull.* (Tokyo), **22**, 2288 (1974).
82. V. Pliška and T. Barth, *Coll. Czech. Chem. Comm.*, **35**, 1576 (1970).
83. B. F. Cain, R. N. Seelye, and G. J. Atwell, *J. Med. Chem.*, **17**, 922 (1974).
84. B. Reiff, S. M. Lambert, and I. L. Natoff, *Arch. Intern. Pharmacodyn. Therap.*, **192**, 48 (1971).
85. J. P. Hellot, G. Lauquin, and H. Pacheco, *Chim. Therap.*, **5**, 65 (1970).
86. J. D. Turnbull, G. L. Biagi, A. J. Merola, and D. G. Cornwell, *Biochem. Pharmacol.*, **20**, 1383 (1971).
87. B. Reiff, *Brit. J. Pharmacol.*, **40**, 135P (1970).
88. G. Pelizza, G. C. Lancini, G. C. Allievi, and G. G. Gallo, *Farmaco, Ed. Sci.*, **28**, 298 (1973).
89. H. W. Kosterlitz, F. M. Leslie, and A. A. Waterfield, *European J. Pharmacol.*, **32**, 10 (1975).
90. J. K. Seydel and R. Abrecht, *European J. Med. Chem.—Chim. Therap.*, **10**, 378 (1975).
91. K. C. James, P. J. Nichols, and G. T. Richards, *European J. Med. Chem.–Chim. Therap.*, **10**, 55 (1975).
92. J. Schmutz, *Arzneim.-Forsch.*, **25**, 712 (1975).
93. W. Morozowich, *J. Chromatog. Sci.*, **12**, 453 (1974).
94. R. M. Carlson, R. E. Carlson, and H. L. Kopperman, *J. Chromatog.*, **107**, 219 (1975).
95. J. M. McCall, *J. Med. Chem.*, **18**, 549 (1975).
96. P. J. Twitchett and A. C. Moffat, *J. Chromatog.*, **111**, 149 (1975).
97. W. J. Haggerty and E. A. Murill, *Res. Develop.*, **25**, 30 (1974).
98. J. J. Kaufman, N. M. Semo, and W. S. Koski, *J. Med. Chem.*, **18**, 647 (1975).
99. P. L. Huyskens and J. J. Tack, *J. Phys. Chem.*, **79**, 1654 (1975).
100. J. Hine and P. K. Mookerjee, *J. Org. Chem.*, **40**, 292 (1975).
101. F. E. Norrington, R. M. Hyde, S. G. Williams, and R. Wootton, *J. Med. Chem.*, **18**, 604 (1975).
102. R. W. Taft, Jr. and I. C. Lewis, *J. Amer. Chem. Soc.*, **80**, 2436 (1958).
103. R. W. Taft, Jr. and I. C. Lewis, *J. Amer. Chem. Soc.*, **81**, 5343 (1959).
104. S. K. Dayal, S. Ehrenson, and R. W. Taft, *J. Amer. Chem. Soc.*, **94**, 9113 (1972).
105. S. Ehrenson, R. T. C. Brownlee, and R. W. Taft, *Progr. Phys. Org. Chem.*, **10**, 1 (1973).
106. C. G. Swain and E. C. Lupton, *J. Amer. Chem. Soc.*, **90**, 4328 (1968).
107. C. Hansch, A. Leo, S. H. Unger, K. H. Kim, D. Nikaitani, and E. J. Lien, *J. Med. Chem.*, **16**, 1207 (1973).

108. N. Kakeya, N. Yata, A. Kamada, and M. Aoki, *Chem. Pharm. Bull.* (*Tokyo*), **17**, 2558 (1969).
109. O. Exner, *Coll. Czech. Chem. Comm.*, **31**, 65 (1966).
110. O. Exner and J. Lakomý, *Coll. Czech. Chem. Comm.*, **35**, 1371 (1970).
111. M. Sjöström and S. Wold, *Chem. Scripta*, **6**, 114 (1974).
112. C. Hansch, *J. Med. Chem.*, **13**, 964 (1970).
113. A. P. G. Kieboom, *Tetrahedron*, **28**, 1325 (1972).
114. R. W. Taft, in *Steric Effects in Organic Chemistry*, Chap. 13, M. S. Newman, ed. (Wiley, New York, 1956).
115. S. H. Unger and C. Hansch, *Progr. Phys. Org. Chem.*, **12**, 91 (1976).
116. A. Verloop, W. Hoogenstraaten, and J. Tipker, in *Drug Design*, Vol. VII, Chap. 4, E. J. Ariëns, ed. (Academic Press, New York, 1976).
117. Z. Simon and Z. Szabadai, *Studia Biophys.*, **39**, 123 (1973).
118. Z. Simon, *Angew. Chem.*, **86**, 802 (1974); EE, **13**, 719 (1974).
119. Z. Simon, *Studia Biophys.*, **51**, 49 (1975).
120. P. Pratesi, L. Villa, and E. Grana, *Farmaco, Ed. Sci.*, **30**, 315 (1975).
121. L. Pauling and D. Pressman, *J. Amer. Chem. Soc.*, **67**, 1003 (1945).
122. C. Hansch and D. Calef, *J. Org. Chem.*, **41**, 1240 (1976).
123. M. Yoshimoto and C. Hansch, *J. Org. Chem.*, **41**, 2269 (1976).
124. M. Yoshimoto and C. Hansch, *J. Med. Chem.*, **19**, 71 (1976).
125. C. Silipo and C. Hansch, *J. Med. Chem.*, **19**, 62 (1976).
126. C. Silipo and C. Hansch, *J. Amer. Chem. Soc.*, **97**, 6849 (1975).
127. M. Yoshimoto, C. Hansch, and P. Y. C. Jow, *Chem. Pharm. Bull.* (*Tokyo*), **23**, 437 (1975).
128. C. Silipo and C. Hansch, *Farmaco, Ed. Sci.*, **30**, 35 (1975).
129. C. Silipo and C. Hansch, *Mol. Pharmacol.*, **10**, 954 (1974).
130. C. Hansch and E. Coats, *J. Pharm. Sci.*, **59**, 731 (1970).
131. J. Fukanaga, C. Hansch, and E. Steller, *J. Med. Chem.*, **19**, 605 (1976).
132. F. Franks, in *Water*, Vol. 4, Chap. 1, F. Franks, ed. (Plenum Press, New York, 1975).
133. C. D. Hubbard and J. F. Kirsch, *Biochemistry*, **11**, 2483 (1972).
134. J. C. McGowan, *J. Appl. Chem.*, **4**, 41 (1954).
135. P. Ahmad, C. A. Fyfe, and A. Mellors, *Biochem. Pharmacol.*, **24**, 1103 (1975).
136. L. J. Mullins, *Chem. Rev.*, **54**, 289 (1954).
137. B. Hetnarski and R. D. O'Brien, *Biochemistry*, **12**, 3883 (1973).
138. B. Hetnarski and R. D. O'Brien, *J. Med. Chem.*, **18**, 29 (1975).
139. B. Hetnarski and R. D. O'Brien, *J. Agric. Food Chem.*, **23**, 709 (1975).
140. M. Charton, *J. Org. Chem.*, **31**, 2991 (1966).
141. H. Sakurai, *J. Org. Chem.*, **35**, 2807 (1970).
142. L. A. Singer and D. J. Cram, *J. Amer. Chem. Soc.*, **85**, 1080 (1963).
143. G. Aloisi, G. Cauzzo, and U. Mazzucato, *Trans. Faraday Soc.*, **63**, 1858 (1967).
144. N. Kucharczyk, B. Kakáč, and V. Horák, *Coll. Czech. Chem. Comm.*, **34**, 2959 (1969).
145. M. H. Hussain and E. J. Lien, *J. Med. Chem.*, **14**, 138 (1971).
146. M. S. Tute, *J. Med. Chem.*, **13**, 48 (1970).
147. H. Nogami, T. Nagai, and S. Wada, *Chem. Pharm. Bull.* (*Tokyo*), **18**, 348 (1970).
148. H. Nogami, T. Nagai, and N. Nambu, *Chem. Pharm. Bull.* (*Tokyo*), **18**, 1643 (1970).
149. H. Umeyama, T. Nagai, H. Nogami, and T. Oguma, *Chem. Pharm. Bull.* (*Tokyo*), **19**, 412 (1971).
150. R. Franke, *Acta Biol. Med. Germ.*, **25**, 789 (1970).
151. K. Nishimura, T. Kawata, K. Asada, and M. Nakajima, *Agric. and Biol. Chem.* (*Japan*), **39**, 867 (1975).
152. W. A. Seth-Paul and A. van Duyse, *Spectrochim. Acta*, **28A**, 211 (1972).
153. J. P. Tollenaere, *Chim. Therap.*, **6**, 88 (1971).

154. J. P. Tollanaere, *J. Med. Chem.*, **16**, 791 (1973).
155. A. Rastelli, P. G. De Benedetti, G. G. Battistuzzi, and A. Albasini, *J. Med. Chem.*, **18**, 963 (1975).
156. T. K. Lin, Y. W. Chien, R. R. Dean, J. E. Dutt, H. W. Sause, C. H. Yen, and P. K. Yonan, *J. Med. Chem.*, **17**, 751 (1974).
157. T. Okano, J. Maenosono, T. Kano, and I. Onoda, *Gann*, **64**(3), 227 (1973); *Chem. Abs.*, **79**, 74695 (1973).
158. N. Kakeya, N. Yata, A. Kamada, and M. Aoki, *Chem. Pharm. Bull.* (*Tokyo*), **18**, 191 (1970).
159. T. C. Bruice, N. Kharasch, and R. J. Winzler, *Arch. Biochem. Biophys.*, **62**, 305 (1956).
160. S. M. Free, Jr. and J. W. Wilson, *J. Med. Chem.*, **7**, 395 (1964).
161. P. N. Craig, *Adv. Chem. Series*, #114, p. 115 (*American Chemical Society*, 1972).
162. R. Reiner, *Arzneim.-Forsch.*, **21**, 2032 (1971).
163. D. R. Hudson, G. E. Bass, and W. P. Purcell, *J. Med. Chem.*, **13**, 1184 (1970).
164. T. Fujita and T. Ban, *J. Med. Chem.*, **14**, 148 (1971).
165. P. N. Craig, *J. Med. Chem.*, **15**, 144 (1972).
166. R. T. Buckler, *J. Med. Chem.*, **15**, 578 (1972).
167. W. H. Lawrence, G. E. Bass, W. P. Purcell, and J. Autian, *J. Dent. Res.*, Part 2, **51**, 526 (1972).
168. B. Tinland, C. Decoret, and J. Badin, *Res. Comm. Chem. Pathol. Pharmacol.*, **4**, 131 (1972).
169. P. N. Craig and C. Hansch, *J. Med. Chem.*, **16**, 661 (1973).
170. H. Cousse, G. Mouzin, and L. D. D'Hinterland, *Chim. Therap.*, **8**, 466 (1973).
171. C. E. Berkoff, P. N. Craig, B. P. Gordon, and C. Pellerano, *Arzneim.-Forsch.*, **23**, 830 (1973).
172. H. Orzalesi, J. Castel, P. Fulcrand, G. Bergé, A.-M. Noël, and P. Chevallet, *Compt. rend., Ser. C*, **279**, 709 (1974).
173. A. Cammarata and T. M. Bustard, *J. Med. Chem.*, **17**, 981 (1974).
174. L. J. Schaad, R. H. Werner, L. Dillon, L. Field, and C. E. Tate, *J. Med. Chem.*, **18**, 344 (1975).
175. J. Thomas, C. E. Berkoff, W. B. Flagg, J. J. Gallo, R. F. Haff, C. A. Pinto, C. Pellerano, and L. Savini, *J. Med. Chem.*, **18**, 245 (1975).
176. G. Grassy, A. Terol, A. Belly, Y. Robbe, J. P. Chapat, R. Granger, M. Fatome, and L. Andrieu, *European J. Med. Chem.-Chim. Therap.*, **10**, 14 (1975).
177. B. Tinland, *Farmaco, Ed. Sci.*, **30**, 423 (1975).
178. F. Darvas, Z. Budai, L. Petócz, and I. Kosóczky, *Res. Comm. Chem. Pathol. Pharmacol.*, **12**, 243 (1975).
179. C. Hansch, C. Silipo, and E. E. Steller, *J. Pharm. Sci.*, **64**, 1186 (1975).
180. C. Daniel and F. S. Wood, *Fitting Equations to Data*, pp. 55, 169, 203 (Wiley-Interscience, New York, 1971).
181. H. Bechgaard and C. Lund-Jensen, *European J. Med. Chem.-Chim. Therap.*, **10**, 103 (1975).
182. K. Bowden and K. R. H. Wooldridge, *Biochem. Pharmacol.*, **22**, 1015 (1973).
183. Y. C. Martin, P. H. Jones, T. J. Perun, W. E. Grundy, S. Bell, R. R. Bower, and N. L. Shipkowitz, *J. Med. Chem.*, **15**, 635 (1972).
184. Y. C. Martin and K. R. Lynn, *J. Med. Chem.*, **14**, 1162 (1971).
185. R. W. Wald and G. Feuer, *J. Med. Chem.*, **14**, 1081 (1971).
186. C. Hansch and W. R. Glave, *J. Med. Chem.*, **15**, 112 (1972).
187. C. Hansch and M. Yoshimoto, *J. Med. Chem.*, **17**, 1160 (1974).
188. C. Hansch, R. N. Smith, and R. Engle, in *Pharmacological Basis of Cancer Chemotherapy*, p. 215 (Williams and Wilkins, Baltimore, Maryland, 1975).
189. F. R. Quinn, J. S. Driscoll, and C. Hansch, *J. Med. Chem.*, **18**, 332 (1975).

190. C. Hansch, A. Vittoria, C. Silipo, and P. Y. C. Jow, *J. Med. Chem.*, **18**, 546 (1975).
191. C. Hansch and C. Silipo, *J. Med. Chem.*, **17**, 661 (1974).
192. T. Higuchi and S. S. Davis, *J. Pharm. Sci.*, **59**, 1376 (1970).
193. J. W. McFarland, *J. Med. Chem.*, **13**, 1192 (1970).
194. A. Cammarata, S. J. Yau, and K. S. Rogers, *J. Med. Chem.*, **14**, 1211 (1971).
195. A. Cammarata, *J. Med. Chem.*, **15**, 573 (1972).
196. S. H. Yalkowsky and G. L. Flynn, *J. Pharm. Sci.*, **62**, 210 (1973).
197. C. Hansch and J. M. Clayton, *J. Pharm. Sci.*, **62**, 1 (1973).
198. T. K. Lin, *J. Med. Chem.*, **17**, 151 (1974).
199. T. K. Lin, *J. Med. Chem.*, **17**, 749 (1974).
200. G. L. Flynn, S. H. Yalkowsky, and N. D. Weiner, *J. Pharm. Sci.*, **63**, 300 (1974).
201. W. P. Purcell, *European J. Med. Chem.*, **10**, 335 (1975).
202. G. M. Furnival and R. W. Wilson, Jr., *Technometrics*, **16**, 499 (1974).
203. A. E. Hoerl and R. W. Kennard, *Technometrics*, **12**, 55 (1970).
204. C. Hansch and T. Fujita, *J. Amer. Chem. Soc.*, **86**, 1616 (1964).
205. J. T. Penniston, L. Beckett, D. L. Bentley, and C. Hansch, *Mol. Pharmacol.*, **5**, 333 (1969).
206. G. L. Flynn and S. H. Yalkowsky, *J. Pharm. Sci.*, **61**, 838 (1972).
207. S. H. Yalkowsky and G. L. Flynn, *J. Pharm. Sci.*, **63**, 1276 (1974).
208. G. L. Flynn, S. H. Yalkowsky, and T. J. Roseman, *J. Pharm. Sci.*, **63**, 479 (1974).
209. S. H. Yalkowsky, T. G. Slunick, and G. L. Flynn, *J. Pharm. Sci.*, **63**, 691 (1974).
210. R. M. Hyde, *J. Med. Chem.*, **18**, 231 (1975).
211. C. Hansch, A. R. Steward, S. M. Anderson, and D. Bentley, *J. Med. Chem.*, **11**, 1 (1968).
212. E. Druckrey, H. Schwarz, and H. Leditschke, *Chim. Therap.*, **7**, 188 (1972).
213. E. Kutter, A. Herz, H. J. Teshemacher, and R. Hess, *J. Med. Chem.*, **13**, 801 (1970).
214. C. Hansch and E. J. Lien, *J. Med. Chem.*, **14**, 653 (1971).
215. C. Hansch and W. J. Dunn III, *J. Pharm. Sci.*, **61**, 1 (1972).
216. H. Glaser, and J. Krieglstein, *N. S. Arch. Pharmakol.*, **265**, 321 (1970).
217. W. E. Kreighbaum, F. A. Grunwald, E. F. Harrison, J. A. LaBudde, and A. A. Larsen, *J. Med. Chem.*, **13**, 247 (1970).
218. M. Yamazaki, N. Kakeya, T. Morishita, A. Kamada, and M. Aoki, *Chem. Pharm. Bull. (Tokyo)*, **18**, 708 (1970).
219. J. Krieglstein, W. Meiler, and J. Staab, *Biochem. Pharmacol.*, **21**, 985 (1972).
220. J. M. Vandenbelt, C. Hansch, and C. Church, *J. Med. Chem.*, **15**, 787 (1972).
221. T. Fujita, *J. Med. Chem.*, **15**, 1049 (1972).
222. W. Müller and U. Wollert, *N. S. Arch. Pharmakol.*, **278**, 301 (1973).
223. O. Gandolfi, A. M. Barbaro, and G. L. Biagi, *Experientia*, **29**, 689 (1973).
224. J. Krieglstein, *Arzneim.-Forsch.*, **23**, 1527 (1973).
225. H. Terada, S. Muraoka, and T. Fujita, *J. Med. Chem.*, **17**, 330 (1974).
226. Y. W. Chien, M. J. Akers, P. K. Yonan, *J. Pharm. Sci.*, **64**, 1632 (1975).
227. D. Gilbert, P. J. Goodford, F. E. Norrington, B. C. Weatherley, and S. G. Williams, *Brit. J. Pharmacol.*, **55**, 117 (1975).
228. Y. W. Chien, H. J. Lambert, and T. K. Lin, *J. Pharm. Sci.*, **64**, 961 (1975).
229. W. J. Dunn III, *J. Med. Chem.*, **16**, 484 (1973).
230. W. J. Dunn III, *J. Pharm. Sci.*, **62**, 1575 (1973).
231. G. Van den Berg, R. F. Rekker, and W. Th. Nauta, *European J. Med. Chem.–Chim. Therap.*, **10**, 408 (1975).
232. R. D. Poretz and I. J. Goldstein, *Biochem. Pharmacol.*, **20**, 2727 (1971).
233. F. G. Loontiens, J. P. van Wauve, R. De Gussem, and C. K. De Bruyne, *Carbohydrate Res.*, **30**, 51 (1973).
234. M. Tichý, *Coll. Czech. Chem. Comm.*, **39**, 935 (1974).

225. R. Franke and W. Schmidt, *Acta Biol. Med. Germ.*, **31**, 273 (1973).
236. M. Yoshimoto, K. N. von Kaulla, and C. Hansch, *J. Med. Chem.*, **18**, 950 (1975).
237. E. J. Lien, *Drug. Intell. Clin. Pharm.*, **4**, 7 (1970).
238. E. J. Lien, R. T. Koda, and G. L. Tong, *Drug. Intell. Clin. Pharm.*, **5**, 38 (1971).
239. K. A. Herzog and J. Swarbrick, *J. Pharm. Sci.*, **60**, 1666 (1971).
240. J. C. Dearden and E. Tomlinson, *J. Pharm. Pharmacol.*, **23**, 73S (1971).
241. N. F. H. Ho and W. I. Higuchi, *J. Pharm. Sci.*, **60**, 537 (1971).
242. H. Machleidt, S. Roth, and P. Seeman, *Biochim. Biophys. Acta*, **255**, 178 (1972).
243. B. C. Sherrill and J. M. Dietschy, *J. Membrane Biol.*, **23**, 367 (1975).
244. A. D. M. Glass and J. Dunlop, *Plant Physiol.*, **54**, 855 (1974).
245. J. B. Houston, D. G. Upshall, and J. W. Bridges, *J. Pharmacol. Exptl. Therap.*, **195**, 67 (1975).
246. R. L. Preston, J. F. Schaeffer, and P. F. Curran, *J. Gen. Physiol.*, **64**, 443 (1974).
247. C. Hansch and W. R. Glave, *Mol. Pharmacol.*, **7**, 337 (1971).
248. M. Rabinovitch and M. J. Destefano, *Exptl. Cell Res.*, **88**, 153 (1974).
249. N. Kurihara, M. Uchida, T. Fujita, and N. Nakajima, *Pest. Biochem. Physiol.*, **4**, 12 (1974).
250. M. S. Roberts, E. J. Triggs, and R. A. Anderson, *Nature*, **257**, 225 (1975).
251. S. Roth and P. Seeman, *Biochim. Biophys. Acta*, **255**, 207 (1972).
252. P. Seeman, A. Staiman, and M. Chau-Wong, *J. Pharmacol. Exptl. Therap.*, **190**, 123 (1974).
253. P. Seeman, *Pharmacol. Rev.*, **24**, 583 (1972).
254. H. Levitan and J. L. Barker, *Science*, **176**, 1423 (1972).
255. J. L. Barker, *Nature*, **252**, 52 (1974).
256. J. L. Barker and H. Levitan, *Brain Res.*, **67**, 555 (1974).
257. J. L. Barker, *Brain Res.*, **92**, 35 (1975).
258. J. L. Barker and H. Levitan, *J. Pharmacol. Exptl. Therap.*, **193**, 892 (1975).
259. W. R. Glave and C. Hansch, *J. Pharm. Sci.*, **61**, 589 (1972).
260. J. P. Liberti and K. S. Rogers, *Biochim. Biophys. Acta*, **222**, 90 (1970).
261. K. Bowden and R. C. Young, *J. Med. Chem.*, **13**, 225 (1970).
262. J. R. Sanborn and T. R. Fukuto, *J. Agric. Food Chem.*, **20**, 926 (1972).
263. K. J. Chang, R. C. Deth, and D. J. Triggle, *J. Med. Chem.*, **15**, 243 (1972).
264. R. L. Jones, T. R. Fukuto, and R. L. Metcalf, *J. Econ. Entomol.*, **65**, 28 (1972).
265. D. Hellenbrecht and K.-F. Müller, *Experientia*, **29**, 1255 (1973).
266. N. Busch, J. Moleyre, R. Bondivenne, J. Lambert, and C. Labrid, *Chim. Therap.*, **8**, 7 (1973).
267. A. H. Lee and R. L. Metcalf, *Pest. Biochem. Physiol.*, **2**, 408 (1973).
268. A. Rey, *Farmaco, Ed. Sci.*, **28**, 766 (1973).
269. J. K. Seydel and O. Wassermann, *Chim. Therap.*, **8**, 427 (1973).
270. A. Rey, *Farmaco, Ed. Sci.*, **29**, 294 (1974).
271. C. Hansch, *J. Org. Chem.*, **35**, 620 (1970).
272. J. J. Zimmerman and J. E. Goyan, *J. Med. Chem.*, **14**, 1206 (1971).
273. I. V. Berezin, A. V. Levashov, and K. Martinek, *FEBS Lett.*, **7**, 20 (1970).
274. C. Hansch, *J. Org. Chem.*, **37**, 92 (1972).
275. V. N. Dorovska, S. D. Varfolomeyev, N. F. Kazanskaya, A. A. Klyosov, and K. Martinek, *FEBS Lett.*, **23**, 122 (1972).
276. R. N. Smith and C. Hansch, *Biochemistry*, **12**, 4924 (1973).
277. J.-J. Béchet, A. Dupaix, and C. Roucous, *Biochemistry*, **12**, 2566 (1973).
278. J. Fastrez and A. R. Fersht, *Biochemistry*, **12**, 1067 (1973).
279. R. N. Smith, C. Hansch, and T. P. Poindexter, *Physiol. Chem. Phys.*, **6**, 323 (1974).
280. V. N. Dorovska, S. D. Varfolomeyev, and K. Martinek, *Biokhimiya*, **38**, 381 (1973).
281. F. Van Wijnendaele and C. K. De Bruyne, *Carbohydrate Res.*, **14**, 189 (1970).

282. H. Kersters-Hilderson, M. Claeyssens, F. G. Loontiens, A. Kryński, and C. K. De Bruyne, *European J. Biochem.*, **12**, 403 (1970).
283. A. Y. Sun and T. Samorajski, *J. Neurochem.*, **17**, 1365 (1970).
284. H. Hayashi and J. T. Penniston, *Arch. Biochem. Biophys.*, **159**, 563 (1973).
285. M. Uchida, N. Kurihara, T. Fujita, and N. Nakajima, *Pest. Biochem. Physiol.*, **4**, 260 (1974).
286. J. W. Bunting and C. D. Myers, *Canad. J. Chem.*, **51**, 2639 (1973).
287. J. W. Bunting and C. D. Myers, *Canad. J. Chem.*, **52**, 2053 (1974).
288. J. W. Bunting, J. Murphy, C. D. Myers, and G. G. Cross, *Canad. J. Chem.*, **52**, 2648 (1974).
289. N. Hennrich, M. Klockow, H. D. Orth, U. Femfert, P. Cichocki, and K. Jany, *Z. physiol. Chem.*, **354**, 1339 (1973).
290. K. Yamamoto, *J. Biochem. (Japan)*, **76**, 385 (1974).
291. A. Cogoli and G. Semenza, *J. Biol. Chem.*, **250**, 7802 (1975).
292. E. A. Coates, *J. Med. Chem.*, **16**, 1102 (1973).
293. C. Hansch, J. Schaeffer, and R. Kerley, *J. Biol. Chem.*, **247**, 4703 (1972).
294. C. Hansch, K. H. Kim, and R. H. Sarma, *J. Amer. Chem. Soc.*, **95**, 6447 (1973).
295. K. S. Rogers and S. C. Yusko, *J. Biol. Chem.*, **247**, 3671 (1972).
296. E. Druckrey and H. Metzger, *J. Med. Chem.*, **16**, 436 (1973).
297. M. Stockdale and M. J. Selwyn, *European J. Biochem.*, **21**, 416 (1971).
298. Y. C. Martin, W. B. Martin, and J. D. Taylor, *J. Med. Chem.*, **18**, 883 (1975).
299. E.J. Lien, M. Hussain, and G. L. Tong, *J. Pharm. Sci.*, **59**, 865 (1970).
300. T. Fujita, *J. Med. Chem.*, **16**, 923 (1973).
301. H. Orzalesi, J. Castel, P. Fulcrand, G. Bergé, A. M. Noël, and P. Chevallet, *Compt. rend., Ser. C*, **279**, 709 (1974).
302. R. W. Fuller, J. Mills, and M. M. Marsh, *J. Med. Chem.*, **14**, 322 (1971).
303. R. W. Fuller and M. M. Marsh, *J. Med. Chem.*, **15**, 1068 (1972).
304. E. Coats, W. R. Glave, and C. Hansch, *J. Med. Chem.*, **13**, 913 (1970).
305. C. Hansch and N. Cornell, *Arch. Biochem. Biophys.*, **151**, 351 (1972).
306. K. H. Mole, D. M. Shepherd, and J. W. Watkins, *Arch. Intern. Pharmacodyn. Therap.*, **216**, 192 (1975).
307. T. Novinson, J. P. Miller, M. Scholten, R. K. Robins, L. N. Simon, D. E. O'Brien, and R. B. Meyer, Jr., *J. Med. Chem.*, **18**, 460 (1975).
308. P. B. Hulbert, *Mol. Pharmacol.*, **10**, 315 (1974).
309. Y. Pocker and M. W. Beug, *Biochemistry*, **11**, 698 (1972).
310. K. Kumaki, S. Tomioka, K. Kobashi, and J. Hase, *Chem. Pharm. Bull. (Tokyo)*, **20**, 1599 (1972).
311. H. F. Schaeffer, R. N. Johnson, E. Odin, and C. Hansch, *J. Med. Chem.*, **13**, 452 (1970).
312. R. C. Allen, G. L. Carlson, and C. J. Cavallito, *J. Med. Chem.*, **13**, 909 (1970).
313. M. L. Fonda, *Arch. Biochem. Biophys.*, **153**, 763 (1972).
314. R. Franke, E. Gäbler, and P. Oehme, *Acta Biol. Med. Germ.*, **32**, 545 (1974).
315. V. N. Shibaev, G. I. Eliseeva, and N. K. Kochetkov, *Biochim. Biophys. Acta*, **403**, 9 (1975).
316. M. Stockdale and M. J. Selwyn, *European J. Biochem.*, **21**, 565 (1971).
317. S. C. Yusko, E. S. Higgins, and K. S. Rogers, *Proc. Soc. Exp. Biol. Med.*, **141**, 10 (1972).
318. G. Schäfer and D. Bojanowski, *European J. Biochem.*, **27**, 364 (1972).
319. H. Terada and S. Muraoka, *Mol. Pharmacol.*, **8**, 95 (1972).
320. W. Draber, K. H. Büchel, and G. Schäfer, *Z. Naturforsch.*, **27B**, 159 (1972).
321. L. S. Yaguzhinsky, E. G. Smirnova, L. A. Ratnikova, G. M. Kolesova, and I. P. Krasinskaya, *J. Bioenerg.*, **5**, 163 (1973).
322. R. Labbe-Bois, C. Laruelle, and J.-J. Godfroid, *J. Med. Chem.*, **18**, 85 (1975).

323. G. Van den Berg, T. Bultsma, R. F. Rekker, and W. Th. Nauta, *European J. Med. Chem.*, **10**, 242 (1975).
324. R. E. McCarty and C. H. Coleman, *Arch. Biochem. Biophys.*, **141**, 198 (1970).
325. Y. Mukohata, T. Yagi, M. Higashida, and A. Matsuno, *J. Bioenerg.*, **4**, 479 (1973).
326. F. J. Carmichael and Y. Israel, *J. Pharmacol. Exptl. Therap.*, **186**, 253 (1973).
327. D. G. O'Sullivan and C. M. Ludlow, *Experientia*, **28**, 889 (1972).
328. R. Franke, *Acta Biol. Med. Germ.*, **30**, 467 (1973).
329. M. Yamazaki, N. Kakeya, T. Morishita, A. Kamada, and M. Aoki, *Chem. Pharm. Bull.* (*Tokyo*), **18**, 702 (1970).
330. G. L. Biagi, M. C. Guerra, and A. M. Barbaro, *Farmaco, Ed. Sci.*, **25**, 755 (1970).
331. G. L. Biagi, M. C. Guerra, A. M. Barbaro, and M. F. Gamba, *J. Med. Chem.*, **13**, 511 (1970).
332. K. H. Büchel, W. Draber, E. Regel, and M. Plempel, *Arzneim.-Forsch.*, **22**, 1260 (1972).
333. B. J. Broughton, P. Chaplen, W. A. Freeman, P. J. Warren, K. R. H. Wooldridge, and D. E. Wright, *J.C.S. Perkin I*, 857 (1975).
334. D. Nardi, E. Massarani, A. Tajana, R. Cappelletti, and M. Veronese, *Farmaco, Ed. Sci.*, **30**, 727 (1975).
335. C. Hansch, K. Nakamoto, M. Gorin, P. Denisevich, E. R. Garrett, S. M. Heman-Ackah, and C. H. Won, *J. Med. Chem.*, **16**, 917 (1973).
336. E. J. Lien, C. T. Kong, and R. J. Lukens, *Pest. Biochem. Physiol.*, **4**, 289 (1974).
337. W. H. Dekker, H. A. Selling, and J. C. Overeem, *J. Agric. Food Chem.*, **23**, 785 (1975).
338. W. Dittmar, E. Druckrey, and H. Urbach, *J. Med. Chem.*, **17**, 753 (1974).
339. G. E. Bass, D. R. Hudson, J. E. Parker, and W. P. Purcell, *J. Med. Chem.*, **14**, 275 (1971).
340. Y. C. Martin, T. M. Bustard, and K. R. Lynn, *J. Med. Chem.*, **16**, 1089 (1973).
341. P. J. Goodford, F. E. Norrington, W. H. G. Richards, and L. P. Walls, *Brit. J. Pharmacol.*, **48**, 650 (1973).
342. R. Cranfield, P. J. Goodford, F. E. Norrington, W. H. G. Richards, G. C. Sheppey, and S. G. Williams, *Brit. J. Pharmacol.*, **52**, 87 (1974).
343. H. R. Munson, Jr., R. E. Johnson, J. M. Sanders, C. J. Ohnmacht, and R. E. Lutz, *J. Med. Chem.*, **18**, 1232 (1975).
344. E. Kutter, H. Machleidt, W. Reuter, R. Sauter, and A. Wildfeuer, *Arzneim.–Forsch.*, **22**, 1045 (1972).
345. C. A. R. Baxter and H. C. Richards, *J. Med. Chem.*, **14**, 1033 (1971).
346. M. W. Miller, H. L. Howes, Jr., R. V. Kasubick, and A. R. English, *J. Med. Chem.*, **13**, 849 (1970).
347. E. J. Lien, M. Hussain, and M. P. Golden, *J. Med. Chem.*, **13**, 623 (1970).
348. E. J. Lien, *J. Med. Chem.*, **13**, 1189 (1970).
349. E. J. Lien, G. L. Tong, J. T. Chou, and L. L. Lien, *J. Pharm. Sci.*, **62**, 246 (1973).
350. M. P. Breen, E. M. Bojanowski, R. J. Cipolle, W. J. Dunn III, E. Frank, and J. E. Gearien, *J. Pharm. Sci.*, **62**, 847 (1973).
351. C. F. Barfknecht, D. E. Nichols, and W. J. Dunn III, *J. Med. Chem.*, **18**, 208 (1975).
352. W. J. Dunn III and J. P. Bederka, Jr., *Res. Comm. Chem. Pathol. Pharmacol.*, **7**, 275 (1974).
353. R. E. Tessel, J. H. Woods, R. E. Counsell, and M. Lu, *J. Pharmacol. Exptl. Therap.*, **192**, 310 (1975).
354. J. C. Dearden and E. Tomlinson, *J. Pharm. Pharmacol.*, **24**, 165P (1972).
355. J. Zaagsma and W. Th. Nauta, *J. Med. Chem.*, **17**, 507 (1974).
356. S. Feldman, M. de Francisco, and P. J. Cascella, *J. Pharm. Sci.*, **64**, 1713 (1975)).
357. M. A. Dinno, T. L. Holloman, and M. Schwartz, *Proc. Soc. Exp. Biol. Med.*, **141**, 397 (1972).

358. B. B. Brodie, J. R. Gillette, and B. N. La Du, *Ann. Rev. Biochem.*, **27**, 427 (1958).
359. Y. C. Martin and C. Hansch, *J. Med. Chem.*, **14**, 777 (1971).
360. C. Hansch, *Drug. Metab. Rev.*, **1**, 1 (1972).
361. H. L. Schmidt, M. R. Möller, and N. Weber, *Biochem. Pharmacol.*, **22**, 2989 (1973).
362. G. M. Cohen and G. J. Mannering, *Mol. Pharmacol.*, **9**, 383 (1973).
363. R. Franke, *Acta Biol. Med. Germ.*, **30**, 151 (1973).
364. R. Franke, *Chem.-Biol. Interactions*, **6**, 1 (1973).
365. C. F. Wilkinson, K. Hetnarski, G. P. Cantwell, and F. J. Di Carlo, *Biochem. Pharmacol.*, **23**, 2377 (1974).
366. K. A. S. Al-Gailany, J. W. Bridges, and K. J. Netter, *Biochem. Pharmacol.*, **24**, 867 (1975).
367. J. A. Durden, Jr., *J. Med. Chem.*, **16**, 1316 (1973).
368. M. E. Wolff and C. Hansch, *Experientia*, **29**, 1111 (1973).
369. J. G. Topliss and E. L. Shapiro, *J. Med. Chem.*, **18**, 621 (1975).
370. H. S. Boudier, J. de Boer, G. Smeets, E. J. Lien, and J. van Rossum, *Life Sci.*, **17**, 377 (1975).
371. A. Cammarata, *J. Med. Chem.*, **15**, 573 (1972).
372. R. F. Borne, R. L. Peden, I. W. Waters, M. Weiner, R. Jordan, and E. A. Coats, *J. Pharm. Sci.*, **63**, 615 (1974).
373. D. H. Rich, P. D. Gesellchen, A. Tong, A. Cheung, and C. K. Buckner, *J. Med. Chem.*, **18**, 1004 (1975).
374. B. Paul, M. F. Chen, and A. R. P. Paterson, *J. Med. Chem.*, **18**, 968 (1975).
375. J. G. Topliss and M. D. Yudis, *J. Med. Chem.*, **15**, 394 (1972).
376. P. N. Craig, H. C. Caldwell, and W. G. Groves, *J. Med. Chem.*, **13**, 1079 (1970).
377. E. Mizuta, J. Toda, N. Suzuki, H. Sugibayashi, K. Imai, and M. Nishikawa, *Chem. Pharm. Bull. (Tokyo)*, **20**, 1114 (1972).
378. M. A. Rogawski, S. Knapp, and A. J. Mandell, *Biochem. Pharmacol.*, **23**, 1955 (1974).
379. E. J. Lien and G. A. Gudauskas, *J. Pharm. Sci.*, **62**, 1968 (1973).
380. E. O. Dillingham, R. W. Mast, G. E. Bass, and J. Autian, *J. Pharm. Sci.*, **62**, 22 (1973).
381. F. R. Murphy, V. Krupa, and G. S. Marks, *Biochem. Pharmacol.*, **24**, 883 (1975).
382. B. J. Broughton, P. Chaplen, P. Knowles, E. Lunt, S. M. Marshall, D. L. Pain, and K. R. H. Wooldridge, *J. Med. Chem.*, **18**, 1117 (1975).
383. C. G. Waringa, R. F. Rekker, and W. Th. Nauta, *European J. Med. Chem.–Chim. Therap.*, **10**, 349 (1975).
384. J. D. Chapman, J. A. Raleigh, J. Borsa, R. G. Webb, and R. Whitehouse, *Internat. J. Radiation Biol.*, **17**, 475 (1972).
385. K. Schwarz, L. A. Porter, and A. Fredga, *Ann. New York Acad. Sci.*, **192**, 200 (1972).
386. S. I. Rapoport and H. Levitan, *Neuroradiology*, **6**, 279 (1974).
387. A. E. Bird, *J. Pharm. Sci.*, **64**, 1671 (1975).
388. G. van den Berg, T. Bultsma, and W. Th. Nauta, *Biochem. Pharmacol.*, **24**, 1115 (1975).
389. T. J. McDonald, R. D. Farley, and R. B. March, *J. Insect Physiol.*, **20**, 1761 (1974).
390. E. J. Lien, J. Kuwahara, and R. T. Koda, *Drug Intell. Clin. Pharm.*, **8**, 470 (1974).
391. H. L. Holmes, *Structure–Activity Relationships for Some Conjugated Heteroenoid Compounds, Catechol Monoethers, and Morphine Alkaloids* (Defense Research Establishment, Suffield, Ralston, Alberta, Canada, 1975).
392. W. B. Neely, W. E. Allison, W. B. Crummett, K. Kauer, and W. Reifschneider, *J. Agric. Food Chem.*, **18**, 45 (1970).
393. I. P. Kapoor, R. L. Metcalf, A. S. Hirwe, J. R. Coats, and M. S. Khalsa, *J. Agric. Food Chem.*, **21**, 310 (1973).
394. H. L. Kopperman, R. M. Carlson, and R. Caple, *Chem.-Biol. Interactions*, **9**, 245 (1974).
395. B. Devaux, B. Duperray, and H. Pacheco, *European J. Med. Chem.*, **9**, 424 (1974).

396. O. Yamada, S. Ishida, F. Futatsuya, K. Ito, H. Yamamoto, and K. Munakata, *Agric. and Biol. Chem.* (*Japan*), **38**, 1235 (1974).
397. A. Fujinami, A. Mine, and T. J. Fujita, *Agric. and Biol. Chem.*, (*Japan*), **38**, 1399 (1974).
398. C. Hansch, R. N. Smith, R. Engle, and H. Wood, *Cancer Chemotherap. Rep.*, Part 1, **56**, 443 (1972).
399. A. H. Khan and W. C. J. Ross, *Chem.-Biol. Interactions*, **1**, 27 (1969).
400. S. M. Kupchan, M. A. Eakin, and A. M. Thomas, *J. Med. Chem.*, **14**, 1147 (1971).
401. C. Hansch, *Cancer Chemotherap. Rep.*, Part 1, **56**, 433 (1972).
402. E. J. Lien and G. L. Tong, *Cancer Chemotherap. Rep.*, Part 1, **57**, 251 (1973).
403. J. A. Montgomery, J. G. Mayo, and C. Hansch, *J. Med. Chem.*, **17**, 477 (1974).
404. G. P. Wheeler, B. J. Bowdon, J. A. Grimsley, and H. H. Lloyd, *Cancer Res.*, **34**, 194 (1974).
405. C. Hansch, *Cancer Chemotherap. Rep.*, Part 2, **4**(4), 51 (1974).
406. J.-L. Montero and J.-L. Imbach, *Compt. rend., Ser. C*, **279**, 809 (1974).
407. V. A. Levin, D. Crafts, C. B. Wilson, P. Kabra, C. Hansch, E. Boldrey, J. Enot, and M. Neely, *Cancer Chemotherap. Rep.*, Part 1, **59**, 327 (1975).
408. B. F. Cain, *Cancer Chemotherap. Rep.*, Part 1, **59**, 679 (1975).
409. T. P. Johnston, G. S. McCaleb, and J. A. Montgomery, *J. Med. Chem.*, **18**, 104 (1975).
410. W. J. Wechter, M. A. Johnson, C. M. Hall, D. T. Warner, A. E. Berger, A. H. Wenzel, D. T. Gish, and G. L. Neil, *J. Med. Chem.*, **18**, 339 (1975).
411. W. B. Neeley, D. R. Branson, and G. E. Blau, *Environ. Sci. Tech.*, **8**, 1113 (1974).
412. C. Hansch, A. Leo, and D. Elkins, *J. Chem. Doc.*, **14**, 57 (1974).

10

A Critical Compilation of Substituent Constants

Otto Exner

10.1. Introduction

The practical usefulness of a correlation equation as well as its significance in theory strongly depend on its range of validity, not only as it is defined but also as its use is restricted in practice by the number of parameters actually known. Hence the tables of parameters are of primary importance and reflect the actual state of a given field. In this chapter

Otto Exner • Institute of Organic Chemistry and Biochemistry, Czechoslovak Academy of Sciences, Prague 6, Czechoslovakia.

attention is focused on constants characterizing the substituent, which are mainly denoted by the letter σ with various subscripts and superscripts, but sometimes also by other symbols (E_s, υ). The correlation equation most frequently has the form (10.1). For the most common equations of this type

$$y = \rho_0 + \sum_{i=1}^{A} \rho_i \sigma_i \tag{10.1}$$

(Hammett, Taft, Yukawa–Tsuno) many tabulations of σ constants are available (e.g. Refs. 1–13), but authors usually select from the literature only one preferred value for each substituent and the extent of the table is always a compromise between two demands: reliability and completeness. Actually neither can be fully satisfied, so that the user cannot really rely on all the data on the one hand, and must always search for additional data on the other. The problem is more urgent than with the tabulation of any other constants, since the individual σ values are not only of varying precision but may even be defined and calculated in quite different ways. It follows that the main problem of this chapter will be the clear definition of how a given σ value was obtained.

10.2. *Definition and Determination of Substituent Constants*

The various kinds of σ constants have been defined by different procedures.[14] Sometimes an equation of the type (10.1) is given and its range of validity is delimited; then the σ constants are empirical parameters that satisfy this relation, although some procedure may be recommended as particularly convenient for their determination (e.g., σ_m, σ_p). In other cases, the constants may be defined on the basis of a certain model, according to theoretical requirements (e.g., σ_I); they then have—more or less exactly—the intended physical meaning, but their possible use in any correlation is *a priori* not secure and must be examined empirically in each case.

Even when the σ constants are defined by an equation of the type (10.1), it never defines them unequivocally by the condition of the best fit only. There are always several degrees of freedom in excess, which are disposed by a convention giving some constants, σ and/or ρ, certain fixed values. The most common is the choice $\sigma = 0$ for hydrogen, thus defining the origin of the scale. Another well-known convention is the choice $\rho = 1$ for the Hammett equation for the dissociation of substituted benzoic acids; it defines the range of the σ scale.† The latter definition accords

†Generally the number of excessive degrees of freedom for equation (10.1) is[15] $A(A+1)$. The possible transformations of σ constants and their consequences have been discussed in detail elsewhere.[15,16]

importance to a particular reaction series and suggests it as a standard reaction for determining new values of σ. In general, these may be obtained by the following procedures.

Primary Standards. If the σ constant is defined unambiguously from experimentally accessible data, or, if from many possible reaction series one is given preference for this purpose, the pertinent model is called the primary standard and the resulting σ's the primary values. In the simplest case of equation (10.1) with $A=1$, it is only necessary to choose a value of ρ, say ρ_1, and the σ values are obtained directly from two experimental results, cf. equation (10.2). The primary values are to be viewed as essentially experimental quantities, since they are determined from experimental results and general constants only; their errors depend on the precision

$$\sigma(\mathrm{X}) = (\log k_{\mathrm{X}} - \log k_{\mathrm{H}})/\rho_1 \qquad (10.2)$$

of the experiment and have nothing to do with the precision of the correlation equation itelf. The primary standard must be chosen carefully with respect to the sensitivity of the model (i.e., an advantageous scaling factor) and its accessibility and theoretical meaning. If these conditions are met, the primary standards have clear advantages and some authors have advocated their use whenever possible.[2]

Secondary Standards. It will always happen that some substituents cannot be studied by means of the primary standard and another reaction series must be applied, provided that its ρ constant has first been determined on the basis of known primary σ values. Any resulting secondary σ value depends on the pertinent reaction series chosen and the primary σ's applied. Its accuracy reflects experimental errors, as well as the precision of the correlation, which is characterized by the correlation coefficient (r), taking into account also the number of data sets (N). The calculation of the secondary values is no longer quite unambiguous since three expressions have been used and advocated,[1,2,8,17,18] viz., (10.3a)–(10.3c).

$$\sigma(\mathrm{X}) = (\log k_{\mathrm{X}} - \overline{\log k})r^2/\rho + \bar{\sigma} \qquad (10.3a)\dagger$$

$$\sigma(\mathrm{X}) = (\log k_{\mathrm{X}} - \log k^0)/\rho \qquad (10.3b)$$

$$\sigma(\mathrm{X}) = (\log k_{\mathrm{X}} - \log k_{\mathrm{H}})/\rho \qquad (10.3c)$$

The inclusion of r, as in (10.3a), or not depends on the underlying statistical model, while the choice between the statistical intercept $\log k^0$ in (10.3b) or the experimental value $\log k_{\mathrm{H}}$ in (10.3c) depends on the precision of the latter. The choice between the expressions (10.3a)–(10.3b) may be rather delicate, and in the literature it is sometimes not even stated which one has actually been used. Fortunately, the results obtained do not differ critically insofar as the correlation is good.

†In (10.3a) the barred quantities are mean values.

Evidently some reactions are more suitable than others and have thus been recommended as secondary standards. They should have high ρ and r values, contain many data sets known with good precision, and if possible be similar to the corresponding primary standard. Sometimes even tertiary values are distinguished[2] if some secondary σ values have been involved in determining ρ. In the writer's opinion this subclassification is not necessary.

Average Values. When several secondary values are available, it may appear natural to take their arithmetic mean, particularly if none of them can be given preference. As to the factors determining these average values and their accuracy, essentially the same arguments apply as for secondary values. The accuracy can be estimated simply from the variance. The average value may be more dependable than the individual secondary values, if it is a mean of several values of comparable reliability; in this case it represents an intermediate step towards *statistical values* (see below). Moreover, averages have sometimes been taken of the values from the same reaction but at different temperatures. These will not be considered average values but simply secondary values with a somewhat improved precision.

Statistical Values. A simple and theoretically well-founded definition of σ constants defines them as values that fit best the entire body of data according to the appropriate equation. The solution is single-valued when certain constants have been conventionally fixed as described above. The statistical problem is not trivial, even for the simplest equations, and a correct solution was achieved only recently.[18,19] Apart from the mathematical difficulties, the main problem is the choice of data. All available data can evidently not be processed, because the procedure would have to be repeated whenever new items appeared. Hence one must select carefully data of comparable reliability within the range of validity of the equation; in addition the data matrix must be reasonably well filled so that short series are to be excluded. Under these conditions valuable results are obtained. Their standard deviations give a general idea of the accuracy of the underlying equation and allow one to classify the substituents into those obeying the correlation better or worse. These deviations are of course essentially different from the errors of the primary values.

Estimated Values. Sometimes missing σ constants may be estimated with reasonable accuracy according to their relationship to other constants of the same type for similar substituents, or to σ constants of a different type for the same substituent. Evidently substituents differing only in a remote part must possess similar values of σ. (This does not apply to steric constants.) For families of substituents with a common first atom (Y), a correlation between the σ constant of the rest of the group (X) and of the whole group (YX) was claimed[8,20] [equation (10.4)]. The equation may be applied for various kinds of constants, σ_i, e.g., σ_I, σ_m, σ_p; the constant σ_j in the second term need not necessarily be of the same kind, but the proper kind must be found empirically together with the value of m. The estimation may become imprecise when X and Y interact strongly; on the other hand it is satisfactory for pure inductive effects, in particular for $Y = CH_2$. In the form (10.4a) it has been widely used.[8]

$$\sigma_i(YX) = \sigma_i(YH) + m\sigma_j(X) \tag{10.4}$$

$$\sigma_I(CH_2X) = \sigma_I(X)/2.8 \tag{10.4a}$$

Relationships between various kinds of σ constants will be discussed after describing the constants involved; see Section 10.3.8.

Corrected Values. One of the last-mentioned correlations was considered to be so generally valid that certain σ constants of experimental origin were corrected slightly in order to obey it exactly.[21] In other cases the experimental σ constants were corrected by a theoretically calculated amount accounting for an extraneous contribution.[22,23] Such corrections are rather problematic. On the one hand their theoretical justification may be criticized,[24] and on the other hand their verification has been attempted by correlations with experimental data from other reaction series. It follows that these series could be simply made into a new standard instead of introducing a correction. Still another type of corrected σ constant represents purely an amendment of scaling.[21,25] This is merely a formal question, which is important only when various kinds of σ constants are compared.

10.3. Types of Substituent Constants

In this section only those types of constants are mentioned that are well documented for a sufficient number of reaction series as well as of substituents.

10.3.1. The Original Hammett Constants, σ_m and σ_p

The Hammett equation is valid for benzene *meta* and *para* derivatives having the functional group in the side chain; there must not be strong conjugation between the side chain and the substituent.[14] The σ constants are unequivocally defined by arbitrarily giving the values $\sigma = 0$ to hydrogen, and $\rho = 1$ to dissociation of substituted benzoic acids in water at 25°C; by this choice this reaction becomes a primary standard.

Certain theoretical problems are connected with the constants $\sigma_{m,p}$ for large and for charged substituents.[14] Some of the substituents in Table 10.1 may be too large, but they were included because the nonvalidity of the

Hammett equation has not been proven. The charged substituents obey the Hammett equation but only with a substantially lowered precision,[14] mainly due to the influence of the ionic strength of the medium. Hence these σ values are collected in a separate section of Table 10.1 and are only orientative.

The primary values, $\sigma_{m,p}$ (denoted pK-W in the tables) are obtained simply from pK values of the substituted and unsubstituted benzoic acids according to equation (10.2). In a critical compilation,[2] their limits of accuracy were estimated to be ± 0.02 σ units at best and more than ± 0.1 at worst. Particular problems are connected with tautomeric compounds.[14] Certain more precise pK determinations have been reported since 1958, but in the writer's opinion the actual accuracy of σ values is somewhat better if pK_X and pK_H have been obtained simultaneously. As secondary standards there were recommended dissociation constants of benzoic acids in 50% (v/v) ethanol[2] (denoted pK-50), or at another concentration (pK-A), or in 80% (w/w) methyl cellosolve[26] (pK-MC); these suffer from some discrepancies concerning the exact values of ρ to be used. The condition of making all the measurements in one laboratory (with one glass electrode) is even more important here. Some values were obtained also from the hydrolysis of benzoic esters (HydrE) under a variety of conditions or from other reactions (OReac). The values from various physical properties (Phys) are not very dependable, since the relevant correlations have never been sufficiently proved. Another emergency solution is to compute σ_m and σ_p from σ_I and σ_R; see Section 10.3.8.

The most difficult problem in this area is the solvent dependence of σ constants, which is striking for hydrogen-bonding substituents such as OH, NH_2, but detectable for many others. The available experimental data do not allow classification according to the different types of solvent, or the determination of specified constants. Hence one must accept the lowered precision and a unified σ constant; for many constants from the literature it is not known to which solvents they are applicable. Until more gas-phase data are available, those from aqueous solutions seem most suitable. Another not fully appreciated problem is temperature dependence, which is marked for acid dissociation.

10.3.2. The Unbiased Values, σ_m^0 and σ_p^0

The constants $\sigma_{m,p}$, insofar as they were determined from primary or common secondary standards, overestimated the mesomeric effects of the donor groups compared to those of acceptor groups.[14] The unbiased values,[7,27] $\sigma_{m,p}^0$, correspond to a restriction of the range of validity to nonconjugated reaction centers, e.g., those insulated by a CH_2 group. Since these values must be equal to common $\sigma_{m,p}$ values for all acceptor groups, no primary standard is possible. As secondary standards, alkaline hydrolysis of substituted ethyl phenylacetates[28] in 60% (v/v) acetone at 25°C, of substituted benzyl benzoates[29] in 70% (v/v) acetone at 25°C, or substituted phenyl tosylates[30] (mean of measurements at three temperatures) were advanced; all these reactions are denoted as HydrE. Dissociation of substituted phenylacetic acids in water (pK-W) is less suitable as a standard because of the low ρ value, although the claim that acceptor substituents exhibit a kind of mesomeric interaction with the CH_2 group,[32,33] was disproved.[31] Many σ^0 values reported are average values[27] (Aver), including those obtained by use of imperfect statistics.[7]

The constants $\sigma^0_{m,p}$ represent progress in the theory, but this is counterbalanced by their statistically less firm basis.[14] For many substituents it is not known whether the published values represent $\sigma^0_{m,p}$ or $\sigma_{m,p}$ or whether these two types actually differ. A recent fundamental statistical analysis[19] did not reveal differences for any *meta*-substituents, or for weak donors in the *para*-position; only for the substituent 4-OMe and still stronger donors is it necessary to make a distinction.

10.3.3. The Dual Values, σ_p^+ and σ_p^-

The dual values[1] express enhanced mesomeric interaction and are to be used when the reaction center is directly conjugated with the substituent,[14] σ_p^+ being defined only for donors, and σ_p^- for acceptors. In the opposite cases, as well as for all *meta*-substituents, they are identical with the normal values. The *meta*-substituents serve also to determine ρ and to retain the common scaling with $\sigma_{m,p}$. Hence all the standard systems in use are secondary in character. For σ_p^+ the solvolysis of substituted t-cumyl chlorides in approximately 90% (empirically adjusted) aqueous acetone at 25°C (CumS) is suitable owing to a high ρ value[12]; otherwise various direct electrophilic substitutions into the benzene nucleus may be applied[12] (ArEl). For σ_p^- the dissociation constants of substituted anilines (pK-An) or *N,N*-dimethylanilines (pK-DM) are approximately equally suitable, while those of phenols (pK-Ph) exhibit some anomalies.[11] Alternatively many nucleophilic substitutions in the benzene nucleus (ArNu) were applied.[34] Statistical analysis[19] confirmed that the constants σ_p^- are identical with σ_p for donors (except for 4-F).

10.3.4. The Aryl Values, σ_a

When extending the term "substitution"[14] one may, for example, regard the pyridine nucleus as benzene with the aza substituent and calculate the pertinent values of σ_m and σ_p. Alternatively constants of the same numerical value may be interpreted as due to substituting phenyl by

3-pyridyl or 4-pyridyl; in this connection they are called aryl values (σ_a) of the pertinent nucleus.[35] The concept may be generalized to various isocyclic and heterocyclic rings, and various types of constants σ_a^0, σ_a^+, σ_a^- may also be defined. The standard reactions are the same as for the corresponding $\sigma_{m,p}$ constants.

The σ_a values must not be mistaken for $\sigma_{m,p}$ of the same group bound to the benzene nucleus, although this distinction is not always sharp in the literature. In addition, σ_a values are often calculated even for groups bearing heteroatoms in position 2. The range of validity of the Hammett equation is thus clearly exceeded, and general validity for these constants cannot be expected.

Another set of constants,[36,37] denoted σ_r, has a meaning very similar to σ_a; the reference substituent is α-naphthyl ($\sigma_r = 0$) and the standard reaction is protonation of hydrocarbons ($\rho = 1$). Since the substitution proceeds directly on the nucleus, these constants correspond to σ_a^+.

10.3.5. The Inductive and Mesomeric Constants, σ_I and σ_R

The inductive constants, σ_I, have been defined on several independent models. Virtually the same sequence of substituents has been obtained, with the possible exception of alkyl groups, whose small constants, σ_I, seem to be controlled by factors other than the actual inductive effect. It was concluded that they lie on a separate scale, and compared to electronegative substituents they are actually equal to zero.[43] Another problem is the proper scaling of σ_I constants from various models in order to be comparable with each other and with the values of $\sigma_{m,p}$. If the latter is intended, the scaling factors must be derived from the benzene series and no strictly primary standard is possible. The original definition[8] of σ_I, as in (10.5), is based on hydrolysis of substituted acetic esters (HydrE) in basic

$$\sigma_I(\mathrm{X}) = 0.181[\log (k_\mathrm{X}/k_\mathrm{H})_\mathrm{B} - \log (k_\mathrm{X}/k_\mathrm{H})_\mathrm{A}] \tag{10.5}$$

and acidic media, hence four experimental values are needed. This is not a primary standard since esters with various alkyl groups were used. The scaling factor was adjusted by reference to dissociation constants of 4-substituted bicyclo[2,2,2]octane-1-carboxylic acids[38] in 50% (v/v) ethanol at 25°C (pK − BCO), which acids are comparable to benzene derivatives. A still better model, especially sensitive, is represented by dissociation constants of 4-substituted quinuclidines[39,40] in water at 25°C (pK-Qui). On the other hand the dissociation of α-substituted *m*- or *p*-toluic acids[21,41] gives σ_I with less precision, but the compounds are more easily accessible. A very simple and useful model[9] defines σ_I from dissociation constants of substituted acetic acids in water at 25°C, equation (10.6) (or similarly at another temperature). This equation is, however, not completely general since acids with intramolecular hydrogen bonds or with very strong steric effects must be excluded. Many σ_I values have also been determined[10]

$$\sigma_I(\mathrm{X}) = (\mathrm{p}K^0 - \mathrm{p}K_\mathrm{X})/3.95 \tag{10.6}$$

from ^{19}F chemical shifts of 3-substituted fluorobenzenes (F-nmr), equation (10.7). The method is of high experimental precision, but the equation is

$$\sigma_I = -\delta_m/7.1 + 0.084 \tag{10.7}$$

somewhat obscure and the results deviate distinctly from the preceding ones.[9] The σ_I constants may also be obtained[21,25,42] from $\sigma_{m,p}$ values; see Section 10.3.8. As a final possibility, some σ_I values may be estimated by use of homologous substituents (Est) according to equation (10.4a) to a good approximation.[8,13,43]

The historically first set of inductive constants, σ^*, was defined by an equation similar to (10.5) but related to substituted formic esters and to the standard $\sigma^*(CH_3) = 0$. Hence the data for the ester XCH_2CO_2R may be used to calculate either σ_I of the substituent X, or σ^* of the whole group CH_2X. These are related by equation (10.8). In addition σ^* values for the

$$\sigma_I(\mathrm{X}) = 0.45\sigma^*(\mathrm{CH_2X}) \tag{10.8}$$

groups CHXY or CXYZ (mostly for branched alkyls) were derived, and have no analogy among σ_I constants.†

These values for branched alkyls were, however, criticized on the grounds that they involve steric and other effects.[43–45] Since σ^* can be converted into σ_I by use of (10.8) and some of the values are of doubtful meaning, their further use cannot be recommended.

The mesomeric constants, σ_R, are defined[8,27] on the basis of σ_p and σ_I as in (10.9). According to the type of σ_p used, the resulting constants are

$$\sigma_R = \sigma_p - \sigma_I \tag{10.9}$$

denoted[11] σ_R^0, σ_R^{Bz}, σ_R^+, or σ_R^-. Their errors originate in the errors of both σ_p and σ_I, and in addition in their imperfect relative scaling. It is this latter factor that causes big differences between literature data[11,16,21] and makes σ_R the least certain of all σ constants. Alternatively σ_R may be obtained from correlations with σ_m, σ_p, and σ_I, see Section 10.3.8, or by a statistical approach referring directly to the Taft equation (10.10). It is possible to

$$y - y^0 = \rho_I\sigma_I + \rho_R\sigma_R \tag{10.10}$$

calculate all the σ_I and σ_R constants simultaneously from the body of experimental data only,[46] but usually the σ_I values are part of the input.[11]

†The σ^* value for the group CXYZ may of course, be used to estimate $\sigma^*(CH_2CXYZ)$ according to (10.4a), and this may be converted into σ_I(CXYZ) according to (10.8). The result is an approximate relation $\sigma_I(\mathrm{X}) = \sigma^*(\mathrm{X})/6.23$.

The results (Stat) must be carefully examined because of the excessive number of degrees of freedom,[15,16] and are less dependable than the others, unlike the statistical $\sigma_{m,p}$ values.

A more direct approach is based on ^{19}F chemical shifts in 3- and 4-substituted fluorobenzenes[10] (F-nmr), as in (10.11). In contradistinction

$$\sigma_R^0 = (\delta_m - \delta_p)/29.5 \quad (10.11)$$

to δ_m and σ_I calculated therefrom, δ_p and σ_R^0 are rather sensitive to the solvent. A still more direct method (ir) is based on the intensity (A) of the transitions ν_{16a} and ν_{16b} in monosubstituted benzenes,[47,48] equation (10.12). Measurement is usually done in tetrachloromethane solution and

$$(\sigma_R^0)^2 = (A - 100)/17{,}600 \quad (10.12)$$

the two bands at 1600 and 1585 cm^{-1} are integrated together. The sign of σ_R^0 must be estimated from disubstituted derivatives or from another source. Although there is a parallelism between the results of individual methods, there are also significant discrepancies. In the author's opinion a single scale of mesomeric effects is an unrealizable objective.

The constants $\Delta\sigma_R^+$ defined[28] in (10.13) have a meaning very similar to σ_R^+. In the original concept[49] σ_p was used instead of σ_p^0 but the change is

$$\Delta\sigma_R^+ = \sigma_p^+ - \sigma_p^0 \quad (10.13)$$

not of much significance. $\Delta\sigma_R^-$ has also been defined[50] but has been less used.

10.3.6. Steric Constants

For steric constants[51] a unified scale applicable to all series has virtually never been postulated. A simple physical meaning belongs to constants υ, which are differences of the van der Waals radii, calculated for a symmetrical substituent and for hydrogen [52] [equation (10.14)]. Hence

$$\upsilon_X = r_X - 1.20 \quad (10.14)$$

they are primary values (vdW); secondary values for unsymmetrical substituents may be determined empirically[24] from correlations of acid-catalyzed esterification and ester hydrolysis data (HydrE). The values υ are correlated rather well with the widely used E_s constants, defined as average values from several reactions of the same type,[8] i.e., acid-catalyzed ester hydrolysis or esterification, equation (10.15). The corrected values E_s^c and

$$E_s = \log k_X - \log k_H \quad (10.15)$$

E_s^0, respectively, were suggested[22,23] to account for the number of α-hydrogen atoms and possible hyperconjugation. The meaning of the corrections is, however, doubtful.[24] Even the basic assumptions of Taft's approach have been serious questioned[44,51] (see also Section 10.3.5), but the E_s values as a relative measure of steric effects in a particular reaction have not been depreciated.

Other quite different reactions have been suggested to evaluate steric effects; they are either little sensitive to polar effects or the latter may be eliminated by a suitable comparative series. Examples are the nucleophilic bromine exchange in alkyl bromides,[44] addition of di-isoamylborane to a carbon–carbon double bond,[53] addition of alkylmagnesium bromides to 3,3-dimethylbutan-2-one,[23] alkylation of cyclic ketimines with alkyl halides,[54] comparison of the acidity of the corresponding 3-substituted and 5-substituted *o*-toluic acids,[55] and comparison[56] of acidities of 2-substituted benzoic acids (steric+polar effects) and 2-substituted pyridinium ions (polar effect only). The last example raises the complex questions connected with the *ortho*-effect. The various *ortho*-constants will not be reported here since they lack generality; see a recent review.[52] The variability of steric effects originates in their dependence on conformation,[54] possible admixture of polar effects,[52] etc. Although E_s values are clearly not additive,[57,58] attempts have been made to calculate the values for complex substituents either from the basic E_s values[57] or from pure structural features.[58]

10.3.7. Various Substituent Constants

Some other types of substituent constants are not listed in the tables because of paucity of data or too special a character. They include various values intermediate in character between σ_p and σ_p^+ or σ_p and σ_p^-, or various other attempts to separate inductive and mesomeric contributions (F–M, $\mathfrak{R}$–$\mathfrak{F}$, σ_i–σ_π. σ_L–σ_D, σ_G–σ_P) which were mentioned previously[14] and are discussed in Chapter 4. More interesting are special constants accounting for conjugation with acceptors and with donors; they were determined from the protonation of substituted pyridine-*N*-oxides.[59] A special scale ($\sigma^\bullet$) to be used in radical reactions was also considered,[60,61] although the evidence is[14] that $\sigma_{m,p}$ or σ_p^+ are satisfactory.

The most elaborate of the special systems concerns substitution on phosphorus.[62–64] The constants σ^ϕ were defined from the dissociation constants of dialkylphosphinic acids, $X_1X_2P(O)OH$, and also separated into the inductive (σ_I^ϕ) and mesomeric (σ_R^ϕ) components by assuming that the alkyl groups exert only an inductive effect. While σ_I^ϕ are related to σ_I, the mesomeric effects are specific for conjugation with phosphorus. All attempts to express σ^ϕ or σ_R^ϕ through known constants failed. The experimental material is ample[62] but the reaction series are not very different in character. In the writer's opinion, even the fundamental assumption that the bulky atom of phosphorus eliminates all steric interactions is not quite securely founded.

Special constants, φ, to express an essentially steric interaction were derived from heats of formation[65] and applied also to reactivities.[66] Steric factors may or need not be involved in the σ_o constants for *ortho*-substituents. The various values are of quite restricted utility and will not be quoted here; see two critical reviews.[52,67]

10.3.8. Relationships between σ Constants of Different Types

Apart from certain correlations,[68] which are merely fortuitous, relationships between σ constants are restricted to σ_I and σ_R on the one hand and various aromatic reactivity parameters on the other.[42] This topic is closely related to the multiparameter correlation equations (Chapter 4) and will be mentioned here only as far as it can be applied to calculation of certain σ values. A common form of the correlations is as in (10.16) and (10.17). Equation (10.16) itself may serve to define σ_R when a and b are known; see (10.9). Equation (10.17) is less suitable for this purposes,

$$\sigma_p = a\sigma_I + b\sigma_R \tag{10.16}$$

$$\sigma_m = c\sigma_I + d\sigma_R \tag{10.17}$$

because of the low value of d, but the two equations together may be used in a statistical approach,[46] searching for the best values of a–d and σ_R simultaneously. Alternatively, the two equations may be solved for σ_I, and σ_R if a–d are known; the values obtained are called aromatic. The σ_I's may be dependable, particularly for substituents with a prevailing I-effect, but σ_R's are more affected by the precision of a–d, hence the values in the literature differ.[11,16,21] For the same reason also, the reverse procedure, i.e., to compute σ_m and σ_p from σ_I and σ_R, is not dependable. However, values that are not otherwise accessible may be obtained by this approach, starting from ^{19}F nmr shifts[10] and (10.7) and (10.11).

The meaning of (10.16) and (10.17) is virtually unchanged in the approach of Swain and Lupton,[69] Dewar and Grisdale,[70] Yukawa and Tsuno,[71] or Charton.[25] In addition to modifications in notation ($\mathfrak{F}$–$\mathfrak{R}$, F–M, σ_i–σ_π, σ_L–σ_D, respectively, instead of σ_I–σ_R), there are differences in the selection of data and the computation process, hence the constants a–d also differ. Similar correlations[69] of σ_p^+ and σ_p^- are only quite approximate and cannot be used at all for calculations.

Another relationship[21] between σ_m and σ_p values (10.18) is restricted to dipolar substituents without a lone pair of electrons (and with a full

$$\sigma_p = 1.14\sigma_m \tag{10.18}$$

octet) in the α position; even carbonyl substituents and polyfluorinated alkyls deviate slightly. The equation may serve to estimate missing values or even to amend a little existing values.[21] The amendment is always within the limits of accuracy, hence it is difficult to say whether it is warranted.

Sometimes the σ constants have even been estimated from the relationship between σ and ρ constants.[72]

10.4. Arrangement of the Tables

The main ideas in arranging the tables were: (*a*) to collect as many data as possible, even those of uncertain reliability, when there are no better, (*b*) to give sufficient information about how each constant was obtained, (*c*) to quote several values for one constant, particularly if these are at variance and/or were obtained by different approaches, (*d*) to recommend a particular value only if the reasons are quite convincing. Generally only the literature concerning actual σ values has been covered; no attempt has been made to search for all the experimental data from which σ constants could be calculated. The detailed selection proceeded along the following lines.

Substituents. The substituents were arranged according to the first atom (H, C, others in a sequence according to the groups of the periodic system); charged substituents were separated. In each group the arrangement proceeded according to the next atoms. A more sophisticated system[6] was abandoned since it separated very similar substituents. Several very complex substituents only were not included, and some very similar ones with equal σ constants were grouped together (e.g., CH_2CO_2R, etc.). In particular, the constants for substituted phenyls were omitted since they may be easily estimated or can be dispensed with if the substituent and side chain are properly defined. For similar reasons the number of substituents of the type $(CH_2)_nX$ was restricted.

For the constants σ_m and σ_p, and σ_I and σ_R, the substituents are essentially the same and are numbered accordingly. In the tables of σ_p^+ and σ_p^- the only substituents included were those whose conjugation in the given direction is either proven or should be considered. For the others σ_p^+ or σ_p^- is *a priori* supposed to be equal to σ_p even if the experimental values differ somewhat.

Constants. These were selected according to the following rules.

(1) The rigorously obtained statistical results and primary values are always given; for σ_I, σ_R, and steric constants, all the independent methods are considered as primary. (The various scales of steric constants have been restricted to the seven most important.)

2. For other values the more reliable and more recent ones are preferred, the preference of the methods being in general in the following order: secondary standards, average values, other reactions, nontypical reactions (e.g., non-benzenoid), physical properties, derived values, and estimates. If several more reliable methods are available, the less reliable are not quoted.

3. If several values from the same approach do not differ substantially, only the most reliable or most recent one is given.

4 Data are omitted that are evidently in error or have already been disproved in the literature.

5. If there is a serious disagreement that cannot be safely resolved, all the values are quoted.

6. The values estimated on the basis of constants of another type are quoted to a limited extent, viz., σ_m, σ_p derived from σ_I, σ_R by using equations (10.16) and (10.17) if there are no better values, σ_R derived from σ_I and σ_p by using (10.9) if these were obtained by a statistical approach or if they are well-known values. Other values of σ_R may be calculated from any pair of values σ_I and σ_p, or values of σ_I and σ_R from any pair σ_m and σ_p.

Reliability and Accuracy. Statistical values are accompanied by their 95% confidence interval. Primary and some secondary values based on dissociation constants are accompanied (*in parentheses*) by the "limit of uncertainty" based on the classification scale of McDaniel and Brown.[2] Of course, the confidence interval has a quite different meaning in these two instances. Other values are arranged approximately in the sequence of decreasing accuracy, but their actual reliability is not known. In some cases the value that seems preferable to the author is printed in bold type. The values in parentheses are particularly uncertain, since they were either denoted as such in the original literature or were obtained by unreliable procedures and are at variance with other values. The values in square brackets relate to particular conditions as indicated by the notes.

References. The references quoted are those in which the constant was calculated, rather than those giving the primary experimental results. If the number quoted is not explicitly given in the reference but had to be calculated by a simple operation, the reference is given in parentheses. Certain review articles are quoted instead of the original literature if their authors either recalculated or critically re-examined the values of constants,[1,2] or if they refer mainly to the work of the reviewer her- or himself (e.g., Ref. 13). Otherwise the original sources were always sought for, and further transferring of values from one review to another was avoided.

Methods. The methods of determining the constants are given in Tables 10.1–10.5 in abbreviations that should be intelligible. Their exact meaning is as follows.

p*K*-W	the p*K*'s of substituted benzoic acids (the ordinary Bz-reactivity) or of phenylacetic acids (σ^0 reactivity) in water
p*K*-50	ditto in 50% (v/v) ethanol
p*K*-A	ditto in aqueous ethanol of another concentration
pK-MC	ditto in 80% (w/w) methyl cellosolve
pK-Av	ditto, average values in several solvents
p*K*-Ac	the p*K*'s of substituted acetic acids in water
p*K*-An	the p*K*'s of substituted anilines in water or aqueous ethanol
p*K*-DM	the p*K*'s of substituted *NN*-dimethylanilines in water or aqueous ethanol
p*K*-Ph	the p*K*'s of substituted phenols in water or aqueous ethanol
p*K*-BCO	the p*K*'s of 4-substituted bicyclo[2,2,2]octane-1-carboxylic acids in aqueous ethanol or methanol
p*K*-Qui	the p*K*'s of 4-substituted quinuclidines in water or 5% ethanol
p*K*	the p*K*'s of all other acids or in other solvents
Stat	values obtained by correct statistics by using at least six observed data sets
←Stat	statistical values based on the value in the left-hand column
Aver	average values obtained by other procedures
HydrE	kinetics of hydrolysis of substituted benzoic esters ($\sigma_{m,p}$; Bz-reactivity), or phenylacetic esters and analogous compounds ($\sigma^0_{m,p}$), or hydrolysis of aliphatic esters or esterification of the corresponding acids (σ_I, E_s)
CumS	kinetics of solvolysis of substituted cumyl chlorides in 90% aqueous acetone
OSolv	other solvolytic reactions
AnReac	kinetics of a reaction of substituted anilines
DMReac	kinetics of a reaction of substituted *NN*-dimethylanilines
PhReac	kinetics of a reaction of substituted phenols
ArNu	kinetics of nucleophilic aromatic substitution
ArEl	kinetics of electrophilic aromatic substitution
ArProt	the equilibrium constants of protonation of aromatic hydocarbons
OReac	kinetics of other reactions within the proper validity range
NBz	kinetics or equilibria of reactions of nonbenzenoid compounds
F-nmr	chemical shifts of ^{19}F in substituted fluorobenzenes; normal solvents are preferred
ir	integrated intensity of the ν_{16a} and ν_{16b} transitions in monosubstituted benzenes; signs in parentheses preceding values of σ^0_R are determined on the basis of other evidence (see p. 447)
Phys	other physical properties

Est(4) Est(4*a*) Est(16) Est(17)	values estimated from the other constants by using the equations (10.4), (10.4*a*), (10.16), or (10.17)
Est	estimated by comparison with similar substituents or with other constants of the same substituent
Cor(16) Cor(17)	previous estimates corrected by applying (10.16) or (10.17) with improved constants
Cor(18)	experimental values corrected by less than 0.03 σ units in order to satisfy (10.18)
CorH	the original experimental values corrected for the contribution of hyperconjugation
vdW	values calculated from the van der Waals radii
BrEx	kinetics of the nucleophilic bromine exchange in alkyl bromides
AddOl	kinetics of the addition of di-isoamylborane to olefins
3-5BA	comparison of the pK values of 3- and 5-substituted 2–methylbenzoic acids
RedCo	reduction of carboxylatopentamminecobalt(III) complexes

Table 10.1. The Hammett Constants, σ_m and σ_p

No.[a]	Substituent	Type[b]	σ_m[c]	Method[d]	Ref.	Type[b]	σ_p[c]	Method[d]	Ref.
1	H	^{0}B	**0.00** ± 0.03	Stat	73				
2	D	^{0}B	-0.000_5	OReac	74	^{0}B	(−0.001)	Est	74
		^{0}B	−0.001	Est	74				
		^{0}B	−0.001	Est(17)	75	^{0}B	−0.001	Est(16)	75
3	Me	^{0}B	**−0.06** ± 0.03	Stat	73	^{0}B	**−0.14** ± 0.03	Stat	73
		B	−0.05(0.01)	p*K*-W	76	B	−0.16(0.01)	p*K*-W	76
		B	−0.07(0.02)	p*K*-W	77	B	−0.16(0.02)	p*K*-W	77
		^{0}	−0.07	HydrE	29	^{0}	−0.12	HydrE	29
						B	[−0.08][e]	OReac	78
5	Et	^{0}B	−0.08 ± 0.12	Stat	73	^{0}B	−0.13 ± 0.09	Stat	73
		B	−0.07(0.1)	p*K*-A	2	B	−0.15(0.02)	p*K*-W	2
						^{0}	−0.13	HydrE	29
6	Pr^n		(−0.07)	Phys	79	B	−0.15	HydrE	80
7	Pr^i	^{0}B	−0.08 ± 0.14	Aver	73	^{0}B	−0.13 ± 0.07	Stat	73
			(−0.07)	Est(17)	81	B	−0.15(0.02)	p*K*-W	2
						^{0}	−0.15	HydrE	29
8	Bu^n		(−0.08)	Phys	79	B	−0.19	HydrE	80
						^{0}	−0.16	Phys	82
9	Bu^i		(−0.07)	Est(17)	81		(−0.14)	Est	—
10	CHMeEt		(−0.08)	Est	—	B	−0.19	HydrE	80
						^{0}	−0.16	Phys	82
11	Bu^t	^{0}B	**−0.09** ± 0.11	Stat	73	^{0}B	**−0.15** ± 0.07	Stat	73
		B	−0.10(0.03)	p*K*-W	2	B	−0.20(0.02)	p*K*-W	2
		B	(0.00)(0.01)	p*K*-W	83	B	−0.18(0.01)	p*K*-W	83
						^{0}	−0.17	HydrE	29
12	n-C_5H_{11}		(−0.08)	Phys	79		−0.15	Est(4)	79
						^{0}	−0.16	Phys	82
13	$CHMePr^n$		(−0.08)	Est	—	B	−0.22	HydrE	80

Table 10.1—continued

No.[a]	Substituent	Type[b]	σ_m[c]	Method[d]	Ref.	Type[b]	σ_p[c]	Method[d]	Ref.
14	CH_2Bu^t		(−0.08)	Est	—	o	−0.10	Aver	33
							(−0.15)	Est	—
15	CMe_2Et		(−0.09)	Est	—	B	−0.21	HydrE	80
19	$CH{:}CH_2$	B	0.08(0.05)	p*K*-W	84	B	−0.08(0.05)	p*K*-W	84
		o	0.01	Est(17)	(10)	o	−0.01	Est(16)	(10)
			(0.05)	Phys	79		−0.02	Est(4)	79
20	CH:CHMe(*E*)		(0.05)	Est	—		(−0.04)	NBz	86
							(0.02)	Est	—
26	C⋮CH	B	0.20(0.1)	p*K*-50	87	B	0.23(0.1)	p*K*-50	87
						o	0.22	Phys	81
						o	(0.16)	Phys	85
27	C⋮CMe		(0.10)	Est	—	o	0.05	Phys	85
							0.09	NBz	86
							(0.12)	Est	—
30	c-C_3H_5	B	−0.07(0.02)	p*K*-W	88	B	−0.21(0.02)	p*K*-W	88
		B	−0.10(0.1)	p*K*-50	89	B	−0.14(0.1)	p*K*-50	89
						o	−0.22(0.1)	p*K*-50	90
32	c-C_4H_7	B	−0.13(0.1)	p*K*-50	89	B	−0.13(0.1)	p*K*-50	89
						B	−0.15(0.05)	p*K*-W	91
33	c-C_5H_9	B	−0.15(0.1)	p*K*-50	89	B	−0.13(0.1)	p*K*-50	89
						B	(−0.02)(0.05)	p*K*-W	91
34	c-C_6H_{11}	B	−0.15(0.1)	p*K*-50	89	B	−0.13(0.1)	p*K*-50	89
						B	(−0.22)(0.05)	p*K*-W	91
37		B	−0.06(0.1)	p*K*-50	89	B	−0.05(0.1)	p*K*-50	89
38		B	−0.10(0.1)	p*K*-50	89	B	−0.08(0.1)	p*K*-50	89

No.	Substituent		Value	Method	Ref.		Value	Method	Ref.
40	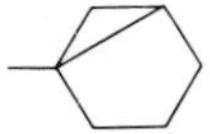					B	0.15(0.05)	pK-W	91
42	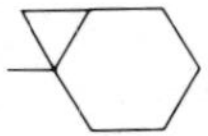					B	0.01(0.05)	pK-W	91
45	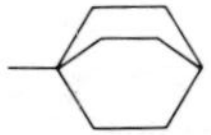					B	−0.25	pK	92
47	1-Adamantyl	0	−0.12	?	93	B	−0.24	pK	92
						0	−0.13	?	93
48	Ph	0B	0.04 ± 0.14	Aver	73	0B	0.05 ± 0.10	Stat	73
		B	**0.05**(0.1)	pK-A	94	B	0.02(0.1)	pK-A	94
		B	0.06(0.05)	pK	2	B	−0.01(0.05)	pK	2
		B	0.06	Cor(17)	21	B	0.03	Cor(16)	21
						0	−0.03	Phys	85
						0	0.04	HydrE	29
49	CH_2Ph	0B	−0.05 ± 0.18	Aver	73	0B	−0.06 ± 0.13	Aver	73
		B	−0.08	pK-Av	21	B	−0.09	pK-Av	21
							(−0.11)	NBz	95
50	$(CH_2)_2Ph$		(−0.07)	Est	—		−0.12	OReac	96
53	$CHPh_2$		(−0.03)	Est	—		(−0.04)	NBz	95
54	CPh_3	B	−0.01	OReac	97	B	0.02	OReac	97
							(0.08)	NBz	95
60	CH:CHPh(*E*)	B	0.03(0.1)	pK-A	98	B	−0.07(0.1)	pK-A	98
							(−0.08)	NBz	95
61	C⋮CPh	B	0.14(0.1)	pK-A	98	B	0.16(0.1)	pK-A	98
65	CH_2CONH_2		0.06	Est	—	B	0.07(0.1)	pK-50	41
68	CH_2CO_2H						(−0.07)	Phys	99
69	$(CH_2)_2CO_2H$		−0.03	OReac	1		−0.07	OReac	1
71	CH_2CN	B	0.16	Cor(18)	21	B	0.18	Cor(18)	21
		0	0.22	Est(17)	(100)	0	0.18	Est(16)	(100)

Table 10.1—continued

No.[a]	Substituent	Type[b]	σ_m[c]	Method[d]	Ref.	Type[b]	σ_p[c]	Method[d]	Ref.
73	$CH(CN)_2$	o	0.53	Est(17)	(100)	o	0.52	Est(16)	(100)
74	$C(CN)_3$	o	0.98	Est(17)	(100)	o	0.99	Est(16)	(100)
75	$C(CN)_2Me$	o	0.60	Est(17)	(100)	o	0.57	Est(16)	(100)
77	CH_2SiMe_3	B	−0.16(>0.1)	p*K*-50	2	B	−0.21(>0.1)	p*K*50	2
		o	−0.17	Est(17)	(10)	o	−0.27	Est(16)	(10)
87	CH_2NH_2	o	−0.03	Est(17)	(10)	o	−0.11	Est(16)	(10)
		o	(−0.07)	Est(17)	(10)	o	(−0.15)	Est(16)	(10)
88	CH_2NMe_2		(0.00)	Est	—	B	0.01	HydrE	102
89	$(CH_2)_2NMe_2$		(−0.04)	Est	—	B	−0.09	HydrE	102
90	$(CH_2)_3NMe_2$		(−0.06)	Est	—	B	−0.13	HydrE	102
91	$(CH_2)_4NMe_2$		(−0.08)	Est	—	B	−0.16	HydrE	102
92	CH_2NHAc		(−0.04)	Est	—	B	−0.05(0.1)	p*K*-50	41
97	CMe_2NO_2		(0.18)	Est(18)	—		0.20	NBz	103
98	$CMe(NO_2)_2$		(0.54)	Est(18)	—		0.61	NBz	103
99	$CEt(NO_2)_2$		(0.56)	Est(18)	—		0.64	NBz	103
100	$C(NO_2)_3$		(0.72)	Est(18)	—		0.82	NBz	103
103	CH_2OH	B	0.01	Est(18)	—	B	0.01(0.1)	p*K*-50	41
							0.01	NBz	104
104	CH(OH)Me						−0.07	NBz	95
105	CH(OH)Ph						−0.03	NBz	95
109	CH_2OR	B	0.02	Cor(18)	21	B	0.02	Cor(18)	21
							0.03	OReac	105
110	CH(OMe)Ph						−0.01	NBz	95
111	$CH(OR)_2$	o	(−0.04)	Est(17)	(100)	o	(−0.05)	Ext(16)	(100)
112	$C(OMe)_3$	o	(−0.03)	Est(17)	(100)	o	(−0.04)	Est(16)	(100)
113	CH_2OPh	B	0.04	Cor(18)	21	B	0.05	Cor(18)	21
115	CH_2OAc		0.04	Est(18)	—	B	0.05(0.1)	p*K*-50	41
116	CH_2OSiMe_3		(−0.04)	Est(18)	—		−0.05	OReac	106

123	CH_2SCF_3	B	0.12(0.1)	p*K*-50	13	B	0.15(0.1)	p*K*-50	13
124	$CH(SCF_3)_2$	o	0.44	Est(17)	(100)	o	0.44	Est(16)	(100)
125	$C(SCF_3)_3$	o	0.51	Est(17)	(100)	o	0.53	Est(16)	(100)
126	CH_2SCN	B	0.12	Est(18)	—	B	0.14(0.1)	p*K*-50	41
127	$CH_2SO{\cdot}CF_3$	o	0.25	Est(17)	13	o	0.24	Est(16)	13
128	$CH_2SO_2{\cdot}R$	B	0.15	Cor(18)	21	B	0.17	Cor(18)	21
129	$CH_2SO_2{\cdot}CF_3$	B	0.29(0.1)	p*K*-50	13	B	0.31(0.1)	p*K*-50	13
130	CH_2F	o	0.11	Est(17)	(100)	o	0.10	Est(16)	(100)
131	CHF_2	o	0.32	Est(17)	(100)	o	0.35	Est(16)	(100)
132	CF_3	oB	**0.46** ± 0.09	Stat	73	oB	**0.53** ± 0.11	Stat	73
		B	0.43(0.1)	p*K*-50	2	B	0.54(0.1)	p*K*-50	2
		B	0.46	Cor(18)	21		0.53	Aver	33
						B	0.53	Cor(18)	21
133	CH_2Cl	B	**0.11**	Cor(18)	21	B	**0.12**	Cor(18)	21
			(0.14)	Phys	79				
134	$CHCl_2$	o	0.31	Est(17)	(100)	o	0.32	Est(16)	(100)
							0.19	NBz	107
135	CCl_3	B	0.40(0.1)	p*K*-MC	21	B	0.46(0.1)	p*K*-MC	21
		o	0.31	Est(17)	(100)	o	0.33	Est(16)	(100)
136	CH_2Br	B	**0.12**	Cor(18)	21	B	**0.14**	Cor(18)	21
			(0.16)	Phys	79				
137	$CHBr_2$	o	0.31	Est(17)	(100)	o	0.32	Est(16)	(100)
138	CBr_3	o	0.28	Est(17)	(100)	o	0.29	Est(16)	(100)
		o	(0.26)	Est(17)	13	o	(0.27)	Est(16)	13
139	CH_2I	B	0.10	Cor(18)	21	B	0.11	Cor(18)	21
			(0.14)	Phys	79				
140	CHI_2	o	0.26	Est(17)	13	o	0.26	Est(16)	13
141	CF_2Cl	o	0.42	Est(17)	(100)	o	0.46	Est(16)	(100)
144	CH_2CF_3		0.16	p*K*-An	109		0.14	p*K*-An	109
148	C_2F_5		0.50	p*K*-An	110		0.52	Est(16)	110
			0.47	Est(17)	110				

Table 10.1—continued

No.[a]	Substituent	Type[b]	σ_m[c]	Method[d]	Ref.	Type[b]	σ_p[c]	Method[d]	Ref.
149	$CF(CF_3)_2$		0.37(0.1)	p*K*-50	110		0.53(0.1)	p*K*-50	110
		o	0.50	Est(17)	110	o	0.52	Est(16)	110
150	C_4F_9		0.52	p*K*-An	110	o	0.52	Est(16)	110
		o	0.47	Est(17)	110				
151	$C(CF_3)_3$	B	0.35(0.1)	p*K*-50	13	B	0.52(0.1)	p*K*-50	13
153	$C(OH)(CF_3)_2$		0.35	p*K*-An	110	o	0.30	Est(16)	110
		o	0.29	Est(17)	110				
154	c-C_4F_7	o	0.49	Est(17)	(109)		0.53	Est(16)	(109)
155	$C(OH)(CF_2)_3$	o	0.49	Est(17)	(109)	o	0.53	Est(16)	(109)
156	CH:CHCHO (*E*)		0.24	NBz	84		0.13	NBz	84
157	CH:CHAc (*E*)		0.21	NBz	84		(−0.01)	NBz	84
158	CH:CHBz (*E*)		0.18	NBz	84		(0.05)	NBz	84
160	$CH{:}CHCO_2R$ (*E*)		0.19	NBz	84		(0.03)	NBz	84
							0.32	NBz	111
161	CH:CHCN (*E* or *Z*)		0.24	NBz	84		0.17	NBz	84
							0.38	NBz	111
162	$CH{:}C(CN)_2$		0.55	p*K*-DM	108		0.70	Est(16)	108
		o	0.55	Est(17)	108				
163	$C(CN){:}C(CN)_2$		0.77	p*K*-DM	108		0.98	Est(16)	108
		o	0.82	Est(17)	108				
164	$CH{:}CHNO_2$ (*E*)	B	0.32(0.02)	p*K*-W	112		0.26(0.02)	p*K*-W	112
			0.28	NBz	84				
165	$CH{:}CHSO_2{\cdot}CF_3$ (*E*)	o	0.44	Est(17)	13	o	0.55	Est(16)	13
169	$CH{:}CHCF_3$ (*E*)	B	0.22(0.1)	p*K*-50	13	B	0.20(0.1)	p*K*-50	13
			0.20	p*K*-DM	108				
170	$CH{:}CHCF_3$ (*Z*)		0.16	p*K*-DM	108				
171	$CF{:}CFCF_3$ (*E*)	o	0.15	Est(17)	108	o	0.17	Est(16)	108

		⁰	0.37	Est(17)	13	⁰	0.44	Est(16)	13
177	C⋮CCF_3	B	0.41(0.1)	p*K*-50	13	B	0.51(0.1)	p*K*-50	13
178	$C_6H_2(NO_2)_3$-2,4,6	B	0.27(0.1)	p*K*-50	113	B	0.31(0.1)	p*K*-50	113
		B	0.43	OReac	114	B	0.41	OReac	114
179	C_6F_5	B	(−0.12)	p*K*	115	B	(−0.03)	p*K*	115
			0.28	ArEl	116		0.26	ArEl	116
		⁰	0.26	Est(17)	115	⁰	0.27	Est(16)	115
180	C_6Cl_5	⁰	0.24	Est(17)	115	⁰	0.24	Est(17)	115
181	CHO	⁰B	0.41 ± 0.13	Aver	73	⁰B	0.47 ± 0.18	Aver	73
		B	0.36(0.03)	p*K*-W	117	B	0.43(0.03)	p*K*-W	117
						B	0.45(0.02)	p*K*-W	118
182	COMe	⁰B	**0.36** ± 0.07	Stat	73	⁰B	**0.47** ± 0.10	Stat	73
		B	0.36	Cor(17)	21	B	0.44	Cor(16)	21
		B	0.38(0.02)	p*K*-W	2	B	0.50(0.02)	p*K*-W	2
183	COEt		(0.38)	Est	—	B	0.48(0.02)	p*K*-W	118
184	$COPr^i$		(0.38)	Est	—	B	0.47(0.02)	p*K*-W	118
185	$COBu^t$		(0.27)	Est	—	⁰B	0.33 ± 0.18	Aver	73
						B	0.32(0.02)	p*K*-W	118
186	COPh	⁰B	**0.36** ± 0.18	Aver	73	⁰B	**0.46** ± 0.10	Stat	73
		B	0.34	Cor(17)	21, 120	B	0.42	Cor(16)	21, 120
187	$COCF_3$	⁰	0.63	Est(17)	13, 110	⁰	0.80	Est(16)	13, 110
189	$CONH_2$		0.28	OReac	1	B	0.31(0.02)	p*K*-W	33
							(0.43)	NBz	95
							(0.36)	Est(4)	79
190	CONHMe	B	0.35(0.1)	p*K*-50	121	B	0.36(0.1)	p*K*-50	121
192	CONHPh						0.41	NBz	95
193	$CONPh_2$						0.35	NBz	95
194	CO_2H	⁰B	**0.35** ± 0.18	Aver	73	⁰B	**0.44** ± 0.18	Aver	73
		B	0.35	Cor(17)	21	B	0.42	Cor(16)	21
		B	0.37	Est	2	B	0.45	Est	2
							0.41	Aver	33

Table 10.1—continued

No.[a]	Substituent	Type[b]	σ_m[c]	Method[d]	Ref.	Type[b]	σ_p[c]	Method[d]	Ref.
195	CO_2R	^{0}B	**0.35** ± 0.10	Stat	73	^{0}B	**0.44** ± 0.09	Stat	73
		B	0.36	Cor(17)	21	B	0.43	Cor(16)	21
		B	0.37(0.1)	p*K*-50	2	B	0.45(0.1)	p*K*-50	2
						B	0.46(0.02)	p*K*-W	33
							[0.58][f]	OReac	119
197	CO_2CHPh_2		0.36	Phys	122		(0.55)	Phys	122
198	COF	^{0}	0.55	Est(17)	13	^{0}	0.70	Est(16)	13
199	COCl	^{0}	0.53	Est(17)	13	^{0}	0.69	Est(16)	13
200	CSNHMe	B	0.30(0.1)	p*K*-50	121	B	0.34(0.1)	p*K*-50	121
201	CH:NPh (*E*)		0.35	NBz	123	B	0.42(0.1)	p*K*-A	124
202	CH:NOH (*E*)		0.22	NBz	125		0.10	NBz	125
203	CN	^{0}B	**0.62** ± 0.05	Stat	73	^{0}B	0.71 ± 0.08	Stat	73
		B	0.61	Cor(18)	21	B	0.69	Cor(18)	21
		B	0.61(0.02)	p*K*-W	77	B	0.65(0.02)	p*K*-W	77
		^{0}	0.65	HydrE	29	B	**0.70**(0.02)	p*K*-W	33
						^{0}	0.71	HydrE	29
						^{0}	[0.65][e]	Aver	126
							[0.88][f]	OReac	119
204	—◁ NH	^{0}	−0.06	Est(17)	101	^{0}	−0.10	Est(16)	101
205	—◁ O	^{0}	0.05	Est(17)	101	^{0}	0.03	Est(16)	101
206	—◁ S	^{0}	0.04	Est(17)	101	^{0}	0.01	Est(16)	101
207	—◁ NMe, O	^{0}	0.09	Est(17)	101	^{0}	0.12	Est(16)	101

211	2-C_4H_3O	B	0.06(0.1)	p*K*-50	127	B	0.02(0.1)	p*K*-50	127
213	2-C_4H_3S	B	0.09(0.1)	p*K*-50	128	B	0.05(0.1)	p*K*-50	128
							0.13	p*K*	111
215	3-C_4H_3S	B	0.03(0.1)	p*K*-50	128	B	−0.02(0.1)	p*K*-50	128
217			0.23	p*K*-An	130				
223		B	0.19(0.1)	p*K*-MC	129	B	0.24(0.1)	p*K*-MC	129
		B	0.07(0.1)[l]	p*K*-50	129	B	0.12(0.1)[l]	p*K*-50	129
224		B	0.31(0.1)	p*K*-MC	129	B	0.34(0.1)	p*K*-MC	129
225		B	0.33(0.1)	p*K*-MC	129	B	0.34(0.1)	p*K*-MC	129
226		B	0.25(0.1)	p*K*-50	131	B	0.39(0.1)	p*K*-50	131
227	$C_{10}H_9Fe$	B	−0.15(>0.1)	p*K*	132	B	−0.18(>0.1)	p*K*	132
			0.00	OReac	133		−0.15	OReac	133
			0.0	OReac	134				
228	1-$C_2HB_{10}H_{10}$(1,2)	B	0.47(>0.1)	p*K*-A	135	B	0.46(>0.1)	p*K*-A	135
	[−C—CH $B_{10}H_{10}$]	B	0.49(>0.1)	p*K*-A	136	B	0.43(>0.1)	p*K*-A	136

Table 10.1—continued

No.[a]	Substituent	Type[b]	σ_m[c]	Method[d]	Ref.	Type[b]	σ_p[c]	Method[d]	Ref.
229	$1\text{-}C_2HB_{10}H_{10}(1,7)$	B	0.25(>0.1)	p*K*-A	136	B	0.33(>0.1)	p*K*-A	136
		o	0.17	Est(17)	135	o	0.15	Est(16)	135
233	HgMe	o	0.43	Est(17)	6	o	0.10	Est(16)	6
235	$HgCF_3$	o	0.29	Est(17)	13	o	0.32	Est(16)	13
236	HgCN	o	0.28	Est(17)	13	o	0.34	Est(16)	13
237	HgOAc	o	0.39	Est(17)	13	o	0.40	Est(16)	13
238	$HgOCOCF_3$	o	0.50	Est(17)	13	o	0.52	Est(16)	13
239	$HgSCF_3$	o	0.39	Est(17)	13	o	0.42	Est(16)	13
240	HgF	o	0.34	Est(17)	13	o	0.33	Est(16)	13
241	HgCl	o	0.33	Est(17)	13	o	0.35	Est(16)	13
243	$6\text{-}B_{10}H_{13}$	o	0.18	Est(17)	137	o	0.24	Est(16)	137
244	$B(OH)_2$	B	−0.01	Cor(17)	21	B	0.12	Corr(16)	21
245	BF_2	o	0.32	Est(17)	109	o	0.48	Est(16)	109
247	$3\text{-}B_{10}H_9C_2H_2(1,2)$	B	0.20(>0.1)	p*K*-A	136	B	0.19(>0.1)	p*K*-A	136
249	SiH_3	o	(0.05)	Est(17)	13	o	0.10	Est(16)	13
							0.10	Phys	139
250	$SiHMe_2$	o	0.01	Est(17)	(138)	o	0.04	Est(16)	138
							0.06	Phys	139
251	$SiMe_3$	B	0.11(0.02)	p*K*-W	83	B	0.00(0.02)	p*K*-W	83
		B	−0.04(0.1)	p*K*-W	140	B	−0.07(0.1)	p*K*-W	140
		B	−0.09	Cor(17)	21	B	−0.07	Cor(16)	21
			0.02	OReac	106		−0.02	OReac	106
							(0.06)	Phys	139
252	$SiMe_2Ph$	o	0.04	Est(17)	(138)	o	0.07	Est(16)	(138)
253	$SiMePh_2$	o	0.10	Est(17)	(138)	o	0.13	Est(16)	(138)
254	$SiPh_3$	B	−0.03	OReac	141	B	0.10	OReac	141
		o	0.16	Est(17)	(138)	o	0.19	Est(16)	(138)

257	$Si(NMe_2)_3$	o	−0.04	Est(17)	(138)	o	−0.04	Est(16)	(138)
258	$SiMe_2OMe$		0.04	OReac	142		−0.02	OReac	142
							0.10	Phys	139
259	$SiMe(OMe)_2$		0.04	OReac	142		0.01	OReac	142
							0.15	Phys	139
260	$Si(OMe)_3$		0.09	OReac	142		0.13	OReac	142
		o	0.06	Est(17)	(138)	o	0.13	Est(16)	(138)
261	$Si(OEt)_3$	o	0.02	Est(17)	(138)	o	0.08	Est(16)	(138)
262	$SiMe_2OSiMe_3$	B	0.00	OReac	106	B	0.01	OReac	106
263	$SiMe(OSiMe_3)_2$	B	−0.03	OReac	106	B	0.02	OReac	106
264	$Si(OSiMe_3)_3$	B	−0.07	OReac	106	B	0.02	OReac	106
265	$SiFMe_2$	B	0.12	OReac	106		0.17	OReac	142
		o	0.13	Est(17)	(138)	o	0.19	Est(16)	(138)
							0.15	Phys	139
266	SiF_2Me	B	(0.29)	OReac	106		0.23	OReac	142
							0.28	Phys	139
267	SiF_3	o	0.54	Est(17)	13	o	0.66	Est(16)	13
							0.51	Phys	139
268	$SiClMe_2$	o	0.16	Est(17)	(138)	o	0.21	Est(16)	(138)
			(0.07)	OReac	142		(0.02)	OReac	142
							0.15	Phys	139
269	$SiCl_2Me$	o	0.31	Est(17)	(138)		0.39	Est(16)	(138)
			(0.09)	OReac	142		(0.08)	OReac	142
							0.42	Phys	139
270	$SiCl_3$	o	0.48	Est(17)	13	o	0.56	Est(16)	13
							0.37	Phys	139
271	$SiBrMe_2$						0.10	OReac	142
272	$SiBr_2Me$						0.29	OReac	142
273	$SiBr_3$	o	0.48	Est(17)	13	o	0.57	Est(16)	13
							0.41	OReac	142
274	GeH_3		(0.0)	Est	—		0.01	Phys	139
275	$GeMe_3$		(0.0)	Est	—	B	0.0(0.1)	p*K*-A	2
							0.01	Phys	139

Table 10.1—continued

No.[a]	Substituent	Type[b]	$\sigma_m{}^c$	Method[d]	Ref.	Type[b]	$\sigma_p{}^c$	Method[d]	Ref.
276	GeEt$_3$		(0.0)	Est	—	B	0.0(0.1)	p*K*-A	2
277	GePh$_3$	B	0.05	OReac	141	B	0.08	OReac	141
278	GeF$_3$	o	0.85	Est(17)	13	o	0.97	Est(16)	13
279	GeCl$_3$	o	0.71	Est(17)	13	o	0.79	Est(16)	13
							0.60	Phys	139
280	GeBr$_3$	o	0.66	Est(17)	13	o	0.73	Est(16)	13
281	SnMe$_3$		(0.0)	Est	—	B	0.0(0.1)	p*K*-A	2
282	SnEt$_3$		(0.0)	Est	—	B	0.0(0.1)	p*K*-A	2
285	NH$_2$	oB	**−0.09** ± 0.05	Stat	73	o	**−0.30** ± 0.11	Stat	73
		B	−0.16(0.1)	p*K*-A	2	o	−0.33	Aver	29
		B	0.00^g	p*K*-W	143	o	−0.24	Aver	33
						B	**−0.57**^{g}	p*K*-W	143
						B	(−0.66)^h(0.1)	p*K*-W	2
						B	−0.52(0.1)	p*K*-50	144
286	NHMe		(−0.30)	p*K*	1	o	−0.46	Phys	85
			−0.1	Est	—	B	−0.64(0.1)	p*K*-50	144
							−0.59	p*K*	1
						B	(−0.84)^h(0.1)	p*K*-W	2
287	NMe$_2$	oB	**−0.10** ± 0.09	Stat	73	o	**−0.32** ± 0.12	Stat	73
		B	−0.15(0.1)	p*K*-50	145	o	−0.24	Aver	33
						o	−0.48	Phys	85
						B	**−0.63**(0.1)	p*K*-50	144
						B	(−0.83)^h(0.1)	p*K*-W	2
288	NHR		(−0.24)	p*K*	1		−0.61	?	20
			(−0.34)	p*K*	1		−0.51	?	20
289	NEt$_2$		−0.15	Est	—	o	−0.53	Phys	85
290	NHPh		−0.12	Est(4)	146	o	−0.27	Phys	85
							−0.29	OReac	96
						B	−0.45	p*K*-50	144

291	NPh_2	B	−0.07(0.1)	p*K*-50	147	B	−0.29(0.1)	p*K*-50	147
						B	−0.26(0.1)	p*K*-50	144
293	$NHCH_2CO_2R$	B	$-0.10^{g}(0.05)$	p*K*-W	143	B	$-0.68^{g}(0.05)$	p*K*-W	143
294	$N(CF_3)_2$	B	0.40(0.1)	p*K*-50	148	B	0.53(0.1)	p*K*-50	148
			0.47	p*K*-An	148		0.51	OReac	111
296	NHCHO	B	0.19(0.1)	p*K*-Av	150	B	0.00(0.1)	p*K*-Av	150
		o	0.19	Est(17)	151	o	0.09	Est(16)	151
297	NHAc	oB	**0.14** ± 0.11	Stat	73	o	0.00 ± 0.13	Aver	73
		B	0.12(0.1)	p*K*-Av	150	o	(0.14)	Aver	33
		o	0.37	Est(17)	13	B	−0.09(0.1)	p*K*-Av	150
							0.13	OReac	111
						o	0.27	Est(16)	13
298	NMeAc	o	0.31	Est(17)	13	o	0.26	Est(16)	13
300	NAc_2	o	0.35	Est(17)	13	o	0.33	Est(16)	13
302	$NH\text{-}COPr^i$	B	0.11(0.1)	p*K*-Av	150	B	−0.10(0.1)	p*K*-Av	150
303	NH-Bz	B	0.02(0.1)	p*K*-MC	150	B	−0.19(0.1)	p*K*-MC	150
			(0.22)	p*K*	1		(0.08)	p*K*	1
			0.12	Phys	122		−0.07	Phys	122
						o	−0.14	OReac	152
304	$NHCOC_6H_4OMe\text{-}p$	B	0.09(0.1)	p*K*-MC	150	B	−0.06(0.1)	p*K*-MC	150
305	$NHCOCF_3$	B	0.30(0.1)	p*K*-Av	150	B	0.12(0.1)	p*K*-Av	150
		o	0.36	Est(17)	13	o	0.27	Est(16)	13
							0.13	OReac	111
306	$NMeCOCF_3$	o	0.41	Est(17)	13	o	0.39	Est(16)	13
307	$NHCOCH_2Cl$	B	0.17(0.1)	p*K*-Av	150	B	−0.03(0.1)	p*K*-Av	150
308	$NHCONH_2$	B	−0.03(0.1)	p*K*-MC	150	B	−0.24(0.1)	p*K*-MC	150
309	HNCONHEt	B	0.04(0.1)	p*K*-50	121	B	−0.26(0.1)	p*K*-50	121
310	$NHCO_2R$	B	0.07(0.1)	p*K*-Av	150	B	−0.15(0.1)	p*K*-Av	150
			0.03	OReac	153		0.04	OReac	153
		B	0.11(0.1)	p*K*-50	154		−0.05	OReac	155
311	$N(CF_3)COF$	o	0.56	Est(17)	148	o	0.56	Est(16)	148
312	$N(COF)_2$	o	0.58	Est(17)	148	o	0.57	Est(16)	148

Table 10.1—continued

No.[a]	Substituent	Type[b]	σ_m[c]	Method[d]	Ref.	Type[b]	σ_p[c]	Method[d]	Ref.
313	NHCSMe	B	0.24(0.1)	p*K*-50	121	B	0.12(0.1)	p*K*-50	121
314	$NHCSNH_2$	0	0.22	Est(17)	149	0	0.16	Est(16)	149
315	NHCSNHEt	B	0.30(0.1)	p*K*-50	121	B	0.07(0.1)	p*K*-50	121
319	NHCN	0	0.21	Est(17)	149	0	0.06	Est(16)	149
320	$NHNH_2$		−0.02	p*K*	1		−0.55	p*K*	1
		0	−0.08	Est(17)	10	0	−0.29	Est(16)	10
322	NHOH		(−0.04)	Phys	1		(−0.34)	Phys	1
323	$NHSO_2Me$	B	0.20(0.1)	p*K*-Av	150	B	0.03(0.1)	p*K*-Av	150
		0	0.32	Est(17)	13		0.21	Est(16)	13
324	$NMeSO_2Me$	0	0.29	Est(17)	13	0	0.24	Est(16)	13
325	$N(SO_2Me)_2$	0	0.47	Est(17)	13	0	0.49	Est(16)	13
326	$NHSO_2Ph$	B	0.16(0.1)	p*K*-Av	150	B	0.01(0.1)	p*K*-Av	150
		B	0.20(0.1)	p*K*-50	154	B	−0.01(0.1)	p*K*-50	154
							(0.17)	OReac	111
327	$NHSO_2CF_3$	0	0.44	Est(17)	13	0	0.39	Est(16)	13
							0.40	OReac	111
328	$NMeSO_2CF_3$	0	0.46	Est(17)	13	0	0.44	Est(16)	13
329	$N(SO_2CF_3)_2$	0	0.75	Est(17)	13	0	0.80	Est(16)	13
330	$\overset{+}{N}\vdots\overset{-}{C}$	0	0.48	Est(17)	151	0	0.49	Est(16)	151
331	N:CHPh	B	−0.08(0.1)	p*K*-Av	150	B	−0.55(0.1)	p*K*-Av	150
						B	0.00(0.1)	p*K*-A	124
332	$N{:}CHC_6H_4OMe$-*p*	B	−0.07(0.1)	p*K*-Av	150	B	−0.54(0.1)	p*K*-Av	150
333	$N{:}C(CF_3)_2$	0	0.29	Est(17)	109	0	0.23	Est(16)	109
334	$N{:}CCl_2$	0	0.21	Est(17)	108	0	0.13	Est(16)	108
336	N:C:O		0.30	OReac	153		0.24	OReac	153
		0	0.27	Est(17)	108	0	0.19	Est(16)	108
			(0.43)	OReac	155		(0.35)	OReac	155

No.	Substituent								
337	N:C:S	B	0.48(0.1)	p*K*-Aver	150, 156	B	0.38(0.1)	p*K*-MC	150, 156
			0.34	p*K*-An	156	0	0.35	Est(16)	148
		0	0.39	Est(17)	148				
338	N:NPh	B	**0.29**(0.1)	p*K*-MC	150	B	**0.33**(0.1)	p*K*-MC	150
			0.32	Phys	122	B	0.31(0.1)	p*K*-50	124
							0.39	Phys	122
339	N:NC$_6$H$_3$OH-2, Me-5	B	0.27(0.1)	p*K*-MC	150	B	0.31(0.1)	p*K*-MC	150
340	N:NCF$_3$	0	0.56	Est(17)	13	0	0.68	Est(16)	13
							0.74	OReac	111
343	N:N$^+$:N$^-$	B	0.27(0.1)	p*K*-Av	150	B	0.15(0.1)	p*K*-Av	150
344	NO	0	0.49	Est(17)	148	0	0.65	Est(16)	148
							(0.12)[h]	p*K*-DM	1
345	N:S:O						0.44	Phys	157
346	NO$_2$	0B	**0.71** ± 0.04	Stat	73	0B	**0.81** ± 0.05	Stat	73
		B	0.70	Cor(18)	21	B	0.80	Cor(18)	21
		B	0.74(0.01)	p*K*-W	76	B	0.78(0.01)	p*K*-W	76, 2
		B	0.71(0.02)	p*K*-W	2	0	0.82	HydrE	29
		0	0.71	HydrE	29	0	[0.73][e]	Aver	126
347	N(CH$_2$)$_2$	0	−0.07	Est(17)	101	0	−0.22	Est(16)	101
350	N(CH$_2$)$_5$						−0.12	p*K*	158
351	—N (N, N; benzo-fused ring)		0.49	p*K*-An	159		(0.51)	p*K*-An	159
352	—N (N=N, CH=N ring)		0.60	p*K*-DM	108		(0.57)	p*K*-DM	108
353	NH—C (N—N, S—N ring)	0	0. 30	Est(17)	149	0	0.19	Est(16)	149

Table 10.1—continued

No.[a]	Substituent	Type[b]	$\sigma_m{}^c$	Method[d]	Ref.	Type[b]	$\sigma_p{}^c$	Method[d]	Ref.
356	PH_2	o	0.06[i]	Est(17)	160	o	0.05[i]	Est(16)	160
357	PMe_2	o	0.05	Est(17)	161	o	0.03	Est(16)	161
358	PEt_2						0.03	OReac	162
359	PPh_2	B	0.11(0.1)	p*K*-50	147	B	0.19(0.1)	p*K*-50	147
		o	0.17	Est(17)	161	B	0.19(>0.1)	p*K*	163
						o	0.16	Est(16)	161
							(0.68)	Phys	164
360	$P(CF_3)_2$	o	0.60	Est(17)	161	o	0.59	Est(16)	161
361	$P(CN)_2$	o	(0.82)	Est(17)	161	o	(0.90)	Est(16)	161
362	$P(NMe_2)_2$	o	−0.03	Est(17)	161	o	−0.06	Est(16)	161
363	P(OMe)Ph						(0.32)	Phys	164
364	$P(OR)_2$	o	0.12	Est(17)	161	o	0.15	Est(16)	161
							(0.33)	Phys	164
365	PF_2	o	0.48	Est(17)	161		0.59	Est(16)	161
			(0.26)	Phys	6	o	(0.61)	Phys	6
366	PClPh						(0.44)	Phys	164
367	$PClNMe_2$		(0.38)	Phys	6		(0.56)	Phys	6
368	PCl_2	o	0.53	Est(17)	161	o	0.61	Est(16)	161
							(0.61)	Phys	164
369	$\overset{+}{P}Ph_2\overset{-}{C}H_2$	o	0.45	Est(17)	165	o	0.54	Est(16)	165
370	$\overset{+}{P}Ph_2\overset{-}{B}Cl_3$	o	0.73	Est(17)	161	o	0.81	Est(16)	161
371	$POMe_2$		0.43	OReac	166	o	0.50[i]	Est(16)	160
		o	0.43[i]	Est(17)	160				
372	$POBu^n_2$	B	0.35(0.1)	p*K*-50	167	B	0.49(0.1)	p*K*-50	167
373	$POPh_2$	B	0.38(0.1)	p*K*-50	147	B	0.53(0.1)	p*K*-50	147
		B	0.42(0.1)	p*K*-50	168	B	0.55(0.1)	p*K*-50	168
		o	0.44[i]	Est(17)	160	B	0.50(>0.1)	p*K*	163
						o	0.51[i]	Est(16)	160

374	$PO(OH)_2$	o	0.36	Est(17)	165	o	0.42	Est(16)	165
375	$PO(OMe)_2$	B	0.42(0.03)	p*K*-W	167	B	0.55(0.03)	p*K*-W	167
		o	0.34^{i}	Est(17)	160	o	0.43^{i}	Est(16)	160
376	$PO(OEt)_2$	B	0.43(0.03)	p*K*-W	167	B	0.53(0.03)	p*K*-W	167
		B	0.55(0.05)	p*K*-W	169	B	0.60(0.05)	p*K*-W	169
		o	0.30^{i}	Est(17)	160	o	0.38^{i}	Est(16)	160
377	$PO(OR)_2$	B	0.38(0.1)	p*K*-50	167	B	0.50(0.1)	p*K*-50	167
378	POF_2		(0.81)	Phys	6		(0.89)	Phys	6
379	$POCl_2$	o	0.78^{i}	Est(17)	160	o	0.90^{i}	Est(16)	160
							(0.43)	Phys	164
380	$PSPh_2$	B	0.29(0.1)	p*K*-50	147	B	0.47(0.1)	p*K*-50	147
		o	0.45^{i}	Est(17)	160	B	0.49(>0.1)	p*K*	163
						o	0.51^{i}	Est(16)	160
381	$PSCl_2$	o	0.70^{i}	Est(17)	160	o	0.80^{i}	Est(16)	160
							(0.39)	Phys	164
383	PF_4	o	0.63	Est(17)	161	o	0.80	Est(16)	161
389	OH	oB	0.02 ± 0.08	Stat	73	o	-0.22 ± 0.12	Stat	73
		B	0.13(0.01)	p*K*-W	76	o	(−0.12)	Aver	33
		B	0.07(0.1)	p*K*-50	170	B	−0.38(0.01)	p*K*-W	76
		o	$[0.04]^{e}$	Aver	27	o	$[-0.13]^{e}$	Aver	27
390	OMe	oB	**0.10** ± 0.03	Stat	73	o	-0.12 ± 0.05	Stat	73
		B	0.11(0.01)	p*K*-W	76	o	−0.15	HydrE	29
		o	$[0.06]^{e}$	Aver	27	B	−0.28(0.01)	p*K*-W	76
			$[0.30]^{f}$	OReac	119	o	$[-0.16]^{e}$	Aver	85, 27
							$[0.05]^{f}$	OReac	119
391	OEt	B	0.1(0.1)	p*K*-W	2	o	−0.14	HydrE	33
						B	−0.24(0.1)	p*K*-W	2
							$[-0.20]^{e}$	Phys	85
392	OR	B	0.1(0.1)	p*K*-W	2	B	−0.35(0.1)	p*K*-W	2

Table 10.1—continued

No.[a]	Substituent	Type[b]	σ_m[c]	Method[d]	Ref.	Type[b]	σ_p[c]	Method[d]	Ref.
395	OPh	B	0.26(0.1)	pK-50	171	0	0.05 ± 0.13	Aver	73
		B	0.25(0.02)	pK-W	2	0	0.05	HydrE	29
						B	0.14(0.1)	pK-50	171
						B	(−0.32)(0.02)	pK-W	2
						0	[−0.05][e]	Phys	85
							−0.10	OReac	96
396	OCH_2Ph						(−0.41)	Phys	1
397	OCH_2CO_2H						−0.18	OReac	111
							(−0.33)	Phys	172
398	OCH_2CO_2R						−0.18	OReac	111
399	OCH_2F	0	0.20	Est(17)	173	0	0.02	Est(16)	173
400	$OCHF_2$	B	0.31(0.1)	pK-MC	174	B	0.18(0.1)	pK-MC	174
401	OCF_3	B	0.35(0.1)	pK-50	13	B	0.32(0.1)	pK-50	13
		B	0.38(0.1)	pK-50	148	B	0.35(0.1)	pK-50	175
402	OCF_2CHF_2	B	0.34(0.1)	pK-50	175	B	0.25(0.1)	pK-50	175
403	OCF_2CF_3		0.48	pK-An	175		0.28	pK-An	175
404	OCF_2CHFCl	B	0.35(0.1)	pK-MC	176	B	0.28(0.1)	pK-MC	176
405	OCH_2Cl	0	0.25	Est(17)	173	0	0.08	Est(16)	173
406	$OCHCl_2$	0	0.38	Est(17)	173	0	0.26	Est(16)	173
407	$OCCl_3$	0	0.43	Est(17)	173	0	0.35	Est(16)	173
408	OAc	B	0.26(0.1)	pK-MC	150	B	0.16(0.1)	pK-MC	150
		B	0.39(0.1)	pK-50	2	B	0.31(0.1)	pK-50	2
							0.08	OReac	111
409	OBz	B	0.21(0.1)	pK-Av	150	B	0.13(0.1)	pK-MC	150
							0.14	OReac	111
410	$OCOCF_3$	0	0.56	Est(17)	177	0	0.46	Est(16)	177
413	OCN	0	0.67	Est(17)	177	0	0.54	Est(16)	177
414	$OSiMe_3$		(0.13)	Est(4)	20		(−0.27)	Est(4)	20

416	ONO_2		(0.55)	Est(4)	20		(0.70)	Est(4)	20
417	$OPO(OMe)_2$						0.04	Phys	164
418	OSOMe		(0.44)	Est(4)	20		(0.45)	Est(4)	20
419	OSO_2Me	B	0.39(0.1)	p*K*-Av	150	B	0.36(0.1)	p*K*-Av	150
420	OSO_2Ph }	B	0.36(0.1)	p*K*-Av	150	B	0.33(0.1)	p*K*-Av	150
	OTs }						(0.20)	Phys	178
421	OSO_2CF_3	B	0.56(0.1)	p*K*-50	13	B	0.53(0.1)	p*K*-50	13
422	OSF_5					B	0.44(0.1)	p*K*	179
423	$\overset{+}{O}Me\cdot\overset{-}{B}Cl_3$	0	0.51	Est(17)	177	0	0.48	Est(16)	177
424	SH	B	0.25(0.1)	p*K*-A	2	B	0.15(0.1)	p*K*-A	2
						0	$[0.06]^e$	Phys	85
425	SMe	0B	**0.14** ± 0.18	Aver	73	0	0.06 ± 0.18	Aver	73
		0	0.12	HydrE	29	0	0.05	HydrE	29
		B	0.15(0.1)	p*K*-50	2	0	$[-0.02]^e$	Phys	85
						B	0.00(0.1)	p*K*-50	2
						B	−0.07	HydrE	180
426	SR		(0.18)	Phys	79	B	0.03(0.1)	p*K*-50	2
			0.14	Est	—	B	0.07(0.1)	p*K*-50	2
428	$SCH{:}CH_2$	0	0.18	Est(17)	177	0	0.14	Est(16)	177
429	SC⋮CH	0	0.26	Est(17)	177	0	0.19	Est(16)	177
430	SPh	0	0.17	Est(17)	181	0	0.13	Est(16)	181
							$[0.01]^e$	Phys	85
							0.18	OReac	96
433	SCH_2F	0	0.23	Est(17)	13	0	0.20	Est(16)	13
434	$SCHF_2$	B	0.33(0.1)	p*K*-MC	174	B	0.36(0.1)	p*K*-MC	174
435	SCF_3	B	0.40(0.1)	p*K*-50	175	B	0.50(0.1)	p*K*-50	175
		B	0.36(0.1)	p*K*-50	13	B	0.42	Cor(18)	21
		B	0.37	Cor(18)	21				
436	SCF_2CHF_2	B	0.38(0.1)	p*K*-50	175	B	0.47(0.1)	p*K*-50	175
		B	0.39	Cor(18)	21	B	0.45	Cor(18)	21

Table 10.1—continued

No.[a]	Substituent	Type[b]	σ_m[c]	Method[d]	Ref.	Type[b]	σ_p[c]	Method[d]	Ref.
437	SAc	B	0.39(0.1)	pK-50	2	B	0.44(0.1)	pK-50	2
		o	0.21	Est(17)	177	B	0.45	Cor(18)	21
						o	0.20	Est(16)	177
438	$SCOCF_3$	o	0.48	Est(17)	177	o	0.46	Est(16)	177
445	SCN		0.53	NBz	123	B	0.52(0.1)	pK-50	2
		o	0.60	Est(17)	177	o	0.58	Est(16)	177
			(0.41)	Est(4)	20		(0.70)	pK	1
446	$SNMe_2$	o	0.12	Est(17)	177	o	0.09	Est(16)	177
447	SOMe	o	0.21	Est(17)	177	o	0.17	Est(16)	177
448	SSMe		0.22	Est(4)	20		0.13	Est(4)	20
449	SSO_2Me		0.43	Est(4)	20		0.54	Est(4)	20
450	SCl	o	0.44	Est(17)	177	o	0.48	Est(16)	177
451	$\overset{+}{S}Me\cdot\overset{-}{B}Cl_3$	o	0.77	Est(17)	177	o	0.82	Est(16)	177
452	SMe:NTs	B	0.65(0.1)	pK-50	185	B	0.70(0.1)	pK-50	185
453	SO·Me	B	0.52(0.1)	pK-50	2	B	0.49(0.1)	pK-50	2
		B	0.52	HydrE	180	B	0.54	HydrE	180
						o	0.57	HydrE	29
454	SO·Ph	o	0.51	Est(17)	181	B	0.46(0.1)	pK-50	182
						o	0.50	Est(16)	181
455	$SO\cdot CHF_2$		0.54	Phys	183		0.58	Phys	183
		o	0.70	Est(17)	13	o	0.76	Est(16)	13
456	$SO\cdot CF_3$	B	0.63(0.1)	pK-50	13	B	0.69(0.1)	pK-50	13
457	$SO\cdot NMe_2$	o	0.29	Est(17)	177	o	0.27	Est(16)	177
458	SO·OMe		0.66	Phys	184	o	0.54	Est(16)	177
		o	0.50	Est(17)	177				
459	SOF	o	0.74	Est(17)	177	o	0.83	Est(16)	177
460	SOCl	o	0.75	Est(17)	177	o	0.82	Est(16)	177

461	SO_2Me	^{0}B	0.68 ± 0.10	Stat	73	^{0}B	**0.73** ± 0.09	Stat	73
		B	**0.64**	Cor(18)	21	B	0.73	Cor(18)	21
		B	0.60(0.1)	p*K*-50	2	B	0.72(0.1)	p*K*-50	2
		^{0}	0.70	HydrE	29	^{0}	0.75	HydrE	29
462	SO_2R	B	0.64	Est	—	B	0.73	Est	—
		^{0}	0.59	Est(17)	177	^{0}	0.68	Est(16)	177
464	SO_2Ph	^{0}	0.59	Est(17)	181	B	0.70(0.1)	p*K*-50	182
		B	0.62	Cor(18)	21	B	0.71	Cor(18)	21
			(0.61)	Phys	79	^{0}	0.76	OReac	105
						^{0}	0.66	Est(16)	181
465	SO_2CH_2F	^{0}	0.66	Est(17)	13	^{0}	0.77	Est(16)	13
466	SO_2CHF_2	B	0.75(0.1)	p*K*-MC	174	B	0.86(0.1)	p*K*-MC	174
467	SO_2CF_3	B	0.76(0.1)	p*K*-50	13	B	0.96(0.1)	p*K*-50	13
		B	(0.79)(>0.1)	p*K*-50	175	B	0.93(0.1)	p*K*-50	175
		B	0.80	Cor(18)	21	B	0.91	Cor(18)	21
468	$SO_2CF_2CHF_2$						1.01	OReac	111
469	$SO_2CF_2CHFCF_3$						1.03	OReac	111
470	SO_2CF_2CHFCl						0.98	OReac	111
471	SO_2CN	^{0}	1.10	Est(17)	177	^{0}	1.26	Est(16)	177
472	SO_2NH_2	^{0}B	0.58 ± 0.13	Aver	73	^{0}B	**0.58** ± 0.12	Stat	73
		B	0.46(0.1)	p*K*-W	2	B	0.57(0.1)	p*K*-W	2
		B	0.53	Cor(18)	21	B	0.60	Cor(18)	21
473	SO_2NMe_2	^{0}	0.48	Est(17)	177		0.63	OReac	111
			0.54	Est	—		0.67	OReac	105
			(0.68)	Phys	186		(0.86)	Phys	186
							0.56	Est(16)	177
474	SO_2NHPh		0.56	Est	—		0.65	OReac	111
475	SO_2OH		(0.55)	Phys	172				
476	SO_2OR		0.71	Phys	184		0.90	Phys	187
477	SO_2OPh						(0.51)	Phys	172
478	SO_2F	B	0.79(0.1)	p*K*-Av	188	B	0.91(0.1)	p*K*-Av	188
		B	0.98(0.1)	p*K*-50	109	B	1.08(0.1)	p*K*-50	109

Table 10.1—continued

No.[a]	Substituent	Type[b]	σ_m[c]	Method[d]	Ref.	Type[b]	σ_p[c]	Method[d]	Ref.
479	SO_2Cl	^{0}	0.92	Est(17)	177	^{0}	1.04	Est(16)	177
							1.22	Phys	187
480	$S(OR_F)_2Ph$	^{0}	0.44	Est(17)	181		0.49	Est(16)	181
481	SF_3	^{0}	0.70	Est(17)	177	^{0}	0.80	Est(16)	177
482	SF_5	B	0.61(0.1)	p*K*-50	189	B	0.68(0.1)	p*K*-50	189
		B	0.59	Cor(18)	21		0.67	Cor(18)	21
483	SeMe	B	0.1(0.1)	p*K*-A	2	B	0.0(0.1)	p*K*-A	2
485	SePh						0.13	OReac	96
486	$SeCF_3$	B	0.32(0.1)	p*K*-50	13	B	0.38(0.1)	p*K*-50	13
487	SeCN		0.61	Est	146		0.66	p*K*	1
488	$SeO{\cdot}CF_3$	^{0}	0.81	Est(17)	13	^{0}	0.86	Est(16)	13
							0.63	Phys	13
489	$SeO_2{\cdot}CF_3$	^{0}	1.08	Est(17)	13	^{0}	1.21	Est(16)	13
490	F	^{0}B	**0.34** ± 0.05	Stat	73	^{0}	**0.15** ± 0.06	Stat	73
		B	0.34(0.01)	p*K*-W	76	^{0}	0.20	HydrE	29
		^{0}	0.33	Aver	126	B	0.06(0.02)	p*K*-W	2
491	Cl	^{0}B	**0.37** ± 0.03	Stat	73	^{0}B	**0.24** ± 0.03	Stat	73
		B	0.37(0.01)	p*K*-W	76	B	0.22(0.01)	p*K*-W	76
		^{0}	0.31	Aver	126		0.28	HydrE	29
		^{0}	[0.38][e]	Aver	126				
492	Br	^{0}B	**0.37** ± 0.04	Stat	73	^{0}B	**0.26** ± 0.04	Stat	73
		B	0.39(0.01)	p*K*-W	76	B	0.22(0.01)	p*K*-W	76
		^{0}	0.32	Aver	126	^{0}	0.30	HydrE	29
		^{0}	[0.38][e]	Aver	126				
493	I	^{0}B	**0.34** ± 0.04	Stat	73	^{0}B	0.28 ± 0.04	Stat	73
		B	0.35(0.01)	p*K*-W	76	B	0.21(0.01)	p*K*-W	76
		^{0}	0.38	Aver	126	^{0}	0.27	Aver	126
494	IO_2	B	0.70(0.1)	p*K*-W	2	B	0.76(0.1)	p*K*-W	2
		B	0.68	Cor(18)	21	B	0.78	Cor(18)	21

495	$I(OAc)_2$	o	0.85	Est(17)	13	o	0.88	Est(16)	13
496	$I(OCOCF_3)_2$	o	1.28	Est(17)	13	o	1.34	Est(16)	13
497	IF_2	o	0.85	Est(17)	13	o	0.83	Est(16)	13
498	ICl_2	o	1.10	Est(17)	13	o	1.11	Est(16)	13
499	IF_4	o	1.07	Est(17)	13	o	1.15	Est(16)	13
	Charged substituents [j]								
500	$CH_2\overset{+}{N}H_3\,I^-$		0.32	p*K*-Ph W	190		0.29	p*K*-Ph W	190
	X^-(?)	o	0.25	Est(17) MeOH/W	(10)		0.25	Est(16) MeOH/W	(10)
501	$CH_2\overset{+}{N}HMe_2\,I^-$		0.40	p*K*-Ph W	190		0.43	p*K*-Ph W	190
502	$CH_2\overset{+}{N}Me_3\,I^-$		0.40	p*K*-Ph W	190	B	0.44	HydrE W	102
							0.46	p*K*-Ph	190
503	$(CH_2)_2\overset{+}{N}H_3\,I^-$		0.23	p*K*-Ph W	190		0.17	p*K*-Ph W	190
504	$(CH_2)_2\overset{+}{N}HMe_2\,I^-$		0.24	p*K*-Ph W	190		0.14	p*K*-Ph W	190
505	$(CH_2)_2\overset{+}{N}Me_3\,I^-$		0.16	p*K*-Ph W	190	B	0.13	HydrE W	102
							−0.01	p*K*-Ph W	190
506	$(CH_2)_3\overset{+}{N}Me_3\,I^-$		0.06	p*K*-Ph W	190	B	0.02	HydrE W	102
							−0.01	p*K*-Ph W	190

Table 10.1—continued

No.[a]	Substituent	Type[b]	σ_m[c]	Method[d]	Ref.	Type[b]	σ_p[c]	Method[d]	Ref.
507	$(CH_2)_4\overset{+}{N}Me_3\ I^-$					B	−0.04	HydrE W	102
510	$NH_3^+\ Cl^-$		0.71	pK-An W(0)[h]	191		0.57	pK-An W(0)[k]	191
			0.64	pK-An W(0.1)[k]	192		0.49	pK-An W(0.1)[k]	192
	X^-(?)		0.86	Aver W	193		0.60	Aver W	193
	X^-(?)		0.98[g]	pK-W W	143				
514	$\overset{+}{N}Me_3\ Cl^-$	o	1.04	OReac W(0)[k]	72	o	0.88	OReac W(0)[k]	72
		B	0.99	pK-W W(0)[k]	194	B	0.96	pK-W W(0)[k]	194
		B	1.02	pK-50 A/W	195	B	0.88	pK-50 A/W	195
			0.95	pK-An W(0)[k]	194		0.89	pK-An W(0)[k]	194
			0.91	pK-An W(0)[k]	191		0.81	pK-An W(0)[k]	191
			0.82	pK-DM W(0)[k]	194		0.80	pK-DM W(0)[k]	194
			0.96	pK-Ph W(0)[k]	194		0.85	pK-Ph W(0)[k]	194
	I^-		0.74	pK-Ph W(0.1)[k]	190		0.66	pK-Ph W(0.1)[k]	190
						B	0.72	HydrE W	102

No.	Substituent								
	Cl^-	B	0.91	OReac A	195	B	0.70	OReac A	195
515	$\overset{+}{N}{:}N\ BF_4^-$	B	1.76	p*K*-W W	196	B	1.91	p*K*-W W	196
						o	2.18	p*K*-W W	196
516	$\overset{+}{P}Me_3\ I^-$		0.74	OReac W/MeOH	166				
		o	0.50	Est(17) DMF	160	o	0.60	Est(16) DMF	160
	BF_4^-	o	0.53	Est(17) DMSO	165	o	0.63	Est(16) DMSO	165
518	$\overset{+}{P}MePh_2\ I^-$					B	1.01	p*K* MeOH/W	163
	X^-(?)	o	0.71	Est(17) DMSO	165	o	0.83	Est(16) DMSO	165
	I^-						1.18	Phys $CDCl_3$	187
521	$\overset{+}{S}Me_2\ TsO^-$	B	1.00	p*K*-W	197	B	0.90	p*K*-W	197
	BsO^-		1.00	p*K*-Ph W	197				
	ClO_4^-	o	0.97	Est(17) MeCN	177	o	1.06	Est(16) MeCN	177
522	$\overset{+}{S}(OR_F)Ph$ $CF_3SO_3^-$	o	1.46	Est(17) $CHCl_3$	181	o	1.62	Est(16) $CHCl_3$	181
523	$\overset{+}{I}Ph\ I^-$	B	0.85	p*K* MeCN/W	198				

Table 10.1—continued

No.[a]	Substituent	Type[b]	σ_m[c]	Method[d]	Ref.	Type[b]	σ_p[c]	Method[d]	Ref.
524	$CH_2CO_2^- M^+$(?)	o	−0.12	Est(17) MeOH/W	(10)	o	−0.18	Est(16) MeOH/W	(10)
528	$CO_2^- K^+$	B	0.09	p*K*-W (0.1)[k]	192	B	−0.05	p*K*-W (0.1)[k]	199
	M^+(?)	B	−0.1	p*K*-W	2	B	0.0	p*K*-W	2
	Na^+	o	−0.19	OReac W(0)[p]	72	o	−0.12	OReac W(0)[k]	72
	K^+					o	−0.05	OReac Dioxan/W	152
	M^+(?)		0.02	Aver W	193		0.11	Aver W	193
	Na^+	o	−0.30	Est(17) MeOH/W	(10)	o	−0.25	Est(16) MeOH/W	(10)
	M^+(?)		0.01[g]	p*K*-An W	143				
529	1-$C_2HB_{10}H_{11}^-$(1,2) Na^+	o	−0.47	Est(17) THF/W	200	o	−0.56	Est(16) THF/W	200
530	1-$C_2HB_{10}H_{10}^{2-}$(1,2) $2Na^+$	o	−0.63	Est(17) THF	200	o	−0.73	Est(16) THF	200
533	$PO_2H^- Na^+$					B	0.14	p*K*-W	201
534	$PO_3H^- Na^+$		0.25	p*K*-W	202		0.17	p*K*-W	202
			0.26	p*K*-50	2, 202		0.31	p*K*-50	2, 202
535	$PO_3^{2-} 2Na^+$		−0.02	p*K*-Ph, W	202				
536	$AsO_3H^- K^+$		0.0	Est(16,17) (?)	146	B	−0.02	p*K*-W	2, 203
538	$O^- M^+$(?)	B	−0.71	OReac A/W	1		−0.52	p*K* MeOH/W	1

		K^+	0	−0.87	OReac W(0)[k]	72	0	−0.82	OReac W(0)[k]	72
		M^+(?)		−0.47	Aver W	193		−0.81	Aver W	193
		M^+(?)	0	−0.46	Est(17) MeOH-W	(10)	0	−0.76	Est(16) MeOH/W	(10)
		K^+					0	−0.75	OReac Dioxan/W	152
540	SO_2^-	Na^+	B	−0.02	p*K*-W (0)[k]	204	B	−0.05	p*K*-W (0)[k]	204
			B	0.22	p*K*-W (0.1)[k]	204	B	0.31	p*K*-W (0.1)[k]	204
541	SO_3^-	K^+	B	0.05	p*K*-W (0)[k]	205	B	0.09	p*K*-W (0)[k]	205
		K^+	B	0.31	p*K*-W (0.1)[k]	205	B	0.37	p*K*-W W(0.1)[k]	205
		K^+		0.39	p*K*-An (0.1)[k]	205		0.35	OReac W(0)[k]	72
		M^+(?)	0	0.26	Est(17) MeOH	177	0	0.30	Est(16) MeOH	177
		Na^+		0.15	Phys crystal	172		0.30	Phys crystal	172

[a] The numbering is common to Tables 10.1, 10.2, 10.4, and 10.5.
[b] 0—Sigma-zero reactivity; B—benzoic-like reactivity; 0B—the validity in both scales has been proved. For the values without a symbol the type to which they belong is not certain.
[c] Values printed in boldface are those that seem preferable to the author; values in parentheses are particularly uncertain (see p. 452).
[d] The abbreviations should be intelligible; for detailed explanation see p. 453.
[e] Special values for nonpolar solvents.
[f] Special values for protonated forms in trifluoroacetic acid as solvent.
[g] Corrected for the tautomeric equilibrium.
[h] Uncorrected for the tautomeric equilibrium, although this may affect the result.
[i] These values are based on σ_I and σ_R, which themselves are calculated according to equations (10.7) and (10.11). They differ from σ_m, σ_p given in the original paper,[160] which were calculated by another procedure.
[j] The solvent is given with the method.
[k] Ionic strength.
[l] Error in the original reference.

Table 10.2. The Dual Constants, σ_p^+ and σ_p^-

No.[a]	Substituent[b]	σ_p^+	Method[c]	Ref.
2	D	−0.001	OSolv	206
3	Me	−0.31	CumS	12
		−0.31	CumS	207
		−0.30	ArEl-Av	208
5	Et	−0.30	CumS	12
		−0.28	ArEl-Av	208
7	Pr^i	−0.28	CumS	12
		−0.28	CumS	207
		−0.26	ArEl-Av	208
11	Bu^t	−0.26	CumS	12
		−0.26	CumS	207
		−0.25	ArEl-Av	208
		[−0.31][d]	ArEl	209
26	$C \vdots CH$	0.18	OSolv	87
30	c-C_3H_5	−0.46	CumS	89
		−0.41	CumS	207
		−0.45	OSolv	90
		−0.52	Phys	210
32	c-C_4H_7	−0.29	CumS	89
33	c-C_5H_9	−0.30	CumS	89
34	c-C_6H_{11}	−0.28	CumS	89
40	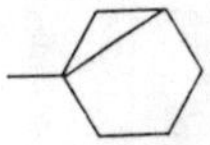	(−0.71)	Phys	210
41	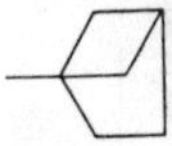	(−0.5)	Phys	210
42	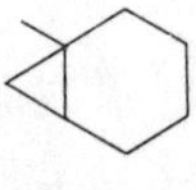	(−0.60)	Phys	210
43	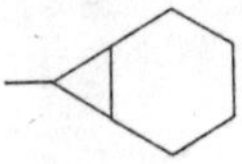	(−0.69)	Phys	210
44	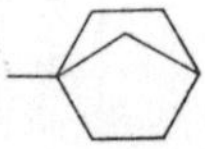	(−0.4)	Phys	210
45	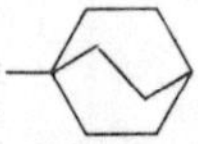	−0.27	CumS	92
46	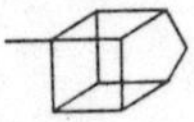	(−0.75)	Phys	210
47	1-Adamantyl	−0.27	CumS	92
48	Ph	−0.18	CumS	12
		[−0.22][d]	CumS	208

Table 10.2—continued

No.[a]	Substituent[b]	σ_p^+	Method[c]	Ref.
48 – *continued*		[−0.34][e]	OSolv	211
		−0.21	OReac	33
		−0.26	ArEl	209
49	CH_2Ph	−0.27	OReac	212
		−0.23	ArEl	209
60	CH:CHPh	−1.0	OSolv	98
61	C⋮CPh	−0.03	OSolv	98
68	CH_2CO_2H	−0.02	ArEl	213
70	CH_2CO_2R	−0.16	ArEl	12
		−0.01	ArEl	213
71	CH_2CN	+0.12	ArEl	213
76	CH_2HgCH_2Ph	(−1.12)	Phys	214
77	CH_2SiMe_3	(−0.66)	Phys	214
81	CH_2SiPh_3	(−0.4)	Phys	215
82	CH_2GePh_3	(−0.6)	Phys	215
83	CH_2SnMe_3	(−0.90)	Phys	214
84	CH_2SnPh_3	(−0.75)	Phys	215
85	CH_2PbPh_3	(−1.08)	Phys	214
86	$(CH_2)_2PbPh_3$	(−0.22)	Phys	214
103	CH_2OH	0.01	ArEl	213
109	CH_2OR	−0.05	ArEl	213
		0.00	ArEl	213
133	CH_2Cl	−0.01	Aver	12
136	CH_2Br	(−0.06)	ArEl	213
179	C_6F_5	0.26	ArEl	116
211	2-C_4H_3O	−0.39	OSolv	127
213	2-C_4H_3S	−0.33	OSolv	216
227	$C_{10}H_9Fe$	−0.7	OSolv	134
		−0.65	Est	132
251	$SiMe_3$	0.02	CumS	12
		−0.03	ArEl	217
281	$SnMe_3$	−0.12	ArEl	217
285	NH_2	−1.3	Aver	12
		−1.31	ArEl-Av	218
		−1.47	OReac	33
287	NMe_2	−1.7	Aver	12
		−1.50	ArEl-Av	218
		−1.67	OReac	33
290	NHPh	−1.4	OReac	12
297	NHAc	−0.6	Aver	12
		(−0.60)	Phys	214
		−0.69	ArEl-Av	218
303	NHBz	−0.6	Aver	12
338	N:NPh	(−0.15)	OReac	61
373	$POPh_2$	0.49	OReac	219
376	$PO(OEt)_2$	0.54	OReac	219
389	OH	−0.92	ArEl-Av	218

Table 10.2—continued

No.[a]	Substituent[b]	σ_p^+	Method[c]	Ref.
390	OMe	−0.78	CumS	12
		−0.78	CumS	207
		−0.76	ArEl-Av	208
		[−0.60][d]	ArEl	209
391	OEt	−0.82	ArEl	220
392	OR	−0.83	ArEl	220
395	OPh	−0.5	Aver	12
		−0.53	OReac	212
		−0.56	ArEl-Av	218
		−0.48	ArEl	220
		[−0.51][d]	ArEl	209
425	SMe	−0.60	CumS	12
		−0.55	ArEl	209
		−0.60	ArEl-Av	221
430	SPh	−0.45	ArEl	209
490	F	−0.07	CumS	12
		−0.02	ArEl-Av	208
491	Cl	0.11	CumS	12
		0.10	ArEl-Av	208
492	Br	0.15	CumS	12
		0.14	ArEl-Av	208
493	I	0.13	CumS	12
		0.13	ArEl-Av	208
	Charged Substituents[f]			
501	$CH_2\overset{+}{N}HMe_2\ SO_4H^-$	0.50	ArEl (AcOH/W)	213
502	$CH_2N^+Me_3\ SO_4H^-$	0.50	ArEl (AcOH/W)	213
538	$O^-\ M^+$	−2.3	Est (?)	33

No.[a]	Substituent[b]	σ_p^-	Method[c]	Ref.
2	D	−0.002	OReac	74
26	C⋮CH	0.52	OReac	222
29	C⋮C·C⋮CH	0.72	OReac	223
30	$c\text{-}C_3H_5$	−0.09	p*K*-An	91
32	$c\text{-}C_4H_7$	−0.07	p*K*-An	91
40	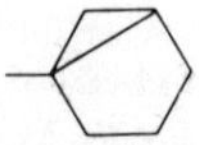	−0.14	p*K*-An	91
42	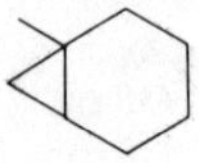	−0.07	p*K*-An	91
48	Ph	0.08	p*K*-An	224
		0.02	AnReac	96
		0.07	AnReac	96

Table 10.2—continued

No.[a]	Substituent[b]	σ_p^-	Method[c]	Ref.
48 – *continued*		0.11	p*K*-Ph	225
		0.30	Est	226
56	C_6H_4Ph-*p*	0.05	AnReac	96
60	CH:CHPh	0.13	p*K*-An	227
		0.15	p*K*-Ph	228
61	C⋮CPh	0.39	p*K*-An	228
		0.36	p*K*-Ph	228
		0.30	AnReac	96
132	CF_3	0.74	p*K*-An	229
		0.64	p*K*-DM	229
		0.62	p*K*-An	110
		0.60	p*K*-DM	110
148	C_2F_5	0.69	p*K*-DM	110
149	$CF(CF_3)_2$	0.68	p*K*-An	110
		0.66	p*K*-DM	110
150	C_4F_9-n	0.73	p*K*-DM	110
153	$C(OH)(CF_3)_2$	0.48	p*K*-An	110
		0.44	p*K*-DM	110
162	$CH:C(CN)_2$	1.20	p*K*-DM	108
163	$C(CN):C(CN)_2$	1.70	p*K*-DM	108
164	$CH:CHNO_2$	0.88	p*K*-Ph	112
169	$CH:CHCF_3(E)$	0.34	p*K*-DM	108
170	$CH:CHCF_3(Z)$	0.29	p*K*-DM	108
181	CHO	1.04	p*K*-Ph	230
		0.94	ArNu	231
182	COMe	0.82	p*K*-An	33
		0.84	p*K*-Ph	230
		0.87	ArNu	231
186	COPh	0.88	p*K*-Ph	124
		0.59	AnReac	96
		0.86	AnReac	232
		0.93	PhReac	232
189	$CONH_2$	0.62	p*K*-An	33
		0.63	OReac	1
		0.61	p*K*-Ph	230
194	CO_2H	0.73	Aver	1
		0.78	p*K*-An	224
195	CO_2R	0.74	p*K*-An	33
		0.66	Aver	1
		0.64	p*K*-Ph	230
196	CO_2CH_2Ph	0.67	p*K*-Ph	224
203	CN	0.99	p*K*-An	33
		1.02	p*K*-An	224
		0.88	p*K*-Ph	230
		1.00	AnNu	231
211	2-C_4H_3O	0.21	p*K*-Ph	127
213	2-C_4H_3S	0.19	p*K*-Ph	128

Table 10.2—continued

No.[a]	Substituent[b]	σ_p^-	Method[c]	Ref.
215	3-C_4H_3S	0.13	p*K*-Ph	128
216	2-C_5H_4N	0.75	p*K*-An	130
217	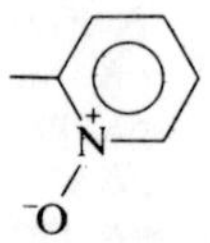	0.27	p*K*-An	130
218	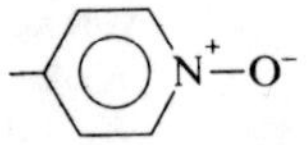	0.33	p*K*-An	130
219	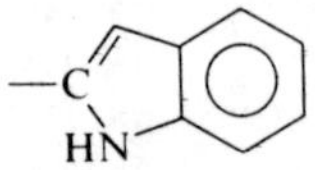	0.39	p*K*-Ph	233
221	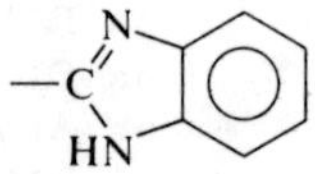	0.48	p*K*-Ph	233
222	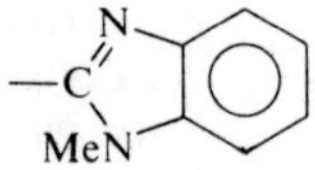	0.58	p*K*-Ph	233
224	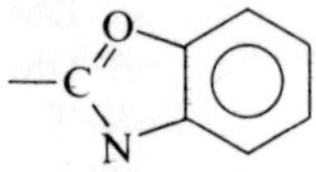	0.68	p*K*-Ph	233
225	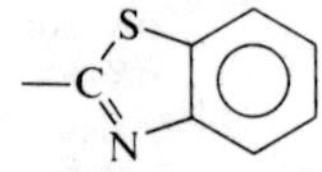	0.65	p*K*-Ph	233
227	$C_{10}H_9Fe$	−0.03	p*K*-An	132
		−0.05	p*K*-Ph	132
228	1-$C_2HB_{10}H_{10}$(1,2)	0.52	p*K*-An	135
244	$B(OH)_2$	0.45	p*K*-An	1
251	$SiMe_3$	0.08	p*K*-An	140
		0.11	p*K*-DM	140
		0.07	p*K*-Ph	140
		0.17	OReac	236
254	$SiPh_3$	0.27	DMReac	141
277	$GePh_3$	0.24	DMReac	141
331	N:CHPh	0.22	p*K*-Ph	124
337	N:C:S	0.34	p*K*-DM	156
338	N:NPh	0.61	p*K*-An	237
		0.69	p*K*-Ph	124
		0.72	PhReac	232
		0.45	AnReac	96
		0.66	ArNu	231
341	N:N(O)Ph	0.60	ArNu	231
342	N(O):NPh	0.77	ArNu	231

Table 10.2—continued

No.[a]	Substituent[b]	σ_p^-	Method[c]	Ref.
343	$N{:}N^+{:}N^-$	0.08	ArNu	231
344	NO	1.46	ArNu	231
		(1.60)	p*K*-Ph	230
		(0.15)	p*K*-DM	238
346	NO_2	1.23	p*K*-An	33
		1.25	p*K*-An	239
		1.28	p*K*-Ph	239
		1.24	p*K*-Ph	230
		1.27	ArNu	231
351	—N< (benzotriazol-2-yl: N=C, N=C fused benzene ring)	0.51	p*K*-An	159
		0.57	p*K*-Ph	159
352	—N(N=N)(HC=N) ring	0.57	p*K*-DM	108
354	—N—CH, N—O—CO (±) (sydnone ring)	0.71	ArNu	234
		(1.0)	Phys	235
355	—N—CH, N—O—C:NH (±) (sydnone imine ring)	(1.1)	Phys	235
359	PPh_2	0.26	p*K*-Ph	147, 168
		0.43	p*K*-DM	240
		(0.39)	Phys	163
371	$POMe_2$	0.72	OReac	166
		0.74	p*K*-DM	240
373	$POPh_2$	0.84	OReac	166
		0.88	p*K*-DM	240
		0.68	p*K*-Ph	147
		(0.99)	Phys	163
376	$PO(OEt)_2$	0.84	p*K*-An	169, 224
		0.75	p*K*-Ph	169, 224
		0.83	OReac	166
380	$PSPh_2$	0.63	p*K*-Ph	147
		(0.89)	Phys	163
		0.76	p*K*-DM	240
425	SMe	0.04	p*K*-An	224
		0.21	p*K*-Ph	241
430	SPh	0.18	AnReac	96

Table 10.2—continued

No.[a]	Substituent[b]	σ_p^-	Method[c]	Ref.
435	SCF_3	0.64	pK-An	175
		0.57	pK-Ph	175
436	SCF_2CHF_2	0.61	pK-An	175
437	SCOMe	0.46	pK-Ph	241
445	SCN	0.59	pK-An	224
		0.60	pK-Ph	241
452	SMe:NTs	1.00	pK-Ph	185
453	SO·Me	0.73	pK-Ph	241
458	SO·OMe	(0.89)	Phys	184
461	SO_2Me	1.05	pK-An	33
		1.13	pK-An	242
		0.98	pK-Ph	242
464	SO_2Ph	0.95	pK-Ph	182
		1.16	ArNu	243
467	SO_2CF_3	1.65	pK-An	175
		1.36	pK-Ph	175
		1.47	pK-Ph	224
472	SO_2NH_2	0.89	pK-An	224
		0.94	pK-An	244
473	SO_2NMe_2	0.99	ArNu	243
476	SO_2OR	(1.06)	Phys	184
478	SO_2F	(1.32)	Phys	184
482	SF_5	0.83	pK-An	189
		0.70	pK-Ph	189
485	SePh	0.13	AnReac	96
	Charged Substituents[f]			
508	2-pyridinium (HN^+) $PO_4H_2^-$	0.75	pK-An (A/W)	130
509	4-pyridinium ($\overset{+}{N}H$) $PO_4H_2^-$	0.65	pK-An (A/W)	130
515	$\overset{+}{N}{:}N\ BF_4^-$	3.43	pK-An (W)	196
		3.04	pK-Ph (W)	196
	X^-(?)	1.87	Est (?)	231
516	$\overset{+}{P}Me_3\ I^-$	1.14	OReac (MeOH/W)	166
		0.95	pK-DM ($CHCl_3$/AcOH)	240
518	$\overset{+}{P}MePh_2\ I^-$	1.28	pK-Ph (A/W)	147
		1.17	pK-DM ($CHCl_3$/AcOH)	240

Table 10.2—continued

No.[a]	Substituent[b]	σ_p^-	Method[c]	Ref.
521	$\overset{+}{S}Me_2\ I^-$	1.16	p*K*-Ph (W)	245
	BsO^-	1.16	p*K*-Ph (W)	197
523	$\overset{+}{I}Ph\ I^-$	0.71	p*K*-Ph (A/W)	198
		(1.06)	p*K*-An (W)	198
527	$CH{:}CHCO_2^-\ M^+(?)$	0.24	ArNu (MeOH)	243
528	$CO_2^-\ K^+$	0.37	p*K*-Ph (W)	199
	$M^+(?)$	0.24	p*K*-Ph (W)	230
	K^+	0.34	OReac (W)	199
	$M^+(?)$	0.32	ArNu (Bz)	34
	$M^+(?)$	0.14	ArNu (MeOH)	243
535	$PO_3^{2-}\ 2Na^+$	−0.16	p*K*-Ph (W)	202
536	$AsO_3H^-\ K^+$	0.19	p*K*-An (W)	203
		(0.71)	p*K*-Ph (W)	203
537	$AsO_3^{2-}\ 2M^+(?)$	0.70	p*K*-Ph (W)	224
540	$SO_2^-\ M^+(?)$	0.08	Phys (A/W)	184
541	$SO_3^-\ K^+$	0.58	p*K*-An (W)	205
	$M^+(?)$	0.40	p*K*-Ph (W)	230
	K^+	0.58	p*K*-Ph (W)	205
	$M^+(?)$	0.19	ArNu (MeOH)	243
	$M^+(?)$	0.24	Est (A/W)	184

[a] The numbering is common to Tables 10.1, 10.2, 10.4, and 10.5.
[b] Substituents are included only for which the conjugation in the given direction has been proved or may be theoretically proposed; for the remaining ones it is assumed that $\sigma_p^{+-} = \sigma_p$.
[c] The abbreviations should be intelligible; for detailed explanation see p. 453.
[d] Special values for polar solvents.
[e] Coplanar phenyl group.
[f] The solvent is given (in parentheses) with the method.

Table 10.3. The Aryl Values, σ_a

System	Position	Type[a]	σ_a	Method[b]	Ref.	σ_a^+	σ_a^-	Method[b]	Ref.
	1		0			0	0		
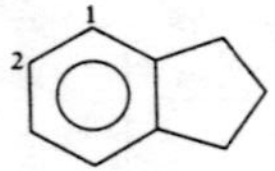	2	0	−0.26	AnReac	1		−0.26	AnReac	1
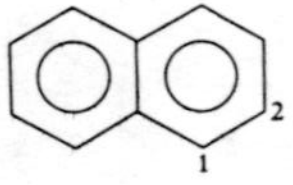	1*[c]	B	0.56*	pK-W	246	−0.45*		ArEl	208
		B	0.12*	pK-A	248	−0.51*		ArProt	208
			[0.17]*[d]	Est	35	−0.35*		ArEl	209
							0.27*	pK-An	(247)
							0.27*	pK-Ph	246
	2	B	0.04	pK-W	249	−0.14		CumS	248
		B	0.04	HydrE	249	−0.28		ArEl	208
		0	0.08	pK-W	249	−0.51		ArProt	208
						−0.14		Aver	249
							0.12	pK-An	249
							0.05	pK-Ph	249
							0.17	Aver	249
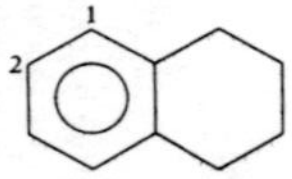	2	0	−0.48	AnReac	1		−0.48	AnReac	1
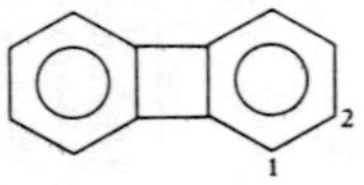	1*					−0.23*		ArEl	209
	2					−0.47		ArEl	209

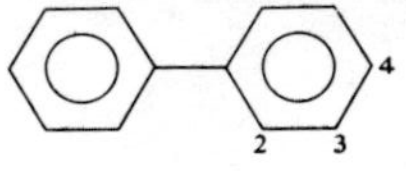	2*	B	0.74*	p*K*-W	52	−0.25*		ArEl	208
						−0.35*		ArProt	208
							0.40*	p*K*-An	(247)
	3	[c]B	0.04	Aver	73	0.11		CumS	208
		B	0.05	p*K*-A	94		(0.13)	p*K*-An	(247)
	4	[c]B	0.05	Stat	73	−0.18		CumS	250
		B	0.02	p*K*-A	94	−0.21		ArEl	208
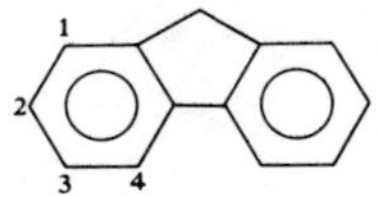	2					−0.30		ArProt	208
							0.02	p*K*-An	224
						−0.49		CumS	250
						−0.54		ArEl-Av	208
							−0.01	p*K*-An	(247)
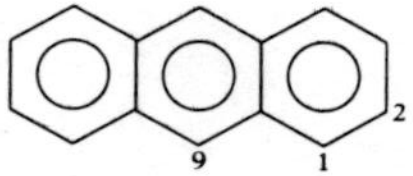	1*	B	0.50*	p*K*-W	(35)		0.33*	p*K*-An	(247)
			[0.28]*[d]	Est	35				
	2	B	0.02	p*K*-W	(35)	−0.30		CumS	250
			[0.14][d]	Est	35		0.27	p*K*-An	(247)
	9*	B	0.55*	p*K*-W	(35)	−0.69*		OSolv	(249)
			[0.44]*[d]	Est	35	−1.25*		ArProt	208
							(0.51)*	p*K*-An	(247)
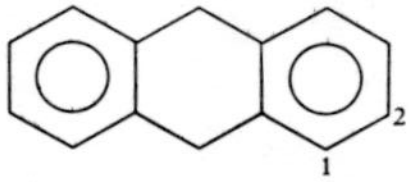	1*					−0.30*		ArEl	209
	2					−0.32		ArEl	209
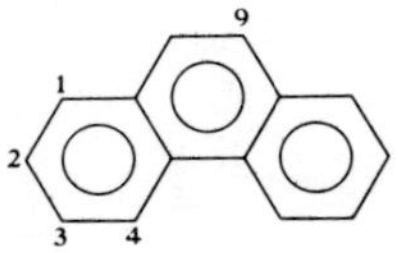	1*					−0.43*		ArEl	208
						−0.55*		ArProt	208
						−0.34*		ArEl	209
							0.33*	p*K*-An	(247)
	2		[0.04][d]	Est	35	−0.12		CumS	250
						−0.33		ArEl	208
						−0.25		ArEl	209
							0.20	p*K*-An	(247)

Table 10.3—continued

System	Position	Type[a]	σ_a	Method[b]	Ref.	σ_a^+	σ_a^-	Method[b]	Ref.
	3		$[0.08]^d$	Est	35	−0.20		CumS	250
						−0.41		ArEl	208
						−0.29		ArEl	209
							0.20	pK-An	(247)
	4*					−0.32*		ArEl	208
						−0.33*		ArEl	209
	9*		$[0.17]^{*d}$	Est	35	−0.45*		ArEl	208
						−0.55*		ArProt	209
						−0.36*		ArEl	208
							0.36*	pK-An	(247)
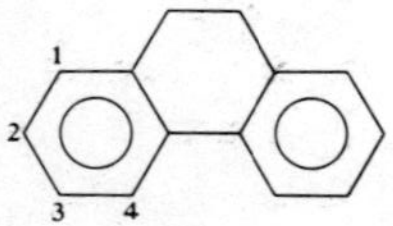	1*					−0.39*		ArEl	209
	2					−0.35		OSolv	249
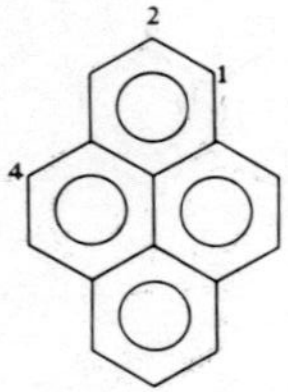	1*					−0.67*		ArEl	208
						−0.71*		ArEl	208
						−1.05*		ArProt	208
	2					−0.22		ArEl	209
	4*					−0.36*		ArEl	209
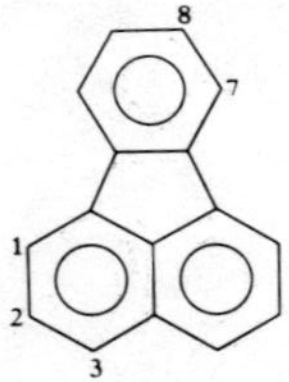	1*					−0.41*		ArEl	208
						−0.28*		ArEl	209
	2					−0.21		ArEl	209
	3*					−0.52*		ArEl	208
						−0.45*		ArEl	209

Structure	Position	Value	Method	Ref.
	7*	−0.46*	ArEl	208
		−0.27*	ArEl	209
	8	−0.49	ArEl	208
		−0.41	ArEl	209
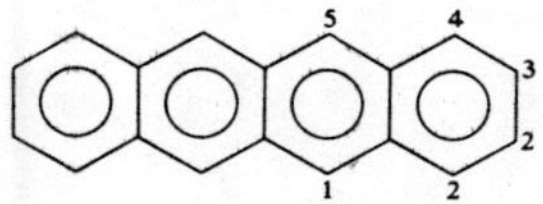 (tetracene, positions 1–5)	5*	−1.40*	ArEl	208
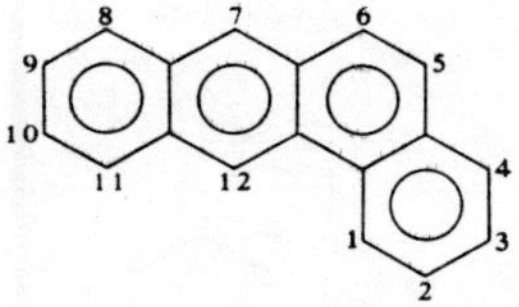 (benz[a]anthracene, positions 1–12)	7*	−1.11*	ArProt	208
	12*	−1.11*	ArProt	208
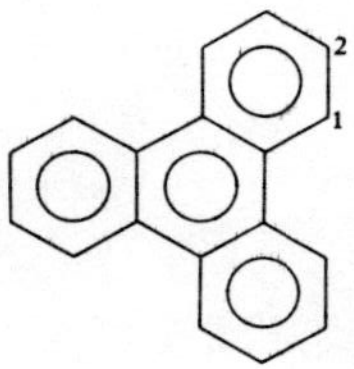 (triphenylene, positions 1, 2)	1*	−0.46*	ArEl	208
		−0.44*	ArProt	208
	2	−0.46	ArEl	208
		−0.35	OSolv	249
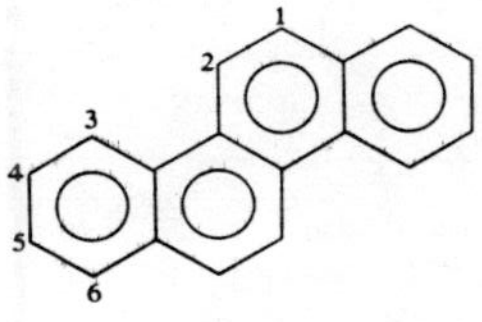 (chrysene, positions 1–6)	6*	−0.59*	ArEl	208
		−0.74*	ArProt	208
		−0.45*	ArEl	209

Table 10.3—continued

System	Position	Type[a]	σ_a	Method[b]	Ref.	σ_a^+	σ_a^-	Method[b]	Ref.
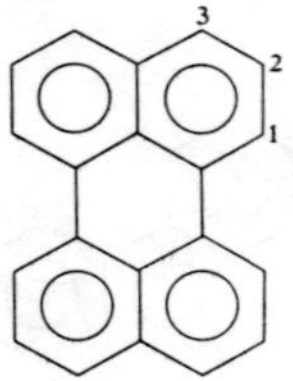	3*					−0.82*		ArEl	208
						−1.27*		ArProt	208
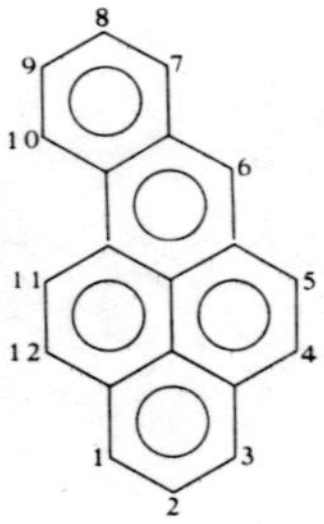	6*					−0.84*		ArEl	208
						−1.51*		ArProt	208
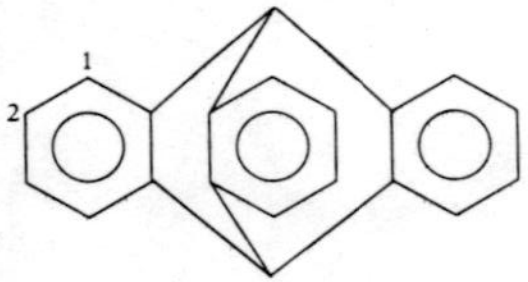	1*					−0.19*		ArEl	209
	2					−0.29		ArEl	209

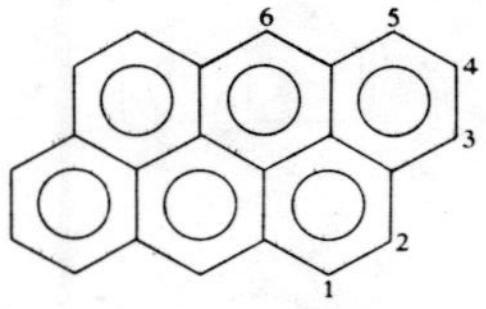	6*					−0.87*	ArEl	208
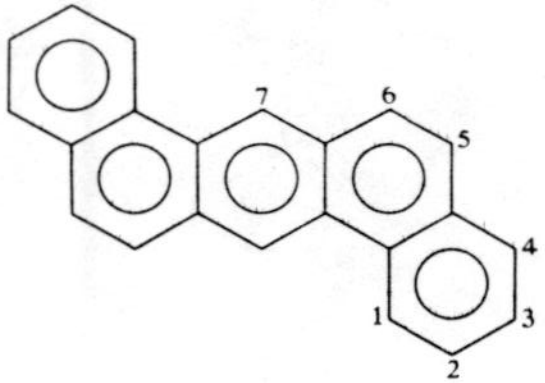	7*					−1.10*	ArProt	208
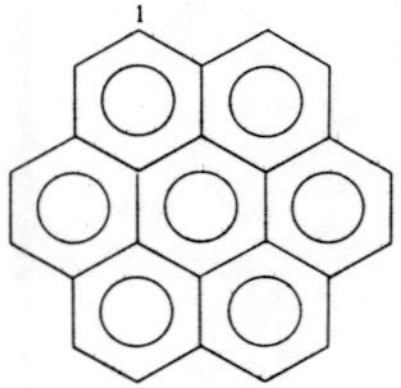	1*					−0.51*	ArEl	208
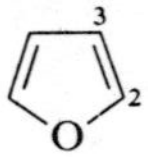	2*	B	1.08*	p*K*-W	158	−0.37*	OReac	158
			[0.67][d]*	Est	35	−0.98*	ArEl	158
		0	0.41*	HydrE	251	−1.32*	Aver[e]	252
		0	0.57*	Aver[e]	252			
		°B	0.32*	HydrE	253			
	3	B	0.25	p*K*-W	35, 158	−0.74	Aver[e]	252
		0	0.42	Aver[e]	252	−0.44	OSolv	253
		B	(0.04)	HydrE	253			

Table 10.3—continued

System	Position	Type[a]	σ_a	Method[b]	Ref.	σ_a^+	σ_a^-	Method[b]	Ref.
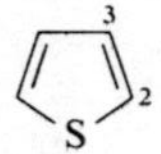 S	2*	B	0.71*	p*K*-W	158	−0.27*		OReac	158
		B	0.31*	p*K*-A	248	−0.84*		ArEl	209
			[0.30][d]*	Est	35	−1.22*		Aver[e]	252
		o	0.50*	HydrE	251		0.58*	OReac	158
		o	0.50*	Aver[e]	252				
		B	(0.03)*	HydrE	253				
	3	B	0.12	p*K*-W	35, 158	−0.49		ArEl	209
		B	0.07	p*K*-A	248	−0.62		Aver[e]	252
		o	0.28	HydrE	251				
		o	0.26	Aver[e]	252				
		B	0.04	HydrE	253				
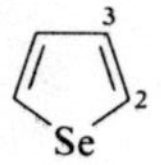 Se	2*	B	0.62*	p*K*-W	252	−1.28*		Aver[e]	252
		B	0.28*	p*K*-A	158				
		o	0.55*	Aver[e]	252				
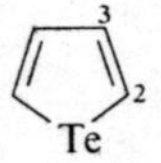 Te	2*	B	0.23*	p*K*-W	216	−1.00*		OSolv	254
						−0.90*		ArEl	254
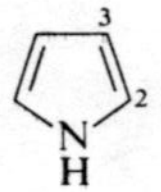 N H	2*	B	−0.25*	p*K*-W	158	−1.61*		OReac	253
		B	−0.58*	HydrE	253	(−1.33)*		Phys	255
	3	B	−0.94	HydrE	253	−1.20		OReac	253
						(−0.54)		Phys	255
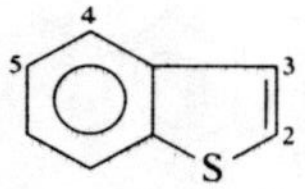 S	2*					−0.61*		ArEl	209
	3					−0.62		ArEl	209

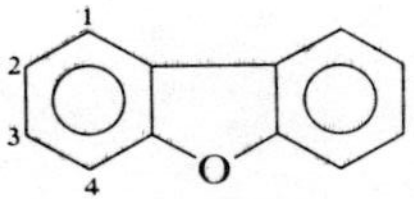 dibenzofuran	1*					−0.24*		ArEl	209
	2					−0.40		ArEl	209
	3					−0.28		ArEl	209
	4*					−0.25*		ArEl	209
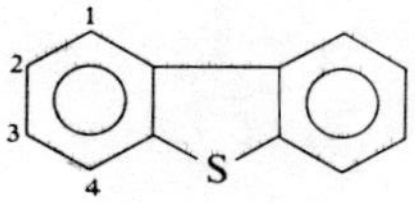 dibenzothiophene	1*					−0.28*		ArEl	209
	2					−0.37		ArEl	209
	3					−0.30		ArEl	209
	4*					−0.29*		ArEl	209
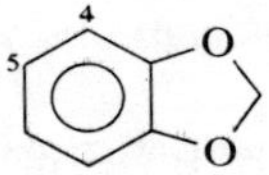 1,3-benzodioxole	5	B	−0.16	pK-W	158				
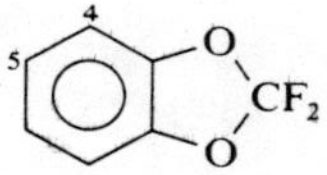 2,2-difluoro-1,3-benzodioxole (CF_2)	5	B	0.36	pK-50	13				
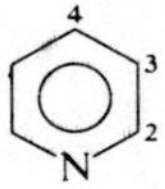 pyridine (N)	2*	B	0.45*	pK-W	158	0.73*		CumS	256
		B	0.75*	HydrE	257		1.00*	ArNu	258
		B	0.88*	OReac	259		0.58*	OReac	158
		o	0.81*	HydrE	260				
	3	B	[0.45][f]	pK-W	261	0.57		CumS	256
		B	0.65	pK-W	158		0.59	ArNu	258
		B	0.62	HydrE	257		0.76	OReac	158
		B	0.74	OReac	259		0.67	pK-Ph	261
		o	0.72	HydrE	260				
		o	[0.53][f]	pK-W	261				
	4	B	[0.76][f]	pK-W	261	1.13		CumS	256
		B	0.67	pK-W	158		1.16	ArNu	258
		B	0.96	HydrE	257		1.26	OReac	158
		B	1.10	OReac	259				
		o	0.95	HydrE	260				
		o	[0.85][f]	pK-W	261				

Table 10.3—continued

System	Position	Type[a]	σ_a	Method[b]	Ref.	σ_a^+	σ_a^-	Method[b]	Ref.
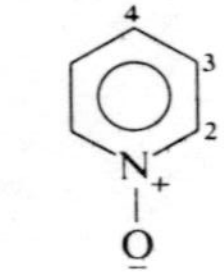	2*	B	(1.5)*	Phys	262		1.50*	ArNu	258
		B	(−0.39)*	OReac	259				
	3	B	1.48	p*K*-W	158	1.18		Phys	158
		B	1.31	OReac	259		1.59	p*K*-Ph	158
		B	0.7	Phys	262		1.18	ArNu	258
	4	B	1.35	p*K*-W	158	0.23		Phys	158
		B	1.14	OReac	259		1.88	p*K*-Ph	158
		B	0.4	Phys	262		1.53	ArNu	258
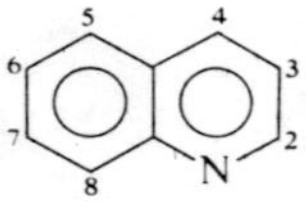	6	B	0.23	p*K*-W	158		0.46	p*K*-Ph	158
	7	B	0.24	p*K*-W	158		0.47	p*K*-Ph	158
$C_{10}H_{10}Fe$	1	B	−0.28	p*K*	132	−1.4		OSolv	134
		B	(−0.20)	p*K*-W	263	−1.54		OReac	262
		B	−0.58	OReac	133				
$C_2H_2B_{10}H_{10}$-(1,2)	1*	B	2.1*	p*K*-50	(200)				
	3*	B	0.20*	p*K*-50	(200)				
Charged systems[g]									
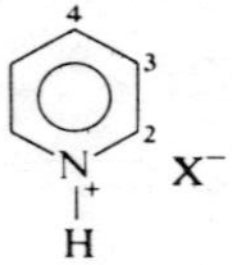	2*	B	3.16^{f}*	p*K*-W	261				
		o	3.41^{f}*	p*K*-W	261				
		B	2.2*	Phys (CF_3CO_2H)	262				
	3	B	2.09^{f}	p*K*-W (W)	261		2.12	p*K*-An (W)	261

		0	2.32^f	p*K*-W	261			
		B	1.9	Phys (CF_3CO_2H)	262			
	4	B	2.34^f	p*K*-W	261	4.0	p*K*-An (W)	158
		0	2.63^f	p*K*-W	261			
		B	1.3	Phys (CF_3CO_2H)	262			
N^+–Me X^-	2*					2.49*	ArNu (MeOH)	258
	3					1.58	ArNu (MeOH)	258
	4					2.32	ArNu (MeOH)	258
N^+–OH X^-	3					2.3	p*K*-An (W)	158
	4					3.9	p*K*-An (W)	158

[a] 0—Sigma-zero reactivity; B—benzoic-like reactivity; ^{0}B—the validity for both types has been proved. For the values without symbol it is not certain to which type they belong.
[b] The abbreviations should be intelligible; for detailed explanations see p. 453.
[c] The *ortho*- and the *peri*-positions and the pertinent values are marked with an asterisk. The Hammett equation is not obeyed for these positions.
[d] These values should express the pure electronic effect of the nucleus, free of steric interaction. They were partly obtained from correlations with quantum chemistry indices, partly by empirical correction.
[e] The values σ_a^0 and σ_a^+ were obtained simultaneously by processing several reaction series by the Yukawa–Tsuno equation.
[f] Corrected for the tautomeric equilibrium.
[g] The solvent is given (in parentheses) with the method.

Table 10.4. *Inductive* (σ_I) *and Mesomeric* (σ_R^0) *Constants*

No.[a]	Substituent	σ_I	Method[b]	Ref.	σ_R^0	Method[b]	Ref.
1	H	(0.01)(0.02)	p*K*-Ac	(9)	0.00	Est	264
		0.00(0.02)	p*K*-Ac	(25)	(−0.03)	ir	(264)
2	D	0.004(0.02)	p*K*-Qui	39	0.00	ir	264
		−0.001	Est	74			
		0.000	OReac	39			
		−0.001	F-nmr	75	-0.000_1	F-nmr	75
3	Me	0.02(0.05)	p*K*-Qui	(40)	(−)0.10	ir	264
		−0.01(0.1)	p*K*-BCO	(265)	−0.05	Cor(16)	21
		−0.04(0.02)	p*K*-Ac	(9)	−0.11	←Stat	11
		−0.05	HydrE	8			
		−0.08	F-nmr	10	−0.15	F-nmr	10
		[0.03][c]	OReac	266			
4	CD_3	0.02(0.05)	p*K*-Qui	39			
5	Et	0.00(0.05)	p*K*-Qui	(40)	(−)0.10	ir	264
		−0.01(0.1)	p*K*-BCO	(265)			
		−0.03(0.02)	p*K*-Ac	(9)			
		−0.05	HydrE	42			
		−0.03	F-nmr	10	−0.14	F-nmr	10
6	Pr^n	−0.02(0.02)	p*K*-Ac	9	(−)0.11	ir	264
7	Pr^i	−0.01(0.05)	p*K*-Qui	(40)	(−)0.12	ir	264
		−0.03(0.02)	p*K*-Ac	9			
		−0.06	HydrE	8			
		−0.03	Est(4*a*)	9			
8	Bu^n	−0.04(0.02)	p*K*-Ac	9	(−)0.12	ir	267
9	Bu^i	−0.03(0.02)	p*K*-Ac	9	(−)0.12	ir	267
		−0.02	Est(4*a*)	9			
10	CHMeEt	−0.03(0.02)	p*K*-Ac	9	(−)0.12	ir	264
		−0.03	Est(4*a*)	9			

11	Bu^t	−0.02(0.05)	p*K*-Qui	(40)	(−)0.13	ir	264
		−0.07(0.02)	p*K*-Ac	9			
		−0.07	HydrE	8			
		0.02	F-nmr	100	−0.17	F-nmr	100
		[−0.02][c]	OReac	266			
12	n-C_5H_{11}	−0.04(0.02)	p*K*-Ac	9	(−)0.12	ir	267
14	CH_2Bu^t	−0.02(0.05)	p*K*-Ac	9	(−)0.09	ir	267
		−0.03	Est(4*a*)	9			
16	n-C_6H_{13}	−0.04(0.02)	p*K*-Ac	9			
17	n-C_7H_{15}	−0.06(0.02)	p*K*-Ac	9			
18	n-C_8H_{17}	(−0.06)	Est	—	(−)0.14	ir	267
19	CH:CH_2	0.08(0.05)	p*K*-Qui	(40)	(−)0.05	ir	264
		0.09(0.02)	p*K*-Ac	9			
		0.06	Est(4*a*)	(3)			
		0.01	F-nmr	10	−0.03	F-nmr	10
20	CH:CHMe(*E*)	0.05(0.02)	p*K*-Ac	9			
		0.06	HydrE	(8)			
21	CH:CMe_2	0.03(0.02)	p*K*-Ac	9			
22	CH_2CH:CH_2	0.01(0.02)	p*K*-Ac	9			
23	CH_2CH:CHMe(*E*)	0.00(0.02)	p*K*-Ac	9			
		−0.02	Est(4*a*)	9			
24	CH_2CH:CMe_2	−0.01(0.02)	p*K*-Ac	25			
25	$(CH_2)_2$CH:CH_2	0.01(0.02)	p*K*-Ac	25			
26	C⋮CH	0.30(0.05)	p*K*-Qui	(40)	(+)0.07	ir	264
		0.35(0.05)	p*K*-Ac	9	0.08	F-nmr	264
27	C⋮CMe	0.21	Est(4*a*)	(8)			
28	CH_2C⋮CH	0.13(0.05)	p*K*-Ac	9			
30	c-C_3H_5	0.01(0.05)	p*K*-Ac	25	(−)0.17	ir	264
		−0.08	F-nmr	101	−0.13	F-nmr	101
31	$CH_2C_3H_5$-c	(−0.01)	Est	—	(−)0.12	ir	267
32	c-C_4H_7	(−0.02)	Est	—	(−)0.12	ir	264
33	c-C_5H_9	−0.03	Est(4*a*)	(8)	(−)0.14	ir	264

Table 10.4—continued

No.[a]	Substituent	σ_I	Method[b]	Ref.	σ_R^0	Method[b]	Ref.
	c-C_6H_{11}	−0.02(0.02)	p*K*-Ac	9	(−)0.13	ir	264
		−0.03	HydrE	8			
	$CH_2C_6H_{11}$-c	−0.05(0.02)	p*K*-Ac	9			
	$(CH_2)_2C_6H_{11}$-c	−0.05(0.02)	p*K*-Ac	25			
	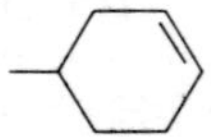	−0.02	Est(4*a*)	(268)			
45	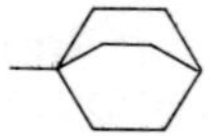				(−)0.17	ir	267
47	1-Adamantyl				(−)0.15	ir	267
48	Ph	0.15(0.05)	p*K*-Qui	(40)	(−)0.10	ir	264
		0.12(0.02)	p*K*-Ac	25	−0.06	Cor(16)	21
		0.10	HydrE	8	−0.11	←Stat	11
		0.09(0.1)	p*K*	41	(−)[0.06][d]	ir	269
		0.08	F-nmr	10	−0.09	F-nmr	10
49	CH_2Ph	0.03(0.02)	p*K*-Ac	25	(−)0.12	ir	264
		0.04	HydrE	42	−0.08	F-nmr	264
					0.00	Cor(16)	21
50	$(CH_2)_2Ph$	−0.01(0.02)	p*K*-Ac	9			
		0.01	HydrE	8			
51	$(CH_2)_3Ph$	−0.01	p*K*	9			
52	$(CH_2)_4Ph$	−0.02	p*K*	9			
53	$CHPh_2$	0.07	Est(4*a*)	(8)	(−)0.11	ir	264
					−0.12	F-nmr	264
54	CPh_3	(0.10)	Est	—	(−)0.13	ir	264
55	CHMePh	0.08(0.05)	p*K*-Ac	25			
57	1-$C_{10}H_7$	0.12(0.02)	p*K*-Ac	9			

58	$CH_2C_{10}H_7$-1	0.07(0.05)	p*K*	9			
59	2-$C_{10}H_7$	0.12(0.02)	p*K*-Ac	9			
60	CH:CHPh(*E*)	0.07	Est(4*a*)	(8)	±0.15	ir	264
61	C⋮CPh	0.22	Est(4*a*)	(8)	0.00	Cor(16)	21
62	C⋮C·C⋮CPh	0.45	p*K*	273			
63	CH_2CHO	(0.10)	Est	—	(−)0.11	ir	264
64	CH_2COMe	0.10	Est(4*a*)	(8)			
65	CH_2CONH_2	0.05(0.02)	p*K*-Ac	9			
66	CH_2CONHR	0.00(0.05)	p*K*-Ac	9			
67	$(CH_2)_2CONH_2$	0.03(0.05)	p*K*-Ac	9			
68	CH_2CO_2H	0.17	Est(4*a*)	(8)			
70	CH_2CO_2R	0.17(0.05)	p*K*-Ac	9			
71	CH_2CN	0.18(0.02)	p*K*-Ac	9	(−)0.09	ir	264
		0.21	Est(4*a*)	9	0.00	Cor(16)	21
		0.26	F-nmr	100	−0.08	F-nmr	100
72	$(CH_2)_2CN$	0.14	OReac	270			
73	$CH(CN)_2$	0.55	F-nmr	100	−0.03	F-nmr	100
74	$C(CN)_3$	0.98	F-nmr	100	0.01	F-nmr	100
75	$C(CN)_2Me$	0.63	F-nmr	100	−0.06	F-nmr	100
77	CH_2SiMe_3	−0.05(0.02)	p*K*-Ac	9	(−)0.20	ir	271
		−0.04	Est(4*a*)	(8)	0.00	Cor(16)	21
		−0.07	F-nmr	10	−0.20	F-nmr	10
78	$(CH_2)_2SiMe_3$	−0.04(0.02)	p*K*-Ac	9			
79	$CH(SiMe_3)_2$				(−)0.24	ir	271
80	$C(SiMe_3)_3$				(−)0.24	ir	271
83	CH_2SnMe_3				±0.26	ir	271
87	CH_2NH_2	0.08	p*K*	9	(−)0.10	ir	272
		0.00	F-nmr	10	−0.15	F-nmr	10
92	CH_2NHAc	0.07(0.02)	p*K*-Ac	9			
93	$CH_2NHCONH_2$	0.07(0.02)	p*K*-Ac	25			
94	$(CH_2)_2NHCONH_2$	0.02(0.02)	p*K*-Ac	25			
95	CH_2N:C:O	(0.13)	Est	—	±0.06	ir	272

Table 10.4—continued

No.[a]	Substituent	σ_I	Method[b]	Ref.	σ_R^0	Method[b]	Ref.
96	$CH_2N{:}C{:}S$	(0.15)	Est	—	±0.07	ir	272
100	$C(NO_2)_3$	0.73	Est(4*a*)	(103)			
101	$CH_2C(NO_2)_3$	0.26	OReac	(103)			
102	$CH_2C(NO_2)_2Hal$	0.24	Est(4*a*)	(274)			
103	CH_2OH	0.05(0.1)	p*K*-BCO	(265)	0.00	ir	264
		0.05(0.05)	p*K*-Ac	9	−0.06	F-nmr	264
		0.10(0.05)	p*K*-Qui	40			
104	CH(OH)Me	0.02(0.05)	p*K*	9			
105	CH(OH)Ph	0.08(0.05)	p*K*-Ac	9			
		0.12	Est(4*a*)	(8)			
106	$CH_2CHMeOH$	−0.01(0.05)	p*K*-Ac	9			
107	CH_2CMe_2OH	−0.04(0.05)	p*K*-Ac	9			
108	$CH(OH)_2$	0.22(0.05)	p*K*-Qui	(40)			
109	CH_2OR	0.10(0.05)	p*K*-Qui	(40)	(−)0.05	ir	264
		0.07(0.05)	p*K*-Ac	9			
		0.08	Est(4*a*)	(8)			
		−0.01	F-nmr	100	−0.06	F-nmr	100
111	$CH(OR)_2$	0.18	Est(4*a*)	(280)	0.00	ir	264
		−0.02	F-nmr	100	−0.03	F-nmr	100
112	$C(OMe)_3$	−0.03	F-nmr	100	0.00	ir	264
		0.26	Est	—	−0.01	F-nmr	100
113	CH_2OPh	0.16	Est(4*a*)	(8)	0.00	Cor(16)	21
114	$CH(OPh)_2$	0.30	Est(4*a*)	(280)			
115	CH_2OAc	0.16(0.05)	p*K*-Qui	(39)	±0.08	ir	272
		0.14(0.05)	p*K*-Ac	9			
117	CH_2OTs	0.23(0.05)	p*K*-Qui	(40)			
118	CH_2SH	0.10(0.05)	p*K*-Ac	9			
119	CH_2SMe	(0.05)	Est	—	(−)0.10	ir	272

120	$CH(SR)_2$	0.15	Est(4*a*)	(280)			
121	CH_2SCH_2Ph	0.06(0.02)	p*K*-Ac	9			
122	$CH(SPh)_2$	0.25	Est(4*a*)	(280)			
123	CH_2SCF_3	0.21	F-nmr	100	−0.06	F-nmr	100
		0.12	Est(17)	—	0.00	Cor(16)	—
124	$CH(SCF_3)_2$	0.44	F-nmr	100	0.00	F-nmr	100
125	$C(SCF_3)_3$	0.49	F-nmr	100	0.04	F-nmr	100
127	$CH_2SO{\cdot}CF_3$	0.26	F-nmr	13	−0.02	F-nmr	13
128	$CH_2SO_2{\cdot}R$	0.21	Est(4*a*)	(8)	±0.09	ir	272
		0.15	Est(17)	—	0.00	Cor(16)	21
129	$CH_2SO_2{\cdot}CF_3$	0.28	F–nmr	13	0.03	F-nmr	13
		0.29	Est(17)	—	0.00	Cor(16)	—
130	CH_2F	0.18	Est(4*a*)	(8)	−0.02	F-nmr	100
		0.12	F-nmr	100			
131	CHF_2	0.32	p*K*	9			
		0.33	Est(4*a*)	(8)			
		0.29	F-nmr	100	0.06	F-nmr	100
132	CF_3	0.42(0.02)	p*K*-Ac	9	(+)0.11	ir	264
		0.43(0.1)	p*K*-BCO	(275)	0.00	Cor(16)	21
		0.41	HydrE	8			
		0.45	Est	11	0.08	←Stat	11
		0.41	F-nmr	10	0.10	F-nmr	110
133	CH_2Cl	0.16(0.05)	p*K*-Qui	(40)	0.00	ir	264
		0.15(0.02)	p*K*-Ac	9	0.00	Cor(16)	21
		0.17	HydrE	8			
		0.17	Est(4*a*)	9			
		0.14	F-nmr	10	−0.03	F-nmr	10
134	$CHCl_2$	0.31	Est(4*a*)	(8)	0.00	ir	264
		0.30	F-nmr	100	0.02	F-nmr	100
135	CCl_3	0.41	Est(4*a*)	(8)	0.00	ir	264
		0.30	F-nmr	100	0.00	Cor(16)	21
					0.03	F-nmr	100

Table 10.4—continued

No.[a]	Substituent	σ_I	Method[b]	Ref.	σ_R^0	Method[b]	Ref.
136	CH_2Br	0.17(0.05)	p*K*-Qui	(40)	0.00	ir	264
		0.18(0.05)	p*K*-Ac	9	±0.02	ir	272
		0.16	Est(4*a*)	9	0.00	Cor(16)	21
		0.16	F-nmr	100	−0.02	F-nmr	100
137	$CHBr_2$	0.30	F-nmr	100	0.00	ir	264
					0.02	F-nmr	100
138	CBr_3	0.26	F-nmr	100	0.00	ir	264
					0.03	F-nmr	100
139	CH_2I	0.18(0.05)	p*K*-Qui	(40)	±0.05	ir	272
		0.16(0.02)	p*K*-Ac	9	0.00	Cor(16)	21
		0.14	Est(4*a*)	9			
		0.14	F-nmr	13	−0.04	F-nmr	13
140	CHI_2	0.26	F-nmr	13	0.00	F-nmr	13
141	CF_2Cl	0.38	F-nmr	100	0.08	F-nmr	100
142	CHClMe	0.16(0.05)	p*K*-Ac	25			
143	CHBrMe	0.20(0.05)	p*K*-Ac	25			
144	CH_2CF_3	0.14(0.02)	p*K*-Ac	9			
145	$CH(CF_3)_2$	0.21	Est(4*a*)	(13)			
146	CH_2CCl_3	0.12(0.02)	p*K*-Ac	9			
147	$(CH_2)_2CCl_3$	0.04(0.05)	p*K*-Ac	9			
148	C_2F_5	0.41	F-nmr	110	±0.08	ir	264
					0.11	F-nmr	110
149	$CF(CF_3)_2$	0.48	F-nmr	110	±0.02	ir	264
					0.04	F-nmr	110
150	C_4F_9-n	0.39	F-nmr	110	0.11	F-nmr	110
151	$C(CF_3)_3$	0.55	F-nmr	13	0.00	F-nmr	13
		0.24	Est(4*a*)	(13)			
152	$CH_2C_3F_7$-n	0.14(0.05)	p*K*-Ac	9			

153	$C(OH)(CF_3)_2$	0.28	F-nmr	110	±0.11	ir	264
					0.02	F-nmr	110
154	c-C_4F_7	0.45	F-nmr	109	0.08	F-nmr	109
155	$C(OH)(CF_2)_3$	0.45	F-nmr	109	0.08	F-nmr	109
159	$CH{:}CHCO_2H$(*E*)	0.16	Est(4*a*)	107			
160	$CH{:}CHCO_2R$(*E*)	(0.18)	Est	—	±0.10	ir	264
162	$CH{:}C(CN)_2$	0.41	F-nmr	108	0.29	F-nmr	108
163	$C(CN){:}C(CN)_2$	0.67	F–nmr	108	0.31	F-nmr	108
164	$CH{:}CHNO_2$(*E*)	0.28	Est(4*a*)	(107)	(+)0.13	ir	264
		0.24	F-nmr	10	0.16	F-nmr	264
165	$CH{:}CHSO_2 \cdot CF_3$(*E*)	0.31	F-nmr	13	0.25	F-nmr	13
166	$CF{:}CH_2$	0.25	Est(4*a*)	(13)			
167	$CH{:}CF_2$	0.19	Est(4*a*)	(13)			
168	$CF{:}CF_2$	0.31	Est(4*a*)	(13)			
169	$CH{:}CHCF_3$(*E*)	0.20	F–nmr	108	0.07	F-nmr	108
170	$CH{:}CHCF_3$(*Z*)	0.12	F-nmr	108	0.05	F-nmr	108
171	$CF{:}CFCF_3$(*E*)	0.30	F-nmr	13	0.14	F-nmr	13
172	$CH{:}CHCl$(*E*)	0.14	Est(4*a*)	(107)			
173	$CH{:}CCl_2$	0.16(0.05)	p*K*-Ac	9			
174	$CCl{:}CCl_2$	0.36	Est(4*a*)	(13)			
175	$CH_2CH{:}CCl_2$	0.03(0.05)	p*K*-Ac	9			
176	$CH{:}CHCCl_3$(*E*)	0.19	Est(4*a*)	(107)			
177	$C \vdots C \cdot CF_3$	0.31	F-nmr	13	0.16	F-nmr	13
178	$C_6H_2(NO_2)_3$-2,4,6	0.26	Est(17)	21	0.00	Cor(16)	21
179	C_6F_5	0.31(0.05)	p*K*-Ac	277	0.02	F-nmr	115
		0.25	F-nmr	115	0.00	Cor(16)	—
180	C_6Cl_5	0.25	F-nmr	115	−0.01	F-nmr	115
181	CHO	0.25	F-nmr	10	(+)0.24	ir	264
					0.27	F-nmr	10

Table 10.4—continued

No.[a]	Substituent	σ_I	Method[b]	Ref.	σ_R^0	Method[b]	Ref.
182	COMe	0.29(0.05)	p*K*-Ac	9	(+)0.22	ir	264
		0.27	HydrE	8	0.05	Cor(16)	21
		0.28	Est	11	0.16	←Stat	11
		0.18	F-nmr	10	0.19	F-nmr	10
					[0.0][d]	ir	269
186	COPh	(0.27)	Est	—	(+)0.19	ir	264
					0.04	Cor(16)	21
					0.17	F-nmr	264
					[0.0][d]	ir	269
187	$COCF_3$	0.59	Est(4*a*)	(13)	0.33	F-nmr	110
		0.47	F-nmr	110			
188	COCN	0.55	F-nmr	10			
189	$CONH_2$	0.33(0.05)	p*K*-Qui	(40)	0.00	Cor(16)	—
		0.27(0.02)	p*K*-Ac	9			
		(0.38)(0.1)	p*K*	41			
		0.21	F-nmr	10			
191	$CONMe_2$	0.31(0.05)	p*K*-Qui	39			
192	CONHPh	0.25(0.05)	p*K*-Ac	9			
194	CO_2H	0.32(0.1)	p*K*-BCO	275	(+)0.29	ir	264
					0.03	Cor(16)	21
195	CO_2R	0.31(0.05)	p*K*-Qui	(40)	(+)0.16	ir	264
		0.32(0.1)	p*K*-BCO	(265)	(+)0.18	ir	264
		0.34(0.05)	p*K*-Ac	9	0.03	Cor(16)	21
		0.30	Est	11	0.14	←Stat	11
		0.11	F-nmr	10	0.19	F-nmr	264
					[0.0][d]	ir	269
198	COF	0.39	F-nmr	10	0.31	F-nmr	10, 109

199	COCl	0.38	F-nmr	13	(+)0.21	ir	264
					0.32	F-nmr	264
					[0.0][d]	ir	269
203	CN	0.57(0.05)	p*K*-Qui	(40)	(+)0.08	ir	264
		0.58(0.1)	p*K*-BCO	38, 276	0.00	Cor(16)	21
		0.57(0.02)	p*K*-Ac	(9)			
		0.59	HydrE	8			
		0.56(0.1)	p*K*	41	0.13	←Stat	11
		0.48	F-nmr	10	0.22	F-nmr	264
204	NH	−0.02	F-nmr	101	−0.08	F-nmr	101
205	O	0.07	F-nmr	101	−0.04	F-nmr	101
206	S	0.07	F-nmr	101	−0.06	F-nmr	101
207	N—Me, O	0.07	F-nmr	101	0.05	F-nmr	101
208	S, S	0.16	Est(4*a*)	(280)			
209	S, S	0.15	Est(4*a*)	(280)			
210	S, S	0.16	Est(4*a*)	(280)			

Table 10.4—continued

No.[a]	Substituent	σ_I	Method[b]	Ref.	σ_R^0	Method[b]	Ref.
211	2-C_4H_3O	0.04(0.1)	pK	9			
		0.17	Est(4*a*)	(278)			
212	3-C_4H_3O	0.10	Est(4*a*)	(278)			
213	2-C_4H_3S	0.21(0.05)	pK	9			
		0.15	Est(4*a*)	(278)			
214	$CH_2C_4H_3S$-2	0.05(0.1)	pK	9			
215	3-C_4H_3S	0.10	Est(4*a*)	(278)			
220	N H	−0.01(0.05)	pK	9			
226	Me N N N Me	0.18	F-nmr	131	0.19	F-nmr	131
228	1-$C_2HB_{10}H_{10}$(1.2)	0.38	F-nmr	135	0.00	F-nmr	135
229	1-$C_2HB_{10}H_{10}$(1,7)	0.19	F-nmr	135	−0.04	F-nmr	135
230	Li				(+)0.14	ir	279
231	ZnPh				(+)0.11	ir	279
232	CdPh				(+)0.10	ir	279
234	HgPh				±0.03	ir	279
235	$HgCF_3$	0.27	F-nmr	13	0.05	F-nmr	13
236	HgCN	0.23	F-nmr	13	0.11	F-nmr	13
237	HgOAc	0.38	F-nmr	13	0.02	F-nmr	13
238	$HgOCOCF_3$	0.48	F-nmr	13	0.04	F-nmr	13

239	$HgSCF_3$	0.37	F-nmr	13	0.05	F-nmr	13
240	HgF	0.34	F-nmr	13	−0.01	F-nmr	13
241	HgCl	0.31	F-nmr	13	0.04	F-nmr	13
242	BPh_2				(+)0.22	ir	279
243	6-$B_{10}H_{13}$	0.12	F-nmr	137	0.12	F-nmr	137
244	$B(OH)_2$	−0.08	F-nmr	10	(+)0.23	ir	279
					0.21	Cor(16)	21
245	BF_2	0.16	F-nmr	109	0.32	F-nmr	10
246	BCl_2				(+)0.30	ir	279
248	$AlPh_2$				(+)0.11	ir	279
249	SiH_3	0.01	F-nmr	13	0.09	F-nmr	13
		0.09	F-nmr	(138)	0.09	F-nmr	(138)
250	$SiHMe_2$	−0.02	F-nmr	(138)	0.06	F-nmr	(138)
251	$SiMe_3$	−0.13(0.05)	p*K*-Ac	9	(+)0.09	ir	272
		−0.12	HydrE	8	0.04	Cor(16)	21
		(−0.15)	p*K*	41			
		−0.10	Est	11	0.06	←Stat	11
		−0.04	F-nmr	(138)	0.05	F-nmr	(138)
252	$SiMe_2Ph$	0.02	F-nmr	(138)	0.05	F-nmr	(138)
253	$SiMePh_2$	0.07	F-nmr	(138)	0.06	F-nmr	(138)
254	$SiPh_3$	0.13	F-nmr	(138)	(0.0)	ir	279
					0.06	F-nmr	(138)
255	$SiMe_2SiMe_3$				±0.04	ir	271
256	$SiMe(SiMe_3)_2$				±0.06	ir	271
257	$Si(NMe_2)_3$	−0.04	F-nmr	(138)	0.00	F-nmr	(138)
260	$Si(OMe)_3$	0.00	F-nmr	(138)	0.13	F-nmr	(138)
261	$Si(OEt)_3$	−0.04	F-nmr	(138)	(+)0.08	ir	272
262	$SiMe_2OSiMe_3$	−0.13(0.05)	p*K*-Ac	9	0.12	F-nmr	(138)
265	$SiFMe_2$	0.08	F-nmr	(138)	0.11	F-nmr	(138)
267	SiF_3	0.42	F-nmr	(138)	0.24	F-nmr	(138)
268	$SiClMe_2$	0.11	F-nmr	(138)	0.10	F-nmr	(138)

Table 10.4—continued

No.[a]	Substituent	σ_I	Method[b]	Ref.	σ_R^0	Method[b]	Ref.
269	$SiCl_2Me$	0.24	F-nmr	(138)	0.15	F-nmr	(138)
270	$SiCl_3$	0.39	F-nmr	13	(+)0.09	ir	272
					0.17	F-nmr	13
273	$SiBr_3$	0.39	F-nmr	(138)	0.18	F-nmr	(138)
275	$GeMe_3$	(0.0)	Est	—	±0.05	ir	271
276	$GeEt_3$	(0.0)	Est	—	(0.0)	ir	279
278	GeF_3	0.74	F-nmr	13	0.23	F-nmr	13
279	$GeCl_3$	0.63	F-nmr	13	0.16	F-nmr	13
280	$GeBr_3$	0.59	F-nmr	13	0.14	F-nmr	13
281	$SnMe_3$	(0.0)	Est	—	±0.07	ir	271
283	$SnPh_3$				(0.0)	ir	279
284	$PbPh_3$				(0.0)	ir	279
285	NH_2	0.12	Est	11	−0.48	←Stat	11
		0.19	Est(4*a*)	281	(−)0.47	ir	264
		0.01	F-nmr	10	−0.48	F-nmr	10
286	NHMe	0.18	Est(4*a*)	281	(−)0.52	ir	264
287	NMe_2	0.06	Est	11	−0.52	←Stat	11
		0.19	Est(4*a*)	281	(−)0.53	ir	264
		0.10	F-nmr	10	−0.54	F-nmr	10
					[−0.13][d]	ir	269
288	NHR	0.17	Est(4*a*)	281	(−)0.53	ir	264
289	NEt_2	0.16	Est(4*a*)	281	(−)0.57	ir	264
					[−0.13][d]	ir	269
290	NHPh				(−)0.50	ir	264
292	$NHCH_2Ph$	(0.22)	Est(4*a*)	281			
293	$NHCH_2CO_2R$	0.26	Est(4*a*)	281	(−)0.44	ir	264
294	$N(CF_3)_2$	0.49	F-nmr	148	±0.13	ir	264
					0.01	F-nmr	148

295	$NHCH_2CF_3$	0.29	Est(4*a*)	281			
296	NHCHO	0.32(0.05)	p*K*-Ac	9	−0.20	F-nmr	151
		0.29	F-nmr	151			
297	NHAc	0.29(0.05)	p*K*-Qui	(40)	(−)0.41	ir	264
		0.26(0.02)	p*K*-Ac	(9)	−0.25	←Stat	11
		0.20(0.1)	p*K*	41	−0.30	F-nmr	264
		0.28	F-nmr	13	−0.23	F-nmr	13
298	NMeAc	0.36	F-nmr	13	(−)0.41	ir	264
					−0.10	F-nmr	13
299	NPhAc	0.22(0.05)	p*K*-Ac	9			
300	NAc_2	0.37	F-nmr	13	−0.04	F-nmr	13
301	NHCOEt	0.25(0.02)	p*K*-Ac	9			
303	NHBz	0.27(0.05)	p*K*-Ac	9			
305	$NHCOCF_3$	0.46	F-nmr	13	−0.19	F-nmr	13
306	$NMeCOCF_3$	0.43	F-nmr	13	−0.04	F-nmr	13
307	$NHCOCH_2Cl$	0.36(0.05)	p*K*-Ac	25			
308	$NHCONH_2$	0.21(0.02)	p*K*-Ac	9			
310	$NHCO_2R$	0.29(0.05)	p*K*-Qui	(40)			
		0.26(0.02)	p*K*-Ac	9			
311	$N(CF_3)COF$	0.56	F-nmr	148	0.00	F-nmr	148
312	$N(COF)_2$	0.58	F-nmr	148	−0.01	F-nmr	148
314	$NHCSNH_2$	0.29	F-nmr	149	−0.13	F-nmr	149
316	$NHCSNMe_2$	0.28(0.05)	p*K*-Ac	25			
317	NHCSSMe	0.42(0.05)	p*K*-Ac	25			
318	NMeCSSMe	0.47(0.05)	p*K*-Ac	25			
319	NHCN	0.37	F-nmr	149	−0.31	F-nmr	149
320	$NHNH_2$	0.14	F-nmr	10	(−)0.49	ir	264
					−0.43	F-nmr	10
321	NHNHPh				(−)0.44	ir	264
322	NHOH				(−)0.22	ir	264
323	$NHSO_2Me$	0.42	F-nmr	13	−0.21	F-nmr	13
324	$NMeSO_2Me$	0.34	F-nmr	13	−0.10	F-nmr	13
325	$N(SO_2Me)_2$	0.45	F-nmr	13	0.04	F-nmr	13

Table 10.4—continued

No.[a]	Substituent	σ_I	Method[b]	Ref.	σ_R^0	Method[b]	Ref.
326	$NHSO_2Ph$	0.32(0.05)	p*K*-Ac	9			
327	$NHSO_2CF_3$	0.49	F-nmr	13	0.10	F-nmr	13
328	$NMeSO_2CF_3$	0.48	F-nmr	13	−0.04	F-nmr	13
329	$N(SO_2CF_3)_2$	0.70	F-nmr	13	0.10	F-nmr	13
330	$\overset{+}{N}\vdots\overset{-}{C}$	0.67(0.05)	p*K*-Ac	25	0.02	F-nmr	151
		0.47	F-nmr	151			
333	$N{:}C(CF_3)_2$	0.35	F-nmr	109	−0.12	F-nmr	109
334	$N{:}CCl_2$	0.29	F-nmr	108	−0.16	F-nmr	108
335	N:C:NPh				(−)0.46	ir	264
336	N:C:O	0.36	F-nmr	148	(−)0.40	ir	264
					−0.17	F-nmr	148
337	N:C:S	0.42	F-nmr	148	(−)0.35	ir	264
					−0.07	F-nmr	148
338	N:NPh	0.19	F-nmr	10, 151	±0.06	ir	264
		0.30	Est(17)	21	0.00	Cor(16)	21
					0.08	F-nmr	151
340	$N{:}NCF_3$	0.44	F-nmr	13	0.24	F-nmr	13
343	$N{:}\overset{+}{N}{:}\overset{-}{N}$	0.42(0.05)	p*K*-Ac	9			
344	NO	0.33	F-nmr	148	±0.07	ir	264
					0.32	F-nmr	148
345	N:S:O				±0.09	ir	264
346	NO_2	0.68(0.05)	p*K*-Qui	(40)	(+)0.17	ir	264
		0.72(0.1)	p*K*-BCO	(265)	0.00	Cor(16)	21
		0.76(0.05)	p*K*-Ac	9			
		0.65	Est	11	0.15	←Stat	11
		0.57	F-nmr	148	0.20	F-nmr	148

347	$N(CH_2)_2$	0.07	F-nmr	101	(−)0.38	ir	264
					−0.29	F-nmr	101
348	$N(CH_2)_3$				(−)0.55	ir	264
349	$N(CH_2)_4$				(−)0.63	ir	264
350	$N(CH_2)_5$				(−)0.47	ir	264
352	—N(N=N—N=CH) (ring: —N, N=N, CH=N)	0.54	F-nmr	108	−0.04	F-nmr	108
353	—NHC(=N—N=...) (ring: —NHC, N—N, S—N)	0.42	F-nmr	149	−0.23	F-nmr	149
356	PH_2	0.07	F-nmr	160	−0.02	F-nmr	160
357	PMe_2	0.08	F-nmr	161	(−)0.08	ir	279
					−0.05	F-nmr	161
359	PPh_2	0.17	F-nmr	161	(−)0.06	ir	279
					−0.01	F-nmr	161
360	$P(CF_3)_2$	0.50	F-nmr	161	0.19	F-nmr	161
361	$P(CN)_2$	(0.74)	F-nmr	161	(0.16)	F-nmr	161
362	$P(NMe_2)_2$	0.00	F-nmr	161	−0.06	F-nmr	161
364	$P(OR)_2$	0.09	F-nmr	161	0.06	F-nmr	161
365	PF_2	0.38	F-nmr	161	0.21	F-nmr	161
368	PCl_2	0.45	F-nmr	161	(+)0.06	ir	279
					0.16	F-nmr	161
369	$\overset{+}{P}Ph_2\overset{-}{C}H_2$	0.36	F-nmr	165	0.18	F-nmr	165
370	$\overset{+}{P}Ph_2\overset{-}{B}Cl_3$	0.65	F-nmr	161	0.16	F-nmr	161
371	$POMe_2$	0.35	F-nmr	160	0.15	F-nmr	160
373	$POPh_2$	0.27(0.05)	p*K*-Ac	282			
		0.30	F-nmr	165	0.12	F-nmr	165
375	$PO(OMe)_2$	0.24	F-nmr	160	(+)0.08	ir	279
					0.19	F-nmr	160

Table 10.4—continued

No.[a]	Substituent	σ_I	Method[b]	Ref.	σ_R^0	Method[b]	Ref.
376	$PO(OEt)_2$	0.35(0.05)	p*K*-Ac	282	0.17	F-nmr	160
		0.21	F-nmr	160			
377	$PO(OR)_2$	0.28(0.05)	p*K*-Ac	282			
379	$POCl_2$	0.65	F-nmr	160	0.25	F-nmr	160
380	$PSPh_2$	0.40	F-nmr	160	0.11	F-nmr	160
381	$PSCl_2$	0.59	F-nmr	160	0.21	F-nmr	160
382	PPh_4				(+)0.03	ir	279
383	PF_4	0.45	F-nmr	161	0.35	F-nmr	161
384	$AsPh_2$				(−)0.07	ir	279
385	$AsPh_4$				(+)0.04	ir	279
386	$SbPh_2$				(−)0.07	ir	279
387	$SbPh_4$				(+)0.06	ir	279
388	$BiPh_2$				(−)0.10	ir	279
389	OH	0.31(0.05)	p*K*-Qui	(40)	(−)0.40	ir	264
		0.25(0.2)	p*K*-BCO	(265)	−0.43	F-nmr	10
		0.22(0.02)	p*K*-Ac	(9)			
		0.25	HydrE	8			
		0.25(0.1)	p*K*	41			
		0.25	F-nmr	10			
390	OMe	0.34(0.05)	p*K*-Qui	(40)	(−)0.43	ir	264
		0.32(0.1)	p*K*-BCO	(265)			
		0.29(0.02)	p*K*-Ac	(9)			
		0.27	Est	11	−0.45	←Stat	11
		0.23	HydrE	8			
		0.25	F-nmr	10	−0.43	F-nmr	10
391	OEt	0.26(0.1)	p*K*	280	(−)0.44	ir	264
392	OR	0.27(0.05)	p*K*-Ac	9			
393	OC_5H_9-c	0.26(0.02)	p*K*-Ac	9			

394	OC_6H_{11}-c	0.30(0.02)	p*K*-Ac	9			
395	OPh	0.42(0.02)	p*K*-Ac	25	(−)0.36	ir	264
		0.38	HydrE	8	−0.34	←Stat	11
		0.39(0.1)	p*K*	41	−0.31	F-nmr	10
		0.37	F-nmr	10			
399	OCH_2F	0.37	F-nmr	173	(−)0.31	ir	173
					−0.35	F-nmr	173
400	$OCHF_2$	0.45	F-nmr	173	(−)0.23	ir	173
					−0.27	F-nmr	173
401	OCF_3	0.55	F-nmr	10	(−)0.19	ir	173
					(−)(0.25)	ir	264
					−0.18	F-nmr	10
405	OCH_2Cl	0.41	F-nmr	173	(−)0.33	ir	173
					−0.33	F-nmr	173
406	$OCHCl_2$	0.49	F-nmr	173	(−)0.25	ir	173
					−0.23	F-nmr	173
407	$OCCl_3$	0.51	F-nmr	173	(−)0.19	ir	173
					−0.16	F-nmr	173
408	OAc	0.41(0.02)	p*K*-Qui	(40)	(−)0.23	ir	264
		0.36(0.1)	p*K*	41	−0.21	F-nmr	10
		0.27	F-nmr	10			
410	$OCOCF_3$	0.65	F-nmr	177	(−)0.23	ir	264
					−0.19	F-nmr	10
411	$OCONMe_2$	0.46(0.05)	p*K*-Ac	25			
412	$OCSNMe_2$	0.50(0.05)	p*K*-Ac	25			
413	OCN	0.80	F–nmr	177	(−)0.27	ir	283
					−0.26	F-nmr	10
415	$ON{:}CMe_2$	0.29(0.05)	p*K*-Ac	9			
416	ONO_2	0.62(0.05)	p*K*-Ac	9			
419	OSO_2Me	0.58(0.05)	p*K*-Ac	25	±0.26	ir	264
420	OSO_2Ph }	0.58	Est	25			
	OTs }	0.59	p*K*	298			

Table 10.4—continued

No.[a]	Substituent	σ_I	Method[b]	Ref.	σ_R^0	Method[b]	Ref.
421	OSO_2CF_3	0.70	F-nmr	13	0.20	F-nmr	13
423	$\overset{+}{O}Me\cdot\overset{-}{B}Cl_3$	0.54	F-nmr	177	−0.06	F-nmr	177
424	SH	0.26(0.05)	p*K*-Ac	9	(−)0.19	ir	264
		0.41	Est	(40)	−0.15	F-nmr	177
		0.19	F-nmr	177			
425	SMe	0.31(0.05)	p*K*-Qui	(40)	(−)0.25	ir	264
		0.25(0.05)	p*K*-Ac	9	[−0.05][d]	Est	269
		0.23	Est	11	−0.20	←Stat	11
		0.13	F-nmr	177	−0.16	F-nmr	177
426	SR	0.25(0.05)	p*K*-Ac	9, 208	(−)0.19	ir	264
		0.13	F-nmr	177	−0.11	F-nmr	177
427	SC_6H_{11}-c	0.31(0.02)	p*K*-Ac	9			
428	$SCH{:}CH_2$	0.21	F-nmr	177	−0.07	F-nmr	177
429	$SC\vdots CH$	0.32	F-nmr	177	−0.13	F-nmr	177
430	SPh	0.30(0.05)	p*K*-Ac	9	(−)0.19	ir	264
		0.20	F-nmr	177	−0.07	F-nmr	177
431	SCH_2Ph	0.27(0.05)	p*K*-Ac	25			
432	$SCPh_3$	(0.12)	p*K*-Ac	25			
433	SCH_2F	0.27	F-nmr	13	−0.07	F-nmr	13
434	$SCHF_2$	0.33	F-nmr	13	0.02	F-nmr	13
					0.00	Cor(16)	21
435	SCF_3	0.44(0.05)	p*K*-Ac	284	0.00	ir	264
		0.42	Est	11	0.04	←Stat	11
		0.42	F-nmr	177	0.00	Cor(16)	21
					0.06	F-nmr	177
437	SAc	0.41(0.05)	p*K*-Ac	25	±0.08	ir	264
		0.21	F-nmr	177	0.00	Cor(16)	21
					0.01	F-nmr	177

438	$SCOCF_3$	0.51	F-nmr	177	0.05	F-nmr	177
439	$SCONH_2$	0.59(0.05)	p*K*	25			
		(0.33)(0.05)	p*K*	9			
440	$SCONMe_2$	0.33(0.05)	p*K*-Ac	25			
441	SCSMe	0.48(0.05)	p*K*-Ac	25			
442	$SCSNMe_2$	0.37(0.05)	p*K*-Ac	25			
443	SCSOEt	0.44(0.05)	p*K*-Ac	25			
444	SCSSEt	0.48(0.05)	p*K*-Ac	25			
445	SCN	0.58(0.05)	p*K*-Ac	9	±0.09	ir	283
		0.51(0.1)	p*K*	41	−0.05	F-nmr	177
		0.63	F-nmr	177			
446	$SNMe_2$	0.15	F-nmr	177	−0.06	F-nmr	177
447	SOMe	0.25	F-nmr	177	−0.08	F-nmr	177
450	SCl	0.40	F-nmr	177	0.08	F-nmr	177
451	$\overset{+}{S}Me\cdot\overset{-}{B}Cl_3$	0.72	F-nmr	177	0.10	F-nmr	177
453	SO·Me	0.50	Est	11	0.00	←Stat	11
		0.49	F-nmr	177	0.00	F-nmr	177
454	SO·Ph	0.52(0.05)	p*K*-Ac	9	±0.06	ir	264
		0.51	F-nmr	181	−0.01	F-nmr	181
455	$SO\cdot CHF_2$	0.65	F-nmr	13	0.11	F-nmr	13
456	$SO\cdot CF_3$	0.69(0.05)	p*K*-Ac	284	0.00	Cor(16)	—
		0.64	Est	11	0.08	←Stat	11
		0.68	F-nmr	177	0.13	F-nmr	177
457	$SO\cdot NMe_2$	0.30	F-nmr	177	0.03	F-nmr	177
458	SO·OMe	0.45	F-nmr	177	0.09	F–nmr	177
459	SOF	0.66	F-nmr	177	0.17	F-nmr	177
460	SOCl	0.68	F-nmr	177	0.14	F-nmr	177
461	SO_2Me	0.64(0.05)	p*K*-Qui	(40)	(+)0.07	ir	264
		0.59(0.05)	p*K*-Ac	9	0.00	Cor(16)	21
		0.59(0.05)	p*K*	9	0.12	←Stat	11
		0.59	HydrE	8	0.16	F-nmr	177
		0.55	F-nmr	177			

Table 10.4—continued

No.[a]	Substituent	σ_I	Method[b]	Ref.	σ_R^0	Method[b]	Ref.
462	SO_2R	0.57(0.05)	p*K*-Ac	9	0.00	Cor(16)	—
		0.50	F-nmr	177	0.18	F-nmr	177
463	$SO_2CH{:}CH_2$	(0.57)	Est	—	±0.03	ir	272
464	SO_2Ph	(0.55)	p*K*	41	(+)0.06	ir	264
		0.52	F-nmr	181	0.00	Corr(16)	21
					0.14	F-nmr	181
465	SO_2CH_2F	0.55	F-nmr	13	0.22	F-nmr	13
466	SO_2CHF_2	0.59	F-nmr	13	0.00	Cor(16)	21
					0.28	F-nmr	13
467	SO_2CF_3	0.72(0.1)	p*K*-Ac	284	0.00	Cor(16)	21
		0.73	F-nmr	13	0.31	F-nmr	13
471	SO_2CN	0.94	F-nmr	177	0.32	F-nmr	177
472	SO_2NH_2	0.44	F-nmr	177	0.00	Cor(16)	21
					0.12	F-nmr	177
473	SO_2NMe_2	0.42	F-nmr	177	0.12	F-nmr	177
476	SO_2OR	0.50	F-nmr	10	(+)0.09	ir	264
478	SO_2F	0.75	F-nmr	177	0.00	Cor(16)	—
					0.26	F-nmr	177
479	SO_2Cl	0.80	F-nmr	177	(+)0.11	ir	264
					0.24	F-nmr	177
480	$S(OR_F)_2Ph$	0.40	F-nmr	181	0.09	F-nmr	181
481	SF_3	0.60	F-nmr	177	0.20	F-nmr	177
482	SF_5	0.57	p*K*-Ac	11	0.06	←Stat	11
		0.53	F-nmr	177	0.00	ir	264
					0.00	Cor(16)	21
					0.08	F-nmr	177
484	SeC_6H_{11}-c	0.38(0.02)	p*K*-Ac	9			
485	SePh	(0.37)	Est	—	(−)0.19	ir	279

486	$SeCF_3$	0.42	F-nmr	13	0.04	F-nmr	13
487	SeCN	0.58(0.05)	p*K*-Ac	9			
488	$SeOCF_3$	0.76	F-nmr	13	0.10	F-nmr	13
489	SeO_2CF_3	0.96	F-nmr	13	0.25	F-nmr	13
490	F	(0.49)(0.05)	p*K*-Qui	(40)	(−)0.34	ir	264
		0.54(0.02)	p*K*-Ac	(9)	−0.59	Cor(16)	21
		0.50	HydrE	8	−0.34	←Stat	11
		0.52	F-nmr	10	−0.32	F-nmr	10
491	Cl	(0.48)(0.05)	p*K*-Qui	(40)	(−)0.22	ir	264
		0.50(0.1)	p*K*-BCO	(265)	−0.36	Cor(16)	21
		0.47(0.02)	p*K*-Ac	(9)			
		0.47	HydrE	8			
		0.46	Est	11	−0.23	←Stat	11
		(0.37)	F-nmr	(27)	−0.18	F-nmr	10
492	Br	(0.51)(0.05)	p*K*-Qui	(40)	(−)0.23	ir	264
		0.49(0.1)	p*K*-BCO	(276)	−0.34	Cor(16)	21
		0.46(0.02)	p*K*-Ac	(9)	−0.16	F-nmr	10
		0.47(0.1)	p*K*	41			
		0.45	HydrE	8			
		0.44	F-nmr	10	−0.19	←Stat	11
493	I	(0.47)(0.05)	p*K*-Qui	(40)	(−)0.22	ir	264
		0.39(0.02)	p*K*-Ac	(9)	−0.16	←Stat	11
		0.41(0.1)	p*K*	41	−0.31	Cor(16)	21
		0.38	HydrE	8	−0.14	F-nmr	10
		0.42	F-nmr	(27)			
494	IO_2				0.00	Cor(16)	21
495	$I(OAc)_2$	0.82	F-nmr	13	0.06	F-nmr	13
496	$I(OCOCF_3)_2$	1.22	F-nmr	13	0.12	F-nmr	13
497	IF_2	0.86	F-nmr	13	−0.03	F-nmr	13
498	ICl_2	1.08	F-nmr	13	±0.12	ir	264
					0.03	F-nmr	13
499	IF_4	1.00	F-nmr	13	0.15	F-nmr	13

Table 10.4—continued

No.[a]	Substituent	σ_I	Method[b]	Ref.	σ_R^0	Method[b]	Ref.
	Charged substituents[e]						
500	$CH_2NH_3^+ X^-$(?)	0.36	p*K*-Ac (W)	9	0.00	ir (D_2O)	264
	Cl^-	0.25	F-nmr (W,A)	10	0.00	F-nmr (W,A)	10
502	$CH_2\overset{+}{N}Me_3 X^-$(?)	0.30	Est(4*a*) (W)	(8)			
510	$NH_3^+ Cl^-$	0.60	p*K*-Ac (W)	9	±0.18	ir (D_2O)	264
	Cl^-	0.58	F-nmr (W,A)	10	±0.33	ir (DMSO)	285
	Cl^-	0.38[f]	p*K*-Ac (W)	286			
511	$\overset{+}{N}H_2Me\ Cl^-$	0.60	p*K*-Ac (W)	9	±0.15	ir (D_2O)	264
512	$\overset{+}{N}H_2R\ Cl^-$	0.60	p*K*-Ac (W)	9			
513	$\overset{+}{N}HMe_2\ Cl^-$	0.70	p*K*-Ac (W)	9	±0.14	ir (D_2O)	264
		(1.25)	p*K* (A/W)	41			
514	$\overset{+}{N}Me_3\ Cl^-$	0.73	p*K*-Ac (W)	9	±0.15	ir (D_2O)	264
	I^-	0.93	p*K*-BCO (A/W)	287	±0.15	ir (DMSO)	285

	Br^-	(0.78)	p*K*-BCO (A/W)	288			
	Br^-	(1.34)	p*K*-BCO (A)	288			
	X^-(?)	0.86	HydrE (W)	8	−0.11	F-nmr (MeOH)	151
	I^-	0.93	F-nmr (W,A)	10			
515	$\overset{+}{N}{:}N\ BF_4^-$				±0.30	ir (D_2O)	264
					±0.29	ir (DMSO)	285
516	$\overset{+}{P}Me_3\ Cl^-$				±0.08	ir (D_2O)	279
	I^-	0.40	F-nmr (DMF)	160	0.20	F-nmr (DMF)	160
	BF_4^-	0.43	F-nmr (DMSO)	165	0.20	F-nmr (DMSO)	165
517	$\overset{+}{P}Bu^n_3\ Cl^-$	0.60	p*K*-Ac (W)	282			
518	$\overset{+}{P}MePh_2\ X^-$(?)	0.60	F-nmr (DMSO)	165	0.23	F-nmr (DMSO)	165
519	$\overset{+}{P}Ph_3\ Cl^-$	0.76	p*K*-Ac (W)	282			
	Br^-				±0.07	ir ($CHCl_3$)	279
520	$O\overset{+}{N}H_3\ Cl^-$	0.47	p*K*-Ac (W)	9			
521	$\overset{+}{S}Me_2\ ClO_4^-$	0.89	F-nmr (MeCN)	177	0.17	F-nmr (MeCN)	177

Table 10.4—continued

No.[a]	Substituent	σ_I	Method[b]	Ref.	σ_R^0	Method[b]	Ref.
522	$\overset{+}{S}(OR_F)Ph$ $CF_3SO_3^-$	1.31	F-nmr ($CHCl_3$)	181	0.31	F-nmr ($CHCl_3$)	181
523	$\overset{+}{I}Ph\ X^-$(?)				±0.28	ir (DMSO)	264
524	$CH_2CO_2^-\ M^+$(?)	0.01	p*K*-Ac (W)	9			
525	$CH_2SO_3^-\ M^+$(?)	0.01	p*K*-Ac (W)	25			
526	$(CH_2)_2SO_3^-\ M^+$(?)	−0.04	p*K*-Ac (W)	25			
528	$CO_2^-\ Na^+$	0.09	p*K*-Qui (W)	(40)			
	M^+(?)	−0.17	p*K*-Ac (W)	9			
	Na^+	−0.12	p*K*-BCO (A/W)	287			
	M^+(?)	0.16[f]	p*K*-Ac (W)	286			
	M^+(?)	−0.49	p*K*-BCO (A/W)	288			
	M^+(?)	−0.85	p*K*-BCO (A)	288			
529	1-$C_2HB_{10}H_{11}^-$(1,2) Na^+	−0.39	F-nmr (THF/W)	200	−0.17	F-nmr (THF/W)	200
530	1-$C_2HB_{10}H_{10}^{2-}$(1,2) $2Na^+$	−0.54	F-nmr (THF)	200	−0.19	F-nmr (THF)	200

531	$\bar{B}Ph_3\ Na^+$				±0.13	ir (DMSO)	279
532	$\bar{B}(OH)_3\ M^+$(?)	−0.36	F-nmr (W,A)	10			
536	$AsO_3H^-\ M^+$(?)	0.01	p*K*-Ac (W)	9			
538	$O^-\ M^+$(?)	−0.16	F-nmr (W,A)	10	−0.60	F-nmr (W,A)	10
	Na^+				(−)0.59	ir (D_2O)	264
539	$S^-\ Na^+$				(−)0.33	ir (D_2O)	264
540	$SO_2^-\ Na^+$				0.00	ir (D_2O)	264
541	$SO_3^-\ K^+$	0.13	p*K*-Ac (W)	9			
	M^+(?)	0.23	F-nmr (MeOH)	177	0.07	F-nmr (MeOH)	177
	Na^+				0.00	ir (D_2O)	264

[a] The numbering is common to Tables 10.1, 10.2, 10.4, and 10.5.
[b] The abbreviations should be intelligible; for detailed explanation see p. 453.
[c] Special values for compounds with several C—C bonds between the substituent and the reaction center.
[d] Special values for substituents twisted out of the plane of the conjugated system.
[e] The solvent is given (in parentheses) with the method.
[f] Extrapolated to zero ionic strength.

Table 10.5. Steric Constants

No.[a]	Substituents	υ (Ref. 24) [υ][b] (Ref. 44)	Method[c]	E_s^d	Ref.	Other scales[e]	Method[c]	Ref.
1	H	0	vdW	+1.24	8	[0][f]	3-5BA	55
		[0][b]	BrEx					
3	Me	0.52	vdW	0.00	8	[0][g]	CorH	290
		[0.35][b]	BrEx			0	AddOl	53
						0	RedCo	291
5	Et	0.56	HydrE	−0.07	8	[−0.27][g]	CorH	289
		[0.38][b]	BrEx			−0.07	AddOl	53
						−0.49	RedCo	291
6	Pr^n	0.68	HydrE	−0.36	8	[−0.56][g]	CorH	290
		[0.42][b]	BrEx			−0.56	AddOl	53
						−0.61	RedCo	291
7	Pr^i	0.76	HydrE	−0.47	8	[−0.85][g]	CorH	290
		[0.62][b]	BrEx			−0.66	AddOl	53
8	Bu^n	0.68	HydrE	−0.39	8	[−0.59][g]	CorH	289
		[0.42][b]	BrEx					
9	Bu^i	0.98	HydrE	−0.93	8	[−1.13][g]	CorH	290
		[0.55][b]	BrEx			−1.06	AddOl	53
10	CHMeEt	1.02	HydrE	−1.13	8	[−1.53][g]	CorH	290
						−1.15	AddOl	53
11	Bu^t	1.24	vdW	−1.54	8	[−2.14][g]	CorH	290
		[1.23][b]	BrEx			−4.22	AddOl	53
12	$n\text{-}C_5H_{11}$	0.68	HydrE	−0.40	8	[−0.60][g]	CorH	289
—	$(CH_2)_2Pr^i$	0.68	HydrE	−0.35	8	[−0.55][g]	CorH	289
—	$CH_2CHMeEt$	1.00	HydrE					
—	$CHEt_2$	1.51	HydrE	−1.98	8	[−2.38][g]	CorH	290
		1.28	HydrE					
14	CH_2Bu^t	1.34	HydrE	−1.74	8	[−1.94][g]	CorH	290
						−1.84	AddOl	53

—	$CHMePr^i$					−1.65	AddOl	53
15	CMe_2Et					−4.72	AddOl	53
16	n-C_6H_{13}	0.73	HydrE	−0.30	289	[−0.45][g]	CorH	289
—	$(CH_2)_3Pr^i$			−0.43	289	[−0.52][g]	CorH	289
—	$(CH_2)_2Bu^t$	0.70	HydrE	−0.34	8	[−0.64][g]	CorH	289
—	$CHEtPr^i$	2.11	HydrE					
—	$CHMeBu^t$	2.11	HydrE	−3.33	8	[−3.73][g]	CorH	289
						−2.43	AddOl	53
—	CMe_2Pr^i					−5.21	AddOl	53
17	n-C_7H_{15}	0.73	HydrE					
—	$CHPr^n_2$	1.54	HydrE	−2.11	8			
—	$CHMeCH_2Bu^t$	1.41	HydrE	−1.85	8	[−2.24][g]	CorH	290
—	CEt_3	2.38	HydrE	−3.8	8	[−4.4][g]	CorH	290
—	CMe_2Bu^t	2.43	HydrE	−3.9	8	[−4.6][g]	CorH	290
						−6.00	AddOl	53
18	n-C_8H_{17}	0.68	HydrE	−0.33	8			
—	$CMe_2CH_2Bu^t$	1.74	HydrE	−2.57	8	[−3.17][g]	CorH	290
—	$CHBu^n_2$	1.56	HydrE					
—	$(CH_2)_2CHEtBu^t$	1.01	HydrE					
—	$CH(CH_2Pr^i)_2$	1.70	HydrE	−2.47	8			
—	$CH(CH_2Bu^t)_2$	2.03	HydrE	−3.18	8			
—	$CMeBu^tCH_2Bu^t$			−4.0	8			
19	$CH{:}CH_2$			[−1.53][h]	(292)			
20	$CH{:}CHMe(E)$			[−1.56][h]	(292)			
—	$CMe{:}CH_2$			[−2.00][h]	(292)			
25	$(CH_2)_2CH{:}CH_2$			−0.33	292			
30	c-C_3H_5					−0.43	RedCo	291
32	c-C_4H_7			−0.06	8			
33	c-C_5H_9			−0.51	8	−0.65	RedCo	291
34	c-C_6H_{11}	0.87	HydrE	−0.79	8	−0.90	RedCo	291
35	$CH_2C_6H_{11}$-c	0.97	HydrE	−0.98	8			
36	$(CH_2)_2C_6H_{11}$-c	0.70	HydrE					
—	$(CH_2)_3C_6H_{11}$-c	0.71	HydrE					

Table 10.5—continued

No.[a]	Substituent	υ (Ref. 24) $[\upsilon]^b$ (Ref. 44)	Method[c]	E_s^d	Ref.	Other scales[e]	Method[c]	Ref.
—	c-C_7H_{13}			−1.10	8			
48	Ph	$[1.66]^h$	HydrE	$[-2.48]^h$	(292)			
		0.57	vdW					
49	CH_2Ph	0.70	HydrE	−0.38	8			
50	$(CH_2)_2Ph$	0.70	HydrE	−0.38	8	$[-0.58]^g$	CorH	289
51	$(CH_2)_3Ph$	0.70	HydrE	−0.45	8			
52	$(CH_2)_4Ph$	0.70	HydrE	−1.76	8			
53	$CHPh_2$	1.25	HydrE	−1.43	293			
54	CPh_3			−4.68	293			
55	CHMePh	0.99	HydrE	−1.19	8			
—	CHEtPh	1.18	HydrE	−1.50	8			
—	$CMePh_2$			−3.55	293			
—	$CEtPh_2$			−4.34	293			
60	CH:CHPh (*E*)			$[-1.82]^h$	(292)			
	[9-R-fluorene structure] R = H			−1.10	293			
	Me			−1.73	293			
	Et			−1.97	293			
	Pr^i			−3.30	293			
	Bu^t			−4.12	293			
	Ph			−3.08	293			

—	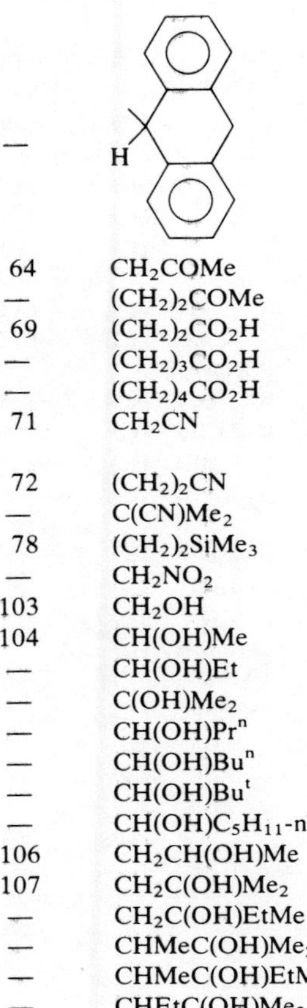			−1.89	293			
64	CH_2COMe			−0.75	289	[−1.08][g]	CorH	289
—	$(CH_2)_2COMe$			−0.89	289	[−1.00][g]	CorH	289
69	$(CH_2)_2CO_2H$			−0.97	289	[−1.08][g]	CorH	289
—	$(CH_2)_3CO_2H$			−0.41	289	[−0.58][g]	CorH	289
—	$(CH_2)_4CO_2H$			−0.32	289	[−0.50][g]	CorH	289
71	CH_2CN	0.89	HydrE	−0.94	294	[−1.41][g]	CorH	289
				−1.14	289	−0.43	RedCo	291
72	$(CH_2)_2CN$			−0.99	289	[−1.08][g]	CorH	289
—	$C(CN)Me_2$			−0.76	295	[−1.49][g]	CorH	289
78	$(CH_2)_2SiMe_3$			−0.12	294			
—	CH_2NO_2			−1.47	289	[−1.85][g]	CorH	289
103	CH_2OH			+0.03	289	[−0.39][g]	CorH	289
104	CH(OH)Me			+0.09	289	[−0.64][g]	CorH	289
—	CH(OH)Et			−0.34	289	[−1.02][g]	CorH	289
—	$C(OH)Me_2$			−0.71	289	[−1.65][g]	CorH	289
—	$CH(OH)Pr^n$			−0.33	289	[−1.02][g]	CorH	289
—	$CH(OH)Bu^n$			−0.31	289	[−1.00][g]	CorH	289
—	$CH(OH)Bu^t$			−2.21	289	[−2.70][g]	CorH	289
—	$CH(OH)C_5H_{11}$-n			−0.34	289	[−1.02][g]	CorH	289
106	$CH_2CH(OH)Me$			−1.07	289	[−1.15][g]	CorH	289
107	$CH_2C(OH)Me_2$			−1.74	289	[−1.73][g]	CorH	289
—	$CH_2C(OH)EtMe$			−1.97	289	[−1.93][g]	CorH	289
—	$CHMeC(OH)Me_2$			−3.11	289	[−3.23][g]	CorH	289
—	CHMeC(OH)EtMe			−3.22	289	[−3.30][g]	CorH	289
—	$CHEtC(OH)Me_2$			−3.76	289	[−3.77][g]	CorH	289

Table 10.5—continued

No.[a]	Substituent	υ (Ref. 24) $[\upsilon]^b$ (Ref. 44)	Method[c]	E_s^d	Ref.	Other scales[e]	Method[c]	Ref.
—	$CH(OH)CH_2OH$			−0.81	289	[−1.45][g]	CorH	289
109	CH_2OMe			−0.19	8	[−0.52][g]	CorH	289
—	$(CH_2)_2OMe$			−0.77	8	[−0.97][g]	CorH	289
—	$(CH_2)_3OMe$			−0.42	289	[−0.60][g]	CorH	289
—	$(CH_2)_4OMe$			−0.34	289	[−0.51][g]	CorH	289
—	CH(OMe)Me			−0.64	289	[−1.29][g]	CorH	289
—	$C(OMe)Me_2$			−1.43	289	[−2.31][g]	CorH	289
—	CH_2OEt			−0.37	289	[−0.75][g]	CorH	289
—	$(CH_2)_2OEt$			−0.97	289	[−1.07][g]	CorH	289
—	$(CH_2)_3OEt$			−0.45	289	[−0.62][g]	CorH	289
—	CH(OEt)Me			−0.69	289	[−1.33][g]	CorH	289
—	CH_2OPr^n			−0.39	289	[−0.77][g]	CorH	289
—	CH_2OBu^n			−0.42	289	[−0.81][g]	CorH	289
—	CH_2OBu^i			−0.47	289	[−0.83][g]	CorH	289
—	$CMe_2OCH_2Bu^t$			−1.43	296			
—	$CMe_2CH_2OBu^t$			−1.57	296			
—	$CMe_2O\cdot OBu^t$			−1.96	296			
113	CH_2OPh			−0.33	289	[−0.66][g]	CorH	289
119	CH_2SMe			−0.34	8	[−0.77][g]	CorH	289
				−0.44	289			
130	CH_2F	0.62	HydrE	−0.24	8	[−0.57][g]	CorH	289
131	CHF_2	0.68	Est	−0.67	8	[−1.33][g]	CorH	289
132	CF_3	0.91	vdW	−1.16	8	[−2.15][g]	CorH	289
						−0.70	RedCo	291
133	CH_2Cl	0.60	HydrE	−0.24	8	[−0.57][g]	CorH	289
						−0.47	RedCo	291
134	$CHCl_2$	0.81	HydrE	−1.54	8	[−2.20][g]	CorH	289
						−0.56	RedCo	291

135	CCl_3	1.38	vdW	−2.06	8	[−3.05][g]	CorH	289
136	CH_2Br	0.64	HydrE	−0.27	8	[−0.60][g]	CorH	289
						−0.49	RedCo	291
137	$CHBr_2$	0.89	Est	−1.86	8	[−2.52][g]	CorH	289
138	CBr_3	1.56	vdW	−2.43	8	[−3.42][g]	CorH	289
139	CH_2I	0.67	HydrE	−0.37	8			
140	CHI_2	0.97	Est					
—	CI_3	1.79	vdW					
—	$(CH_2)_2Cl$			−0.90	8			
—	$(CH_2)_3Cl$			−0.48	289	[−0.64][g]	CorH	289
142	CHClMe			−0.50	289	[−1.18][g]	CorH	289
—	$(CH_2)_2Br$			−1.27	289	[−1.32][g]	CorH	289
143	CHBrMe			−0.69	289	[−1.33][g]	CorH	289
—	$CBrMe_2$	1.39	HydrE					
—	CBr_2Me	1.46	HydrE					
—	$(CH_2)_2I$			−1.02	289	[−1.20][g]	CorH	289
—	CHIMe			−1.36	289	[−1.92][g]	CorH	289
203	CN					[0.15][f]	3-5BA	55
253	$SiMe_3$	1.40	vdW					
287	NH_2					[0.22][f]	3-5BA	55
299	NHAc					[0.27][f]	3-5BA	55
348	NO_2			−1.28	297	[0.25][f]	3-5BA	55
				Est				
389	OH					[0.09][f]	3-5BA	55
390	OMe					[0.12][f]	3-5BA	55
391	OEt					[0.13][f]	3-5BA	55
482	SF_5			−1.67	297			
				vdW				
490	F	0.27	vdW					
491	Cl	0.55	vdW			[0.20][f]	3-5BA	55
492	Br	0.65	vdW			[0.22][f]	3-5BA	55
493	I	0.78	vdW			[0.36][f]	3-5BA	55

Table 10.5—continued

No.[a]	Substituent	υ (Ref. 24) $[\upsilon]^b$ (Ref. 44)	Method[c]	E_s^d	Ref.	Other scales[e]	Method[c]	Ref.
	Charged Substituents (water as solvent)							
500	$CH_2\overset{+}{N}H_3\,X^-$(?)			−2.30	289	[−2.41][g]	CorH	289
—	$(CH_2)_2\overset{+}{N}H_3\,X^-$			−1.82	289	[−1.80][g]	CorH	289
502	$CH_2\overset{+}{N}Me_3\,X^-$(?)			−2.89	289	[−3.02][g]	CorH	289
	Cl^-			−1.56 (Est)	295			
505	$(CH_2)_2\overset{+}{N}Me_3\,Br^-$			−1.99	289	[−2.00][g]	CorH	289
506	$(CH_2)_3\overset{+}{N}Me_3\,Br^-$			−1.35	289	[−1.43][g]	CorH	289
525	$CH_2SO_3^-\,K^+$			−1.25	289	[−1.51][g]	CorH	289
526	$(CH_2)_2SO_3^-\,K^+$			−0.33	289	[−0.52][g]	CorH	289

[a]The numbering is common to Tables 10.1, 10.2, 10.4, and 10.5; the substituents included in Table 10.5 only are not numbered.
[b]Values valid for the nucleophilic substitution at a sp^3-hybridized carbon.
[c]The abbreviations should be intelligible; for detailed explanations see p. 453.
[d]All the values were obtained by the method of ester hydrolysis unless otherwise noted.
[e]The range of validity is unknown; the scaling is essentially similar to that for E_s except under *f*.
[f]The scaling is related to H = 0 and is not comparable with the E_s scale.
[g]The experimental values have been corrected for assumed hyperconjugation; the resulting values are not comparable with other E_s constants.
[h]Purely formal values expressing the mesomeric rather than the steric effect.

References

1. H. H. Jaffé, *Chem. Rev.*, **53**, 191 (1953).
2. D. H. McDaniel and H. C. Brown, *J. Org. Chem.*, **23**, 420 (1958).
3. V. A. Palm, *Uspekhi Khim.*, **30**, 1069 (1961); EE, 471.
4. Yu. A. Zhdanov and V. I. Minkin, *Korrelatsionnyi analiz v organicheskoi khimii* (Izd. rostovskogo universiteta, Rostov, 1966).
5. N. B. Chapman and J. Shorter, eds., *Advances in Linear Free Energy Relationships*, (Plenum, London, 1972).
6. C. Hansch, A. Leo, S. H. Unger, K. H. Kim, D. Nikaitani, and E. J. Lien, *J. Med. Chem.*, **16**, 1207 (1973).
7. H. van Bekkum, P. E. Verkade, and B. M. Wepster, *Rec. Trav. chim.*, **78**, 815 (1959).
8. R. W. Taft, in *Steric Effects in Organic Chemistry*, Chap. 13, M. S. Newman, ed. (Wiley, New York, 1956).
9. M. Charton, *J. Org. Chem.*, **29**, 1222 (1964).
10. R. W. Taft, E. Price, I. R. Fox, I. C. Lewis, K. K. Andersen, and G. T. Davis, *J. Amer. Chem. Soc.*, **85**, 709, 3146 (1963).
11. S. Ehrenson, R. T. C. Brownlee, and R. W. Taft, *Progr. Phys. Org. Chem.*, **10**, 1 (1973).
12. H. C. Brown and Y. Okamoto, *J. Amer. Chem. Soc.*, **80**, 4979 (1958).
13. L. M. Yagupol'skii, A. Ya. Il'chenko, and N. B. Kondratenko, *Uspekhi Kkim.*, **43**, 64 (1974); EE, 32.
14. O. Exner, in Ref. 5, p. 1.
15. O. Exner, *Coll. Czech. Chem. Comm.*, **41**, 1516 (1976).
16. O. Exner and K. Kalfus, *Coll. Czech. Chem. Comm.*, **41**, 569 (1976).
17. C. K. Hancock, *J. Chem. Educ.*, **42**, 608 (1965).
18. S. Wold and M. Sjöström, *Chem. Scripta*, **2**, 49 (1972).
19. M. Sjöström and S. Wold, *Chem. Scripta*, **6**, 114 (1974).
20. M. Charton, *J. Org. Chem.*, **28**, 3121 (1963).
21. O. Exner, *Coll. Czech. Chem. Comm.*, **31**, 65 (1966).
22. C. K. Hancock, E. A. Meyers, and B. J. Yager, *J. Amer. Chem. Soc.*, **83**, 4211 (1961).
23. V. A. Palm, *Grundlagen der quantitativen Theorie der organische Reaktionen*, Chap. 10 (Akademie Verlag, Berlin, D.D.R., 1971).
24. M. Charton, *J. Amer. Chem. Soc.*, **97**, 1552 (1975).
25. M. Charton, personal communication.
26. W. Simon *et al.*, *Zusammenstellung von scheinbaren Dissoziationskonstanten im Lösungsmittelsystem Methylcellosolve/Wasser*, Vols. I–III (Juris-Verlag, Zürich, 1959–1963).
27. R. W. Taft, *J. Phys. Chem.*, **64**, 1805 (1960).
28. Y. Yukawa, Y. Tsuno, and M. Sawada, *Bull. Chem. Soc. Japan*, **39**, 2274 (1966).
29. Y. Yukawa, Y. Tsuno, and M. Sawada, *Bull. Chem. Soc. Japan*, **45**, 1198 (1972).
30. V. M. Maremäe and V. A. Palm, *Reakts. spos. org. Soedinenii*, **1** (2), 85 (1964).
31. M. Sjöström and S. Wold, *Acta Chem. Scand.* **B30**, 167 (1976).
32. A. J. Hoefnagel, J. C. Monshouwer, E. C. G. Snorn, and B. M. Wepster, *J. Amer. Chem. Soc.*, **95**, 5350 (1973).
33. A. J. Hoefnagel and B. M. Wepster, *J. Amer. Chem. Soc.*, **95**, 5357 (1973).
34. S. M. Shein and L. A. Kozorez, *Reakts. spos. org. Soedinenii*, **3**, (4), 45 (1966); EE, 315.
35. Y. Otsuji, M. Kubo, and E. Imoto, *Bull. Univ. Osaka Prefect., Ser.* A, **7**, 61 (1959); *Chem. Abs.*, **54**, 24796 (1960).
36. A. Streitwieser, *Molecular Orbital Theory for Organic Chemists* (Wiley, New York, 1960).
37. T. M. Krygowski, *Bull. Acad. polon. Sci., Sér. Sci. chim.*, **19**, 49 (1971).

38. J. D. Roberts and W. T. Moreland, *J. Amer. Chem. Soc.*, **75**, 2167 (1953).
39. J. Paleček and J. Hlavatý, *Coll. Czech. Chem. Comm.*, **38**, 1985 (1973).
40. C. A. Grob and M. G. Schlageter, *Helv. Chim. Acta*, **57**, 509 (1974); E. Ceppi and C. A. Grob, *Helv. Chim. Acta*, **57**, 2332 (1974); W. Eckhardt and C. A. Grob, *Helv. Chim.Acta*, **57**, 2339 (1974).
41. O. Exner and J. Jonáš, *Coll. Czech. Chem. Comm.*, **27**, 2296 (1962).
42. R. W. Taft and I. C. Lewis, *J. Amer. Chem. Soc.*, **80**, 2436 (1958); **81**, 5343 (1959).
43. V. A. Palm, Ref. 23, Chap. 7.
44. M. Charton, *J. Amer. Chem. Soc.*, **97**, 3691, 3694 (1975).
45. C. D. Ritchie, *J. Phys. Chem.*, **65**, 2091 (1961).
46. S. Ehrenson, *Tetrahedron Lett.*, 351 (1964).
47. R. T. C. Brownlee, A. R. Katritzky, and R. D. Topsom, *J. Amer. Chem. Soc.*, **88**, 1413 (1966).
48. A. R. Katritzky and R. D. Topsom, *Angew. Chem. Internat. Edn.*, **9**, 87 (1970), GE, **82**, 106 (1970).
49. Y. Yukawa and Y. Tsuno, *Bull. Chem. Soc. Japan*, **32**, 971 (1959).
50. M. Yoshioka, K. Hamamoto, and T. Kubota, *Bull. Chem. Soc. Japan*, **35**, 1723 (1962).
51. J. Shorter, in Ref. 5, p. 71.
52. M. Charton, *Progr. Phys. Org. Chem.*, **8**, 235 (1971).
53. R. Fellous and R. Luft, *J. Amer. Chem. Soc.*, **95**, 5593 (1973).
54. A. Babadjamian, M. Channon, R. Gallo, and J. Metzger, *J. Amer. Chem. Soc.*, **95**, 3807 (1973).
55. D. Peltier, *Compt. rend.*, **241**, 57 (1955).
56. D. H. McDaniel and H. C. Brown, *J. Amer. Chem. Soc.*, **77**, 3756 (1955).
57. T. Fujita, C. Takayama, and M. Nakajima, *J. Org. Chem.*, **38**, 1623 (1973).
58. B. I. Istomin and V. A. Baransky, *Org. Reactivity*, **11**, 963 (1975); RE 963.
59. J. H. Nelson, R. G. Garvey, and R. O. Ragsdale, *J. Heterocyclic Chem.*, **4**, 591 (1967).
60. A. Streitwieser and C. Perrin, *J. Amer. Chem. Soc.*, **86**, 4938 (1964).
61. T. H. Fisher and A. W. Meierhoefer, *Tetrahedron*, **31**, 2019 (1975).
62. T. A. Mastryukova and M. I. Kabachnik, *Uspekhi Khim.*, **38**, 1751 (1969); EE, 795.
63. T. A. Mastryukova and M. I. Kabachnik, *Reakts. spos. org. Soedinenii*, **7**, 573 (1970); EE, 255
64. T. A. Mastryukova and M. I. Kabachnik, *J. Org. Chem.*, **36**, 1201 (1971).
65. B. I. Istomin and V. A. Palm, *Reakts. spos. org. Soedinenii*, **9**, 433 (1972).
66. A. J. Talvik and V. A. Palm, *Org. Reactivity*, **11**, 287 (1974); RE, 285.
67. V. M. Maremäe, *Reakts. spos. org. Soedinenii*, **2** (3), 13 (1965); **4**, 96 (1967); EE, 38.
68. I. A. Koppel, *Reakts. spos. org. Soedinenii*, **2**(2), 26 (1965).
69. C. G. Swain and E. C. Lupton, *J. Amer. Chem. Soc.*, **90**, 4328 (1968).
70. M. J. S. Dewar and P. J. Grisdale, *J. Amer. Chem. Soc.*, **84**, 3539, 3548 (1962).
71. Y. Yukawa and Y. Tsuno, *Nippon Kagaku Zasshi*, **86**, 873 (1965).
72. V. M. Nummert, *Org. Reactivity*, **11**, 621 (1975); RE, 617.
73. M. Sjöström and S. Wold, *Chem. Scripta*, **9**, 200 (1976).
74. A. Streitwieser and H. S. Klein, *J. Amer. Chem. Soc.*, **85**, 2759 (1963); A. Streitwieser and J. S. Humphrey, *J. Amer. Chem. Soc.*, **89**, 3767 (1967).
75. W. R. Young and C. S. Yannoni, *J. Amer. Chem. Soc.*, **91**, 4581 (1969).
76. T. Matsui, H. C. Ko, and L. G. Hepler, *Canad. J. Chem.*, **52**, 2906 (1974).
77. J. M. Wilson, N. E. Gore, J. E. Sawbridge, and F. Cardenas-Cruz, *J. Chem. Soc.* (*B*), 852 (1967).
78. H. Kloosterziel and H. J. Backer, *J. Amer. Chem. Soc.*, **74**, 5806 (1952).
79. M. Charton, *J. Org. Chem.*, **30**, 552 (1965).
80. R. L. Herbst and M. E. Jacox, *J. Amer. Chem. Soc.*, **74**, 3004 (1952).
81. M. Charton, *J. Chem. Soc.*, 1205 (1964).

82. C. Laurence and B. Wojtkowiak, *Bull. Soc. chim. France* 3833 (1971).
83. J. M. Wilson, A. G. Briggs, J. E. Sawbridge, P. Tickle, and J. J. Zuckerman, *J. Chem. Soc.* (A), 1024 (1970).
84. G. B. Ellam and C. D. Johnson, *J. Org. Chem.*, **36**, 2284 (1971).
85. M. Berthelot, C. Laurence, and B. Wojtkowiak, *Bull. Soc. chim. France* 662 (1973).
86. M. Charton and H. Meislich, *J. Amer. Chem. Soc.*, **80**, 5940 (1958).
87. J. A. Landgrebe and R. H. Rynbrandt, *J. Org. Chem.*, **31**, 2585 (1966).
88. J. Šmejkal, J. Jonáš, and J. Farkaš, *Coll. Czech. Chem. Comm.*, **29**, 2950 (1964).
89. R. C. Hahn, T. F. Corbin, and H. Shechter, *J. Amer. Chem. Soc.*, **90**, 3404 (1968).
90. Y. Kusuyama and Y. Ikeda, *Bull. Chem. Soc. Japan*, **46**, 204 (1973).
91. Yu. S. Shabarov, T. P. Surikova, and R. Ya. Levina, *Zhur. org. Khim.*, **4**, 1175 (1968); EE, 1131.
92. T. J. Broxton, G. Capper, L. W. Deady, A. Lenko, and R. D. Topsom, *J.C.S. Perkin* II, 1237 (1972).
93. H. Alper, E. C. H. Keung, and R. A. Partis, *J. Org. Chem.*, **36**, 1352 (1971).
94. W. Polaczkowa, N. Porowska, and B. Dybowska, *Bull. Acad. polon. Sci., Sér. Sci. chim.*, **8**, 537 (1960).
95. W. F. Little, C. N. Reilley, J. D. Johnson, and A. P. Sanders, *J. Amer. Chem. Soc.*, **86**, 1382 (1964).
96. L. M. Litvinenko, *Izvest. Akad. Nauk S.S.S.R., Ser. Khim.*, 1737 (1962); EE, 1653.
97. R. A. Benkeser and R. B. Gosnell, *J. Org. Chem.*, **22**, 327 (1957).
98. J. K. Kochi and G. S. Hammond, *J. Amer. Chem. Soc.*, **75**, 3452 (1953).
99. P. Zuman, *Substituent Effects in Organic Polarography* (Plenum, New York, 1967).
100. W. A. Sheppard, *Tetrahedron*, **27**, 945 (1971).
101. R. G. Pews, *J. Amer. Chem. Soc.*, **89**, 5605 (1967).
102. J. H. Smith and F. M. Menger, *J. Org. Chem.*, **34**, 77 (1969).
103. J. Hine and W. C. Bailey, *J. Org. Chem.*, **26**, 2098 (1961).
104. W. K. Kwok, R. A. M. O'Ferrall, and S. I. Miller, *Tetrahedron*, **20**, 1913 (1964).
105. N. Bodor and A. Kövendi, *Rev. Roumaine Chim.*, **11**, 413 (1966); *Chem. Abs.*, **65**, 7011 (1966).
106. Z. Plzák, F. Mareš, J. Hetflejš, J. Schraml, Z. Papoušková, V. Bažant, E. G. Rochow, and V. Chvalovský, *Coll. Czech. Chem. Comm.*, **36**, 3115 (1971).
107. J. Hine and W. C. Bailey, *J. Amer. Chem. Soc.*, **81**, 2075 (1959).
108. W. A. Sheppard, *Trans. New York Acad. Sci.* [2] **29**, 700 (1967).
109. W. A. Sheppard and C. M. Sharts, *Organic Fluorine Chemistry*, (Benjamin, New York, 1969).
110. W. A. Sheppard, *J. Amer. Chem. Soc.*, **87**, 2410 (1965).
111. E. B. Lifshits, L. M. Yagupol'skii, D. Ya. Naroditskaya, and E. S. Kozlova, *Reakt. spos. org. Soedinenii*, **6**, 317 (1969); EE, 133.
112. R. Stewart and L. G. Walker, *Canad. J. Chem.*, **35**, 1561 (1957).
113. D. J. Glover, *J. Org. Chem.*, **31**, 1660 (1966).
114. H. E. Ruskie and L. A. Kaplan, *J. Org. Chem.*, **30**, 319 (1965).
115. W. A. Sheppard, *J. Amer. Chem. Soc.*, **92**, 5419 (1970).
116. R. Taylor, *J.C.S. Perkin* II, 253 (1973).
117. A. A. Humffray, J. J. Ryan, J. P. Warren, and Y. H. Yung, *Chem. Comm.*, 610 (1965).
118. K. Bowden and M. J. Shaw, *J. Chem. Soc.* (*B*), 161 (1971).
119. P. E. Peterson, D. M. Chevli, and K. A. Sipp, *J. Org. Chem.*, **33**, 972 (1968).
120. W. N. White, R. Schlitt, and D. Gwynn, *J. Org. Chem.*, **26**, 3613 (1961).
121. T. Nishiguchi and Y. Iwakura, *J. Org. Chem.*, **35**, 1591 (1970).
122. W. F. Little, C. N. Reilley, J. D. Johnson, K. N. Lynn, and A. P. Sanders, *J. Amer. Chem. Soc.*, **86**, 1376 (1964).
123. C. Weiss, W. Engewald, and H. Müller, *Tetrahedron*, **22**, 825 (1966).

124. J. J. Ryan and A. A. Humffray, *J. Chem. Soc.* (*B*), 842 (1966).
125. P. Cecchi, *Ricerca sci.*, **28**, 2526 (1958).
126. L. A. Cohen and S. Takahashi, *J. Amer. Chem. Soc.*, **95**, 443 (1973).
127. F. Fringuelli, G. Marino, and A. Taticchi, *J. Chem. Soc.* (B), 2304 (1971).
128. F. Fringuelli, G. Marino, and A. Taticchi, *J. Chem. Soc.* (B), 1595 (1970).
129. V. F. Bystrov, Zh. N. Belaya, B. E. Gruz, G. P. Syrova, A. I. Tolmachev, L. M. Shulezhko, and L. M. Yagupol'skii, *Zh. obshchei Khim.*, **38**, 1001 (1968); EE, 963.
130. A. R. Katritzky and P. Simmons, *J. Chem. Soc.*, 1511 (1960).
131. H. L. Nyquist and B. Wolfe, *J. Org. Chem.*, **39**, 2591 (1974).
132. A. N. Nesmeyanov, E. G. Perevalova, S. P. Gubin, K. I. Grandberg, and A. G. Kozlovsky, *Tetrahedron Lett.*, 2381 (1966).
133. Š. Toma, *Coll. Czech. Chem. Comm.*, **34**, 2771 (1969).
134. T. G. Traylor and J. C. Ware, *J. Amer. Chem. Soc.*, **89**, 2304 (1967).
135. M. F. Hawthorne, T. E. Berry, and P. A. Wegner, *J. Amer. Chem. Soc.*, **87**, 4746 (1965).
136. L. I. Zakharkin, V. N. Kalinin, and I. P. Shepilov, *Doklady Akad. Nauk S.S.S.R.*, **174**, 606 (1967); EE, 484.
137. Z. Plzák, B. Štíbr, J. Plešek, and S. Heřmánek, *Coll. Czech. Chem. Comm.*, **40**, 3602 (1975).
138. J. Lipowitz, *J. Amer. Chem. Soc.*, **94**, 1582 (1972).
139. Vo-Kim-Yen, Z. Papoušková, J. Schraml, and V. Chvalovský, *Coll. Czech. Chem. Comm.*, **38**, 3167 (1973).
140. R. A. Benkeser and H. R. Krysiak, *J. Amer. Chem. Soc.*, **75**, 2421 (1953).
141. R. A. Benkeser, C. E. DeBoer, R. E. Robinson, and D. M. Sauve, *J. Amer. Chem. Soc.*, **78**, 682 (1956).
142. F. Mareš, Z. Plzák, J. Hetflejš, and V. Chvalovský, *Coll. Czech. Chem. Comm.*, **36**, 2957 (1971).
143. E. P. Serjeant, *Austral. J. Chem.*, **22**, 1189 (1969).
144. W. G. Herkstroeter, *J. Amer. Chem. Soc.*, **95**, 8686 (1973).
145. J. C. Howard and J. P. Lewis, *J. Org. Chem.*, **31**, 2005 (1966).
146. M. Charton, *J. Org. Chem.*, **30**, 557 (1965).
147. E. N. Tsvetkov, D. I. Lobanov, M. M. Makhamatkhanov, and M. I. Kabachnik, *Tetrahedron*, **25**, 5623 (1969).
148. F. S. Fawcett and W. A. Sheppard, *J. Amer. Chem. Soc.* **87**, 4341 (1965).
149. J. C. Kauer and W. A. Sheppard, *J. Org. Chem.*, **32**, 3580 (1967).
150. O. Exner and J. Lakomý, *Coll. Czech. Chem. Comm.*, **35**, 1371 (1970).
151. L. G. Vaughan and W. A. Sheppard, *J. Amer. Chem. Soc.*, **91**, 6151 (1969).
152. S. S. Gitis, A. V. Ivanov, A. Ya. Kaminskii, and Z. A. Kozina, *Reakts. spos. org. Soedinenii*, **3**(3), 142 (1966); EE, 264.
153. M. Kaplan, *J. Chem. and Eng. Data*, **6**, 272 (1961).
154. H. K. Hall, *J. Amer. Chem. Soc.*, **78**, 2570 (1956).
155. F. H. Brock, *J. Org. Chem.*, **24**, 1802 (1959).
156. P. Kristián, K. Antoš, D. Vlachová, and R. Zahradník, *Coll. Czech. Chem. Comm.*, **28**, 1651 (1963).
157. N. C. Collins and W. K. Glass, *Spectrochim. Acta*, **30A**, 1335 (1974).
158. H. H. Jaffé and H. L. Jones, *Adv. Heterocyclic Chem.*, **3**, 209 (1964).
159. I. Čepčiansky and J. Majer, *Coll. Czech. Chem. Comm.*, **34**, 72 (1969).
160. W. Prikoszovich and H. Schindlbauer, *Chem. Ber.*, **102**, 2922 (1969).
161. J. W. Rakshys, R. W. Taft, and W. A. Sheppard, *J. Amer. Chem. Soc.*, **90**, 5236 (1968).
162. B. I. Stepanov, A. I. Bokanov, and B. A. Korolev, *Zhur. obshchei. Khim.*, **37**, 2139 (1967); EE, 2029.
163. G. P. Schiemenz, *Angew. Chem.*, **78**, 605, 777 (1966); EE, **5**, 595, 731 (1966).
164. H. L. Retcofsky and C. E. Griffin, *Tetrahedron Lett.*, 1975 (1966).

165. A. W. Johnson and H. L. Jones, *J. Amer. Chem. Soc.*, **90**, 5232 (1968).
166. R. W. Bott, B. F. Dowden, and C. Eaborn, *J. Chem. Soc.*, 4994 (1965).
167. E. N. Tsvetkov, D. I. Lobanov, L. A. Izosenkova, and M. I. Kabachnik, *Zhur. obshchei Khim.*, **39**, 2177 (1969); EE, 2126.
168. J. J. Monagle, J. V. Mengenhauser, and D. A. Jones, *J. Org. Chem.*, **32**, 2477 (1967).
169. L. D. Freedman and H. H. Jaffé, *J. Amer. Chem. Soc.*, **77**, 920 (1955).
170. J. D. Roberts and W. T. Moreland, *J. Amer. Chem. Soc.*, **75**, 2267 (1953).
171. B. M. Wepster, cf. Ref. 150.
172. P. J. Bray and R. G. Barnes, *J. Chem. Phys.*, **27**, 551 (1957).
173. I. W. Serfaty, T. Hodgins, and E. T. McBee, *J. Org. Chem.*, **37**, 2651 (1972).
174. R. Pollet, R. van Poucke, and A. de Cat, *Bull. Soc. chim. belges*, **75**, 40 (1966).
175. W. A. Sheppard, *J. Amer. Chem. Soc.*, **85**, 1314 (1963).
176. R. van Poucke, R. Pollet, and A. de Cat, *Bull. Soc. chim. belges*, **75**, 573 (1966).
177. W. A. Sheppard and R. W. Taft, *J. Amer. Chem. Soc.*, **94**, 1919 (1972).
178. J. G. Traynham and G. A. Knesel, *J. Org. Chem.*, **31**, 3350 (1966).
179. J. R. Case, R. Price, N. H. Ray, H. L. Roberts, and J. Wright, *J. Chem. Soc.*, 2107 (1962).
180. C. C. Price and J. J. Hydock, *J. Amer. Chem. Soc.*, **74**, 1943 (1952).
181. L. J. Kaplan and J. C. Martin, *J. Amer. Chem. Soc.*, **95**, 793 (1973).
182. H. H. Szmant and G. Suld, *J. Amer. Chem. Soc.*, **78**, 3400 (1956).
183. L. N. Sedova, L. Z. Gandel'sman, L. A. Alekseeva, and L. M. Yagupol'skii, *Zhur. obshchei Khim.*, **39**, 2057 (1969); EE, 2011.
184. B. J. Lindberg, *Arkiv Kemi*, **32**, 317 (1970).
185. Á. Kucsman, F. Ruff, S. Sólyom, and T. Szirtes, *Acta Chim. Acad. Sci. Hung.*, **57**, 205 (1968).
186. T. Hayashi and R. Shibata, *Bull. Chem. Soc. Japan*, **34**, 1116 (1961).
187. G. P. Schiemenz, *Angew. Chem.*, **80**, 559 (1968); EE, **7**, 544 (1968).
188. K. Kalfus, M. Večeřa, and O. Exner, *Coll. Czech. Chem. Comm.*, **35**, 1195 (1970).
189. W. A. Sheppard, *J. Amer. Chem. Soc.*, **84**, 3072 (1962).
190. J. Epstein, R. E. Plapinger, H. O. Michel, J. R. Cable, R. A. Stephani, R. J. Hester, C. Billington, and G. R. List, *J. Amer. Chem. Soc.*, **86**, 3075 (1964).
191. A. V. Willi, *Z. phys. Chem. (Frankfurt)*, **27**, 233 (1961).
192. A. V. Willi and W. Meier, *Helv. Chim. Acta*, **39**, 318 (1956).
193. J. Hine, *J. Amer. Chem. Soc.*, **82**, 4877 (1960).
194. A. V. Willi, *Z. phys. Chem. (Frankfurt)*, **26**, 42 (1960).
195. J. D. Roberts, R. A. Clement, and J. J. Drysdale, *J. Amer. Chem. Soc.*, **73**, 2181 (1951).
196. E. S. Lewis and M. D. Johnson, *J. Amer. Chem. Soc.*, **81**, 2070 (1959).
197. F. G. Bordwell and P. J. Boutan, *J. Amer. Chem. Soc.*, **78**, 87 (1956).
198. F. M. Beringer and I. Lillien, *J. Amer. Chem. Soc.*, **82**, 5141 (1960).
199. A. V. Willi and J. F. Stocker, *Helv. Chim. Acta*, **38**, 1279 (1955).
200. L. I. Zakharkin, *Pure Appl. Chem.*, **29**, 513 (1972).
201. L. D. Quin and M. R. Dysart, *J. Org. Chem.*, **27**, 1012 (1962).
202. H. H. Jaffé, L. D. Freedman, and G. O. Doak, *J. Amer. Chem. Soc.*, **75**, 2209 (1953).
203. D. Pressman and D. H. Brown, *J. Amer. Chem. Soc.*, **65**, 540 (1943).
204. B. J. Lindberg, *Acta Chem. Scand*, **24**, 2852 (1970).
205. H. Zollinger, W. Büchler, and C. Wittwer, *Helv. Chim. Acta*, **36**, 1711 (1953).
206. A. Streitwieser and H. S. Klein, *J. Amer. Chem. Soc.*, **86**, 5170 (1964).
207. L. B. Jones and V. K. Jones, *Tetrahedron Lett.*, 1493 (1966).
208. L. M. Stock and H. C. Brown, *Adv. Phys. Org. Chem.*, **1**, 35 (1963).
209. R. Baker, C. Eaborn, and R. Taylor, *J.C.S. Perkin II*, 97 (1972).
210. N. A. Clinton, R. S. Brown, and T. G. Traylor, *J. Amer. Chem. Soc.*, **92**, 5228 (1970).
211. D. S. Noyce and S. A. Fike, *J. Org. Chem.*, **38**, 2433 (1973).

212. R. Taylor, *J. Chem. Soc. (B)*, 1450 (1971).
213. A. J. Cornish and C. Eaborn, *J.C.S. Perkin II*, 874 (1975).
214. W. Hanstein, H. J. Berwin, and T. G. Traylor, *J. Amer. Chem. Soc.*, **92**, 829 (1970).
215. G. D. Hartman and T. G. Traylor, *J. Amer. Chem. Soc.*, **97**, 6147 (1975).
216. F. Fringuelli, G. Marino, and A. Taticchi, *J.C.S. Perkin II*, 158, 1738 (1972).
217. O. Buchman, M. Grosjean, and J. Nasielski, *Helv. Chim. Acta*, **47**, 2037 (1964).
218. S. Clementi and P. Linda, *J.C.S. Perkin II*, 1887 (1973).
219. R. B. Davison, Thesis, Pittsburgh, 1965, cf. Ref. 164.
220. K. V. Seshadri and R. Ganesan, *Tetrahedron*, **28**, 3827 (1972).
221. S. Clementi and P. Linda, *Tetrahedron*, **26**, 2869 (1970).
222. C. Eaborn, A. R. Thompson, and D. R. M. Walton, *J. Chem. Soc. (B)*, 859 (1969).
223. C. Eaborn, A. R. Thompson, and D. R. M. Walton, *J. Chem. Soc. (B)*, 357 (1970).
224. K. C. Tseng, *Hua Hsueh Hsueh Pao*, **32**, 107 (1966); *Chem. Abs.*, **65**, 8736 (1966).
225. N. Porowska, W. Polaczkowa, and S. Kwiatkowska, *Roczniki Chem.*, **44**, 375 (1970).
226. H. H. Szmant and C. M. Harmuth, *J. Amer. Chem. Soc.*, **86**, 2909 (1964).
227. H. Veschambre and A. Kergomard, *Bull. Soc. chim. France*, 336 (1966).
228. H. Veschambre, G. Dauphin, and A. Kergomard, *Bull. Soc. chim. France*, 134, 2846 (1967).
229. J. D. Roberts, R. L. Webb, and E. A. McElhill, *J. Amer. Chem. Soc.*, **72**, 408 (1950).
230. L. A. Cohen and W. M. Jones, *J. Amer. Chem. Soc.*, **85**, 3397 (1963).
231. J. Miller and A. J. Parker, *Austral. J. Chem.*, **11**, 302 (1958).
232. J. J. Ryan and A. A. Humffray, *J. Chem. Soc. (B)*, 1300 (1967).
233. J. Durmis, M. Karvaš, and Z. Maňásek, *Coll. Czech. Chem. Comm.*, **38**, 215 (1973).
234. T.-L. Chan, J. Miller, and F. Stansfield, *J. Chem. Soc.*, 1213 (1964).
235. E. V. Borisov, L. E. Khodolov, and V. G. Yashunskii, *Reakts. spos. org. Soedinenii*, **7**, 704 (1970); EE,314.
236. F. Mares and A. Streitwieser, *J. Amer. Chem. Soc.*, **89**, 3770 (1967).
237. M. Syz and H. Zollinger, *Helv. Chim. Acta*, **48**, 383 (1965).
238. M. M. Fickling, A. Fischer, B. R. Mann, J. Packer, and J. Vaughan, *J. Amer. Chem. Soc.*, **81**, 4226 (1959).
239. A. I. Biggs and R. A. Robinson, *J. Chem. Soc.*, 388 (1961).
240. G. P. Schiemenz, *Angew. Chem.*, **78**, 145 (1966); EE, **5**, 129 (1966).
241. F. G. Bordwell and P. J. Boutan, *J. Amer. Chem. Soc.*, **78**, 854 (1956); **79**, 717 (1957).
242. F. G. Bordwell and G. D. Cooper, *J. Amer. Chem. Soc.*, **74**, 1058 (1952).
243. J. Miller, *Austral. J. Chem.*, **9**, 61 (1956).
244. H. Zollinger and C. Wittwer, *Helv. Chim. Acta*, **39**, 347 (1956).
245. S. Oae and C. C. Price, *J. Amer. Chem. Soc.*, **80**, 3425 (1958).
246. L. K. Creamer, A. Fischer, B. R. Mann, J. Packer, R. B. Richards, and J. Vaughan, *J. Org. Chem.*, **26**, 3148 (1961).
247. J. J. Elliot and S. F. Mason, *J. Chem. Soc.*, 2352 (1959).
248. C. C. Price, E. C. Mertz, and J. Wilson, *J. Amer. Chem. Soc.*, **76**, 5131 (1954).
249. A. Fischer, J. Packer, J. Vaughan, A. F. Wilson, and E. Wong, *J. Org. Chem.*, **24**, 155 (1959).
250. H. C. Brown and T. Inukai, *J. Amer. Chem. Soc.*, **83**, 4825 (1961).
251. Y. Otsuji, Y. Koda, M. Kubo, M. Furukawa, and E. Imoto, *Nippon Kagaku Zasshi*, **80**, 1300 (1959).
252. N. N. Zatsepina, Yu. L. Kaminskii, and I. F. Tupitsyn, *Reakts. spos. org. Soedinenii*, **6**, 778 (1969); EE, 333.
253. G. T. Bruce, A. R. Cooksey, and K. J. Morgan, *J.C.S. Perkin II*, 551 (1975).
254. F. Fringuelli, G. Marino, and A. Taticchi, *Gazz. Chim. Ital.*, **102**, 534 (1972).
255. L. W. Deady, R. A. Shanks, and R. D. Topsom, *Tetrahedron Lett.*, 1881 (1973).

256. T. J. Broxton, G. L. Butt, L. W. Deady, S. H. Toh, R. D. Topsom, A. Fischer, and M. W. Morgan, *Canad. J. Chem.*, **51**, 1620 (1973).
257. A. D. Campbell, S. Y. Chooi, L. W. Deady, and R. A. Shanks, *Austral. J. Chem.*, **23**, 203 (1970).
258. M. Liveris and J. Miller, *J. Chem. Soc.*, 3486 (1963).
259. D. M. Dimitrijević, Ž. D. Tadić, M. M. Mišić-Vuković, and M. Muškatović, *J.C.S. Perkin II*, 1051 (1974).
260. L. W. Deady and R. A. Shanks, *Austral. J. Chem.*, **25**, 431 (1972).
261. J. H. Blanch, *J. Chem. Soc.* (*B*), 937 (1966).
262. A. R. Katritzky and F. J. Swinbourne, *J. Chem,. Soc.*, 6707 (1965).
263. E. M. Arnett and R. D. Bushick, *J. Org. Chem.*, **27**, 111 (1962).
264. R. T. C. Brownlee, R. E. J. Hutchinson, A. R. Katritzky, T. T. Tidwell, and R. D. Topsom, *J. Amer. Chem. Soc.*, **90**, 1757 (1968).
265. H. D. Holtz and L. M. Stock, *J. Amer. Chem. Soc.*, **86**, 5188 (1964).
266. H. Kwart and T. Takeshita, *J. Amer. Chem. Soc.*, **86**, 1161 (1964).
267. T. J. Broxton, D. G. Cameron, R. D. Topsom, and A. R. Katritzky, *J.C.S. Perkin II*, 256 (1974).
268. C. F. Wilcox and S. S. Chibber, *J. Org. Chem.*, **27**, 2210 (1962).
269. T. B. Grindley, K. F. Johnson, A. R. Katritzky, H. J. Keogh, C. Thirkettle, R. T. C. Brownlee, J. A. Munday, and R. D. Topsom, *J.C.S. Perkin II*, 276 (1974).
270. G. W. Stevenson and D. Williamson, *J. Amer. Chem. Soc.*, **80**, 5943 (1958).
271. N. C. Cutress, A. R. Katritzky, C. Eaborn, D. R. M. Walton, and R. D. Topsom, *J. Organometal. Chem.*, **43**, 131 (1972).
272. A. R. Katritzky, R. F. Pinzelli, M. V. Sinnott, and R. D. Topsom, *J. Amer. Chem. Soc.*, **92**, 6861 (1970).
273. I. M. Koshkina, L. A. Remizova, E. V. Ermilova, and I. A. Favorskaya, *Reakts. spos. org. Soedinenii*, **7**, 944 (1970); EE, 427.
274. L. A. Kaplan and H. B. Pickard, *J. Org. Chem.*, **35**, 2044 (1970).
275. F. W. Baker, R. C. Parish, and L. M. Stock, *J. Amer. Chem. Soc.*, **89**, 5677 (1967).
276. C. F. Wilcox and C. Leung, *J. Amer. Chem. Soc.*, **90**, 336 (1968).
277. V. M. Vlasov and G. G. Yakobson, *Uspekhi Khim.*, **43**, 1642 (1974); EE, 781.
278. P. A. Ten Thije and M. J. Janssen, *Rev. Trav. chim.*, **84**, 1169 (1965).
279. J. M. Angelelli, R. T. C. Brownlee, A. R. Katritzky, R. D. Topsom, and L. Yakhontov, *J. Amer. Chem. Soc.*, **91**, 4500 (1969).
280. I. Minamida, Y. Ikeda, K. Uneyama, W. Tagaki, and S. Oae, *Tetrahedron*, **24**, 5293 (1968).
281. F. E. Condon, R. T. Reece, D. G. Shapiro, D. C. Thakkar, and T. B. Goldstein, *J.C.S. Perkin II*, 1112 (1974).
282. D. J. Martin and C. E. Griffin, *J. Org. Chem.*, **30**, 4034 (1965).
283. D. Martin and W. M. Brause, *Chem. Ber.*, **102**, 2508 (1969).
284. V. V. Orda, L. M. Yagupol'skii, V. F. Bystrov, and A. U. Stepanyants, *Zhur. obshch. Khim.*, **35**, 1628 (1965); EE, 1631.
285. N. C. Cutress, T. B. Grindley, A. R. Katritzky, M. V. Sinnott, and R. D. Topsom, *J.C.S. Perkin II*, 2255 (1972).
286. I. Koppel, M. Karelson, and V. Palm, *Org. Reactivity*, **11**, 101 (1974).
287. C. F. Wilcox and J. S. McIntyre, *J. Org. Chem.*, **30**, 777 (1965).
288. C. D. Ritchie and E. S. Lewis, *J. Amer. Chem. Soc.*, **84**, 591 (1962).
289. I. V. Talvik and V. A. Palm, *Reakts. spos. org. Soedinenii*, **8**, 445 (1971).
290. V. A. Palm, *Trudy Konferentsii po Problemam Primeneniya Korrelatsionnykh Uravnenii v Organicheskoi Khimii*, Vol. 1, p. 3 (Tartu State University, Tartu, 1962).
291. R. T. M. Fraser, *Nature*, **205**, 1207 (1965).
292. C. G. Evans and J. D. R. Thomas, *J. Chem. Soc.* (*B*), 1502 (1971).

293. K. Bowden, N. B. Chapman, and J. Shorter, *J. Chem. Soc.*, 3370 (1964).
294. W. A. Pavelich and R. W. Taft, *J. Amer. Chem. Soc.*, **79**, 4935 (1957).
295. S. S. Biechler and R. W. Taft, *J. Amer. Chem. Soc.*, **79**, 4927 (1957).
296. W. H. Richardson, R. S. Smith, G. Snyder, B. Anderson, and G. L. Kranz, *J. Org. Chem.*, **37**, 3915 (1972).
297. C. Hansch, *J. Org. Chem.*, **35**, 620 (1970).
298. J. Hine and O. B. Ramsay, *J. Amer. Chem. Soc.*, **84**, 973 (1962).

Index

This index is a guide to the main entries for important topics. Individual compounds, reactions, etc. are not usually mentioned.